NO TRENCHES IN TOWN
NO-DIG 92 PARIS

COMPTES RENDUS DE LA CONFERENCE INTERNATIONALE NO-DIG 92 PARIS
PARIS LA VILLETTE / LA FRANCE / 12-14 OCTOBRE 1992

Pour une Ville sans Tranchée

Rédigé par
JEAN-PIERRE HENRY & MICHEL MERMET
FSTT – Comité Français pour les Travaux sans Tranchée, Paris, la France

A.A. BALKEMA / ROTTERDAM / BROOKFIELD / 1992

PROCEEDINGS INTERNATIONAL CONFERENCE NO-DIG 92 PARIS
PARIS LA VILLETTE / FRANCE / 12-14 OCTOBER 1992

No Trenches in Town

Edited by
JEAN-PIERRE HENRY & MICHEL MERMET
FSTT – French Society for Trenchless Technology, Paris, France

A.A. BALKEMA / ROTTERDAM / BROOKFIELD / 1992

The texts of the various papers in this volume were set individually by typists under the supervision of each of the authors concerned.

Published by
A.A. Balkema, P.O. Box 1675, 3000 BR Rotterdam, Netherlands
A.A. Balkema Publishers, Old Post Road, Brookfield, VT 05036, USA

ISBN 90 5410 085 0

Printed in the Netherlands

No Trenches in Town, Henry & Mermet (eds) © *1992 Balkema, Rotterdam. ISBN 90 5410 085 0*

Table of contents
Table de matières

1.3 *Pipes*

1.4 *Detection and geotechnical investigations*

2 *Rehabilitation*

2.1 *Techniques*

2.2 *Internal inspection*

3 *Strategic and legal aspects*

Late papers

No Trenches in Town, Henry & Mermet (eds) © 1992 Balkema, Rotterdam. ISBN 90 5410 085 0

Foreword

The 9th International Congress on Trenchless Technology is being held in Paris on the prestigious and symbolic site of the Cité des Sciences de la 'La Villette'. It is the first time that this event has been held in France.

Our new Society, FSTT, only two years old, was charged by the international body, ISTT, and its Chairman, Ted Flaxman, with the organisation of this event. We are very happy and proud of this sign of confidence.

The Conference will have two principal aims. Firstly it will focus on the latest developments in the technology on an international level; it should be a platform for technicians, engineers and scientists to express themselves, to exchange experiences and to break through the language barriers and diminish distances. This is the justification for its being an international scientific and technical meeting at the highest level.

The second aim of the Conference is to make trenchless methods better known in France, to draw attention to the advantages of trenchless solutions for the construction and repair of the ever more important pipelines under our streets and roads.

These solutions have proved successful over many years in other countries and have been recently tested in France. Several contractors have made great efforts to buy the equipment and to train their employees in its proper use.

The Conference documents will be a record of the exchange of expertise and a fitting reminder of an important event. They will detail essential information about the evolution of the various techniques, whilst the proficiency and reputations of the authors of the papers as well as the importance of the topics indicate the value of the Conference. We thank them sincerely.

The disruption and damage associated with the street works are no longer necessary. Solutions to the problems exist! They have been proved to be efficient! They are working in France! However, they must be made more widely known and applied. Moreover, the techniques can be further improved and be suitable for more and more contractors.

If these trenchless methods are specified where appropriate, it will break the absurd circle in which the problems are denounced without efforts being made to solve them.

Although this is a scientific and technical conference it will not only appeal to specialists, but also to those who think that technology, well applied, can reduce damage and better protect the environment. It will appeal to those who prefer to act rather than to passively observe.

I wish to thank once more and warmheartedly the authors of the papers and all those who have contributed to the success of this Conference by their dedication, their talents and their enthusiasm.

Jean-Pierre Henry
Président du Comité Scientifique
de la FSTT

Michel Mermet
Président du Comité Français
pour les Travaux sans Tranchée (FSTT)

No Trenches in Town, Henry & Mermet (eds) © 1992 Balkema, Rotterdam. ISBN 90 5410 085 0

Préface

Le 9ème Congrès international des 'Techniques sans tranchée' se tient à Paris sur le site prestigieux et symbolique de la Cité des Sciences de la Villette. Pour la première fois ce Congrès se tient en France.

Après seulement deux années d'existence, notre jeune association la FSTT, s'est vu confier l'organisation de cet évènement par l'association internationale, ISTT et son Président, Ted Flaxman. Nous sommes heureux et fiers de cette marque de confiance.

Ce Congrès vise deux objectifs primordiaux. Faire le point des progrès les plus récents de ces techniques au niveau international. Permettre aux techniciens, ingénieurs et scientifiques de s'exprimer, d'échanger et franchir à cette occasion les obstacles de la langue ou de l'éloignement. C'est la raison d'être, classique, d'une rencontre internationale scientifique et technique de ce niveau.

L'autre objectif de ce Congrès est de faire connaître ces solutions en France. Faire connaître les avantages des solutions dites 'sans tranchée' pour la construction et la réparation des canalisations toujours plus nécessaires dans le sous-sol de nos rues et de nos routes.

Ces solutions ont fait leur preuve depuis assez longtemps à l'étranger. Plus récemment en France nous les avons testées. Des entreprises ont fait l'effort de s'équiper et de former les hommes.

Le présent document, sera la trace de cet échange et la mémoire de ce Congrès. Il réunit des informations essentielles pour le progrès de ces techniques. La compétence et la notoriété des auteurs de ces communications tout comme l'importance des sujets qu'ils ont choisis, situent très haut l'intérêt réel de cette rencontre. Qu'ils en soient sincèrement remerciés.

La gêne et les nuisances de tous ordres provoquées par les 'travaux des rues et des routes' ne sont donc plus une fatalité. Des solutions existent. Elles ont fait leurs preuves. Elles sont opérationnelles en France. Il faut le faire savoir et s'en servir. C'est d'ailleurs ainsi que nous pourrons perfectionner encore ces techniques et donc les utiliser de plus en plus.

C'est surtout en les utilisant à bon escient que l'on sortira du cercle absurde où l'on dénonce le problème sans même avoir essayé la solution.

Ainsi ce Congrès bien que scientifique et technique ne s'adresse pas seulement aux spécialistes, mais à ceux qui pensent que le progrès technologique bien maîtrisé peut réduire les nuisances et mieux protéger l'environnement; à tous ceux qui préfèrent essayer d'agir que regarder.

Je tiens à remercier à nouveau et chaleureusement les auteurs des communications et tous ceux qui par leur dévouement, leur talent et leur enthousiasme auront permis le succès de ce Congrès.

Jean-Pierre Henry
Président du Comité Scientifique
de la FSTT

Michel Mermet
Président du Comité Français
pour les Travaux sans Tranchée (FSTT)

1 Trenchless installations

1.1 Microtunneling

No Trenches in Town, Henry & Mermet (eds) © 1992 Balkema, Rotterdam. ISBN 90 5410 085 0

A variable diameter microtunnelling system

N. Bristow
Markham & Co. Ltd, Chesterfield, UK

The paper describes a recent innovative adaptation of an existing well proven microtunnelling system to allow a wide range of final tunnel diameters to be produced by a single microtunnelling system. The technology on which the system is based is the highly reliable, low risk Super Mini system which originated in Japan and has already been developed to suit the much more varied European conditions. The key technical features of the basic machine are outlined, illustrated by some of the practical experiences seen to date on interesting or unusual jobs. The jobs described highlight the versatility of the system and the particular details of the technology that lead to the very low risks inherent in the system that allow it to be used in potentially difficult situations. The paper then outlines the technical and commercial background to the development of the variable diameter innovation. The diameter can be varied over a wide range and the technology that achieved this will be described. Finally, the results of the initial trials and production results are given following its use on a major contract in the UK.

INTRODUCTION: The Super Mini microtunnelling system developed originally in Japan some years ago and adapted substantially for use in the more varied ground conditions in Europe, is now well established. It has proved to be a very fast and reliable system capable of undertaking jobs that might otherwise have been considered risky under roads, rivers and railway lines for example. Here in Paris the system has been used successfully on many job sites and some of these jobs have been described in previous papers. However, one disadvantage that the system had, in common with some other systems, is that a significant capital investment must be made by contractor or client for a machine that can basically only produce one size of finished tunnel. Consequently, an innovative adaptation of the system has now been designed, manufactured and proven that allows a single microtunnelling system to produce a wide range of final tunnel diameters for a modest additional capital cost.

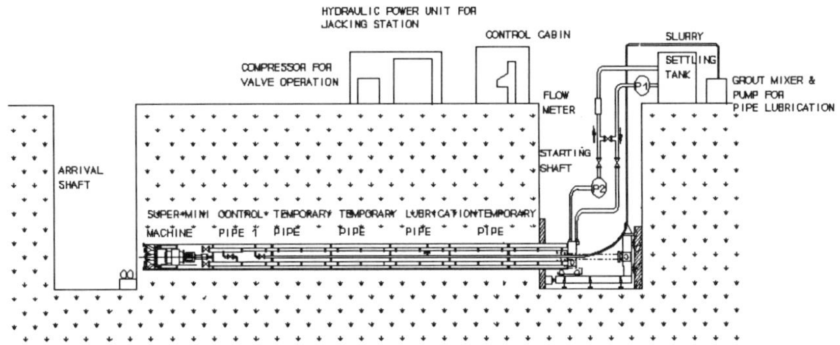

EQUIPMENT LAYOUT FOR SUPER-MINI OPERATION

The paper will briefly highlight the key technical features of the basic system including the particular details of the technology that lead to the very low risks inherent in its use. It will then go on to show how these same features enable the variable diameter innovation to be produced and will outline the technical and commercial background to the development. Finally the results of the initial trials and production results are given following its first use on a major project in the UK.

1 BASIC SUPER MINI SYSTEM

The Super Mini system is a slurry based microtunnelling system in a mature stage of development such that a single system configuration can deal with very varied ground types - clays, sands, silt, gravels and boulders for example, and ground conditions that include those that would normally be considered bad, that is very shallow, or very deep, below water table or in unstable ground. One of the most important features is the temporary pipe system which gives it many of its key performance features:-

- reliable and rapid services connections
- high rates of progress
- accurate line and level control
- long drivage distances
- low thrust forces on permanent pipes
- full retractability of the machine

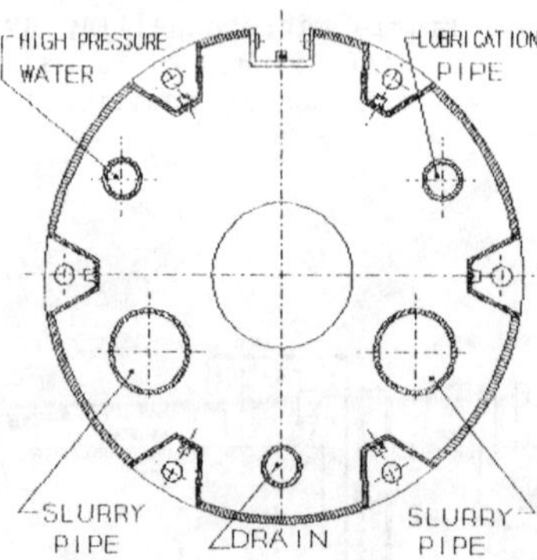

Each temporary pipe contains all slurry and service pipes with connection being made and sealed automatically as the new pipe is fitted. The electric cable carrying all power, control and signal lines is continuous from surface to machine and is laid in a slot in the top of the temporary pipe which is then protected with a steel cover. There is therefore no risk resulting from water ingress into cable connectors for example. High rates of progress are achieved because pipe changeover time is minimal, only a few minutes, and tunnelling rates of 42m per shift have been achieved. Even though the permanent pipe installation requires in effect a second pass, the overall progress rate on most jobs is considerably faster than with a more conventional system. Indeed, as I will show later, the two-pass nature of the system is one of the features that allows it to be adapted to variable pipe sizes.

The accurate line and level control and the long drivage distance capability results from the inherent stiffness of the temporary pipe column. Only low thrust forces are needed on the permanent pipes because the tunnelling has already been finished and a clean, straight, lubricated hole is already prepared in the ground for the permanent pipes to push into as they push the temporary pipes out, and this operation is carried out at a progress rate of about 80m per shift.

The bolted column of temporary pipes allows the system to be put into reverse at will from the control station enabling anything from slight pulling back to relieve thrust or torque build up to full retraction should a drive have to be aborted for any reason. During this latter process grout injection can take place through the slurry lines to refill the void such that it gives full support to the ground but also allows re-excavation later. With this ultimate recourse available, high risk jobs can be confidently undertaken.

DEVELOPMENT OF THE REAMER SYSTEM

The temporary pipe feature described above has, however, two potential disadvantages. Firstly, there is an additional capital cost directly resulting from the need to have sufficient temporary pipes available for the drive lengths to be undertaken, although once acquired, the temporary pipes need little or no maintenance and last almost indefinitely. Secondly, the temporary pipes, along with the tunnelling machine itself are fixed in diameter and are therefore only capable of producing a pipeline of one outside diameter. But with the addition of a relatively low cost additional piece of equipment that itself can have a range of diameters of cutting edge, the second pass inherent in the Super Mini system can be used to good effect to back ream the pipeline to any size from the original nominal size upwards up to a limit of about two times its diameter. This is the innovation that has now been designed, manufactured and used successfully on a major contract in the UK. The reamer unit only comprises the powered cutting module fitted with whatever size of cutter required plus a simple additional jacking station to assist in pushing the reamer back to the start shaft as the permanent pipes are installed. All the other capital equipment including the control can for the machine, the surface control station and the slurry pumping and treatment plant remain in use in their original locations.

The sequence of operation is shown in the figures below and simply entails removing the Super Mini machine (but not its control can) at the normal arrival shaft, replacing it with the Reamer Unit, and back reaming to the original starting shaft.

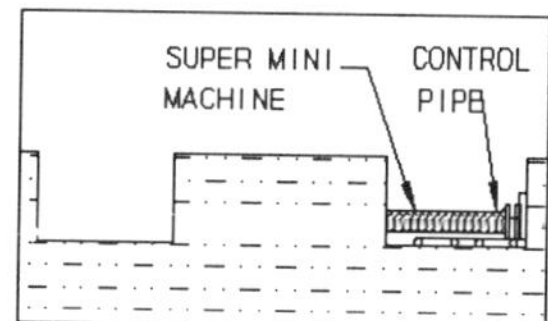

1) LAUNCHING OF SLURRY SHIELD.

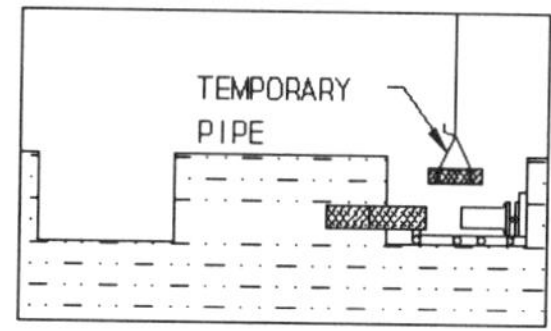

2) INSTALLATION OF TEPORARY PIPES

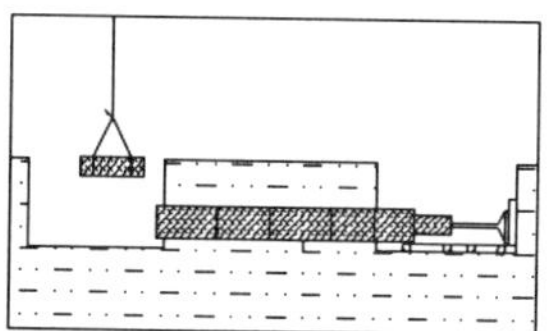

3) RECOVERY OF SLURRY SHIELD

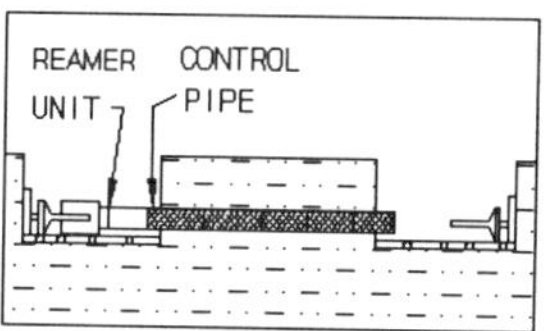

4) INSTALLATION OF REAMER UNIT.

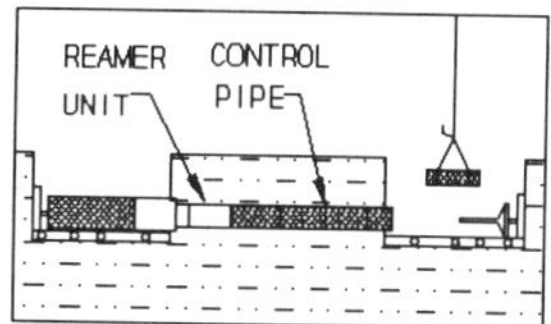

5) INSTALLATION OF FIRST PIPE SECTION.

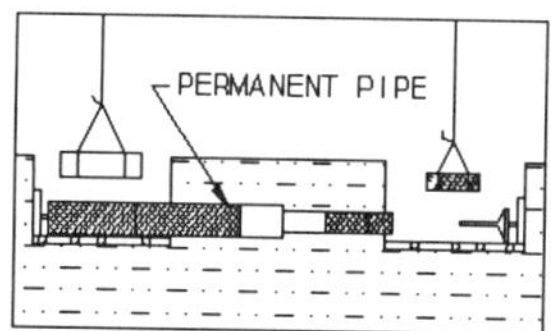

6) INSTALLATION OF PERMANENT PIPE SECTION.

Slurry flow within the reamer and the general design of the cutting unit are based very much on the proven Super Mini itself. All control and slurry pipes and cables are used exactly as for the Super Mini but are now running ahead of the reamer and are shortening as the reamer progresses. Progress rates are rapid partly because no steering control is needed as the reamer has to follow the previously excavated tunnel line. Ideally permanent pipes of the same individual length as the temporary pipes should be installed to minimise problems of coordination between starting and arrival shafts, but this is not essential - indeed the first job to use the reamer installed 2.5m concrete pipes behind 2.0m temporary pipes. As all services are carried in the temporary pipes, the permanent pipes are installed free of any tunnelling equipment and can therefore be equipped with "furniture" or additional linings without risk of damage. The reamer module is designed around a central power unit with a bolted on cutting disc so that the diameter is easily changed. We believe that for example a nominal 500mm inside diameter Super Mini system could be

equipped with a reamer for 600mm, 700mm, 800mm or 900mm pipes
or any standard size in between 500mm and 900mm.

FIRST EXPERIENCES WITH THE REAMER

The reamer was first used on a major new wastewater project in the South of England early this year. The contractor, Miller Markham, owned a 500 mm Super Mini system and had to produce three lengths of microtunnel (96m, 102m and 40m) at 690 mm internal diameter as part of a total of 3km of pipeline works, the rest being undertaken partly in open cut, partly in conventional pipejack at 1000mm and partly by microtunnelling at 1000mm. Being the first use of such an innovative system it was no surprise to find some shortcomings with the equipment. On the first drive some difficulty was experienced with slurry loss during the reamer launch and subsequently. The drive was completed, however, but somewhat slowly. For the second drive the launch procedure was improved reducing the slurry loss and allowing the drive to be completed at a reasonable rate of progress. Between the second and third drives, some minor but significant modifications were made to the reamer which virtually eliminated the slurry loss problem and on the third drive the performance of the reamer exceeded that of the Super mini in its first pass through the same ground.

CONCLUSION

A truly variable diameter microtunnelling system now exists, proven and reliable, with all the potential advantages and attributes required to undertake all kinds of projects ranging from straightforward jobs in good ground to difficult work in poor ground conditions in high risk situations.

No Trenches in Town, Henry & Mermet (eds) © 1992 Balkema, Rotterdam. ISBN 90 5410 085 0

The excavation behaviour of soils in Trenchless Tunnelling Techniques

W.G.M.van Kesteren & F. Bisschop
Delft Hydraulics/WL Delft, Netherlands

ABSTRACT: The possibilities for optimalization of Trenchless Tunnelling techniques are based on knowledge obtained from dredging research. Of primary importance is the ability to control the excavation behaviour in order to optimize the mixing, transport and deposition.

1. INTRODUCTION

The paper is aimed at the application of the fundamental knowledge of excavation processes obtained from dredging research for improvement and development of trenchless tunnelling techniques. In the present trenchless tunnelling techniques four processes can be distinguished: excavation, mixing, transport and deposition (figure 1). Tuning the used excavation technique by changing the operational parameters of a TBM, according to the known soil parameters, the whole production-line can be improved.

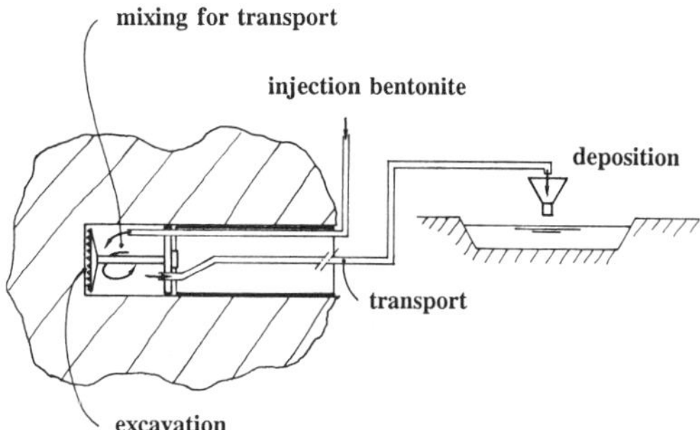

Figure 1: Production-line of TBM

2. EXCAVATION PROCESS

Excavation techniques can be divided into mechanical and hydraulic techniques. Mechanical techniques are the cutting of soils with cutting tools like blades, teeth and pick-points. The main hydraulic excavation technique is hydraulic erosion by jetting.

From a energetically point of view mechanical excavation techniques are preferable to hydraulic excavation techniques. In the case of hydraulic excavation techniques the excavation and mixing process coincide and conditions for transport are difficult to control.

2.1 Mechanical excavation techniques

In the case of mechanical excavation three different failure modes can be distinguished in all soils, including rock (figure 2 and van Kesteren et al., 1992):

- ductile:
  This failure mode is characterized by a continuous deformation of the soil. The production is characterized by a continuous chip (figure 3).
- localization of deformation in a shear plane:
  These shear planes appear periodically from the cutting edge of the cutting tool. The production is characterized by a continuous chip weakened by discrete shear planes. In the case of the cutting of sand the process is accompanied by dilatation (figure 4 and Os et al., 1987).
- localization of deformation in cracks (brittle behaviour):
  This failure mode is characterized by crack propagation initiated by a stress concentration at the border of the plastic zone around the cutting edge (figure 5).

Because most tunnelling operations are executed under water the cut soil will be saturated or nearly saturated with water. The pore water is of primary importance for the mechanical behaviour of the soil during cutting. It influences the kind of failure mode by the degree of saturation, permeability, compressibility of the grain skeleton and cavitation due to a pressure-drop. Other important soil parameters are the fracture toughness and the strength at the brittle-ductile transition.

The three basic cutting processes can be disturbed

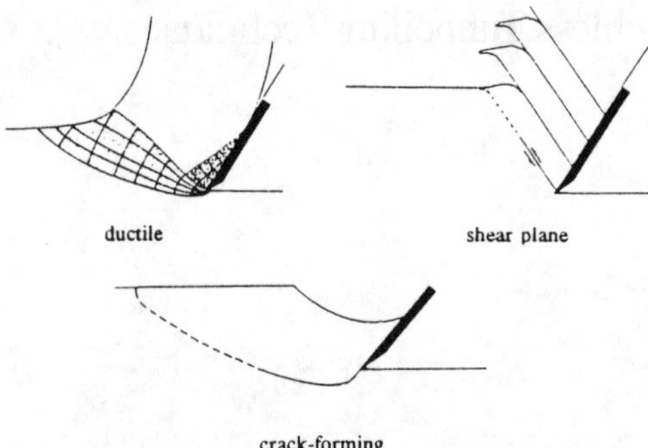

Figure 2: Failure modes in soil cutting

Figure 3: Ductile behaviour during the cutting of clay

Figure 4: Localization of deformation in a shear plane during the cutting of sand

by discontinuities in the soil like inhomogenities, fractures, bedding planes and joints. This results in fragmented production-units.

Theoretically all these failure modes can occur in any soil and therefore the soil failure behaviour cannot be uniquely referred to as either ductile or brittle. The occurance of each failure mode and their mutual transition is not only determined by soil parameters but also by operational parameters like cutting velocity, cutting depth, cutting angle and waterdepth (figure 6).

2.2 Hydraulic processes

In tunnelling hydraulic excavation mainly takes place with the aid of jets. The hydraulic excavation process is determined by the erosion process. The erosion process by a turbulent flow can be separated into two stages:
Failure of the soil near at the interface between soil and water and lateral transport into the turbulent flow.

Dilatation of the soil is necessary to mobilize the soil for transport. The pressure gradient due to the dilatation, over the interface, increases the resistance against erosion. This increase is determined by the flow velocity, permeability, relative density and stiffness of the grain skeleton. The same effect is recognised considering the stability of a slope under water with a porewaterflow inwards the soil (van Rhee e.a., 1991 en Breusers, 1977).

The lateral transport capacity depends on the degree of turbulence, concentration of solids and the flow velocity of water to the interface. This determines the mixing process.

3. RELATION OF EXCAVATION PROCESS WITH MIXING, TRANSPORT AND DEPOSITION

In an optimal functioning tunnelling proces the excavation technique, mixing, transport and deposition have to be tuned to each other. It is practically impossible in tunnelling to work under different soil conditions with one and the same tunnelling technique. But fundamental knowledge about mechanical as well as hydraulic excavation techniques makes the application of one TBM-system under different soil conditions more suitable. Also the efficiency of the whole tunnelling proces in a certain soil can be improved.

3.1 Relation of excavation process with the mixing process

Optimalization of the mixing process and transport can take place by considering the internal age distribution function. This function relates the production-rate at the intake (excavation) and the discharge (transport). This function is determined by the type of mixing procedure and dimensions of the production-units.

During mixing the following processes are important for the internal age distribution function:

- desintegration of the production-units caused by softening due to swelling, erosion by hydraulic impact and collisions and failure into smaller lumps due to collisions.
- segregation of the production-units in the mixture due to the difference in density between production-units and mixing-fluid (consisting of porewater and

added mixing agent), including the effect of the concentration of solids in the mixture (hindered settling) and the rheological properties of the mixing fluid.

flowconditions in the mixing room: a viscous flow causes another kind of distribution of the production-units than a turbulent flow.

The mixing process is not only important for the internal age distibution function but is also important for the stability of the tunnelfront. For the stability of a tunnelfront it is import to balance earth pressure and porewater pressure. To achieve this a homogeneous mixture of high density and viscosity under pressure is required. Due to a higher density and viscosity the power necessary to create a homogeneous mixture increases. The increase of the density and viscosity causes only a problem: separation of soil and mixing agent for deposition needs more effort.

For the control of the stability of a tunnelfront it is also important to know whether the groundwater flows to the mixing room or into the soil in front of the TBM. This depends on the pressure in the mixing room and groundwater pressure. The flow of water into the mixing-room causes a decrease of the stability of the tunnelfront. On the other hand if water flows into the soil around the TBM the stability of the tunnelfront increases. This can be compared with the stability of a slope under water Rhee (et al., 1991). The flow of water into the soil in front of the TBM can cause a "blow out" at the surface. Therefore it is necessary to check the porewater pressure and vertical pressure gradient in front of the tunnel.

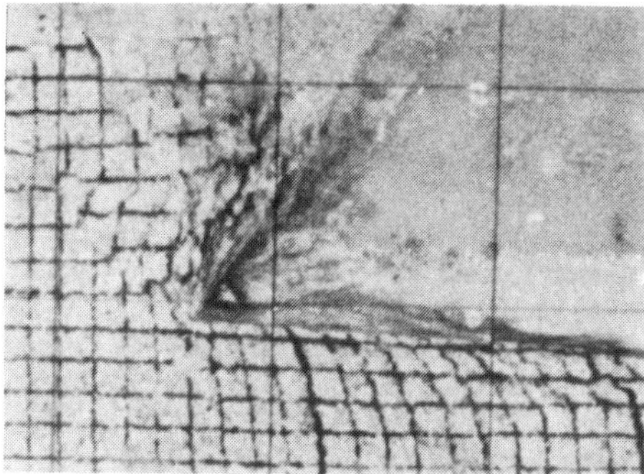

Figure 5: Brittle behaviour of rock (Cools, 1990)

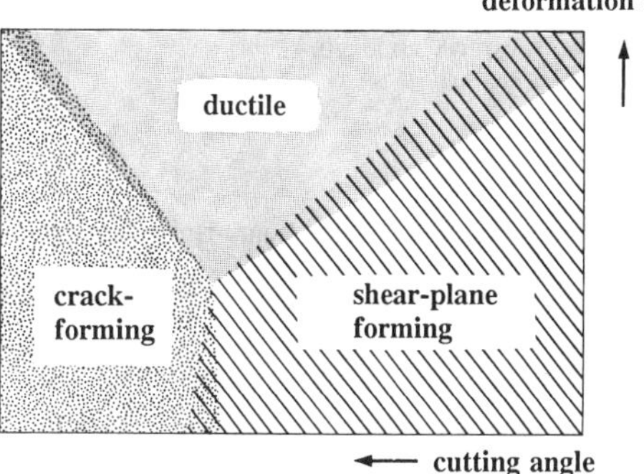

Figure 6: Deformation rate versus cutting angle

3.2 Relation excavation process to transport and deposition

The relation of the excavation process to transport and deposition is determined by the mode of transport chosen. In general the transport part in trenchless tunnelling consists of conveying of bulk solids horizontally and upwards. A major distinction in the mode of transport is:

* discontinuous
* continuous without a fluid as transport medium
* continuous with a fluid as transport medium

The conveyor selection is determined by material characteristics, capacity requirements and length of travel combined with lift. When a conveyor selection and mode of deposition is made only the material characteristics can be varied, or definitely will vary when the tunnelling proces is proceeding. Therefore in order to be able to optimize the trenchless tunnelling process, the material characteristics must be controlled. This can be done by adjusting the excavation proces to the mixing process before and during transport in order to fullfill the boundary conditions for the chosen mode of transport and deposition.
The material characteristics, which are of primary importance for conveyor selection are:

* size (fines, granular, lumps)
* flow behaviour
* abrasiveness
* watercontent
* mechanical behaviour:
  - compressibility
  - permeability
  - failure behaviour in compression and tension

As outlined in the previous chapter the physics involded in the cutting process is independent of soil type ranging from mud, sand, gravel, clay to rock. It is therefore possible, when different kind of soils are encountered during the trenchless tunneling proces, to adjust the cutting and mixing proces in such a way that the conditions for transport can be fullfilled. The only limitation can be the required torque and available power.
A major selection item is if in the continuous mode a fluid is used as transport medium or not.
In the case of application af a fluid it is important to

combine the material parameters to the parameters of the processes involded in fluidal transport i.e:

* fall velocity
* deposition velocity
* desintegration (see paragraph 3.1)

In case of soils, like sand and gravel, which behave cohesionless within the processtime of mixing and transport, the slurry conditions during transport will be determined by the occurence of segregation and the forming of a movable bed. In case of soils, like silt and clay, which have an time dependent apparent cohesion and soils with a true cohesion like heavily overconsolidated clays and rock, the slurry conditions are olso determined by the desintegration process of the production-units (see paragraph 3.1).
Given the length of transport and the deposition method, the dimensions of the production-units can be adjusted to the slurry conditions.

4. CONCLUSIONS

For optimalization and development of new Trenchless Tunnelling techniques it is important to tune the excavation technique on mixing, transport and deposition. The application of fundamental knowledge of dredging of soils, varying from mud, clay and sand to rock can improve existing Trenchless Tunnelling techniques.

The two main excavation techniques are hydraulic and mechanical techniques. For hydraulic techniques erosion plays an important role and for mechanical techniques the cutting process determined by the failure modes: ductile, shear plane and crack forming. Besides soil parameters the operational parameters as cutting depth, cutting velocity, water pressure and flow velocity determine these processes. These parameters can be tuned in such a way that in almost all soil conditions the excavation process can be optimal adjusted to the mixing and transport mechanism.

To control and optimize the mixing process the internal age distribution function can be used.

The ability to control the mode of deformation and failure during excavation is the first important instrument in optimizing the production line of excavation, mixing, transport and deposition. The second important instrument is the control of the mixng process. Although the mixing process often is used for the stability of the tunnelfront, it has also a major function in the link between excavation and transport. In the case of hydraulic excavation techniques the excavation and mixing process coincide and therefore are difficult to control.

REFERENCES

Cools, P.M.C.B.M. 1990. Research activities on mechanical rock cutting and dredging at Delft Hydraulics. *Proceedings VIth Intern. Congress Intern. Assoc. of Engin. Geology.* Amsterdam, 6-10 August.

Breusers, H.N.C. 1977. Hydraulic excavation of sand. *Proc. Int. Course Modern Dredging.* The Hague. 5-10 June.

Kesteren, W.G.M. van. 1992. Pore water behaviour in dredging processes. *XIIIth World Dredging Congress.* Bombay. India. April.

Os, A.G. van and Leussen, W. van. 1987. Basic research on cutting in saturated sand. *Journal of Geotechnical Engineering.* Vol. 113. No. 12. December.

Rhee, C. van and Bezuijen, A. 1991. The influence of seepage on the stability of a sandy slope. *Paper submitted to the Journal of the Geotechnical Engineering Division of the ASCE.*

No Trenches in Town, Henry & Mermet (eds) © 1992 Balkema, Rotterdam. ISBN 90 5410 085 0

Proposition of a method for the study of the microtunneling machine behaviour and of the soil-structures interaction

S.Quebaud, E.Morel & J.-P. Henry
Laboratory of Mechanics of Lille, France

ABSTRACT : a research project on the microtunneling machines has been launched in France in 1990 ; it lead to an experimentation on a Val-de-Marne microtunneling site. The experience consisted of a study of the machine driving, with a parameters measure, and of a study of pipes and joints behaviour, with sticked gages on the pipes. Data were recorded with a data acquisition station. Results allow us to see some phenomena, like flexion, viscosity, otherthrust, and lubrication influence. Other experiments will have to be done to carry on this study ; instrumentation, data following-up and geotechnical reconnaissence will be considered with attention.

1 INTRODUCTION

The microtunneling system has been used in Japan for twenty years. It was used for the first time in France in 1989 and now sites follow sites.

However, some questions remain, especially about :

- the microtunneling system driving according to the soils nature ;
- the behaviour of pipes under the action of various loadings.

For these reasons, a research project has been launched in 1990 involving two research laboratories (the laboratory of Mechanics of Lille and the laboratory of Civil Engineering of Bordeaux), a construction firm (entreprises Léon Ballot BTP) and the Water and Sewer Service (D.S.E.A.) of the Val-de-Marne department ; as a result of an experimental program on a Val-de-Marne microtunneling site has been launched.

In the long term, this project should allow to use the microtunneling machine as a tool for the soil reconnaissance.

2 EXPERIMENTATION

2.1 Site description

The chosen site for experimentation was Bry-sur-Marne : a insufficient, damaged 200 millimeter-diameter sewer had to be replaced by a 300 millimeter-diameter sewer, on a length of 1542 meters, with 1200 meters realized by a microtunneling machine.

The D.S.E.A . chose the Léon Ballot's firm with a RVS 100 A microtunneling auger system, made by Decon Engineering Company (UK), under Soltau's licence (FRG). It was interesting to study the machine behaviour in clay soils. In addition, the small size of the machine allowed to bore from 2000 millimeter-diameter shafts, constituted by prefabricated concrete rings. These rings can support a slab and a standard inspection hole.

The piping is made by special elements, for a microtunneling use only. The chosen pipes trademarks are :

- Bonna, reinforced concrete pipes ;
- Eternit, asbestos-cemente pipes ;
- Naylor-Denlok, vitrified clay pipes.

3 natures of soil were initially retained on the site lengthwise section to carry out experimentations : green clay, sands and marls. We chose to experiment four drives in these soils, but due to a delay in the program of works, the experimental drive in sands was replaced by a second experience in marls.

2.2 The process

Some boreholes have been realized on the last two drives in order to characterize soils with Atterberg limits determination, and triaxial tests.

During experimentations, some measures of driving parameters have been made ; the different signals are :

- torque moment ;
- thrust ;
- head jacks stroke ;
- machine inclination ;
- rolling ;
- x and y axis deviation.

The aim of this signals study was to recognize soils during the machine was driven through.

A metallic ring with 4 hollow cylinders arranged

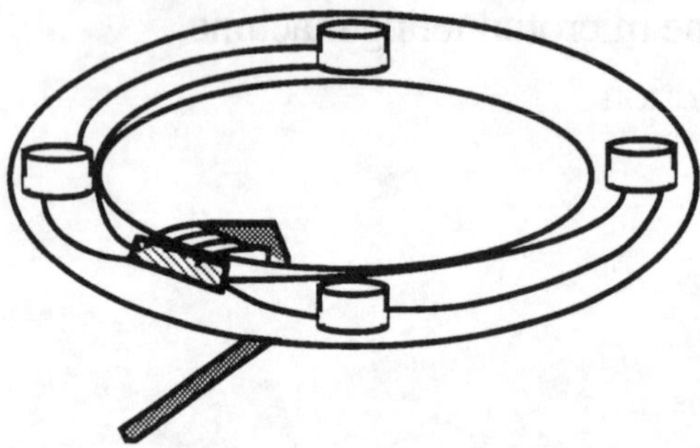

figure 1 : metallic ring with hollow cylinders

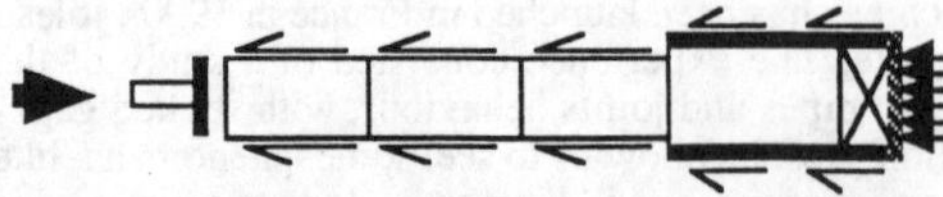

figure 2 : principle of loadings transmission during the boring of pipes

every 90 degrees has been put between the machine and the first pipe. Each cylinder was equipped with 2 gages, which are connected 2 by 2 (full bridge). This mechanism permitted to measure loadings directly behind the microtunneling system and then to know thrust in front of the machine, depending on the nature of soil (see figure 1).

Considering that the pipes price is quite high in comparison with the total cost of the bargain, it is necessary to optimize them. In fact, they are overdesigned, and generally tested with a press. A study of the real behaviour had not been done till today. The matters are :

- the joint efficency when the piping is not in a straight line ;
- the knowledge of ground pressure distribution along the piping.

In order to answer these questions, some pipes were equipped with strain gages (4 longitudinal gages, 2 transversal gages). The number of equipped pipes for a drive was equal to 3, that is to say 18 gages for a drive.

The gages position was chosen so that :

- we could study the joints behaviour : the gages were pasted on different pipes generating lines ;
- we could specify the thrust continuations : the experimented pipes were spaced along the piping.

The strain gages were connected by cables to a data acquisition station piloted by a computer. The acquisition program allowed to record the gages responses.

The control panel was linked to the acquisition system with a cable in order to record the driving parameters. All data recorded on the site were transferred on a other computer for their analysis.

2.3 Data processing

Concerning the pipes, the study concerned the longitudinal gages responses for a first time, in order to observe the pipes behaviour when they undergo thrust loadings. The different processing steps were :

- signals initialization : the data acquisition program does it when gages are connected to the station, but we could see a difference on curves : a possible reason could be the gages thermic dilatation, due to the temperature difference between road and shaft bottom, and due to the temperature increase inside the pipe (the casing overheats itself under the auger action). These 2 phenomena could induce a temperature variation of several degrees ;
- data correction : the curves low points represented in fact a thrust stop (the thrust jacks strokes are equal to 200 millimeters, that is to say that every 200 millimeters, the thrust is stopped in order to shorten jacks) : these values were useless for the study ;
- strain average value calculation of the bored pipe : in order to observe its behaviour alongside the drive, we determined the average value of the 4 gages responses. This was a simplist method, but it permitted a good approximation. We consider the half-bored pipe strain ;
- with this average value, we could calculate the loading drilled on the pipe.

To analyze ring data, the steps were almost the same as for the pipes : the difference was in the meaning of chanels response : in fact, the gages connection gave a global strain value for 2 cylinders, for each chanel (on the contrary, the 3-quarter bridge for the pipes gages gives local strains). To determinate the loading corresponding to a known strain, we have made a compression test on two cylinders before the first experimentation, in order to obtain a calibration curve. The average of the 2 chanels responses gave the average loading drilled on the ring. This value allowed the study of loadings transmission alongside the pipe train, as it represented the head loading ; we could know the machine reaction depending on the soil nature too.

3 RESULTS

3.1 The pipes

The thrust loading analysis allowed to study the pipe during the boring. This behaviour was scheduled on figure 2 : when a pipe is bored, it undergoes a equaling loading to the thrust value. When the pipe is bored little by little, the soil reaction acts on the pipe and prevent it from advancing. The thrust pressure must then be increased in order to compensate this reaction. When the following pipe is bored, the loading is transmitted to the bored pipes by the joints ; the surface which is submitted to the soil reaction increases.

We can see that a pipe globally undergoes the same thrust loading during its boring. On the other hand, the thrust pressure increases with the bored pipes number : the farther the pipe from the machine, the greater loading it undergoes. For this reason, manufacturers design their pipes with the maximal thrust capacity of the machine. However, this behaviour is verified if we suppose that :

- we have an homogeneous soil ;
- we bore without lubrication, and the boring is in a straight line.

In fact, the soil homogeneity means a constant loading in front of the machine ; on the contrary, the loading in the pipes will change. Concerning the lubrication, it plays an important part because it reduces the soils reaction, and then reduces the thrust pressure. Its efficiency depends on soils nature. After this analysis, we verified if results were in agreement with theory. If we follow the schedules for the processing, we can draw the average loading Fm as a function of bored pipe meter. Fm is calculated with the average strain εm and the pipes characteristics.

We can see that the curve of figure 3 decreases linearly, then suddenly increases, and then decreases again. We can explain these phenomena :

- the curve sudden increase is due to a soil changing : in fact, we could see on the site that the limons change to marls, which have better geotechnical characteristics than the others. These results were confirmed by the triaxial tests : the sample which came from the bore hole realized at 18 meters from the beginning shaft showed a more resistant soil than the sample coming from the bore hole realized at 9 meters from the same shaft ;
- the decrease is maybe due to the fact that the piping is not in a straight line : the joints dissipate the thrust energy in the ground.

The results show some others phenomena :

- the flexion phenomenon on the figure 4 given by longitudinal gages responses pasted on a Naylor-Denlok's pipe : we can clearly see that the response of 2 gages indicates a more compressed state, on the contrary of the 2 others ; this graph represents the pipe turning. The study of this phenomenon will allow a better understanding of joints behaviour ;
- the viscosity phenomenon, shown on figure 5 : a longitudinal gage response increases as the experimental Eternit pipe is bored, and becomes constant at the end : it means that thrust loading which acts on the pipe becomes more and more important, the compression is directly linked to the thrust loading : it is a typical viscosity phenomenon. Its study is interesting because its influence must be more and more important with an increasing machine diameter .

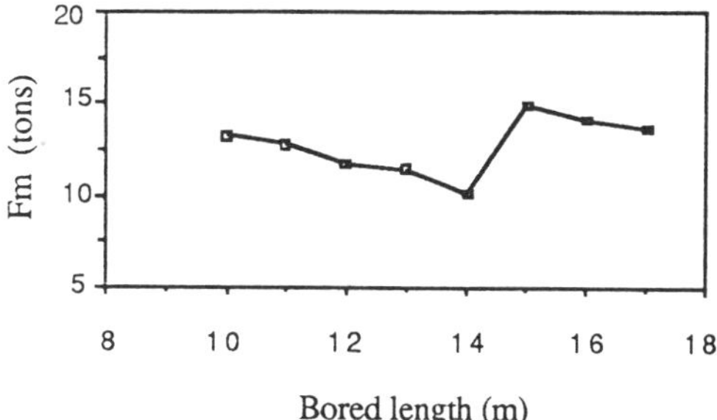

figure 3 : average loading recorded during the Bonna drive realization

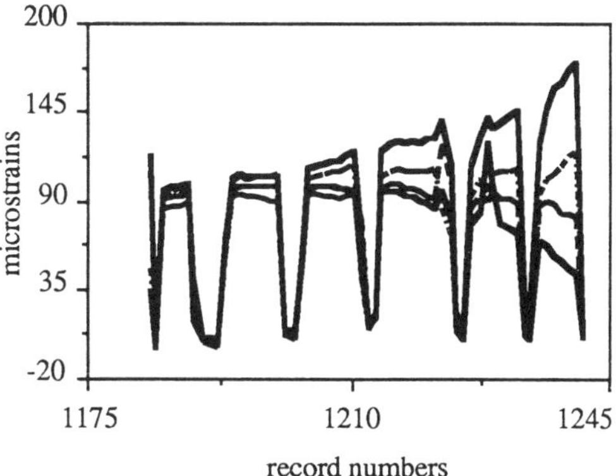

figure 4 : flexion phenomena observed during the Naylor-Denlok drive realization

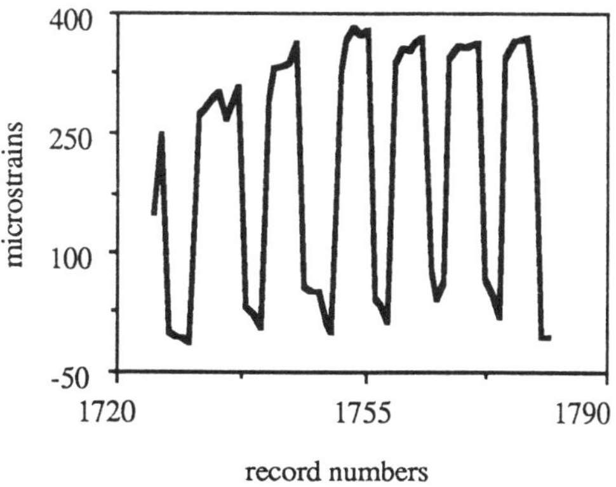

figure 5 : viscosity phenomena observed during the Eternit drive realization

This viscosity phenomenon has been shown too with the triaxial tests, realized with different speeds : the parameters evolution is quite significant.

At last, transversal gages should have allowed to study the ground pressure on the pipes, and also the torsion phenomena. This study could not be carried out, because the gages response was to weak to be analyzed. We can however verify they were in working order : in fact, their response is on the contrary of longitudinal gages response.

3.2 The ring

We have obtained a few results, only because one of both chanels gave values during the Eternit drive : it was due to the cables tearing. However, we could observe a viscosity phenomenon with these results, as for the pipes.

3.3 The driving parameters

We could see some phenomena :
- the overthrust phenomenon, when the thrust jacks stop : during the Eternit drive, where that value was equal to 18-20 tons, these picks could reach 110 tons : this phenomenon must be due to a synchronization problem of starting between thrust jacks and auger. These high values oblige firms to overdesign the pipes.
- the figure 6 shows the lubrication influence during the boring ; the 2 graphs can be divided into 3 phasis :
- phasis 1 : the thrust increases, due to the successive pipes additions : the torque increase is surely due to a too important rotation speed, which provokes the filling of the auger, and some difficulty to evacuate excavated soil ;
- phasis 2 : in opposition to the first phasis, we bore with lubrication ; the thrust becomes constant (homogeneous soil), and the torque decreases, because of a better soil excavation (the auger is little by little totally lubricated) ;
- phasis 3 : when the lubrication is stopped, the thrust and torque values come back to a linear response with the first phasis : the soil reaction and the fillng of the auger acts again.

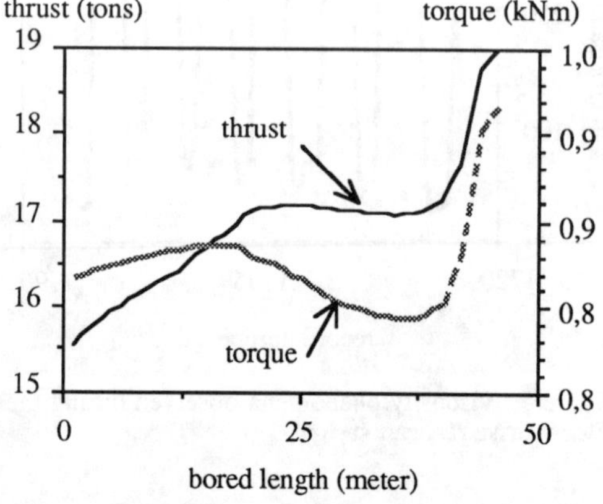

figure 6 : lubrication influence observed during the Eternit drive realization

The other parameters have not been studied, because they were not very important ; the more important could have been the bored length parameter : it allowed the determination of advancing speed, in order to study the viscosity phenomenon (which depends on time).

4 RECOMMENDATIONS

Though the experimentation on the site of Bry-sur-Marne has been carried out, the results show the necessity to carry on our study on new sites. We have to think of the instrumentation we will use on future sites, to the data follow-up, and to the geotechnical reconnaissance.

About the instrumentation, we have to :
- convince the manufacturers of microtunneling systems to equip their machines with sensors in order to get a lot of driving parameters ;
- study the influence of the gages position on a pipe, and realize tests on a compression bench ;
- put the measure systems in the pipes, in order to limit the clutter due to cables.

About the data follow-up, it has to be standardized, in order to be useful to several workers, and even to lead to a database creation.

About the geotechnical reconnaissance, which is important, we must be careful to :
- the bore holes campaign, in order to avoid some problems, like the meeting of an other sewer or buried cables ;
- the follow-up of some phenomenon like the subsidence ;
- the possible correlation between the microtunneling technology and the penetrometric and pressiometric tests, which could be interesting.

5 CONCLUSION

After the recorded data analysis, we have established that an important data ratio could not be studied. However, the values allow us to observe the instrumented equipment behaviour, and to verify that the operative procedure is well chosen.

This research project is just beginning and we can say with obtained results that we will be able to modelize the microtunneling machine behaviour during the boring, and optimize then its driving. The pipes design should profit by this study. Moreover, the lauching of the National Project about microtunneling machines should accelerate this study, and should allow :
- the knowledges spreading ;
- the technical recommendations publication ;
- the tuning of a french machine.

No Trenches in Town, Henry & Mermet (eds) © 1992 Balkema, Rotterdam. ISBN 90 5410 085 0

A new position detection system for microtunneling using a mounted optical fiber gyroscope

Kurosawa Tomohiro, Kawabata Kazuyoshi & Nojiri Yoshihiko
NTT, Japan

ABSTRACT : NTT has developed a new horizontal position detection system with a highly accurate optical fiber gyroscope for use in microtunneling systems. Continuous and reliable position detection over curved lines allows highly accurate curved and long-distance driving, and has expanded the range of working environments to which the microtunneling system is suited.

The new position detection system uses a highly accurate optical fiber gyroscope mounted on the driving machine to detect the minute changes in the angle and direction of driving. The horizontal driving line is calculated by a combination of this rate of change, readings from an accelerationmeter attached to the driving machine and a highly accurate distancemeter installed in the rear-end jacking unit. Because the driving speed is extremely slow, the optical fiber gyroscope loaded on the driving machine must be accurate enough to measure rotation on the earth. It has a diameter of 150mm.

With the system, the 300mm-diameter driving machine of the microtunneling system can drive long-distance and curved lines while maintaining a high level of efficiency and accuracy. Its successful application under actual working conditions has widened the range of environment microtunneling.

1. INTRODUCTION

Increasing social dependance on information has brought a sharp increase in demand for advanced communications services such as facsimile and data transmission. Providing such new services beyond the traditional telephone has placed greater emphasis on equipment and system reliability. As part of the drive to achieve this, NTT is now actively moving its communication transmission cables underground at a rate of 1,500 km a year. Underground cable laying has conventionally involved

Table 1: Comparison of position detecting methods

Method	Safety	Curved lines	Accuracy	Overall
Laser(photonic)	very good	fair	very good	fair
Angle of turn	very good	very good	good	good
Gyroscope	very good	very good	very good	very good
Electromagnetic	fair	very good	good	good

Table 2: Comparison of gyroscope methods

	Type of gyroscope		
	Gyrocompass	Rate integral gyroscope	Optical fiber gyroscope
Working principle	Mounting a weight on the gyroscope creates precession by gravity, causing, the rotation axis of the gyroscope to indicate north	Rate is detected from precession by the changes of direction	Speed of turn is detected using the sagnac effect — the passing speed of light changes as the loop of optical fiber cable turns
Accuracy	± 0.2 °	0.01～ 2 ° /hr	0.1 ～0.01° /hr
Startup time	About 2 hrs	About 15 mins	Several seconds
Reliability	Vulnerable to vibration	Vulnerable to vibration	Light, resistant to vibration and shock
Suitability	fair	good	very good

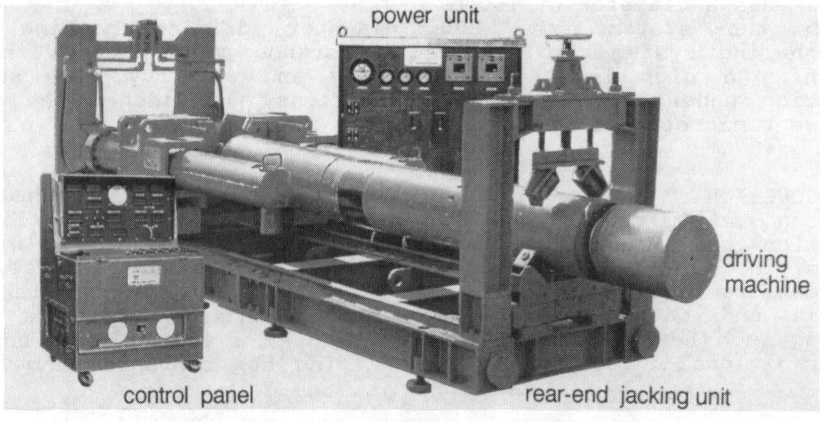

figure 1

digging trenches from the surface, but escalating cost and worsening working conditions spurred the search for an alternative.

The solution has been NTT's development of a trenchless microtunneling system called the Ace Mole series. There are four Ace Mole models in the series - the PC10, DC15, PL30 and DL35 - each suited to different soil conditions and pipe diameters. The PL30 system is designed in particular for laying 300mm diameter steel pipes in relatively soft soils with an N value of up to 15. It uses a pressure driving, non-soil discharge method of propulsion, allowing it to drive over long distances(up to 300m) and perform

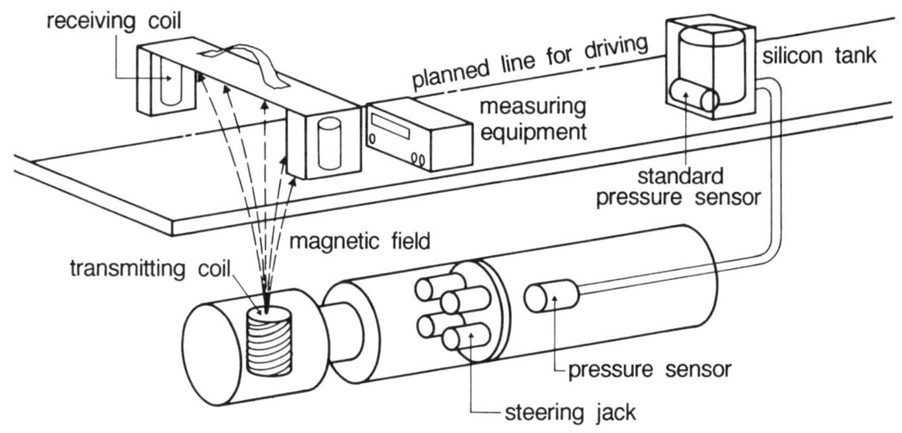

figure 2

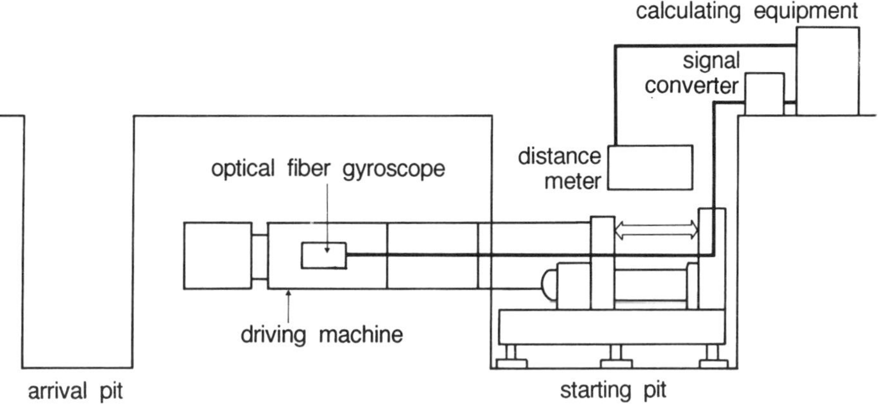

figure 3

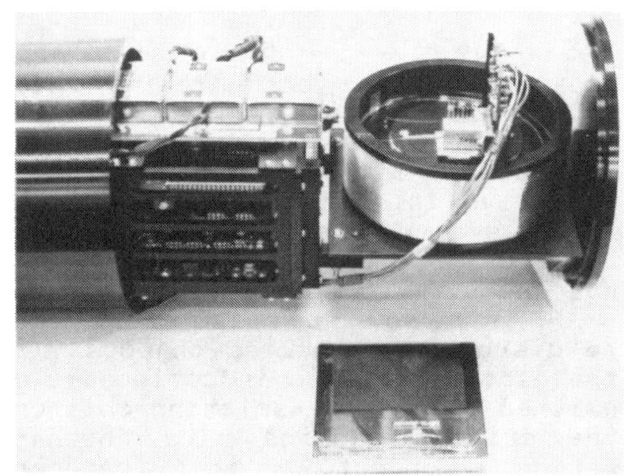

figure 4

curved-line work with a minimum radius of 100m. Fig. 1. shows the elements of the PL30 system : a driving machine, rear-end jacking unit, power unit and control panel.

2. CONVENTIONAL POSITION DETECTION SYSTEMS.

Accurate position detection is a major factor in the PL30's ability to drive over curves and long distances. As Fig.2 shows, the horizontal position of driving is calculated from the strength of the magnetic field generated by an electronic coil mounted on the driving machine, as measured from the surface above. The depth of driving beneath the surface is

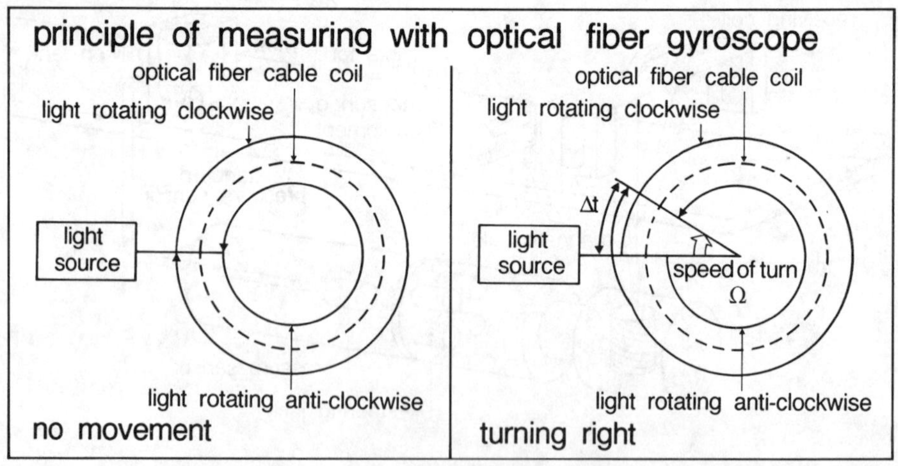

figure 5

figure 6

assessed by sensors which measure the difference between the pressure at the driving machine and the pressure on the ground. This approach is referred to as the electromagnetic induction method ; its depends on the principle that when a magnetic core crosses over a circle of conducting wire placed still in a magnatic field, an electrical voltage is created in the circle. In practical application, the electromagnetic force generated by the transmitting coil creates an electrical voltage in the receiving coil on the surface ; the size or variation in this voltage allows the horizontal position of the transmitting coil to be measured.

Although the electromagnetic induction method is generally effective in allowing the Ace Mole PL30 system to operate long-distance and curved-line work, it has encountered some problems :

. accuracy of measurement can be disrupted by buried objects. If the electromagnetic induction field emitted by the transmitting coil on the driving machine hits another buried object, it becomes dispersed, disrupting the reading obtained on the surface.

. accurate measurement is limited to a depth of about six

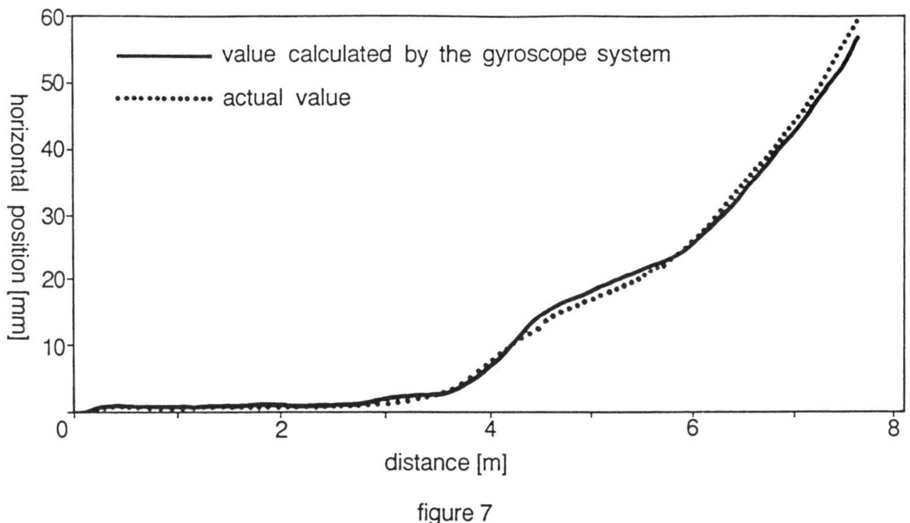

figure 7

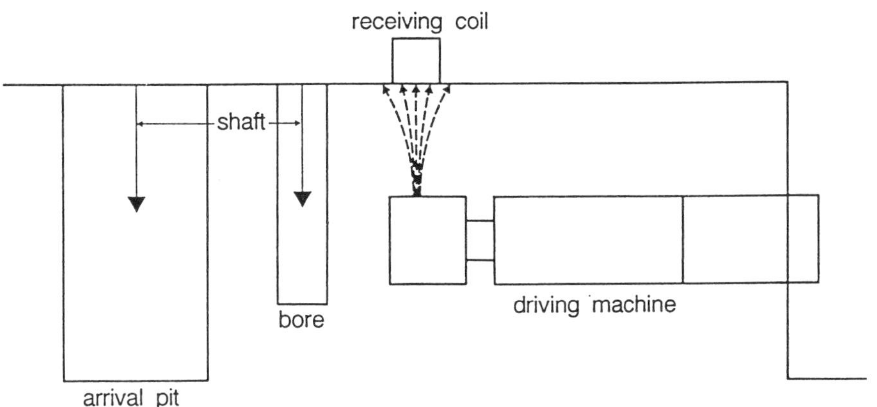

figure 8

meters. As the driving machine moves deeper underground, the electromagnetic induction field emitted by the transmitting coil becomes weaker. Boosting the strength of the field would only exacerbate the problem above if it hit another buried object. Six meters is the maximum depth at which accurate readings can be obtained using the most appropriate strength of electromagnetic field.

. measurement is difficult when driving under a river or a railway.Because the strength of the emitted field is measured at the surface, a river is highly unsuitable. Railways cause similar problems to buried objects, distorting the strength of the field.

. accuracy is impeded because measurement is not continuous along the entire length of the driving route.

. other problems include the danger of taking measurements on a road.

3. DEVELOPING A NEW APPROACH TO POSITION DETECTION.

The following three approaches to position detection where considered in terms of safety, suitability for curved-line work, and accuracy.

Laser method (photonic

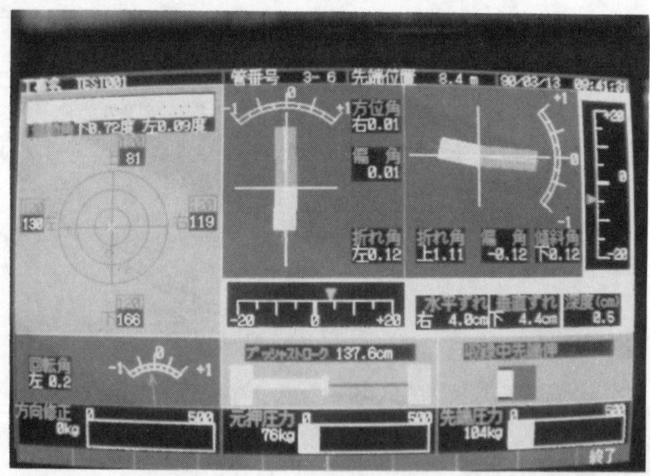

figure 9

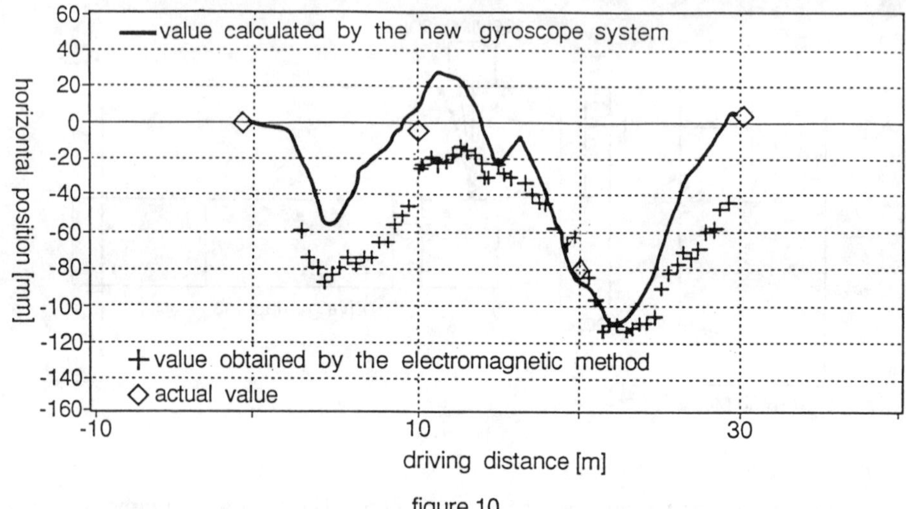

figure 10

measurement)
A laser target is mounted on the driving machine ; its position is detected by tracing it from the starting pit.

. Angle of turn method
An angle of turn mechanism is mounted on the driving machine, detecting its direction. Position is calculated from the angle of turn and distance driven.

. Gyroscope method
A gyroscope is mounted on the driving machine. The driving machine's direction is detected by the gyroscope, and the position is calculated from the direction and the distance driven.

As table 1. shows, the gyroscope method was judged superior to the others in terms of safety, curved-line suitability and measuring accuracy. The development team therefore moved on to refining the method. Table 2. shows the gyroscope variations considered. The optical fiber gyroscope was finally selected for use in Ace Mole systems for its accuracy of detection, its quick response, and its reliability.

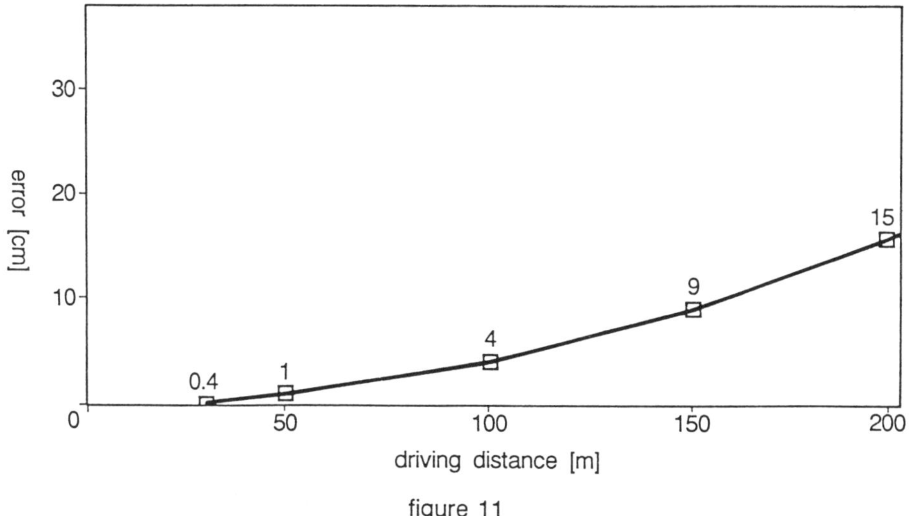

figure 11

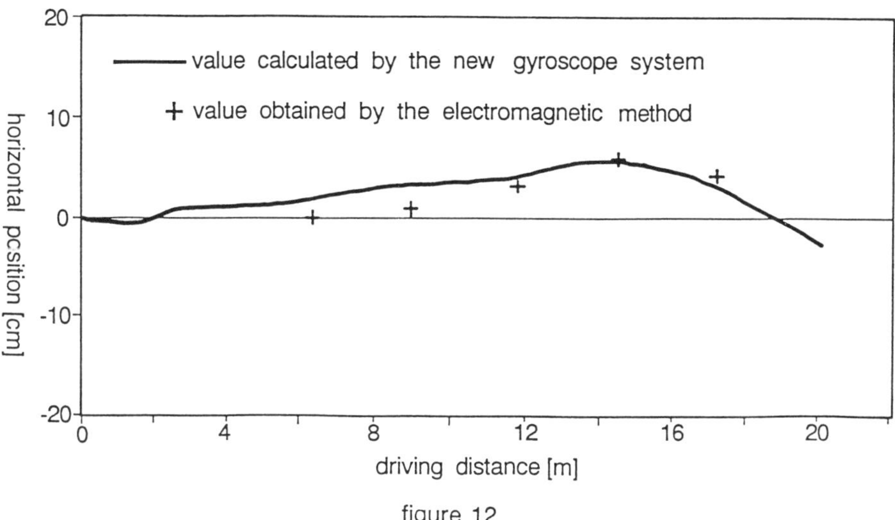

figure 12

4. OUTLINE OF THE POSITION DETECTION SYSTEM USING AN OPTICAL FIBER GYROSCOPE.

Figure 3. shows the structure of the new position detection system. It consists of an optical fiber gyroscope, a signal converter, a distance meter and a calculating equipment.

The optical fiber gyroscope measures the speed and angle of the gyroscope's rotation. As Fig. 4. illustrates, it consists of a coil around which the optical fiber cable is wound, and a unit that processes the signals emitted by the coil. Equal lengths of optical fiber cable are wound clockwise and anti-clockwise around the gyroscope's coil, and the point at which the optical signals enter and leave each cable is the same.

The gyroscope's measuring principle is presented in outline form in Fig. 5. When the gyroscope is not moving, meaning that the driving machine is not moving either, the difference in arrival time (Δ t) is zero. However, if the gyroscope coil begins rotating in a clockwise direction, indicating that the driving machine too is turning to the right, the value for Δt becomes positive. The speed of turn by the gyroscope (Ω) can then be established with the following formula :

$$\Omega = \frac{c^2 \times \Delta t}{4 \times S \times N}$$

Ω = speed of turn by the gyroscope
Δt = difference in arrival time
c = speed of light
S = surface area of coil
N = number of fibers wound on the gyroscope's coil

Once the value for Ω is known, the angle of turn by the gyroscope's coil in any specified period of time can be understood. The calculating device uses the angle of turn by the driving machine and the distance travelled to identify its movement.

The driving machine of the Ace Mole PL30 system moves forward very slowly. To be effective in position detection, the gyroscope mounted on the driving machine must be able to identify movement in the order of 0.001° - 0.0001°/second. A highly-accurate optical fiber gyroscope able to work to this degree of sensitivity was therefore used.

5. BASIC RESEARCH INTO SYSTEM EFFECTIVENESS.

To establish the basic feasibility of the new system, an experiment was performed as in Fig. 6. A highly accurate optical fiber gyroscope was fixed to a wagon which was then propelled on rails along an already measured route. The horizontal position readings calculated by the system were monitored. The wagon-mounted gyroscope was driven 8m along the rails five times. It produced differences between calculated position and measured position at an average of 7mm, and at one time as close as 3mm.

Fig. 7. shows the results for the closest run. Based on the error of 3mm, it was calculated that the error over a 50m driving course could be about 5cm. However, the measuring equipment used and the method of calculating distance travelled by the wagon may have affected the accuracy of this system. To gain appropriate readings under conditions of actual use, the gyroscope was mounted on the driving machine of an Ace Mole PL30 system and tested during driving.

6. TESTING UNDER ACTUAL USE CONDITIONS.

The new position detection method was tested with a PL30 system at NTT's Telecommunication Field Systems R+D Center in Tsukuba. Fig. 8. shows the overall testing set up, and Fig. 9. gives results recorded by the calculating device.

To confirm the accuracy of the new method's position detection, three independent tests were made : measurement with the electromagnetic induction method ; sinking a shaft from the surface into the bore to measure directly ; and measuring the horizontal position at the arrival pit.

Fig. 10. gives the results. The solid line is the value given by the new gyroscope system ; (+) shows the result obtained by the electromagnetic induction method ; and (◇) is the measurement at the bore and arrival pit. The values from the gyroscope and electromagnetic methods differ by a maximum of 5cm along the driving length. However, the values produced by the electromagnetic method already contain a margin of error, so the gyroscope's values were then compared with the measurements at the bore and arrival pit. Increased weight was given to the measurement at the arrival pit, as this was considered the more accurate. The readings from the system showed an error of just 4mm at the arrival point.

Fig. 11. shows how the figure of a 4mm error over 30m was used to estimate the accuracy of the gyroscope method over distances in the order of 50-100m. Accuracy was judged adequate for application of the new method to actual working sites. For driving distances over

100m, the system's readings can be adjusted by electromagnetic testing at the end of each 100m stretch.

7. PERFORMANCE IN USE.

Further tests on the new gyroscope system were carried out on an Ace Mole PL30 assignment in Amagasaki, Hyogo Prefecture, at a site with an N value of five.

In this case, cross-measurement at the bore and arrival pit were not possible. Confirmation of position values was made by comparison wih electromagnetic readings.

The results are shown in Fig. 12. The solid line represents the value measured by the gyroscope system, and (+) values are taken by the electromagnetic system. The values from the two methods differed by a maximum of 2-3cm along the route. Taking into account the error factor of the electromagnetic method, the system was judged accurate enough for the job.

8. CONCLUSIONS.

Compared with electromagnetic position detection, the new gyroscopic method has shown reliability and accuracy in both laboratory tests and under actual use conditions. It also offers an effective answer to the problem with conventional methods of horizontal position detection of non-continuous monitoring of the driving machine's location. As shown by the "angle of movement" in Fig.9., conventional methods are unable to identify the direction in which the driving machine is pointing. The new gyroscopic system allows the operator to control the direction of the driving machine, using the extremely useful information transmitted automatically from the tunnel. The new system can be expected to produce a new level of driving accuracy when it enters regular service. This enables the PL30 system to drive long-distance and curved lines while maintaining a high level of efficiency and accuracy.

9. FUTURE MOVES.

The current version of the gyroscopic position detection system uses automated processes to locate the driving vehicle and transmite data to a human operator. We are now researching ways to expand the applicability of the system versions by replacing the human operator with artificial intelligence, thereby making the system entirely automatic.

No Trenches in Town, Henry & Mermet (eds) © 1992 Balkema, Rotterdam. ISBN 90 5410 085 0

Un nouveau système de détection de position pour le micro-tunnelier utilisant un gyroscope embarqué à fibre optique

Kurosawa Tomohiro, Kawabata Kazuyoshi & Nojiri Yoshihiko
NTT, Japon

RESUME : NTT a développé un nouveau système de détection de position hrozontal avec un gyroscope à fibre optique à haute résolution pour l'utilisation dans des systèmes de micro-tunneliers.Une détection de position continue et fiable sur des lignes courbes permet une conduite sur des courbes et à longue distance avec une grande précision, et a ainsi étendu le domaine d'application pour lequel le système de micro-tunnelier est utilisé.

Le nouveau système de détection de position utilise un gyroscope à fibre optique de haute résolution embarqué dans le tunnelier, lequel détecte les changements d'angle et de direction dans le guidage de l'ordre de la minute. La trajectoire horizontale est calculée par une combinaison de ces taux de changement lus à partir d'un accéléromètre monté dans la machine et d'un distancemètre à haute résolution installé sur les vérins de poussée. Par le fait que la vitesse d'avancement est extrêmement lente, le gyroscope à fibre optique embarqué dans la machine doit être suffisamment précis pour mesurer la rotation en place. Il a un diamètre de 150mm.

Avec le système, la tête du micro-tunnelier de 300mm de diamètre peut être contrôlée sur de longues distances et sur des lignes courbes tout en maintenant un haut niveau d'efficacité et de précision. Les applications fructueuses de ce système, dans des conditions réelles de travaux, ont élargi les possibilités d'application du micro-tunnelier.

1. INTRODUCTION.

L'accroissement de la dépendance sociale sur l'information a apporté une brutale augmentation dans la demande pour des services de Communications avancées telles que la télécopie et les transmissions de données. La fourniture de tels services à travers le téléphone traditionnel a mis une plus grande accentuation sur les équipements et sur la fiabilité des systèmes. Pour permettre de réaliser ceci, NTT est actuellement entrain de changer ses transmissions par câbles souterrains à une moyenne de 1.500 Km par an. La pose de câbles souterrains utilisait habituellement des tranchées mais les coûts croissants et les pires conditions de travail ont stimulé les recherches pour trouver une alternative.

La solution qu'a développée NTT pour un système de travaux sans tranchée a été appelée "la série Ace Mole". Il y a quatre modèles Ace Mole dans la série : PC10, DC15, PL30 et DL35, chacun étant adapté à des conditions de sols différentes et à des diamètres de conduits différents. Le système PL30 est conçu en particulier pour poser des tuyaux d'acier de 300mm de diamètre dans des sols relativement mous avec une valeur de N jusqu'à 15. Il permet, grâce à sa méthode de propulsion, de piloter sur de longues distances (sup. à 300 m) et d'effectuer des trajectoires courbes avec un rayon minimum de 100m. La Fig. 1. montre les éléments du système PL30 : un berceau de poussée, l'unité de vérins, les groupes hydrauliques et le panneau de contrôle.

2. SYSTEME CONVENTIONNEL DE DETECTION DE POSITION.

La détection précise de position est une des particularités du système PL30 dont les possibilités d'effectuer des courbes sur des longues distances. Comme le montre la Fig. 2., la position horizontale de la tête est calculée à partir d'un champ magnétique généré par une bobine électronique montée sur la tête, et ceci à partir de la surface. La profondeur de la tête par rapport à la surface est assurée par des capteurs qui mesurent la différence de pression à la tête du tunnelier et à la surface. Cette approche est appelée la méthode à induction électro-magnétique. elle est basée sur le principe que lorsque un noyau magnétique traverse un câble conducteur circulaire placé également dans un champ magnétique, un courant électrique est créé dans le conducteur circulaire. Dans les applications pratiques, les forces électro-magnétiques engendrées par la bobine émettrice créée un courant électrique dans la bobine réceptrice en surface. L'amplitude, les variations dans les tensions permettent de mesurer la position dans le plan horizontal de la bobine émettrice.

Bien que la méthode d'induction électro-magnétique généralement efficace pour permettre au système Ace Mole du PL30 pour effectuer des opérations courbes et à longue distance, elle a rencontré quelques difficultés :

. la précision de la mesure peut être perturbée par des objects enterrés. Si le champ d'induction électro-magnétique émis par la bobine émettrice sur la tête de la machine rencontre un autre objet enterré, il devient dispersé, perturbant la lecture obtenue à la surface.

. les mesures précises sont limitées à des profondeurs approximatives de 6m. Lorsque la tête du micro-tunnelier est plus profonde, le champ d'induction électromagnétique émis par la bobine émettrice devient plus faible. Augmenter l'intensité du champ reviendrait à accroître le problème précédent s'il y a un objet enterré. Six mètres est la profondeur maximale pour laquelle des lectures précises peuvent être obtenues, utilisant une intensité appropriée du champs électro-magnétique.

. la mesure est difficile lorsque les travaux ont lieu sous une rivière ou sous voie ferrée. Par le fait que l'intensité du champs émis est mesurée à la surface, une rivière est particulièrement défavorable. Les voies ferrées causent des problèmes semblables aux objets enterrés créant des distorsions de l'intensité du champs.

. la précision est limitée parce que la mesure n'est pas continue le long du tracé.

. des problèmes de sécurité peuvent apparaître pour la prise de mesures sur la chaussée.

3. DEVELOPPEMENT D'UNE NOUVELLE APPROCHE POUR LA DETECTION DE POSITION.

Les trois approches suivantes pour la détection de position ont été étudiées en termes de sécurité, faisabilité pour des tracés courbes, et de précision.

. Méthode Laser

Une cible laser est montée sur la tête de la machine ; sa position est détectée par repérage dans le puits d'entrée.

. Méthode d'angle de rotation

Un mécanisme de mesure d'angle de rotation est monté sur la tête de la machine pour détecter la direction. La position est calculée à partir de l'angle de rotation et de la distance parcourue.

. Méthode gyroscopique

Un gyroscope est placé sur la tête de la machine. La direction de la tête est détectée par le gyroscope et la position est calculée à partir de la direction et de la distance parcourue.

Comme le montre le tableau 1., la méthode gyroscopique a été jugée supérieure aux autres méthodes en termes de sécurité, faisabilité d'une courbe et de précision de mesures. Le propos a donc été d'affiner la méthode. Le tableau 2. montre les différents types de gyroscopes. Le gyroscope à fibres optiques a finalement été sélectionné pour être utilisé dans le système Ace Mole pour sa précision de détection, sa réponse rapide et sa fiabilité.

METHODE	SECURITE	TRACE COURBE	PRECISION	CONCLUSION
Laser (photonique	très bon	moyen	très bon	moyen
Angle de rotation	très bon	très bon	bon	bon
Gyroscope	très bon	très bon	très bon	très bon
Electromagnétique	moyen	très bon	bon	bon

Tableau 2. : Comparaison de méthodes gyroscopiques

Type de Gyroscope

	Gyrocompas	Gyroscope à intégration de vitesse	Gyroscope à fibre optique
Principe de fonctionnement	Mettre un poids sur un gyroscope créée une précession par gravité, causant une rotation des axes du gyroscope pour indiquer le nord	La vitesse est détectée à partir de la précession par le changement de direction.	La vitesse de rotation est détectée en utilisant l'effet sagnac. La vitesse de la lumière change lorsque la boucle de la fibre optique tourne.
Précision	plus ou moins 0,2°	0,01 - 1°/hr	0,1 - 0.01°/hr
Temps de démarrage	environ 2 heures	environ 15 minutes	plusieurs secondes
Fiabilité	Vulnérable aux vibrations	Vulnérable aux vibrations	résistant aux vibrations et aux chocs.
Adaptabilité	Moyenne	bonne	très bonne

4. DESCRIPTION DU SYSTEME DE DETECTION DE POSITION UTILISANT UN GYROSCOPE A FIBRE OPTIQUE.

La Fig. 3. montre la structure du nouveau système de détection de position. Il comprend un gyroscope à fibre optique, un convertisseur de signal, un système de mesure de distance et un équipement de calcul.

Le gyroscope à fibre optique mesure la vitesse et l'angle de rotation du gyroscope. Comme il est illustré dans la Fig. 4., il consiste en une bobine autour de laquelle une fibre optique est enroulée et une unité qui exécute le signal émis par la bobine. Des longueurs identiques de fibre optique sont bobinées dans le sens des aiguilles d'une montre et dans le sens contraire autour de la bobine du gyroscope et le point où les signaux entrent et sortent sur chaque câble est le même.

Le principe de mesure du gyroscope est présenté de façon schématique dans la Fig. 5. Lorsque le gyroscope n'est pas en mouvement, c'est-à-dire lorsque la tête de la machine n'est pas en mouvement, la différence dans les temps d'arrivée (Δ t) est égale à zéro. Cependant, si la bobine du gyroscope commence à tourner dans le sens des aiguilles d'une montre indiquant que la tête de la machine est en train de tourner à droite, la valeur de Δ t devient positive. La vitesse de rotation donnée par le gyroscope (Δt) peut être établie à partir de la formule suivante :

: vitesse de Ω rotation du gyroscope

: différence dans les temps d'arrivée

: vitesse de la lumière

: surface de la bobine

: quantité de fibres enroulées sur la bobine du gyroscope.

Ainsi, la valeur de est connue, l'angle de rotation de la bobine du gyroscope pendant une période de temps spécifiée peut être appréhendée. Le système de calcul utilise l'angle de rotation de la tête de la machine et la distance parcourue pour identifier son mouvement.

La machine Ace Mole PL 30 avance très lentement. Pour avoir une détection de position efficace, le gyroscope monté sur la tête doit être capable d'identifier des mouvements de l'ordre de 0,001° - 0,0001°/seconde. Un gyroscope à fibre optique de haute résolution capable de travailler avec cette précision a alors été utilisé.

5. RECHERCHE DE BASE SUR L'EFFICACITE DU SYSTEME.

Pour établir la faisabilité de base du nouveau système, un montage expérimental a été réalisé comme le montre la Fig. 6. Un gyroscope àfibre optique de haute sensibilité a été fixé à un chariot qui a été guidé sur des rails le long d'un tracé pré-défini. Les lectures de la position horizontale calculée par le système ont été enregistrées. Le gyroscope monté sur le chariot a été manoeuvré cinq fois sur le banc. Ceci a produit des différences entre la position calculée et la position mesurée d'une valeur moyenne de 7mm et une fois inférieure à 3mm.

La Fig. 7. montre les résultats pour le meilleur passage. En se basant sur l'erreur de 3mm, il a été calculé que l'erreur sur un tracé de 50m serait d'environ 5cm. Cependant, l'équipement de mesure utilisé et la méthode de calcul de distances effectuée par le chariot peuvent avoir affecté la précision de ce système. Pour obtenir des lectures appropriées dans des conditions réelles d'utilisation, le gyroscope a été monté sur une tête de micro-tunnelier Ace Mole PL30 et testée durant le percement.

6. ESSAIS EN CONDITIONS RELLES.

La nouvelle méthode de détection de position a été testée avec le système PL 30 sur un chantier expérimental dans le Centre de Recherches de Tsukuba. La Fig. 8. montre le schéma expérimental et la Fig. 9. donne les résultats enregistrés par le système de calcul.

Pour confirmer la précision de cette nouvelle méthode de détection de position, trois essais indépendants ont été effectués: mesure avec la méthode d'induction électro-magnétique ; enfoncement d'une aiguille de la surface dans le micro-tunnel pour une mesure directe ; et mesure de la position horizontale depuis l'arrivée.

La Fig. 10. donne les résultats. La ligne continue est la valeur donnée par le nouveau système gyroscopique; (+) montre les résultats obtenus par la méthode d'induction électro-magnétique et (◊) est la mesure directe et au puits d'arrivée. Les valeurs obtenues par le gyroscope et la méthode électro-magnétique diffèrent d'environ 5cm au maximum le long du tracé. Cependant, les valeurs données par la méthode électro-magnétique contiennent toujours une marge d'erreurs dont les valeurs gyroscopiques ont été comparées avec les mesures de positionnement direct du puits et du positionnement du point d'arrivée.

Un poids plus important a été donné à la mesure au puits d'arrivée car elle est considérée comme étant la plus précise. Les lectures des systèmes ont montré une erreur de juste 4mm au point d'arrivée.

La Fig. 11. montre comment l'hypothèse de 4mm d'erreur sur 30m a été utilisée pour estimer la précision de la méthode gyroscopique sur des distances de 50-100m. La précision a été jugée adéquate pour l'application de la nouvelle méthode à des travaux réels. Pour des tracés supérieurs à 100m, le système de lecture peut être réajusté par un test électro-magnétique à la fin de chaque tranche de 100m.

7. PERFORMANCE EN UTILISATION.

Des tests complémentaires avec le nouveau système gyroscopique ont été effectués sur le système Ace Mole PL30, à Amagasaki, Hyogo Prefecture, dans un site avec une valeur de N = 5.

Dans ce cas, des mesures directes de positionnement de la canalisation et d'arrivée au puits n'ont pas été possibles. La confirmation des valeurs de position ont été faites par comparaison avec des lectures électro-magnétiques.

Les résultats sont montrés dans la Fig. 12. La ligne continue représente la valeur mesurée par le système gyroscopique et (+) les valeurs données par le système électro-magnétique. Les valeurs données par les deux méthodes diffèrent d'un maximum de 2-3cm le long du tracé. En prenant en compte l'incertitude de la méthode électro-magnétique, le nouveau système a été jugé suffisamment précis pour l'exploitation.

8. CONCLUSIONS.

Comparée avec la méthode de détection deposition électro-magnétique, la nouvelle méthode gyroscopique a montré fiabilité et précision tant en essais de laboratoire qu'en conditions réelles d'utilisation. Il offre également une réponse réelle au problème avec des méthodes conventionnelles de détection de position horizontale et d'enregistrements non continus de localisation de la tête de la machine. Comme il est montré par l'angle de mouvement dans la Fig. 9., les méthodes conventionnelles sont incapables d'identifier la direction vers laquelle la tête de la machine se dirige. Le nouveau système gyroscopique permet à l'opérateur de contrôler la direction de la tête en utilisant l'information pratique transmise automatiquement à partir du tunnel. Il est permis d'attendre du nouveau système de produire un nouveau niveau de précision de conduite lorsqu'il rentrera en service régulier. Celui-ci permet au système PL 30 d'effectuer des longues distances et des lignes coubes tout en mantenant un haut niveau d'efficacité et de précision.

9. DEVELOPPEMENTS FUTURS.

La version ordinaire du système gyroscopique de détection de position utilise un processus automatique pour positionner la tête et transmettre les données à un opérateur. Nous recherchons actuellement des pistes pour étendre l'applicabilité des versions de ce système en remplaçant un opérateur humain par de l'intelligence artificielle, c'est-à-dire, en rendant le système complètement automatique.

(voir les figures dans la traduction Anglaise du texte).

No Trenches in Town, Henry & Mermet (eds) © 1992 Balkema, Rotterdam. ISBN 90 5410 085 0

On a common fuzzy control system for the micro-tunnelling – Introduction of control configured construction machine

Tomoji Takatsu, Hideyuki Takeda & Atsushi Sugiyama
Public Works Research Institute, Ministry of Construction, Tsukuba City, Ibaraki, Japan

ABSTRACT : In this paper, we study an efficient development of the micro–tunnelling system in order to cope increase of the number of micro– tunnelling works.
Firstly, we indicate the need to establish a standard for the efficient system, as development. Secondly, we propose a virtual machine and a fuzzy theory applied on the machine controlling system which is the standard model. Finally, the performance of the controller is shown by actual construction works.

Key words – Fuzzy Control, Adaptive Control, Control Configured Construction Machine

1 INTRODUCTION

The 40 millions people use the sewerage in Japan. Although the percent of sewered population has been increased in these years, the ratio of sewered population is lower than any other advanced country's as from figure 1 and figure 2. [Uehara, 1987] This survey shows us the needs to promote the construction of more swerages in our country. There are two types of method to construct sewer culverts. The one is the open cut method. This method causes traffic blockage because it is conducted at busy and narrow streets, Which in fact is the nature of Japan. The other one is the pipe pushing method which creates less traffic at construction sites.

Recently, we have come to use a micro–tunnelling system at a place, such as at a railroad and at a road for constructing culverts, pipes and etc. under the ground. (The micro– tunnelling is one of pipe jacking methods which applies pipes of inner diameter with less than 800mm in Japan.) The reason of adopting the micro–tunnelling system is that we construct culverts and pipes in the depth of life–lines at 5 meters or deeper. We do this because some pipe–lines have been already buried in the underground at the depth of less than 5 meters. The open cut method is costly , takes time and requires lar er s ace when di in more than 5 meters, so it is not advantageous to apply in that case. Moreover the number of construction has been increased at rural cities. At rural cities,it is inclined to use the pipes of which the diameter is below 800 mm. The first time we used the micro tunnelling system was back in the age of the ancient Rome when sewer pipes was placed under the ground. In Japan, we use the micro–tunnelling system for the first time in 1958 at Amagasaki city in Hyogo prefecture.

Now, we apply the system not only in constructing sewer pipes but also in extending telephone cables. The modern micro– tunnelling system consists of only one power transmission to run various digging mechanism. The transmitter transmits the power from the cylinder to the cutter head. The micro–tunnelling system needs a control system for correcting bended pipes. The reason why it is needed is mainly because workers cannot enter into the pipes for placing them to the right position when the pipes are off the line from the place which is originally planned to be set. It is necessary for the machine operator to be very skillful in operating the control mechanism, because controlling the operation is differed depending upon the length of pipe line, the variation of the soil and so on. It takes time for young construction workers to master the micro–tunnellin controllin s stem. Besides,

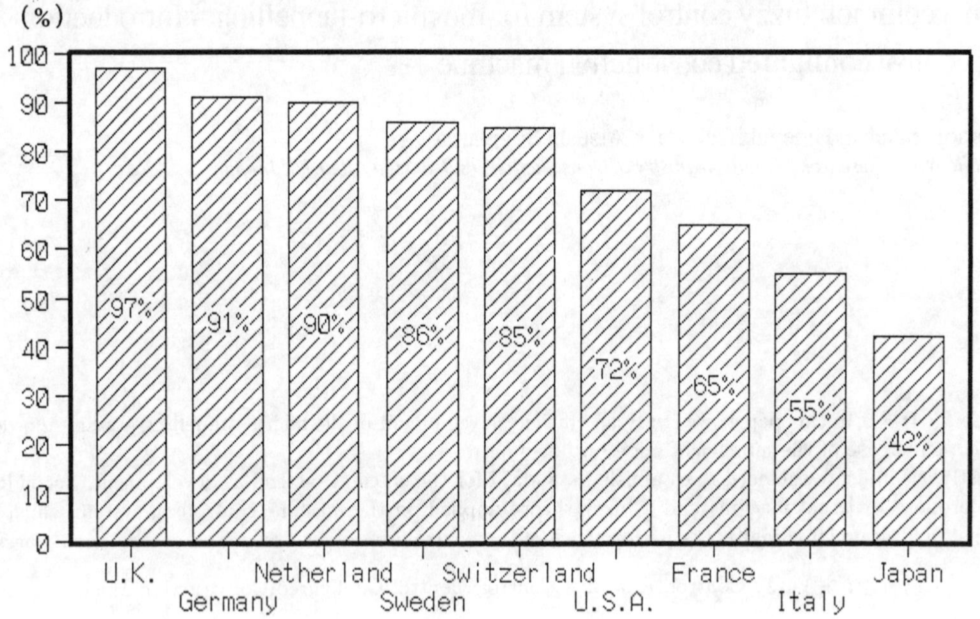

Fig. 1 Percentage of sewered diffusion

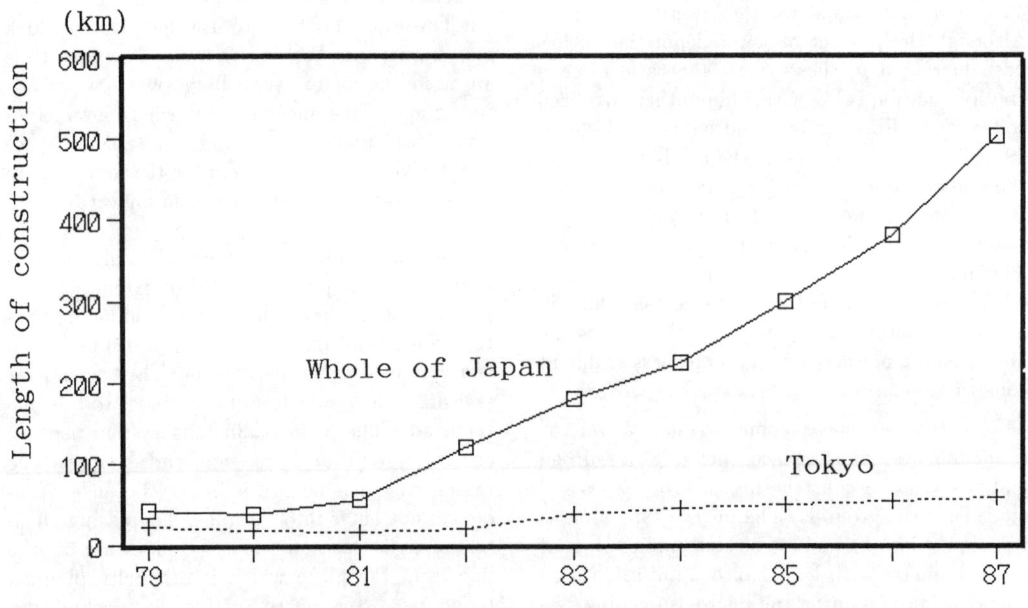

Fig. 2 Transition of Micro-tunnelling

Table 1: Compression of research expenses

	Market (billion yens)	Research expenses (percent)	Research expenses (billion yens)
Elector industrial	4000	5(%)	200
Motor Industry	3600	3(%)	108
Micro-tunnelling	75	0.5(%)	0.38

there is shortage of workers in construction industries. So,constructors and makers have been developing a control system of micro-tunnelling which does not need so much manpower.

The ministry of construction has started a joint project of "AI control system for the micro-tunnelling" with 12 private companies. It's term is from 1988 to 1991.

2 NEEDS OF DEVELOPMENT

First of all, we compare the scale between the factory industry and the sewage construction using the micro- tunnelling system. Scales of markets are shown in table 1. [Databook, 1988][Newspaper, 1990] The table also shows the percentages of research expenses for the micro-tunnelling, which is 4/1000 times as much as that of the electorindustrial. This implies that we must have more efficient researches and developments for the micro-tunnelling.

Secondly, we consider the comparison of the construction work and factory industry. There are few difference between the construction work and factory industry. One of case is the size of works to be operated. We must consider the kilometer order length in micro-tunnelling. On the other hand, the factory operates the meter order length. The next one is the characteristics of works to be operated. At factory industry, each tools only operates a few types of work which have little dispersion in to the equality control. On the other hand, we must work with various objects for micro-tunnelling under unfixed conditions compared to that of the factory industry. The third is the relation between works and tools (machines). The construction machines need to be carried to do the work, but in factory production, generally

Table 2: Comparison among factory, construction and micro-tunnelling

	Factory	Construction	Micro-tunnelling
Size of works	Meters	Kilometer	100 Meters
Characteristics of Works	Constant	Not Constant	Constant
Movement of Tools	Stand	Move	Move
Movement of works	Move	Stand	Move/Stand
Handling and Work	Separated	Not Separated	Separated

the tools are fixed for manufacturing. The fourth one is the relation between handling and work. The handling and work are separated at the factory industry. At the construction works, they cannot be separated clearly. [Muramatsu T. and Cho K., 1989]

Under those circumstances, technologies used in the factory industry are not appropriate in construction industry if there is no modification is made. We ought to develop a new technology for construction works.

Finally, we show some characteristics of the micro-tunnelling. We consider the length of the line is to be about 100 meters. Conditions of soil are not constant and it often changes. The works are constant, such as piping sewage and placing telephone cables and so on. It implies that the prefabrication construction has been advanced in the micro-tunnelling. The machine head moves in construction work. The handling and the work are separated.

From the above consideration in mind, the system for the micro- tunnelling must be developed by the original approach. We would define some standard for micro-tunnelling in order to make the development system efficient. Because the characteristics of micro-tunnelling is considered same with general construction though it is not, and the development budget is less than those of Elector industry, Motor industry and so on. The standardization should be decided in the aspect of hardware, application software and developmental environment.

3 AIMS OF DEVELOPMENT

We set aims of this development as follows,

1. The system needs the skillful operator. This is called deskillful system which is the one of important key words at the 9th International Symposium on Automation and Robotics in Construction, June 1992.

   This implies that we have an advanced concept for the Automation and Robotics in Construction

2. The ability of the control system is to keep deviations from the normal position.

   (a) Vertical 30mm

   (b) Horizontal 50mm

   The aims is decided by the result of questionnaires for 140 cities and contractors in Japan at 1988. If there is the ability to realize above at the control system, it will be pass almost check in Japan.

4 COMMON CONTROLLER

4.1 FUNDAMENTAL CONCEPT

There are some standardization for the computer systems. The one is the hardware standard. This should be considered from a CPU level to a connector level. The standardization of programming language such as FORTRAN 77 is the next be considered. The third one is the standardization of OS (for example UNIX, TRON and so on). Protocols, MAP (Manufacturing Automation Protocol) might be considered also.

First of all, we discuss I/O standard connectors. The reason is that we can easily exchange measuring instruments with the application of the standard so that we can finish the construction work by the due date.

We secondly research and develop an application software [Takatsu and Takeda,1989] which is a sort of adaptive control system. The motivation of the development is that the system can be used in any micro–tunnelling system with any soil conditions.

Thirdly, a standardization of hardware is defined. The most simple way to realize the standarization of hardware is obtained when all of micro–tunnelling systems use a same hardware. It takes a lot of time and money for the development of the standard hardwares. To utilize conventional hardwares, there are two ways. The one is to use the whole system, OS, programming language and so on. However, the existed systems include functions like a graphic function which is not needed to the micro–tunnelling system, so that it is hard to select the best system appropriate for the micro–tunnelling.

Thus, we propose a concept of a virtual machine (see figure 3). In general, the virtual machines are not functioned with real time and not for multitask purpose because it is simulated by work stations. We request the virtual machine to be for multitask and work on real time. The virtual machine needs multitask function because the measurement and some control are separately implemented, but the programs to do each work should run at the same time in order not to miss any data. The real time property are required so as to perform jobs in the field. The virtual machine we try to develop is not a simple emulator but it should be a highly qualified control system. There are two types of systems. The one is a work stations in a laboratory, and the other is a micro–computer at the work site. The former has some simulators and data base. The simulator is used to train operators and to develop new controller. The data base supply data to the simulator and support a new development. A micro–computer at work site is only loaded with data base. We would propose to use IC cards in order to send the data from field systems to laboratory systems.(See Fig. 3)

4.2 FUZZY CONTROL

Professor Zadeh has established Fuzzy logic at 1962. The Fuzzy logic is one of set theory of which characteristic function is defined as follows.

$$\mu(x) = \begin{cases} 1 & (x \in A : aset) \\ 0 & (x \notin A : aset) \end{cases}$$

The range of characteristic function at Fuzzy logic is

$$\mu(x) = [0,1] \in \mathcal{R}(x \in A : aset)$$

and is called the membership function. We can operate the feeling of "A is big", "B is rather big" and so on. The feeling like this cannot be operated by ordinary set theory.

The Fuzzy control is application of the Fuzzy logic. The Fuzzy control is combined the Fuzzy logic with IF THEN rule. IF THEN rule is convenience to express human operations and judgments. Moreover, the feeling of human is explained by the Fuzzy logic. The Fuzzy control is very convenience to transport human skills to computer program.

The reason why we use the Fuzzy control is not only to transport human skills to computer program but also easy to change the program.

4.3 REAL SYSTEM

In. this section, we introduce about a real system which is developed by the concept of standard discussed the previous section.

First of all, we must have an actual machine. A fuzzy computer is selected as the actual machine to be used, because micro– tunnelling attribute is still unclear as already being discussed and machine operation is described by the language. Moreover, the operation patterns are same basically, even if the machines are different and membership function and its role are same even though the fuzzy application is differed. Therefore, it implies that we need only one control system for the micro tunnelling using the fuzzy logic. Secondly, we define the I/O, the communication protocol and so on.

Now, a few companies participate in this standardization project. They finish testing the system in the laboratories field and real constructions. We show the data of real constructions by the fuzzy controller.(see fig.8)

4.4 FIELD TESTS

4.4.1 BASE MACHINES

This Common Fuzzy Controller is adapted three type machines. The machines are follows,

1. Pressurized slurry type (One–process type) for gravel (1st Case)

2. Pressurized slurry type (One–process type) for sandy clay (2nd Case)

3. Jack–in method (One–process type) (3rd Case)

Table 3: Results of field tests

	1st Case	2nd Case	3rd Case
Deviation (vertical)	22mm	21mm	27mm
Deviation (horizontal)	8mm	9mm	43mm
Length	56m	108m	60m

4.4.2 RESULT OF TESTS

The result of field test is shown by table 3. From the table 3, it proved that :

1. The common Fuzzy controller can be applied at least 3 types micro–tunnelling machines.

2. The deviation from the planning line is fitted the aim of development.

This is implies that the common Fuzzy controller makes the micro–tunnelling machines easy–operating ones i.e. *Deskillful system.*

5 DISCUSSIONS

5.1 FUZZY CONTROLLER

In this paper, we propose some standardization in the micro– tunnelling system which are derived from the potentiality of development and the characteristics of construction works. Standardization may slowdown a development of new technology. It may be true at the factory industry, but it is not demerit in construction field. What is important is that it is possible to realize it at present technology. The standardization gives us very convenient merits as follows;

1. We can make some application programs used in micro– tunnelling systems.

   This makes the development cost lower and the quality of the program higher.

2. We can easily exchange parts of the system.

   This makes the construction term accurate.

3. We can easily build up the data base.

   This realizes an efficient new system development.

The performance of the fuzzy controller reach the goal whose values are 50mm horizontal error and 30mm vertical error. This goal is authorized by a committee in the Japan Micro– tunneling Association whose chairman is Dr. Nagata and co–chairman is Dr. Takeshita.

We are is award the prize for this concept of the Common Fuzzy Controller from *Japan Construction Mechanization Association* in May 1992.

5.2 ADVANCED RESEARCHES

5.2.1 OTHER APPROACH

The weak point of the Fuzzy control is that there is no guarantee of the stability, the controllerbility and the observability. These properties are important at the system and control theory.

So, we investigate another control technique though the Fuzzy control has the advantage of easy programming. The control technique is the adaptive control derived from the stochastic system theory which is one of the most advanced theory.

The adaptive control is made up of the parameter identification, the state estimation and the optimal control. In this paper, we would like to discuss the parameter identification and the state estimation, because the estimation process is importance for the adaptive control.

5.2.2 PROBLEM FORMULATION

Although there exists a lot of representations of the system dynamics and the observation mechanism, the phenomena of the micro–tunnelling are expressed as the discrete equation, in this paper. The reason is that the micro-tunnelling system often includes digital computers so that these digital computers take a share in the measurement. Then the system dynamics and the observation mechanism are described as follows:

System:

$$\begin{aligned} z(k+1) &= A(k)z(k) + G(k)w(k) \\ z(0) &= z_0 \quad (k = 0, 1, 2, \cdots) \end{aligned} \tag{1}$$

Observation:

$$\begin{aligned} y(k) &= H(k)z(k) + R(k)v(k) \\ & (k = 0, 1, 2, \cdots) \end{aligned} \tag{2}$$

Parameter:

$$\begin{aligned} \theta(k+1) &= C(k)\theta(k) + S(k)\mu(k) \\ \theta(0) &= \theta_0 \quad (k = 0, 1, 2, \cdots) \end{aligned} \tag{3}$$

where $z(\bullet)$, $y(\bullet)$ and $\theta(\bullet)$ are n, l and n^2 vectors $A(\bullet)$, $G(\bullet)$, $H(\bullet)$, $R(\bullet)$, $C(\bullet)$ and $S(\bullet)$ are $n \times m_1$, $l \times n, l \times m_2$ $n^2 \times n^2$ and $n^2 \times m_3$ matrices
$\theta(\bullet) = [a_{11}(\bullet), a_{12}(\bullet) \cdots a_{nn}(\bullet)]'$
and where $a_{ij} = \{A(\bullet)\}_{ij}$, "$'$" implies the transposition.

$z(\bullet)$ is the system state value which includes the drift of the cutter head and so on. $w(\bullet)$, $v(\bullet)$, and $\mu(\bullet)$ are mutually independent white Gaussian noise. We may regard the parameter of $\theta(\bullet)$ as the set of muddles.

5.2.3 ESTIMATOR

There are a lot of approaches of estimator of the system state and the parameter. In this paper, we find the minimum variance estimator because this estimator are derived easier than any other estimators.

Now, let the form of the estimators as

$$\begin{aligned} \hat{z}(k+1|k+1) &= \hat{z}(k+1|k) + K_z(k+1) \\ &\{y(k+1) - H(k+1)\hat{z}(k+1|k)\} \\ \hat{z}(0) &= \hat{z}_0 \qquad (k = 0, 1, 2, \cdots) \end{aligned} \tag{4}$$

$$\begin{aligned} \hat{\theta}(k+1|k+1) &= \hat{\theta}(k+1|k) + K_\theta(k+1) \\ &\{y(k+1) - H(k+1)\hat{z}(k+1|k)\} \\ \hat{\theta}(0) &= \hat{\theta}_0 \qquad (k = 0, 1, 2, \cdots) \end{aligned} \tag{5}$$

where $\hat{\bullet}(i|j) = E\{\bullet(i)|Y^j\}$
and where the operator E implies the expectation and Y^j is the set of $\{y(k)|k = 0, 1, \cdots, j\}$.

The purpose of this investigation is to find $K_z(k+1)$ and $K_\theta(k+1)$ to minimize the trace of estimation error martices;

$$\begin{aligned} & P_z(k+1|k+1) \\ &:= E\{[z(k+1) - \hat{z}(k+1|k+1)] \\ & \quad [z(k+1) - \hat{z}(k+1|k+1)]'|Y^{k+1}\} \end{aligned} \tag{6}$$

$$\begin{aligned} & P_\theta(k+1|k+1) \\ &:= E\{[\theta(k+1) - \hat{\theta}(k+1|k+1)] \\ & \quad [\theta(k+1) - \hat{\theta}(k+1|k+1)]'|Y^{k+1}\} \end{aligned} \tag{7}$$

$K_z(k+1)$, $K_\theta(k+1)$, $\hat{z}(k+1|k)$ and $\hat{\theta}(k+1|k)$ are calculated by the following equations:

$$\frac{\partial}{\partial K_z \ k+1} tr.\{P_z(k+1|k+1)\} = 0 \tag{8}$$

$$\frac{\partial}{\partial K_\theta(k+1)} tr.\{P_\theta(k+1|k+1)\} = 0 \quad (9)$$

$$K_z(k+1) = M(k)H'(k+1) \\ [H(k+1)M(k)H'(k+1) + R(k)R'(k)]^{-1} \quad (10)$$

$$K_\theta(k+1) = \hat{C}(k|k)M'_{z\theta}(k)H'(k+1) \\ [H(k+1)M(k)H'(k+1) + R(k)R'(k)]^{-1} \quad (11)$$

$$M(k) = [\hat{A}(k|k)P_z(k|k) + \hat{Z}(k|k)P'_{z\theta}(k|k)]\hat{A}'(k|k) \\ + \ M_{z\theta}(k)\hat{Z}'(k|k) - \zeta(k)\zeta'(k) + G(k)G'(k) \quad (12)$$

$$M_{z\theta}(k) = \hat{A}(k|k)P_{z\theta}(k|k) + \hat{Z}(k|k)P_\theta(k|k) \quad (13)$$

$$\hat{z}(k+1|k) = \hat{A}(k|k)\hat{z}(k|k) + \zeta(k) \quad (14)$$

$$\hat{\theta}(k+1|k) = \hat{C}(k|k)\hat{\theta}(k|k) + \xi(k) \quad (15)$$

$$P_z(k+1|k+1) = [I - K_z(k+1)H(k)]M(k) \quad (16)$$

$$P_\theta(k+1|k+1) \\ = \ \hat{C}(k|k)P_\theta(k|k)\hat{C}'(k|k) \\ -K_\theta(k+1)h(k+1)M_{z\theta}(k)\hat{C}'(k|k)S(k)S'(k) \quad (17)$$

$$P_{z\theta}(k+1|k+1) \\ = \ [I - K_z(k+1)H(k)][I - K_z(k+1)H(k)]' \\ M_{z\theta}(k)\hat{C}'(k|k) \\ +K_z(k+1)R(k+1)R'(k+1)K'_\theta(k+1) \\ -[I - K_z(k+1)H(k)]\zeta(k)\xi'(k) \quad (18)$$

where

$$\hat{Z}(k|k) = \begin{bmatrix} \hat{z}(k|k) & 0 & \cdots & \cdots & 0 \\ 0 & \hat{z}(k|k) & 0 & \cdots & 0 \\ \vdots & 0 & \ddots & \ddots & \vdots \\ \vdots & \vdots & \ddots & \ddots & 0 \\ 0 & 0 & \cdots & 0 & \hat{z}(k|k) \end{bmatrix}$$

$$P_{\alpha\beta}(i|j) \\ := \ E\{[\alpha(i) - \hat{\alpha}(i|j)][\beta(i) - \hat{\beta}(i|j)|Y^j]\} \quad (19)$$

$$\zeta(k) = P_{Az'}(k|k) \quad (20)$$

$$\xi(k) = P_{C\theta'}(k|k) \quad (21)$$

The simulation study is shown in references [Takatsu and Takeda, 1989]. The salient features of the proposed adaptive controller are demonstrated by simulation experiments.

5.3 CCCM

Most advanced concept is the Control Configured Construction Machine (CCCM). The fundamental concept is to design the construction machine which the computer control the machines easier. Moreover, the system without the controller need not to be stable. The system include the controller need to be stable.

The first thing we do is the work analysis. We analyse the motion of men and machine in order to change the process, machine and so on better. The control configered construction machine is exist depend on work analysis.

If any construction machine is designed by this concept, the controller can be designed as stable, controllerble and observable. This guarantee the safety and development of construction works.

Acknowledgment

The authors wish to express their sincere thanks to Dr. Nagata who is the professor of Nippon University, Dr. Takeshita who is professor of Ritsumeikan University, joint research companies and the Japan Micro–tunneling Association.

Special thanks should be extended to members of the construction equipment division, the construction method division and the sewage works division for their continuous encouragement and valuable discussions.

REFERENCES

Databook, 1988, "Nippon kokusei zue", Kokuseisya.

Muramatsu, T. and Cho, K., 1989, "A concept of control system for construction robot", IEEE IROS '89, pp.128– 135.

Newspaper, Japan Sewage Newspaper, 1990 Jan. 28.

Takatsu, T. and Takeda, H., 1989, "On an application of the adaptive control to micro–tunnelling", Preprints of ISCIE '89–5, pp.161–162.

Takatsu, T. and Takeda, H., 1989, "On an application of the modern control theory to the micro–tunnelling", No–dig '90 Osaka Pre–Conference, pp.1.1.1–1.1.5, Japan Society for Trenchless Technology (JSTT).

Takatsu, T. and Takeda, H., 1989, "A study of the parameter identification with a parameter's dynamics", SICE Preprints of Dynamical System Theory Symposium, pp.274–277

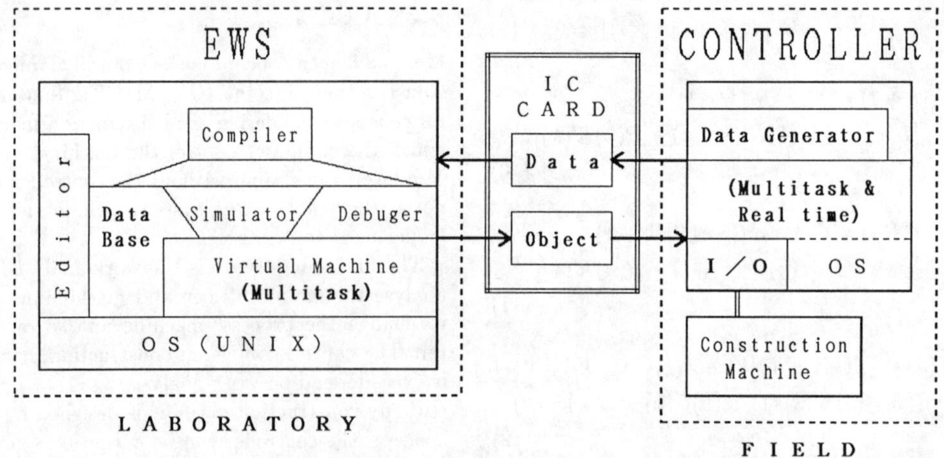

Fig. 3 Architecture of the controller

Uehara, K., 1987,"Syoukoukei suisin kouhou no genjyou", Tisitsu to tyousa, 1987 April, pp.30–37.

Saitou, K. and Ishibasi, N., 1985, "E de miru gesuidou no suisin kouhou", Sankaidou.

Takatsu, T. and Takeda, H., 1990, "A concept of strategy for a standard in the micro–tunnelling system", ISCIE, IEEE and ASME Japan–U.S.A. Symposium on Flexible Automation –A Pacific Rim Conference–, pp.493–496.

Kenji CHO, Takatsu, T. and Takeda, H., 1990, "In case of standard on control system for micro–tunnelling", JSTT, NO–DIG 90 OSAKA, pp.k.1.1–k.1.7

No Trenches in Town, Henry & Mermet (eds) © 1992 Balkema, Rotterdam. ISBN 90 5410 085 0

Une contribution pour un système de controle flou pour micro-tunnelier Introduction sur une machine, d'un contrôle asservi

Tomoji Takatsu, Hideyuki Takeda & Atsushi Sugiyama
Public Works Research Institute, Ministry of Construction, Tsukuba City, Ibaraki, Japon

RESUME. Dans cet article, nous étudions un développement efficace des systèmes de micro-tunneliers en vue de faire face à un nombre croissant de travaux sans tranchée. Tout d'abord, nous indiquons le besoin d'établir un standard pour définir un système efficace. Ensuite, nous proposons une machine virtuelle et une théorie floue appliquée au système de contrôle des machines qui est le modèle standard. Enfin, la performance de l'asservissement est montrée par des travaux en cours.

1. INTRODUCTION.

40 Millions de personnes utilisent le raccordement aux réseaux d'assainissement au Japon. Bien que le pourcentage de la population raccordée au réseau soit en augmentation ces dernières années, le pourcentage est inférieur à celui des autres pays développés, comme montré dans les Fig. 1 et 2. (Uehara, 1987). Ce résumé nous montre la nécessité de promouvoir les constructions de plus de réseaux d'assainissement dans notre pays. Il y a deux types de méthodes de construction des réseaux enterrés d'assainissement. Le premier est la méthode des tranchées. Cette méthode perturbe le traffic car il est conduit dans des rues étroites et très chargées, ce qui est l'une des caractéristiques du Japon. L'autre méthode est celle du pousse-tube qui créée moins de perturbation de traffic sur les sites de travaux.

Récemment, nous avons été amenés à utiliser un système de micro-tunneliers à des endroits tels que chemins de fer ou voiries pour construire des réseaux enterrés, tuyaux, etc... en souterrains (le micro-tunnelier est l'une des méthodes de pousse-tube qui pose des tuyaux d'un diamètre intérieur inférieur à 800 mm au Japon. La raison pour adopter le système de micro-tunnelier est que nous construisons des réseaux et conduits à la profondeur de 5 m ou plus. Nous avons fait ainsi parce que des réseaux ont déjà été enterrés à des profondeurs inférieures à 5 m. La méthode de tranchée est coûteuse, prend du temps et nécessite plus d'espace lorsque l'on creuse à des profondeurs supérieures à 5 m, aussi n'est-il pas avantageux de l'utiliser dans ce cas. De plus, le nombre de constructions a augmenté dans les petites villes . Dans les petites villes, il est plutôt choisi d'utiliser des conduits de diamètre intérieur inférieur à 800mm. La première fois que nous avons utilisé le micro-tunnelier remonte très loin dans le temps de l'ancienne Rome, lorsque les réseaux d'assainissement ont été placés en souterrains. Au Japon, nous avons utilisé le système de micro-tunnelier pour la première fois en 1958 à Amagasaki dans la Préfecture de Hyogo.

Maintenant, nous appliquons ce système non seulement dans la construction de réseaux d'assainissement mais également dans la pose de câbles téléphoniques. Le système moderne de micro-tunneliers consiste à avoir un seul système de transmission de puissance pour mettre en marche différents

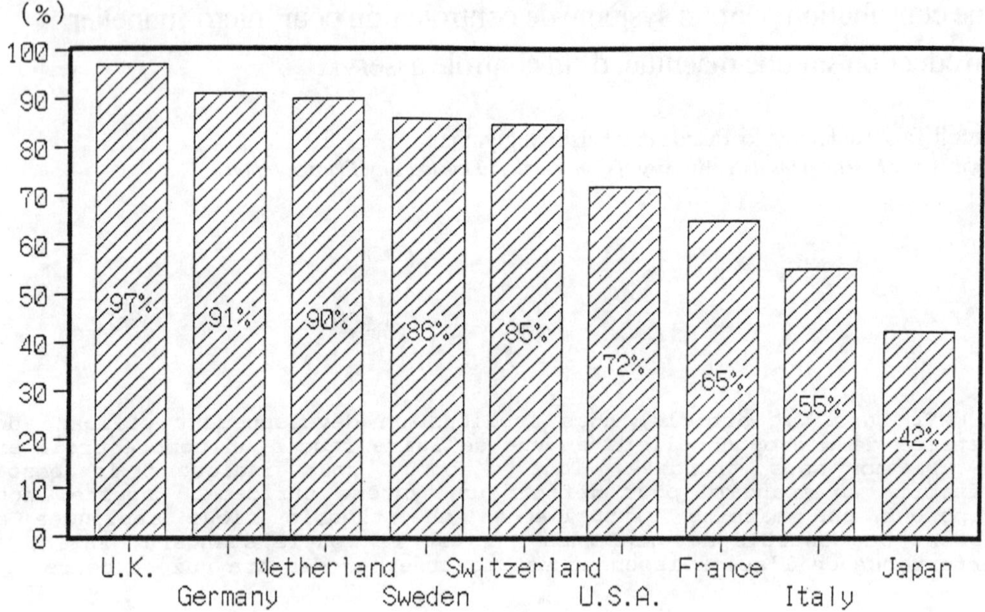

Fig. 1 Percentage of sewered diffussion.

mécanismes de creusement. Le système transmet la puissance du cylindre à la tête de coupe. Le système de micro-tunnelier nécessite un système de contrôle pour corriger les tuyaux courbés. Les raisons de cette nécessité sont principalement dûes au fait que les ouvriers ne peuvent entrer dans les tuyaux pour les placer dans la bonne position lorsque ceux-ci ne sont pas dans la position théorique du tracé. Il est nécessaire pour l'opérateur de la machine soit très expérimenté pour piloter le système de contrôle car les opérations vont dépendre de la longueur des tuyaux poussés, de la variation de la nature du sol, etc... Cela prend du temps pour de jeunes ouvriers pour posséder la connaissance du contrôle des micro-tunneliers.
En outre, il y a un manque de manoeuvres dans les travaux publics. Aussi les Entreprises et les Fabricants ont développé un système de contrôle pour les micro-tunneliers qui ne nécessite plus autant de connaissances.

Le Ministère de la Construction a démarré un projet coordonné de "système de contrôle AI" pour les micro-tunneliers avec 12 sociétés privées. Il s'étend de 1988 à 1991.

2. NECESSITES DE DEVELOPPEMENT.

En premier lieu, nous comparons l'échelle de grandeur entre deux grandes industries et les constructions d'assainissement utilisant le système de micro-tunnelier. Echelle et marchés sont donnés par le tableau I (Databook 1988, Newspaper 1990). Le tableau montre également le pourcentage de dépenses de recherche pour les micro-tunneliers qui est le 4/°°° de celui de l'industrie électrique. Ceci implique que nous devons avoir des recherches et des développements plus efficaces pour les micro-tunneliers.

Deuxièmement, nous regardons la comparaison entre les travaux de génie civil et l'industrie. Il y a peu de différence entre les travaux de génie civil et l'industrie de fabrication. Le premier cas est la taille de l'ouvrage à réaliser. Nous devons considérer la longueur en ordre de kilomètres en matière de micro-tunneliers. D'un autre côté, l'industrie travaille sur des commandes de l'ordre du mètre. Le suivant est la caractéristique de l'ouvrage. Dans l'industrie de transformation, chaque outil n'effectue que peu de variantes de

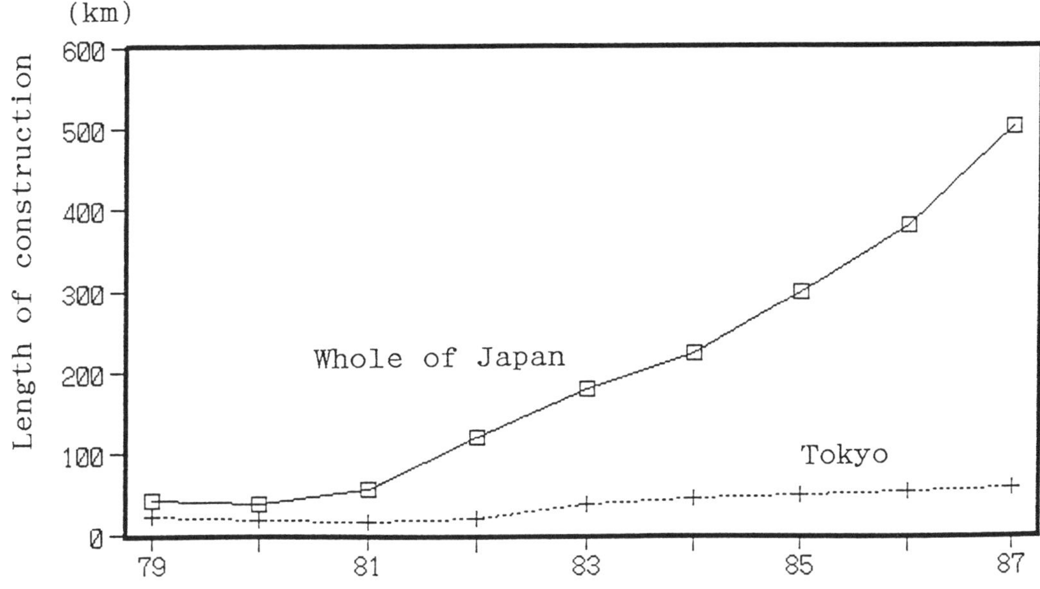

Fig. 2 Transition of micro-tunneling.

travail, lesquelles ont peu de dispersions dans le contrôle de qualité. D'un autre côté, nous devons travailler avec des objets variés pour le micro-tunneliers sous des conditions non précises comparées à celles de l'industrie de fabrication. Le troisième est la relation entre les travaux et les outils (machines). Les machines de construction nécessitent d'être construites pour faire le travail mais dans l'industrie de production, les outils sont généralement imposés pour la fabrication. Le quatrième est la relation entre la manutention et les travaux. Manutention et travaux sont séparés dans l'industrie de transformation. Sur les chantiers, ils ne peuvent pas être séparés clairement (Muramatsu T. et Cho K., 1989).

Selon ces remarques, les technologies utilisées dans l'industrie de transformation ne sont pas applicables aux travaux de génie civil si l'on ne procède pas à des modifications. Nous avons à développer une nouvelle technologie pour les travaux de construction.

Finalement, nous montrons quelques caractéristiques des micro-tunneliers. Considérons une longueur de tracé d'environ 100m. Les conditions de sol ne sont pas

Table 1 Compression of rechearch expenses.

	Market (billion yens)	Research expenses (percent)	Research expenses (billion yens)
Elector industrial	4000	5(%)	200
Motor Industry	3600	3(%)	108
Micro-tunnelling	75	0.5(%)	0.38

constantes et changent souvent. Les travaux sont constants tels que tuyaux d'assainissement, mise en place de câbles téléphoniques... Cela implique que la construction pré-fabriquée se soit développée au micro-tunnelier. La tête de la machine bouge dans les travaux de construction, la manutention et le travail sont séparés.

Tenant compte de ce qui précède, le système pour les micro-tunneliers doit être développé avec une approche originale. Nous définirons quelques standards pour le micro-tunnelier en vue de rendre le système développé efficace. Par le fait que les caractéristiques de micro-tunnelage sont considérées être les mêmes qu'en construction générales bien qu'elles ne le

Table 2 Comparison among factory, construction and micro-tunneling.

	Factory	Construction	Micro-tunnelling
Size of works	Meters	Kilometer	100 Meters
Characteristics of Works	Constant	Not Constant	Constant
Movement of Tools	Stand	Move	Move
Movement of works	Move	Stand	Move/Stand
Handling and Work	Separated	Not Separated	Separated

soient pas, que le budget de développement est inférieur à celui de l'industrie électrique, l'industrie des moteurs, etc..., la standardisation devra être décidée en termes de hardware, software et environnement de développement.

3. DEMANDES DE DEVELOPPEMENT.

Nous posons les besoins de développement comme suit :

1. Le système nécessite un opérateur expérimenté. Ceci est appelé "deskillful system" qui est l'un des mots-clés importants du IXe Symposium International sur l'Automatique et la Robotique dans la construction. Juin 1992.
Ceci implique que nous avons des concepts avancés pour l'automatisation et la robotique dans la construction.

2. La capacité du système de contrôle est de limiter des déviations par rapport à la position normale.
(a) Verticale 30 mm
(b) Horizontale 50 mm
Ces buts résultent de questionnaires provenant de 140 villes et Entrepreneurs au Japon en 1988.

4. ASSERVISSEMENT GENERAL.

4.1. Concept fondamental.

Il y a des standards pour les systèmes de calcul. Le premier est le standard hardware. Celui-ci doit être considéré à partir d'un niveau CPU vers le niveau connecteur. Le standard du langage de programmation tel que le FORTRAN 77 est le suivant à prendre en compte. Le troisième est la standardisation de OS (par exemple UNIX, TRON, etc...) Des procotoles (tels que MAP - Manufacturing Automatic Protocol) doivent être également pris en considération.

En premier lieu, nous discutons le standard de connection I/O. La raison en est que nous pouvons facilement échanger les instruments de mesure avec l'application de ce standard et ainsi nous pouvons terminer les travaux de construction à la date désirée.

En second lieu, nous recherchons et développons une application software (Takatsu et Takeda, 1989) qui est une sorte de système de contrôle adaptatif. La motivation de ce développement est que le système peut être utilisé dans n'importe quel système de micro-tunneliers avec n'importe quelles conditions de sols.

En troisième lieu, une standardisation du hardware est définie. La façon la plus simple de réaliser la standardisation du hardware est obtenue lorsque tous les systèmes de micro-tunneliers utilisent le même hardware. Cela coûte du temps et de l'argent pour le développement de standards hardware. Pour utiliser les hardwares conventionnels, il y a deux façons. La première est d'utiliser l'ensemble complet, OS, langage programmé, etc... Cependant, les systèmes existants incluent des fonctions telles que des fonctions graphiques qui ne sont pas nécessaires pour des micro-tunneliers. Aussi est-il très difficile de sélectionner le meilleur système pour le micro-tunnelier.

Aussi, nous proposons un concept de machine virtuelle (voir Fig. 3). En général, la machine virtuelle ne fonctionne pas en temps réel et en multi-taches parce que cela est simulé par des stations de travail. Nous demandons à la machine virtuelle d'être multi-taches et de travailler en temps réel. La machine virtuelle a besoin de fonctions multi-tâches parce que les mesures et divers contrôles sont appliqués séparément, mais pour faire chaque travail, le programme doit fonctionner en même temps afin de ne pas perdre de données. La propriété de temps réel est nécessaire pour effectuer le travail sur chantier. La machine virtuelle que nous essayons de développer n'est pas un simple émulateur mais elle doit être un système de contrôle hautement qualifié. Il y a deux types de systèmes. Le premier est une station de travail dans un laboratoire et l'autre un micro-ordinateur sur chantier. Le "formateur" a quelques simulateurs et des bases de données. Le simulateur est utilisé pour entraîner les opérateurs et pour développer de nouveaux asservissements. La base de donnée fournit les données au simulateur et est un support à de nouveaux développements. Le micro-ordinateur sur chantier est seulement chargé avec la base de données. Nous proposons d'utiliser les cartes IC en vue d'envoyer les données du système sur chantier au système en laboratoire (voir Fig. 3.).

4.2. CONTROLE FLOU.

Le Professeur Zadeh a établi une logique floue en 1962. La logique floue est l'une des théories dont la fonction caractéristique est définie comme suit.

$$\mu(x) = \begin{cases} 1 & (x \in A : aset) \\ 0 & (x \notin A : aset) \end{cases}$$

La plage de caractéristiques en logique floue est :

$$\mu(x) = [0, 1] \in \mathcal{R}(x \in A : aset)$$

et est appelée la fonction adhérente. Nous pouvons simuler le sentiment tel que "A est grand", "B est légèrement plus grand" etc... Un sentiment comme celui-ci ne peut être obtenu par une théorie ordinaire.

L'asservissement flou est une application de la logique floue. L'asservissement flou combine la logique floue avec la règle IF THEN. La règle IF THEN convient pour exprimer la réflexion et les jugements humains. De plus, la réflexion humaine est expliquée par la logique floue. L'asservissement flou convient très bien au transfert de la connaissance humaine au programme informatique.

La raison pour laquelle nous utilisons l'asservissement flou n'est pas seulement le transfert de la connaissance en programme mais également pour changer facilement celui-ci.

4.3. SYSTEME REEL.

Dans cette section, nous faisons une introduction au système réel qui est développé à partir du concept de standard discuté précédemment.

En premier lieu, nous devons avoir une machine performante. Un programme flou est sélectionné pour la machine car l'attribut micro-tunnelier est encore non clairement défini comme il a été mentionné précédemment et les opérations de la machine sont décrites par le langage. De plus, les modèles des opérations sont essentiellement identiques bien que les machines soient différentes et que la fonction adhérente et son rôle sont les mêmes, même si les applications floues sont différentes. Par conséquent, cela implique que nous avons besoin seulement d'un système de contrôle pour le micro-tunnelier utilisant la logique floue. Deuxième nous définissons le I/O, le protocole de communication, etc...

Actuellement, quelques sociétés participent à ce projet de standardisation. Elles terminent le système en laboratoire et en constructions réelles.

4.4. ESSAIS EN PLACE.

4.4.1. Machines de base.

L'asservissement flou est adapté sur trois types de machines qui sont les suivantes:

1. Pression de boue pour des graves (premier cas)

2. Pression de boue pour argile

Table 3 Results of field tests.

	1st Case	2nd Case	3rd Case
Deviation (vertical)	22mm	21mm	27mm
Deviation (horizontal)	8mm	9mm	43mm
Length	56m	108m	60m

sablonneuse (second cas)

3. Méthode de pousse-tube (troisième cas).

4.4.2. RESULTATS DES TESTS.

Les résultats des tests in-situ sont donnés par le tableau 3. A partir du tableau 3., il est prouvé que :

1. L'asservissement flou peut être appliqué au moins à trois types de micro-tunneliers.

2. Les déviations du tracé théorique correspondent aux objectifs de développement.

Ceci implique que l'asservissement flou rend les micro-tunneliers facilement utilisables, c'est-à-dire "Deskillful system".

5. DISCUSSIONS.

5.1. ASSERVISSEMENT FLOU.

Dans cette étude, nous proposons quelques standardisations dans les systèmes micro-tunneliers qui sont déduites des potentialités de développement et des caractéristiques des travaux. La standardisation peut ralentir le développement de nouvelles technologies. Cela peut être vrai dans l'industrie de transformation mais ce n'est pas un handicap en génie civil. Ce qui est important, c'est de savoir si cela est réalisable avec la technologie actuelle. La standardisation nous donne des cadres agréables comme suit :

1. Nous pouvons faire quelques programmes d'applications utilisés dans les systèmes de micro-tunneliers.
Ceci rend le développement moins cher et la qualité du programme est supérieure.

2. Nous pouvons facilement échanger des parties du système.
Ceci permet de respecter les délais.

3. Nous pouvons facilement construire une base de données.
Ceci réalise un nouveau système efficace de développement.

La performance de l'asservissement flou atteint son objectif avec les valeurs d'erreurs horizontales de 50mm et verticales de 30mm. Cet objectif est autorisé par un Comité dans l'association des micro-tunneliers Japonais dont le Président est le Dr Nagata et le vice-Président le Dr Takeshita.

5.2. RECHERCHES AVANCEES.

5.2.1. AUTRE APPROCHE.

Le point faible de l'asservissement flou est le fait qu'il n'y a pas de garantie de contrôler la stabilité, l'asservissement et les observations. Ces propriétés sont importantes pour le système et la théorie du contrôle.

Aussi, nous recherchons une autre technique d'asservissement puisque l'asservissement flou a l'avantage d'une programmation aisée. La technique de contrôle et le contrôle adaptatif dérivés de la théorie des systèmes stochastiques qui est l'une des théories les plus avancées.

Le contrôle adaptatif est constitué d'identifications de paramètres, de l'estimation de l'état et du contrôle optimal. Dans cette étude, nous discuterons l'identification du paramètre et l'estimation de l'état car le processus d'estimation est important pour. le contrôle adaptatif.

5.2.2. FORMULATION DU PROBLEME.

Bien qu'il existe de nombreuses représentations de la dynamique des systèmes, le phénomène du micro-tunnelier est exprimé dans une équation discrète ci-après. La raison en est que le système micro-tunnelier comprend souvent des ordinateurs donc ces ordinateurs digitaux prennent des points dans la mesure. La dynamique du système et le mécanisme d'observation sont décrits comme suit :

System :

$$
\begin{aligned}
z(k+1) &= A(k)z(k) + G(k)w(k) \qquad (1)\\
z(0) &= z_0 \quad (k = 0, 1, 2, \cdots)
\end{aligned}
$$

Observation:

$$
\begin{aligned}
y(k) &= H(k)z(k) + R(k)v(k) \qquad (2)\\
&\quad (k = 0, 1, 2, \cdots)
\end{aligned}
$$

Paramètre :

$$
\begin{aligned}
\theta(k+1) &= C(k)\theta(k) + S(k)\mu(k) \qquad (3)\\
\theta(0) &= \theta_0 \quad (k = 0, 1, 2, \cdots)
\end{aligned}
$$

where $z(\bullet)$, $y(\bullet)$ and $\theta(\bullet)$ are n, l and n^2 vectors $A(\bullet)$, $G(\bullet)$, $H(\bullet)$, $R(\bullet)$, $C(\bullet)$ and $S(\bullet)$ are $n \times m_1$, $l \times n, l \times m_2$ $n^2 \times n^2$ and $n^2 \times m_3$ matrices
$\theta(\bullet) = [a_{11}(\bullet), a_{12}(\bullet) \cdots a_{nn}(\bullet)]'$
and where $a_{ij} = \{A(\bullet)\}_{ij}$, "′" implies the transposition.

$z(\bullet)$ is the system state value which includes the drift of the cutter head and so on. $w(\bullet)$, $v(\bullet)$, and $\mu(\bullet)$ are mutually independent white Gaussian noise. We may regard the parameter of $\theta(\bullet)$ as the set of muddles.

5.2.3. ESTIMATEUR.

Il existe de nombreuses approches d'estimateurs pour l'état du système et le paramètre. Ci-après on trouvera l'estimateur basé sur la variante minimum car cet estimateur est facilement déduit que les autres.

Soit la forme de l'estimateur suivante :

$$
\begin{aligned}
\hat{z}(k+1|k+1) &= \hat{z}(k+1|k) + K_z(k+1)\\
&\quad \{y(k+1) - H(k+1)\hat{z}(k+1|k)\} \qquad (4)\\
\hat{z}(0) &= \hat{z}_0 \qquad (k = 0, 1, 2, \cdots)
\end{aligned}
$$

$$
\begin{aligned}
\hat{\theta}(k+1|k+1) &= \hat{\theta}(k+1|k) + K_\theta(k+1)\\
&\quad \{y(k+1) - H(k+1)\hat{z}(k+1|k)\} \qquad (5)\\
\hat{\theta}(0) &= \hat{\theta}_0 \qquad (k = 0, 1, 2, \cdots)
\end{aligned}
$$

where $\hat{\bullet}(i|j) = E\{\bullet(i)|\,Y^j\}$
and where the operator E implies the expectation and Y^j is the set of $\{y(k)|k = 0, 1, \cdots, j\}$.

The purpose of this investigation is to find $K_z(k+1)$ and $K_\theta(k+1)$ to minimize the trace of estimation error martices;

$$
\begin{aligned}
&P_z(k+1|k+1)\\
&:= E\{[z(k+1) - \hat{z}(k+1|k+1)]\\
&\qquad [z(k+1) - \hat{z}(k+1|k+1)]'|\,Y^{k+1}\} \qquad (6)
\end{aligned}
$$

$$
\begin{aligned}
&P_\theta(k+1|k+1)\\
&:= E\{[\theta(k+1) - \hat{\theta}(k+1|k+1)]\\
&\qquad [\theta(k+1) - \hat{\theta}(k+1|k+1)]'|\,Y^{k+1}\} \qquad (7)
\end{aligned}
$$

$K_z(k+1)$, $K_\theta(k+1)$, $\hat{z}(k+1|k)$ and $\hat{\theta}(k+1|k)$ are calculated by the following equations:

$$\frac{\partial}{\partial K_z\; k+1} tr.\{P_z(k+1|k+1)\} = 0 \qquad (8)$$

$$\frac{\partial}{\partial K_\theta(k+1)} tr.\{P_\theta(k+1|k+1)\} = 0 \qquad (9)$$

$$
\begin{aligned}
K_z(k+1) &= M(k)H'(k+1)\\
&\quad [H(k+1)M(k)H'(k+1) + R(k)R'(k)]^{-1} \qquad (10)
\end{aligned}
$$

$$
\begin{aligned}
K_\theta(k+1) &= \hat{C}(k|k)M'_{z\theta}(k)H'(k+1)\\
&\quad [H(k+1)M(k)H'(k+1) + R(k)R'(k)]^{-1} \qquad (11)
\end{aligned}
$$

$$
\begin{aligned}
M(k) &= [\hat{A}(k|k)P_z(k|k) + \hat{Z}(k|k)P'_{z\theta}(k|k)]\hat{A}'(k|k)\\
&+ M_{z\theta}(k)\hat{Z}'(k|k) - \zeta(k)\zeta'(k) + G(k)G'(k) \qquad (12)
\end{aligned}
$$

$$M_{z\theta}(k) = \hat{A}(k|k)P_{z\theta}(k|k) + \hat{Z}(k|k)P_\theta(k|k) \qquad (13)$$

$$\hat{z}(k+1|k) = \hat{A}(k|k)\hat{z}(k|k) + \zeta(k) \qquad (14)$$

$$\hat{\theta}(k+1|k) = \hat{C}(k|k)\hat{\theta}(k|k) + \xi(k) \qquad (15)$$

$$P_z(k+1|k+1) = [I - K_z(k+1)H(k)]M(k) \qquad (16)$$

$$
\begin{aligned}
&P_\theta(k+1|k+1)\\
&= \hat{C}(k|k)P_\theta(k|k)\hat{C}'(k|k)\\
&\quad -K_\theta(k+1)h(k+1)M_{z\theta}(k)\hat{C}'(k|k)S(k)S'(k) \qquad (17)
\end{aligned}
$$

$$
\begin{aligned}
&P_{z\theta}(k+1|k+1) \\
&\quad = [I - K_z(k+1)H(k)][I - K_z(k+1)H(k)]' \\
&\qquad M_{z\theta}(k)\hat{C}'(k|k) \\
&\qquad + K_z(k+1)R(k+1)R'(k+1)K_\theta'(k+1) \\
&\qquad - [I - K_z(k+1)H(k)]\zeta(k)\xi'(k) \qquad (18)
\end{aligned}
$$

where

$$
\hat{Z}(k|k) = \begin{bmatrix} \hat{z}(k|k) & 0 & \cdots & \cdots & 0 \\ 0 & \hat{z}(k|k) & 0 & \cdots & 0 \\ \vdots & 0 & \ddots & \ddots & \vdots \\ \vdots & \vdots & \ddots & \ddots & 0 \\ 0 & 0 & \cdots & 0 & \hat{z}(k|k) \end{bmatrix}
$$

$$
\begin{aligned}
&P_{\alpha\beta}(i|j) \\
&\quad := E\{[\alpha(i) - \hat{\alpha}(i|j)][\beta(i) - \hat{\beta}(i|j)|\, Y^j]\}
\end{aligned} \qquad (19)
$$

$$\zeta(k) = P_{Az'}(k|k) \qquad (20)$$

$$\xi(k) = P_{C\theta'}(k|k) \qquad (21)$$

L'étude de la stimulation est montrée en références (Takatsu et Takeda, 1989). Les points forts du contrôle adaptatif proposé sont démontrés par des expériences de simulation.

5.3. CCCM.

Le concept le plus avancé est le "Control Configured Construction Machine - CCCM". Le concept fondamental est de concevoir une machine que l'ordinateur peut contrôler plus facilement. En outre, le système sans asservissement n'a pas besoin d'être stable. Le système avec l'asservissement lui doit être stable.

La première chose que nous avons faite est le travail d'analyse. Nous analysons le mouvement de l'homme et de la machine en vue d'améliorer le process, la machine, etc...

Si une machine est conçue suivant ces concepts, le contrôleur peut être conçu comme stable, contrôlable et observable. Ceci garantit la sécurité et le développement des travaux.

Remerciements.

L'Auteur désire exprime ses sincères remerciements au Dr Nagata qui est Professeur à l'Université Japonaise, Dr Takeshita qui est Professeur à l'université Ritsumeikan, les entreprises partenaires et l'Association Japonaise de micro-tunneliers.

Des remerciements particuliers doivent être adressés aux membres de la Division Construction, de la Division Méthodes et de la Division Assainissement pour leurs encouragements continus et discussions fructueuses.

1.2 Other techniques

No Trenches in Town, Henry & Mermet (eds) © 1992 Balkema, Rotterdam. ISBN 90 5410 085 0

In situ replacement of a sewage pipe by 'Tube-Munching' microtunneling technique at Champigny-sur-Marne

M. Darras
DSEA, Val de Marne, France

ABSTRACT: A 200/230 diameter EU stoneware pipe from Avenue du Général de Gaulle at CHAMPIGNY SUR MARNE was replaced in situ with a 450/640 diameter reinforced concrete pipe using the tube-eater method for microtunnelling. The difficulties encountered in the work were not foreseeable since it seemed to be relatively well known what the material around the lines was, i.e. superimposed hard core put on in accordance with.

1. INTRODUCTION

The "Tube-munching" microtunneling technique was used to replace a length og gravity-driven sewage piping running under the Avenue du Général de Gaulle in Champigny-sur-Marne (Val de Marne department of France). This in situ replacement of a 200 mm clayware pipe was the second experience of its kind in France.

2. SITUATION EXISTING BEFORE THE OPERATION

The existing pipe that had to be replaced ran along an access road to national highway 303 over a length of 117.76 m. It was in severely degraded condition and was inadequate for evacuating the waste waters from the seven buildings feeding it.

3. TRADITIONAL CONSTRUCTION OPTION

The only remedy to the shortcomings of this pipe was to rebuild. The traditional approcah would have required:
- major human and physical means ;
- shielding and support wall, trenches and tunnels ;
- demolition of the existing works (including roadway, filler, walls, and so on) and reconstruction of them afterward ;
- detouring the traffic ;
- bothersome and hazardous measures affecting the local residents.

The estimated time required was twelve weeks.

4. "MODERN" RECONSTRUCTION ALTERNATIVE

The modern alternative is a trenchless technique in which the new pipe is either:

- laid next to the old one;

- or the old pipe is replaced directly by the new one, "in situ".

The estimated cost of an in situ reconstruction with various types of machinery was of the order of the traditional approach, while a parallel reconstruction running next to the old pipe would have cost more, as the inlets from the surface would have to be rearranged.

The expected time estimate was eleven weeks.

In consideration of which, the choice fell to the in situ replacement, which was nicknamed the "tube-munching" approach.

5. THE TUBE-MUNCHING APPROACH WITH ISEKI MACHINERY AND BONNA PIPES

The preliminary analysis of the project brought out a few particular features of this job that called for caution:

- same water course ;
- passages through manholes ;
- incoming brench connections ;
- the portion of the pipe near the highway support wall, i.e. where it passes through the wall foundation (see figure 1).

The microtunneler had to pass through three manholes. When these were inspected closely, it was found that they were made of milestone rubble masonry having mechanical properties exceeding the performance levels of the machinery to be used, and likely to cause same guidance problems.

The seven inlet connections located along the route were examined and it was found that they were encased in concrete of variable thicknesses, but compatible with the machine capacities.

Last but not least of the worries was the proximity of the underpass support wall running alongside the pipe.

Figure 2 below hinted that the pipe would out through the wall foundation plane at about 60 m from the machine pit, and that the new pipe would come to within about 30 cm of the wall foundation.

Finally the pipe to be replaced was assumed to lie in a known geologically and geotechnical environment of filler.

The following decisions were made as a consequence of the above points.

a) The 200 mm i.d. clayware sewage pipe would be replaced in situ with a 450 mm i.d. pipe of reinforced concrete.

b) In order to maintain the water course of the old pipe, a slurry of fly ash mixed with cement would be injected in it and in the manholes, where passages would first be cut out for the machine. The purpose of this was to keep:

   1) the microtunneler from centering in on the old pipe, by following the "easiest path" ;

   2) the machine from deviating for lack of resistance as it passed through the manholes.

c) To maintain the effluent flow while work was in progress, the waste water would be pumped to the surface and piped to an inspection manhole beyond the job site.

d) A EURO-ISEKI shield of the UNCLEMOLE TCC 500ID/660DD model will be used.

e) Reinforced concrete pipes of the ROCLA Microtunnel line would be used. These pipes are of high-performance concrete, made by the Tuyaux BONNA company.

f) The pipes would be jacked in two sections.

6. ACTUAL EXECUTION OF THE WORKS

Considering the dense utilization of the underground space by other utilities, the machinery pit was placed in such a way as to reduce the distance between pits from 130 m to 118 m, and the pipes were jacked through in a single 118 m drive. The work pit measured 3.35 m x 2.70 m and this was dug at the low point of the existing pipe. The 2.50 m x 1.00 m outlet pit was located at the other end, and work proceeded in the upstream direction.

- The filler around the pipe, which we had assumed was homogeneous and executed according to the rules of the trade, were found to be heterogeneous and executed at different periods, going, against all good common sense. A few exploratory drillings would have been enlightening, but an extrapolation was made on the basis of information from a previous microtunnelling project in the same region. The machinery behaved differently, depending on what kind of terrain it was passing through, as a few comparative parameters will show:

- instataneous velocities of 42 mm/min. through the first 60 m and 58 mm/min. through the last 50 m ;

- average torque of 75 % through the first 60 m and 52 % through the last 50 m ;

- thrust force of 47 metric tonnes per linear meter jacked in over the first 60 m and 97 tonnes per linear meter through the last 50 m (see figure 3).

Though the impossibility of getting reliable documents on existing works was deplorable, it must be said that, even if a reliable dossier of executed works existed, it still would not have solved the problem of the foreign matter in the filler. This plethora of foreign matter (nails, wood, concrete blocks, iron ingot) were not just accidentally forgotten, but were purposely left behind by the companies that had executed the previous works - e.g. the wooden formworks they did not know what to do with, the leftover concrete at the end of the day, formwork nails, and so on.

- The losses of sludge after the pipejacking was recommenced are probably due to a loosening up of the earth after the concrete block was excavated.

- The clogging of the spoil-ridding circuit was certainly due to the wooden blocks and PVC pipe scrap encountered.

These second-order difficulties could therefore have been avoided if they had been taken into account from the start. As concerns the geotechnical analysis, though, this would have revealed the heterogeneity of the filler but would not have hinted at the presence of all of the foreign matter.

This experimental job site also provided an opportunity to analyze job costs. Expressed in percentages, the breakdown is as follows.

A) Breakdown by resources

1	Labor	27.31 %
2	Rentals	43.47 %
3	Materials	13.36 %
4	Consumables	2.82 %
5	Subcontracting	3.89 %
6	Hand tools	1.51 %
7	Miscellaneous	8.14 %
	Total	100.00 %

B) Breakdown by job

1	Instal./removal/misc.	18.02 %
2	Special connections	38.37 %
3	Microtunneling	43.61 %
	Total	100.00 %

The financial situation at the end of the job revealed a 12 % overrun of the initial estimate, or 16.5 % compared with the traditional approach.

The exexution time was nine weeks.

7. PROBLEMS ENCOUNTERED

The following problems were encountered.

a) At pipe no. 6 about 12 m in, the spoil-ridding circuit became clogged by wooden wedges that had been left behind in the filler of the old pipe, along with PVC debris, nails and fly ash.

b) Approximately in the region

where the course crosses the base of the support wall (about 60 m in), the machine struck concrete blocks and shut down on one of them. This block had to be excavated from the surface.

1) When this block was excavated, it was found to be very large and it contained a large piece of metal resembling an iron ingot. As the machine had advanced several meters through this block before coming to a halt, it lost two of its cutting teeth, and these had to be replaced. We even found a shovel in the block !

2) When the jacking began again, losses of slurry were observed.

3) Another problem of seel tightness was observed at this particular point, which still has not been satisfactorily explained.

c) At pipes nos. 33 (about 66 m in), 34 (68 m) and then 48 (96 m) and 49 (98 m), more clogging of the spoil-ridding circuit was observed. This was again caused by wooden wedges, nails, PVC debris and ash.

   It should be noted that the fly ash complicates the sludge treatment process as it makes frequent cleaning and regular draining necessary.

8. POST-MORTEM ANALYSIS

The difficulties we encountered were not the ones we had expected.

The anticipated underground environment contained no hint or indication of the problems that were actually to occur. Only the following problems were unavoidable :

- the inability of the machinery to grind up the wood and PVC tubing debris properly;
- the poor properties of the fly ash used in making the slurry.

As concerns the other problems, they might have been dealt with better if the Prime Contractor and other subcontractors had reliable information concerning the underground environment of the existing pipe.

- The only available information was a few drawings dating from 1966, at the time the SNCF Paris outer belt railway level crossing was eliminated and the underpass on the Paris-Basel line was constructed at Champigny-sur-Marne. These drawings do not seem to contain verifications, and so no accurate localization of the underground works could be made. This is why the machine ran into the support wall foundations earlier in the job then expected.
- It should be noted that the microtunneling part of the job cost only 43.61 % of the total, while the special connections cost 28.37 %, which is relatively high. The way the flow of effluents was maintained should certainly be studied more carefully to lower the overall costs of the operation.
- Equipment rentals (43.47 %) also amplifies the cost of the job.

9. CONCLUSION

Despite the problems encountered, this experimental job site was finished within a total of nine weeks. Of course more people had to be employed, and therefore more hours than were initially planned. It globally confirmed the advantages of the trenchless approach over the traditional one.

It also:

- confirmed the ground-machinery interaction, as the equipment passed through the two types of filler encountered ;
- opened the path to other direction of research, namely:
  . adapting the machinery to microtunnel through certain other types of material (e.g. wood and PVC) ;

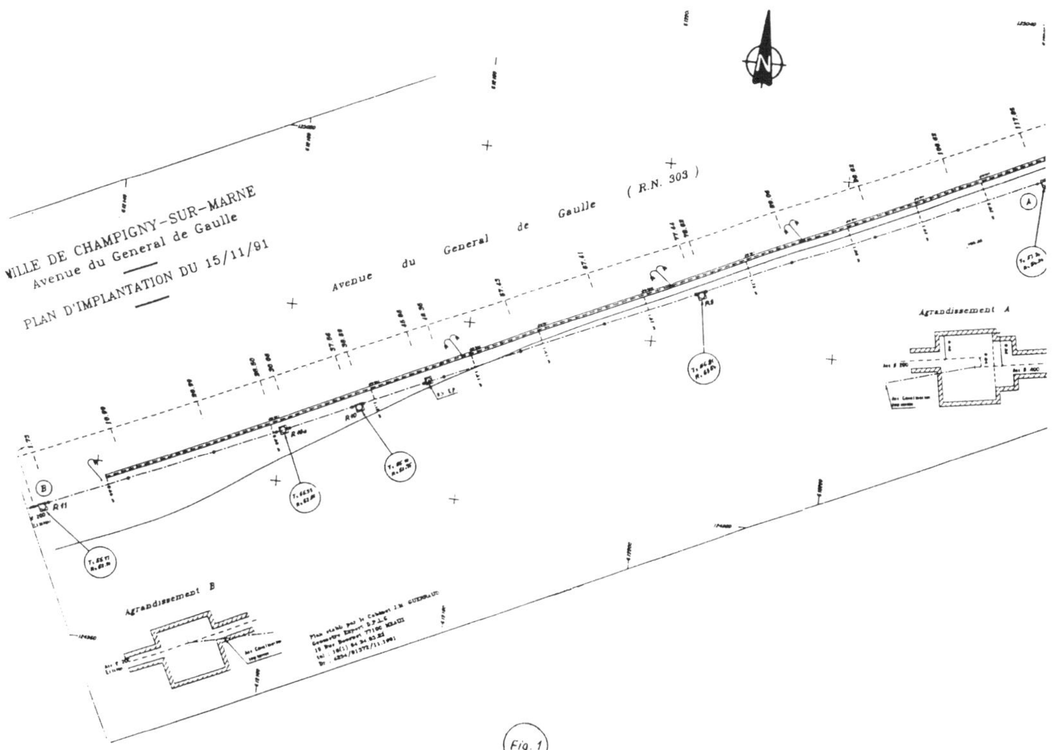

VILLE DE CHAMPIGNY-SUR-MARNE
Avenue du General de Gaulle

PLAN D'IMPLANTATION DU 15/11/91

Avenue du General de Gaulle (R.N. 303)

Agrandissement A

Agrandissement B

Fig. 1

_ *Remplacement in situ de la canalisation EU ⌀ 200* _

_ *Avenue du Général de Gaulle à Champigny sur Marne* _

Profil en long du mur de soutènement et

Coupe longitudinale de la canalisation

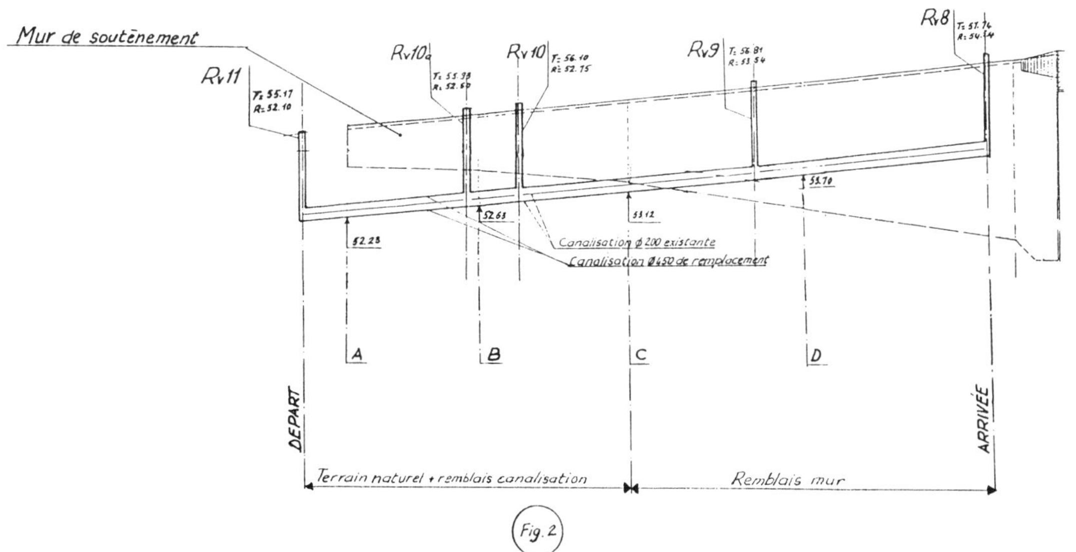

Fig. 2

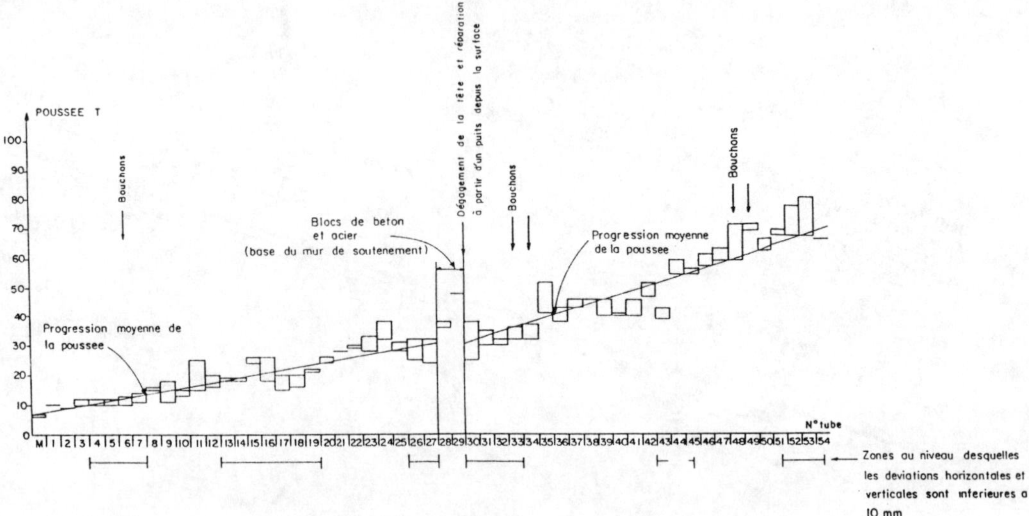

REMPLACEMENT DE CANALISATION A CHAMPIGNY-SUR-MARNE

POUSSEE

Fig.3

. using materials with better properties in the spoil-ridding circuits (e.g. exclusion of ash) ;

. using pipes adapted to this type of job.

The importance of a better knowledge of the terrain around the pipe was also brought out, whence the need for adequate geological ans geotechnical investigation.

And lastly, this experience serves as a warning to all those involved in civil works, of the importance of developing reliable dossiers of executed works, and correctly managing theses dossiers afterward. There is no way to get a good idea of an existing structure unless documents exist that give a faithful account of its history.

No Trenches in Town, Henry & Mermet (eds) © 1992 Balkema, Rotterdam. ISBN 90 5410 085 0

Contribution of Gaz de France towards the achievement of rapid and inconspicuous construction works

B.Calinaud
STG-DEGS, EDF-GDF, France

P.Le Testu & S.Stringhetta
CERSTA, DETN, GDF, France

ABSTRACT : Construction works involving the repair, maintenance and installation of underground networks generally demand large-scale trenching. These generate a certain inconvenience which the general public and local authorities find difficult to tolerate since they have to bear the social costs of such operations. Within this context, Gaz De France has devoted time and effort to developing the utilization of techniques which reduce hindrance to the environment. Whether such engineering work concerns pipe renewal or pipe extension, the various techniques employed range from the re-utilization of existing pipes as sleeves (lining and pipe insertion), to small-scale remote excavation techniques or the installation of new pipes by the steerable horizontal drilling method.
This paper aims to describe a work site, the original aspect of which is the association of a steerable horizontal drilling process for laying a polyethylene pipe on the one hand, with a vertical vacuum excavation process on the other, along with a pneumatic mole drilling process and a remote working technique for connecting customer service lines. After outlining the work site context, the report specifies the selected technical options and gives a report of the works progress.
Finally, an overall cost assessment, allowing for social costs that are more difficult to quantify due to their qualitative aspect, provides a comparison between the cost of a work site of this type and that of a conventional work site.

1 INTRODUCTION

Construction works involving the repair, maintenance and installation of underground networks generally demand large-scale trenching which does not fit in easily with the urban landscape. The public and local authorities are becoming increasingly less tolerant of the inconvenience caused by work-sites and future regulations may place heavier constraints on the digging of trenches.
In order to solve these problems, with which all holders of underground concessions are confronted, Gaz de France has developed new techniques for **rapid and inconspicuous construction works**.

2 CONTEXT

The population supplied by Gaz De France amounts to 38 million, i.e roughly three-quarters of the French population, via a distribution network of a total length of 115 000 km. The gas is distributed under low pressure at 21 mbar or medium pressure at 4 bar.
The oldest pipelines supplying the network are in cast iron under low pressure (13%) and steel under low and medium pressure (45%) and the most recent are in polyethylene (or PE) under medium pressure (33%).
Over 3000 km of pipeline are laid each year in urban areas, of which 90% are in PE.

The extension of the networks, essential to the development of towns, traditionally involves trenching which is proving to be less and less compatible with the urban context and with the quality of city life and environment.

Trenching work effectively entails increasingly heavy social costs while available space for the community is ever more limited.

For users, trenches mean a hindrance to pedestrian and road traffic and therefore a waste of time, increased petrol consumption and more frequent accidents; **for industry and commerce** they are synonymous with loss of profit; **for local authorities** they are responsible for increased costs due to the resulting weakening of the roadway and to the loss of earnings due to a drop in the income derived from the renting out of certain spaces (e.g: parking spaces). Finally, trenching deteriorates the environment through the noise and pollution it generates.

However, although social costs, by definition are borne by the community, they have repercussions on the concessionary firm, who also bears the brunt of the direct costs of trenching (60 to 70% of the total cost of the work). The inconvenience caused by trenching has a penalizing effect on the firm's image likely to entail the loss of potential clients.

An analysis of the above elements has prompted Gaz de France to develop techniques which minimize trenching activities and time, thus meeting the following two-fold objective: improved service for customers and a reduction in direct costs and social costs related to such activities.

3 TRENCHLESS TECHNIQUES

The main techniques used by Gaz de France to meet the above objective are:

- lining,
- pipe insertion,
- steerable horizontal drilling,
- "top-of-the-keyhole" techniques.

Lining is used to maintain cast iron networks (18 000 km) in operation. It consists in inserting in the existing pipeline a resin-coated sleeve which is bonded to the walls of the latter by pressurization. Lining also increases the service life of the network with an increased degree of safety in so far as it limits the risk of fracture of the cast iron. This option, when costing less than 60 % of the cost of identical pipeline renewal, is particularly suited to urban pipelines that that will still be operated under low pressure 10 or 15 years hence.

The technique of pipe insertion into obsolete pipelines of a polyethylene pipe was developed from 1980 onwards, at the same time as the use of such material was generally extended to distribution networks. When the diameter of the inserted PE pipe is less than that of the old pipeline, it allows the conversion of the network from low pressure to medium pressure. In the case where low pressure is maintained, the operation consists in cutting out and enlarging the pipe required to be rehabilitated. Pipe insertion constitutes a complete renewal of the network at a current cost approximately equal to 85% of the cost of laying a new network.

The above techniques play a major role in the renovation of obsolete pipelines and serve to limit trenching to two excavation pits (introduction and receiving ends) plus, in the case of pipe insertion, excavations for service pipe reconnections. As an example, 75% of renovated pipelines, since 1985, have been fitted with pipe insertions or, in a lower proportion, with linings.

The steerable horizontal drilling method consists in laying a PE pipe in accordance with the following principle: a drill head, sunk by a drill, penetrates the soil to create a small-diameter (50 mm) pilot hole from a start excavation pit to a finishing excavation pit of small overall dimensions (1 m 2) by means of drill rods.

Once in the start pit, the pipe is attached to the drill rods preceded by a reamer, and then drawn as far as the finishing pit. Depending on the case, excavation may be carried out by percussion method or using mud (water and bentonite) to cut the ground. Comparitive studies carried out on different excavation methods and equipment served to substantiate the choice of drilling method which depends on the specific worksite characteristics (ground formation, length of pipe installation...). Furthermore, certain installations were assessed by measuring the pulling forces sustained by the tube and by analyzing samples taken from PE pipes whose installation history was well known. From 1988 onwards, roughly 30 km of Gas pipeline were installed using these processes.

So-called **"operations-in-keyholes"** were developed during the eighties. They contrast with conventional "in contact" techniques which demand openings of sufficiently large size as to enable operators to work in satisfactory safety conditions. In this case the operator uses remote techniques to work on the pipe from the surface using an excavation pit with a diameter of no more than 60 cm. These techniques involve the use of a vacuum excavator and a set of remote controlled working tools which can be applied both to repair and maintenance work (leak repairs, suppression of lead service pipes...) and to new installation work (service pipe tapping on steel or PE mains).

4 EXAMPLE OF RAPID AND INCONSPICUOUS CONSTRUCTION WORKS

Each of the techniques described above has resulted in a significant reduction in the volume of construction work. The urge to pursue this course of action has prompted Gaz De France to combine two such techniques for a new project which will divide by 10 the road surface area to be demolished. The originality of the work site outlined below is to associate pipe installation by horizontal drilling and the performance of service pipe-main pipe connections using a vertical drilling process and a set of remote-controlled working tools.

4.1 The work-site environment

The work-site is situated in a private housing estate in the Paris area in a road measuring 8m wide (width of sidewalks : roughly 1,50m). As part of the rehabilitation of utilities including a switch to medium pressure, provision had been made to install a polyethylene pipeline having a diameter of 110mm over a length of 100m at a depth comprised between 0,80 and 1,00m through the existing Telecom, LV electrical power supply and water networks parallel to the roadway and the sewage pipes perpendicular to it.
There are 12 service pipes required to be replaced by polyethylene pipes, diameter 12 mm, located on either side of the road (8 on the even-numbered side and 4 on the odd-numbered side at average intervals of 12 m.

The sub-strata consists of backfill to a depth of 1 m and natural clayey soil below that.

4.2 Installation technique for the main pipe

The polyethylene pipe 0 110 m is laid using a steerable jet cutting drill system designed by the firm FLOWMOLE.

With this process the head is fitted with water + bentonite mud jets which cut the soil in front of the tool. The reamer which has a diameter of 150 mm, is also provided with jets designed to loosen the soil and lubrify the drawn tube.

Drilling is guided by the position of the slanted head operated with effect from the drill, and by the orientation given to the jets. It is localized along the three axes from the surface by means of a detector which receives the appropriate signals from the radio emitter installed in the drill head.
This system demands a drilling machine and a truck to transport the water tanks, the hydraulic power unit and the injection pumps. It is operated by three people.

4.3 "Operation-in-keyholes" techniques for service-main connections

Remote working techniques used for service pipe connections require the following equipment:
- a vacuum excavator
- a small-diameter pneumatic mole
- a set of tools for "operations-in keyholes";

The **vacuum excavator** carries out the vertical drilling of the small openings (diameter roughly equal to 60cm) necessary for installing the tapping tee on the main pipeline, and of the narrow trenches required to install the service lines supplying the suscribers on the even-numbered side of the road (service line of slight length).

Trenching by vacuum process consists in simultaneously:
- breaking up the earth by means of a blow gun comprising a compressed air-supplied steel tube producing a powerful and fine jet of air,
- sucking up the broken earth (subsequently used for backfilling) into a storage tank, using a flexible hose.

To implement this process, the paving must naturally be cut away beforehand.
The use of a vacuum excavator means that excavations are scattered and of slight bulk while the surrounding area remains clean since the excavated earth is stored in the extractor truck. Moreover, as regards the various underground concession holders, this technique is non-agressive and is environmentally friendly.

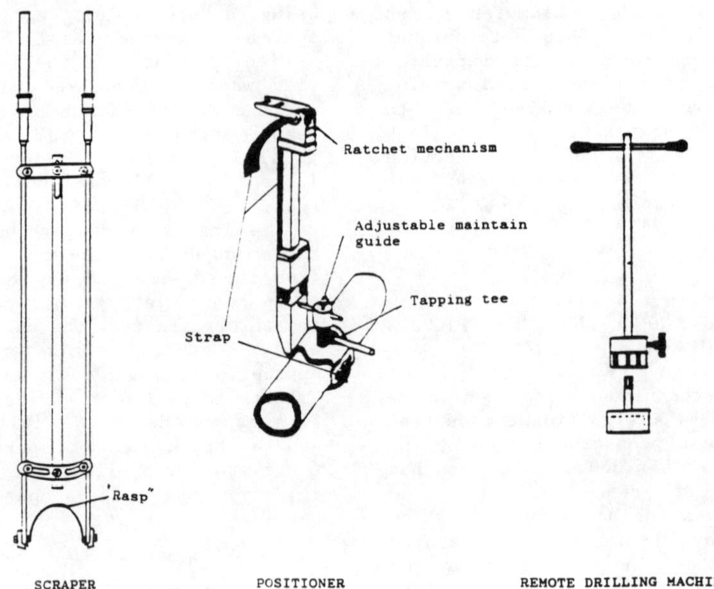

Figure 1.

The **pneumatic mole**, which is a percussion drill with earth compactor, does the horizontal drilling required to lay the service line on the odd-numbered side of the road (length of service line approximately 8 m). This means that the roadway can be crossed without hindrance to road traffic. This technique, which has been used by Gaz de France for over 15 years, is now in current use for this type of application.

The **"operation-in-keyholes"** tooling, designed by Gaz de France in collaboration with equipment manufacturing firms, is used to fasten the tapping tee to the main pipe.

The scraper: consists of a blade actuated by two articulated sleeves which removes the superficial oxydized deposit from the pipe over the width of the tapping tee.
Scraping is followed by a degreasing operation with a brush.

The positioner: used to lower the tapping tee until it is in contact with the pipe and then maintaining it there by means of a strap during the electrofusion operation. Previously, at the top of the excavation, the pipeline routed to the subscriber will have been fixed to the tee by an electrowelded joint.

The remote drilling machine: used for the following:

- to remove the cap from the tapping tee,
- to screw the perforator plug incorporated in the tee so as to pierce the pipe without producing any gas release,
- to unscrew the plug and thus fill the service line with gas,
- to replace and tighten the cap.

Once these operations are finished, an under-pressure tightness check based on washing with soap and rincing will complete the intervention.

4.4 Work Progress

Work is scheduled to last a week, preliminary studies and exploration not included.

Preliminary explorations on existing networks were conducted on the basis of concessionary drawings and observation of in-situ indications (key-operated man-holes, inspection ports...).

The explorations were completed by a summary geological survey of the digging location which included two extra test borings at the places presumed to be the start and finish of the steerable drilling.

This phase made it possible to define the optimal layout for the pipe to be installed.

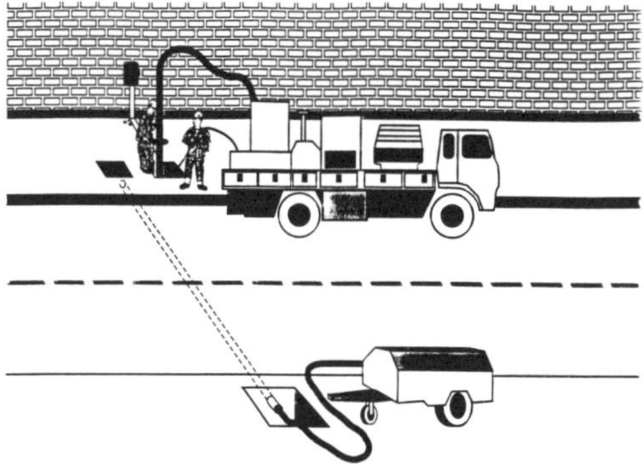

Figure 2. Installation of services with vacuum excavator, impact mole and small hole tools.

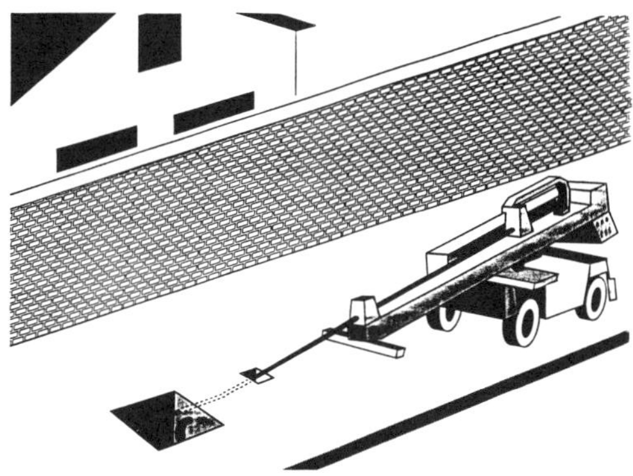

Figure 3. Installation of main pipe with steerable horizontal drilling.

The work schedule was as follows:

1st Day: setting up of work site, marking out of the existing pipes and of the pipe route, installation of signalling and marking system, start excavation pit for steerable drilling, pneumatic mole drilling for the four crossings

2nd day: preparation of "subscriber's" service box, steerable drilling terminal excavation pit and final service excavations,

3rd day: connection of main pipe to network, tapping tee installation and backfilling,

4th day: end of tapping operations and backfilling.

These operations were carried out by a steerable drilling team (3 persons during 1 day), a civil engineering firm, 3 Gaz de France teams (2 X 2 persons during 3 days, 1 extra team on the 3rd day).

A few technical points involving the work-site should be emphasised:

The digging location being particularly favorable, pipe laying by steerable drilling proved to be relatively quick (100m/day) and progressed without notable incident other than the necessity of lowering the drilling device, by a few dozen centimeters, to pass under an obstacle which exploration had failed to identify.

The simultaneous performance of service openings and of steerable horizontal drilling meant that it was possible to visually follow the progression of the drill head and punctually check the location indicated by the detector.

The first two meters of PE pipe that were pulled after having passed through 100 meters of soil, were sampled and analysed. Pressure resistance testing gave satisfactory results.

Apart from a pneumatic mole which slightly deviated from course, the pre-defined drilling paths were adhered to and the time required for each pipe crossing did not exceed 1h 30.

Two sets of "operation-in-keyhole" tools were necessary to complete the service connections.

4.5 Work Assessment

The direct cost of the work site was noticeably higher than the cost of a conventional work site in spite of the gain in time, backfill and surfacing material due to a significant reduction in trenching operations. There were several reasons for this:

- a site of this type demands a large amount of varied equipment which remains immobilized for four days,
- such equipment represents a relatively heavy investment, particularly for the excavation engine,
- this type of work site is still of an experimental nature, which raised the price.

The social cost of the work site was practically nil :
- the working time was divided by three in relation to a conventional site,
- the cut out paving surface was ten times smaller, which implies a longer service life for the roadway and a larger usable space for users,

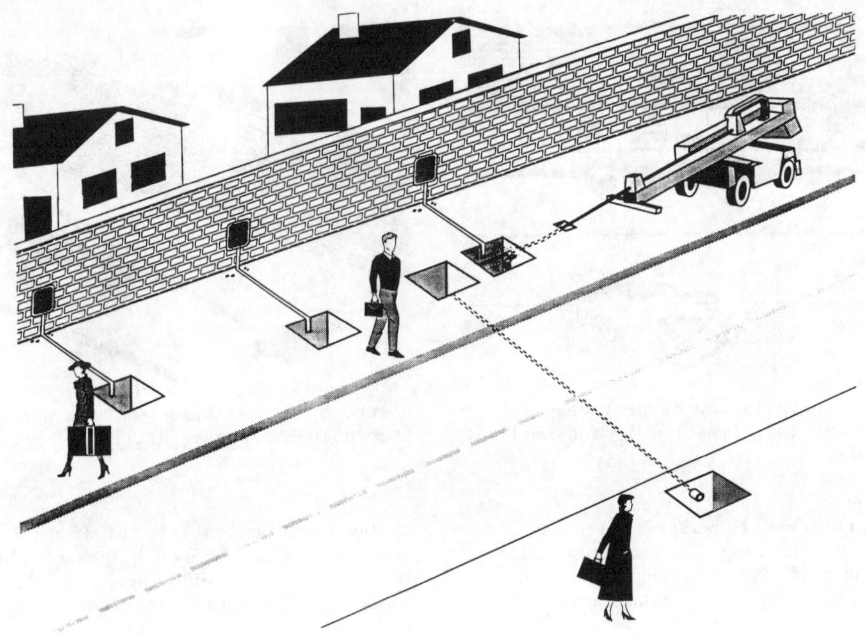

Figure 4. Association of horizontal and vertical drilling.

- no interruption of the road traffic was made necessary,
- the various excavations were small-size and were backfilled as soon as the work was completed, thereby minimizing inconvenience in terms of road surface and time,
- local residents did not undergo any inconvenience. They were able to leave their garage without difficulty and circulate normally on the sidewalks...
- the local environment was protected.

5 CONCLUSION AND PROSPECTS :

The achievement of rapid and inconspicuous engineering work is one of the targets that Gaz De France has set itself in order to satisfy its clients, guarantee safety and maintain good relations with the public and local authorities. Although the technique described herein represents considerable progress in the field concerned, it is still only used on a marginal basis and offers a considerable potential for research and development.

A research programme has recently been undertaken on this theme. It should help to improve the above techniques and generalize their application by making them more profitable.

No Trenches in Town, Henry & Mermet (eds) © 1992 Balkema, Rotterdam. ISBN 90 5410 085 0

Contribution de Gaz de France à la réalisation de Travaux Rapides et Discrets

B.Calinaud
STG-DEGS, EDF-GDF, France

P.Le Testu & S.Stringhetta
CERSTA, DETN, GDF, France

Résumé: Les travaux de pose et d'exploitation des réseaux en zone urbaine induisent généralement des terrassements importants. Ceux-ci génèrent des nuisances de moins en moins tolérées par le public et les collectivités locales qui en supportent les coûts sociaux.

Dans ce cadre, Gaz De France s'est attaché à développer l'utilisation de techniques réduisant les gênes à l'environnement. Que ce soit des travaux de renouvellement ou d'extension, ces différentes techniques vont de la réutilisation des conduites existantes comme fourreaux (chemisage et tubage), à l'exécution de travaux du haut de fouille de dimensions réduites ou à la pose de conduites neuves par forage horizontal dirigé.

L'objet de cette communication est la description d'un chantier dont l'originalité est d'associer d'une part un procédé de forage horizontal dirigé pour la pose de la conduite polyéthylène, d'autre part un forage vertical par aspiration, un fonçage par fusée et une technique de travail à distance pour la réalisation des branchements d'abonnés. Après une présentation du contexte du chantier, les options techniques retenues sont explicitées, avant de dépeindre le déroulement du chantier.

Un bilan économique global,tenant compte des coûts sociaux plus difficiles à quantifier du fait de leur aspect qualitatif, a permis de comparer le coût d'un tel chantier avec le coût d'un chantier traditionnel.

1 INTRODUCTION

Les travaux de pose et d'exploitation des réseaux souterrains nécessitent généralement des terrassements importants qui s'intègrent difficilement dans le paysage urbain. Les nuisances causées par les chantiers sont en effet aujourd'hui de moins en moins tolérées par le public et les collectivités locales. Demain, la réglementation pourrait imposer des contraintes plus fortes sur la réalisation de tranchées.

Pour résoudre ces problèmes, auxquels tout concessionnaire du sous-sol se trouve confronté, Gaz De France a mis au point des techniques permettant l'exécution de ***travaux rapides et discrets.***

2 CONTEXTE

La population desservie par Gaz De France s'élève à 38 millions, soit environ les trois quarts de la population française, ceci par l'intermédiaire d'un réseau de distribution d'une longueur totale de 115 000 km. Le gaz est distribué en basse pression à 21 mbar ou en moyenne pression à 4 bar.

Les conduites alimentant le réseau sont constituées pour les plus anciennes de fonte en basse pression (13%) et d'acier en basse et moyenne pression (45%) et pour les plus récentes de polyéthylène (ou PE)en moyenne pression (33%).

Plus de 3000 km de canalisations sont posés chaque année en zone urbaine, dont 90% en PE.

L'extension des réseaux, indispensable à l'expansion des villes, induit traditionnellement des terrassements qui apparaissent de moins en moins compatibles avec le contexte urbain et la qualité de la vie et de l'environnement citadins.

Les travaux de terrassement représentent en effet des **coûts sociaux** d'importance croissante à mesure que l'espace collectif disponible dans la ville est compté.

Pour les usagers, les tranchées signifient la perturbation de la circulation piétonne et routière, donc perte de temps, consommation accrue de carburant et fréquence plus grande des accidents ; **pour les industries et les commerces**, elles sont synonymes de perte de profit ; **pour les collectivités locales**, elles sont responsables de surcoûts dûs à la fragilisation de la chaussée qu'elles entraînent ou de manques à gagner causés par la diminution de revenus issus de la location de

certains espaces aux usagers (places de parking par exemple). Enfin, les terrassements détériorent l'environnement par le bruit et la pollution qu'ils génèrent.
Mais, si les coûts sociaux sont, par définition, supportés par la société, ils se répercutent sur l'entreprise maître d'ouvrage, qui, par ailleurs, fait les frais des coûts directs des tranchées (60 à 70 % du coût total des travaux) : les nuisances occasionnées par les terrassements pénalisent l'image de marque de l'entreprise qui risque ainsi de perdre des clients potentiels.
L'analyse de ces différents points a conduit Gaz De France à développer des techniques minimisant les terrassements et les durées d'intervention répondant au double objectif suivant : l'amélioration du service rendu à ses clients et la réduction des coûts directs et sociaux de ses travaux.

3 DES TECHNIQUES A TERRASSEMENT REDUIT

Les principales techniques employées par Gaz De France pour répondre à l'objectif cité ci-avant sont :
- le chemisage
- le tubage
- le forage horizontal dirigé
- les techniques de travail du haut de la fouille

Le chemisage est utilisé pour le maintien en exploitation des réseaux en fonte grise cassante (18 500 km). Il consiste à introduire dans la canalisation existante une gaine souple de résine collée sous pression aux parois de celle-ci. Le chemisage permet d'augmenter la durée de vie du réseau avec une sécurité garantie dans la mesure où il pallie au risque de cassure de la fonte. Cette solution, économiquement intéressante pour un coût inférieur à 60% du coût de renouvellement à l'identique, est particulièrement adaptée aux conduites urbaines encore exploitées en basse pression dans les 10 ou 15 ans à venir.
Le tubage des canalisations anciennes par insertion d'un tube en polyéthylène s'est développé à partir de 1980, parallèlement à la généralisation de l'emploi de ce matériau pour les réseaux de distribution. Lorsque le diamètre du PE inséré est inférieur à celui de l'ancienne canalisation, il permet de convertir le réseau basse pression en moyenne pression. Dans le cas d'un maintien en basse pression, le tubage est réalisé par découpe et élargissement de la conduite à rénover. Le tubage constitue un renouvellement complet du réseau avec un coût aujourd'hui sensiblement égal à 85% du coût de la pose d'un réseau neuf.
Ces deux techniques occupent une place prépondérante dans la rénovation des anciennes canalisations et limitent les terrassements à deux fouilles (d'introduction et de réception) auxquelles s'ajoutent, dans le cas du tubage, les fouilles de reprises de branchement. A titre d'exemple, 75% des canalisations rénovées depuis 1985 à Paris ont été tubées ou, en moindre proportion, chemisées.
Le forage horizontal dirigé permet de poser une conduite PE selon le principe suivant : une tête de forage pénètre dans le sol poussée par une foreuse (ou Drill) et crée un trou pilote de petit diamètre (50mm) d'une fosse de départ vers une fosse d'arrivée de faibles encombrements (1m2) au moyen d'un train de tiges creuses. Une fois dans la fosse d'arrivée, la conduite est accrochée au train de tiges, précédée d'un réaléseur,puis tirée jusqu'à la fosse de départ. Le creusement s'effectue selon les cas, par percussion, ou à l'aide d'un liquide (eau et bentonite) découpant le sol. Des études comparatives des différents modes de creusement et matériels ont été menées constituant un support pour le choix du type de forage à utiliser selon les spécificités du chantier (terrains traversés, longueur de pose...). De plus, certains systèmes ont été évalués notamment par la mesure des efforts de traction subis par le tube et l'analyse d'échantillons de PE dont l'historique de pose était parfaitement connu. Depuis 1988, environ 30 km de conduites de Gaz ont été posés par ces procédés.
Les techniques d'intervention dites **du haut de la fouille** ont été élaborées à partir des années 80. Elles s'opposent aux techniques classiques "au contact" qui nécessitent des ouvertures dont la dimension doit être suffisamment grande pour que les intervenants travaillent dans des conditions satisfaisantes de sécurité. Ici, les agents agissent à distance sur la conduite depuis la surface, avec une excavation dont le diamètre n'excède pas 60 cm. Ces techniques mettent en oeuvre un engin de terrassement par aspiration (ETPA) et un outillage de travail à distance et permettent de réaliser les travaux d'exploitation (réparation de fuites, suppression des branchements plomb....) comme les travaux neufs (pose de prises de branchement sur acier ou PE).

4 UN EXEMPLE DE TRAVAUX RAPIDES ET DISCRETS

Les techniques décrites précédemment permettent chacune des réductions significatives du volume des terrassements. La volonté d'aller plus loin dans cette voie a conduit Gaz De France à additionner deux d'entre elles pour la réalisation de travaux neufs, divisant ainsi par 10 la surface de revêtement démoli. Le chantier dont la description suit présente en effet l'originalité d'associer la pose de la conduite par forage horizontal et l'exécution des raccordements branchements-conduite à l'aide d'un forage vertical et d'un outillage de travail à distance.

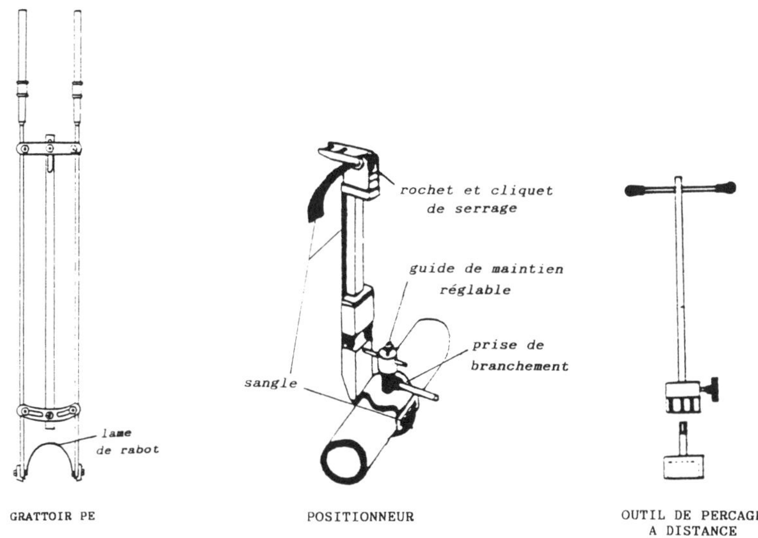

Figure 1.

4.1 Cadre du chantier :

Le chantier se situe dans un quartier pavillonnaire de la région parisienne sur une rue d'une largeur de 8m (dimension des trottoirs : environ 1,50 m). Dans le cadre d'un renouvellement de réseau avec passage en moyenne pression, il s'agit de mettre en place une canalisation en polyéthylène de diamètre 110mm sur une longueur de 100m à une profondeur comprise entre 0,80 et 1,00m au travers des réseaux existants : PTT, électricité BT et eau parallèlement à la chaussée et assainissement perpendiculairement à celle-ci.
Le nombre de branchements à renouveler avec des conduites polyéthylène de diamètre 20mm s'élève à 12, situés de part et d'autre de la rue (8 côté pair, 4 côté impair) distants de 12m en moyenne.
Le sous-sol est constitué de remblai jusqu'à 1m puis de terrain naturel argileux en dessous.

4.2 Technique de pose de la conduite principale :

La conduite polyéthylène ø110mm est posée grâce à un système de forage dirigé à jets découpants mis en oeuvre par la société FLOWMOLE.
Dans ce procédé, la tête est munie de jets de liquide (eau+bentonite) qui assurent une découpe du sol devant l'outil. L'aléseur de diamètre 150mm, est aussi équipé de jets dont le rôle est d'ameublir le sol et de lubrifier le tube tiré.
Le forage est dirigé grâce à la position du méplat de la tête commandée depuis la foreuse et à l'orientation donnée aux jets. Il est localisé selon les 3 axes depuis la surface avec un détecteur, qui reçoit les signaux de l'émetteur radio embarqué dans la tête.
Ce système nécessite une foreuse et un camion transportant les réservoirs d'eau, l'unité hydraulique de puissance et les pompes d'injection. Son fonctionnement est assuré par trois personnes.

4.3 Techniques "de haut de fouille" pour le raccordement branchement-conduite :

Les techniques de travail à distance utilisées pour le raccordement des branchements mettent en oeuvre le matériel suivant :

- un engin de terrassement par aspiration
- une fusée de faible diamètre
- un outillage de haut de fouille

L' engin de terrassement par aspiration réalise le forage vertical des petites ouvertures (diamètre environ égal à 60cm) nécessaires à la pose de la prise de branchement sur la canalisation principale et des tranchées étroites d'installation des conduites de dérivation alimentant les abonnés sur le côté pair de la rue (longueur de branchement faible).
Le procédé de terrassement par aspiration consiste à provoquer simultanément :
- l'émiettement des terres au moyen d'une soufflette composée d'un tube acier alimenté en air comprimé produisant un jet d'air puissant et fin,
- l'aspiration des terres émiettées (utilisées ultérieurement pour le remblai) vers une cuve de stockage, par l'intermédiaire d'un flexible.

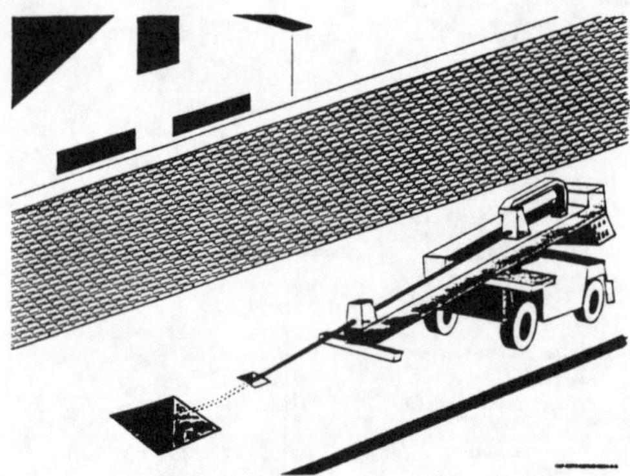

Figure 2. Pose de la conduite principale par forage horizontal dirigé.

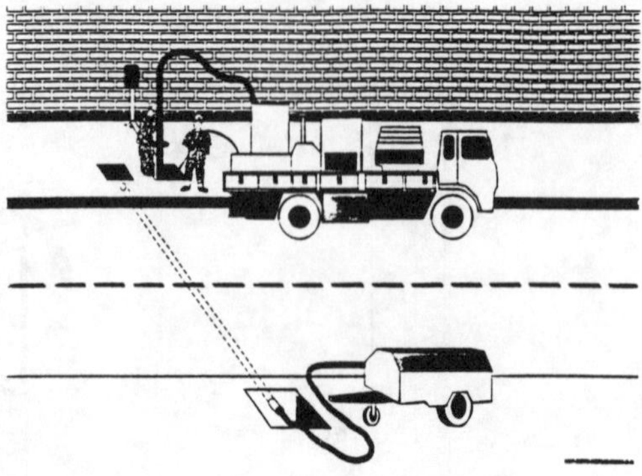

Figure 3. Pose des branchements avec un camion aspirateur, une fusée et un outillage "du haut de la fouille".

La mise en oeuvre de ce procédé nécessite naturellement la découpe préalable du revêtement.

L'utilisation de l'engin de terrassement par aspiration permet la réalisation de fouilles éparses et de faible encombrement avec des abords propres puisque les déblais sont entreposés dans le camion aspirateur. Par ailleurs, cette technique est non-agressive vis à vis des différents concessionnaires du sous-sol et préserve l'environnement.

La fusée pneumatique, agissant par percussion avec compression des terres, effectue le forage horizontal nécessaire à la pose de la conduite de branchement sur le côté impair de la rue (longueur de branchement avoisinant les 8m). Elle permet la traversée sous la chaussée sans perturbation pour la circulation routière. Cette technique, employée par Gaz De France depuis plus de 15 ans, est aujourd'hui d'utilisation courante pour ce type d'applications.

L'outillage du haut de la fouille, développé par Gaz De France en collaboration avec des entreprises de fabrication de matériel, est utilisé pour la fixation de la prise de branchement sur la conduite principale.

Le gratteur, constitué d'une lame actionnée par deux manches articulés enlève la couche superficielle oxydée de la conduite sur la largeur de la prise.

Ce grattage est suivi d'un dégraissage au pinceau.

Le positionneur permet de descendre la prise au contact de la conduite, puis de la maintenir sur celle-ci à l'aide d'une sangle durant l'électrosoudage. Préalablement, en haut de la fouille, la canalisation allant chez l'abonné aura été fixée sur la prise avec un manchon électrosoudable.

L'électrosoudage accompli, le positionneur est retiré. Un essai d'étanchéité est effectué à partir du"coffret abonné" puis le client est raccordé.

La machine à percer à distance réalise :
- le retrait du capuchon de la prise,
- le vissage du bouchon perforateur incorporé à la prise qui perce la conduite sans dégagement de gaz,
- le dévissage de ce bouchon et donc la mise en gaz du branchement,
- la remise en place et le serrage du capuchon.

Ces opérations terminées, un contrôle d'étanchéité en charge par savonnage puis rinçage de l'ensemble prise-conduite va clore l'intervention.

4.4 Déroulement du chantier :

La durée du chantier a été d'une semaine, études et reconnaissances préliminaires non comprises.

Les reconnaissances préalables des réseaux en place ont été menées à partir des plans des concessionnaires et de l'observation des indications in situ (bouches à clef, regards,...).

Deux sondages complémentaires, aux endroits présumés du départ et de l'arrivée du forage dirigé ont complété ces reconnaissances par une étude géologique sommaire du terrain à traverser.

Cette phase a permis de définir le tracé optimal de la conduite à poser.

Le calendrier du chantier a été le suivant :

1er jour : installation sur le chantier, marquage des concessionnaires et du parcours de la conduite à poser, mise en place de signalisation, balisage,
réalisation de la fouille de départ du forage dirigé,
fonçage à la fusée pour les quatre traversées,

2ème jour : préparation des "coffrets abonnés",
réalisation de la fouille d'arrivée du forage dirigé et des dernières fouilles de branchement,
pose de la conduite principale par forage dirigé,

3ème jour : raccordement de la conduite principale au réseau,

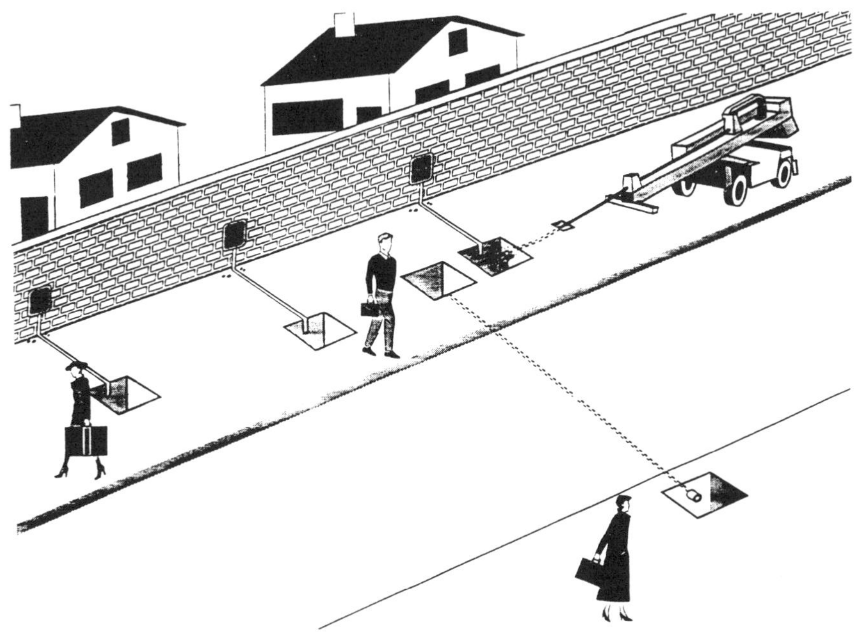

Figure 4. Association de techniques de forage horizontal et de forage vertical

pose des prises de branchements et remblais,
4ème jour : fin des branchements et remblais.
Ces operations ont été réalisées avec une équipe de forage dirigé (3 personnes durant 1 jour), une entreprise de travaux publics, 3 équipes de Gaz De France (2x2 personnes pendant 3 jours, 1 équipe supplémentaire le 3 ème jour).
Quelques points techniques sont à souligner concernant ce chantier.
Le terrain étant particulièrement favorable, la pose par forage dirigé a été relativement rapide (100m/jour) et s'est déroulée sans incident notoire, si ce n'est le recul de quelques dizaines de centimètres du dispositif de forage pour le passage sous un obstacle non identifié au cours des reconnaissances.
La réalisation simultanée des ouvertures de branchement et du forage horizontal dirigé a permis de suivre la progression de la tête de visu et de vérifier ponctuellement la localisation donnée par le détecteur.
Les deux premiers mètres de PE tirés, ayant traversé les 100 mètres du forage, ont été prélevés et analysés. La mesure de leur résistance en pression a donné des résultats satisfaisants.
A l'exception d'une fusée légèrement déviée, les trajectoires définies pour les fonçages ont été respectées et la durée de chacune des traversées n'a pas excédé 1h30.
Deux jeux d'outillage "de haut de fouille" ont été nécessaires pour accomplir dans les délais l'ensemble des branchements.

4.5 Bilan du chantier :

Le coût direct du chantier a été sensiblement supérieur au coût d'un chantier classique, malgré les gains de temps, la réduction des matériaux de remblai utilisés et la diminution importante des terrassements qu'il a permis d'obtenir. Ceci pour plusieurs raisons :
- un tel chantier nécessite une concentration d'équipements variés qui se trouvent immobilisés pendant quatre jours,
- ces équipements représentent un investissement relativement lourd, notamment l'engin de terrassement,
- l'utilisation de ce matériel requiert un personnel d'un niveau technique supérieur,
- ce type de chantier revêt encore un caractère expérimental, ce qui en augmente le prix.

Le coût social du chantier a été quasiment nul :

- la durée du chantier a été divisée par trois par rapport à un chantier classique,
- la surface de revêtement découpée a été dix fois inférieure, ce qui implique une durée de vie plus grande de la chaussée et un espace utilisable plus important pour les usagers,
- aucune interruption de la circulation routière n'a été nécessaire,
- les différentes excavations réalisées, de petite taille, ont été remblayées sitôt le travail terminé, d'où un dérangement de surface et de durée minimales,
- les riverains n'ont subi aucune

nuisance; en particulier, ils ont pu sortir sans problème de leur garage, circuler normalement sur les trottoirs...
- le site a été préservé.

5 CONCLUSION ET PERSPECTIVES

Rendre les travaux rapides et discrets est un des objectifs que s'est fixé Gaz De France, soucieux de satisfaire ses clients, de garantir la sécurité et d'entretenir de bonnes relations avec le public et les collectivités locales. Si la technique décrite ci-avant représente un progrès considérable dans cette voie, elle reste encore utilisée de façon marginale et présente de nombreuses potentialités de recherche et développement.

Un programme de recherche portant sur ce thème a été engagé. Il devrait permettre d'améliorer ces techniques et d'en généraliser l'emploi afin d'en accroître la rentabilité.

No Trenches in Town, Henry & Mermet (eds) © 1992 Balkema, Rotterdam. ISBN 90 5410 085 0

Directionally drilled crossing of Dauphin Island, Mobile Bay, Alabama

Neil Smith
Costain Land & Marine, Inc., Houston, Tex., USA

ABSTRACT: The design and construction of three, bundled, 305 mm diameter, 1190 m long pipelines beneath Dauphin Island, Alabama, was completed in December 1991. Construction was achieved by horizontal directional drilling with entry and exit points in water depths of 4.5 m and 6.7 m, respectively. The selection of the staging vessels (self elevating working platforms), drilling fluid supplies' logistics, pipe laying operations and installation procedures are described to provide an insight to the problems and solutions of constructing drilled crossings in this extraordinary environment.

INTRODUCTION

The first directionally drilled pipeline installation was constructed in the United States over 20 years ago. Since that time, the advantages of the directional drilling technique have been recognized by the pipeline industry. The technology has been applied as the most environmentally responsible and economical method for installing river crossings on many projects worldwide.

The application of the technology to the offshore pipeline industry has predominantly been limited to shore approaches. However, a small number of water-to-water crossings have been constructed on projects where the economical advantages have been recognized or where conventional dredging or post-lay burial techniques have been impractical.

In 1991, Costain Land & Marine designed and constructed two crossings of the Mobile Bay Ship Channel (with drilled lengths of 480 m and 590 m) from a sea-going work barge in water depths which exceeded those of any previously constructed water-to-water crossing. Having successfully completed those crossings, Costain Land & Marine was contracted by BP Exploration,Inc. to install the directionally drilled crossing of Dauphin Island.

BP Exploration's development of Mobile Block 821 for the production and transport of gas included the construction of a 305 mm diameter pipeline beneath Dauphin Island. This pipeline delivers gas from their production platform, located offshore Alabama in the Gulf of Mexico, to onshore facilities. In order to accommodate future requirements, two additional 305 mm diameter pipelines were installed at the crossing of the island (the bundle consisted of two 12.75 inch o.d. x 0.500 inch w.t. and one 12.75 inch o.d. x 0.562 inch w.t. pipelines, all to API 5L Grade B).

By selecting directional drilling as the method of construction, BP Exploration realized the economical and environmental advantages it provides compared with conventional installation procedures. Directional drilling mitigated the construction impact to the environmentally sensitive island, maintained the islands stability, provided enhanced pipeline burial depths(particularly in the surf zones), gave a shorter and more predictable construction schedule, and minimized the risks associated with winter weather conditions.

STAGING VESSELS

With the construction of the crossing being undertaken during the winter, the weather sensitivity of the equipment to be used was of prime importance. The impact of the weather was carefully considered during the pre-construction planning phase in order that the risks associated with inclement weather could be minimized.

A stable working platform was required from which to conduct the drilling operations. The drilling rig has only limited capacity to accommodate the yaw, pitch and roll tendencies of offshore vessels. Even small motion of the staging vessel can potentially over stress the drill pipe and the rig's mechanical components. Although the crossings that had been constructed in Mobile Bay earlier in the year had been drilled from a sea-going (120 m x 30 m) work barge, it was not considered economically feasible to utilize such a large vessel on this project. The likelihood of experiencing weather downtime due to winter sea states, even when drilling from a relatively stable floating platform, was also considered prohibitive. Hence, other options were investigated.

Staging the drilling rig on a medium sized (55 m x 16 m) flat barge ballasted to the sea floor was considered. However, this option was regarded as impractical due to the possibility of causing damage to the barge, the difficulty in obtaining the requisite insurance, and the probability of water over topping the deck during only moderate seas.

The third option, which was finally selected, was to stage the drilling rig on a "self elevating working platform" (SEWOP). These self propelled lift boats are able to jack-up on self contained legs to provide a stable working platform. Drilling operations could be carried out from such vessels in all but storm conditions.

The selection of the class of SEWOP to be used was dependent on the depth of water in the vicinity of the proposed site location and the availability of the vessels. Since only relatively small SEWOP's (class 150 and 105) would be able to maneuver in the water depths which varied between 3 m and 4.5 m, at least two were required to stage the drilling rig and its control module.

Stability analyses of a class 150 SEWOP were performed to determine factors of safety against overturning and sliding, and to calculate anticipated stresses which might be imposed due to the combined effects of the maximum pulling capacity of the drilling rig (250 tons) and environmental loadings (wind and waves). These analyses proved that two class 150 SEWOP's, attached in tandem, would provide a suitable staging platform for the drilling rig.

In order to provide the drilling operator with an adequate view of the drilling rig, it was necessary to use a third, smaller, SEWOP (class 105) on which to mount the control module. This vessel was sited alongside, but perpendicular to, that staging the drilling rig.

The other equipment associated with the drilling rig (drill pipe, mud pumps and mud tank) were placed on a 55 m x 16 m flat barge which was moored on the leeward side of the SEWOP's. In order to protect the SEWOP's from damage by the flat barge, three 914 mm diameter pipe piles were designed as fenders between the vessels. A second 55 m x 16 m flat barge was also utilized to hold tanks of premixed drilling fluid (bentonite) which was used during the operation. This second flat barge was moored against that with the drill pipe, pumps and mud tank.

DRILLING FLUIDS

Throughout the drilling operations (pilot hole drilling, pre-reaming and pullback) drilling fluid was pumped in order to stabilize the drilled hole, to carry cuttings away from the face, and to reduce the friction on the drill pipe and, ultimately, on the pipeline bundle as it was pulled into the drilled hole.

The collection, containment and subsequent disposal of drilling fluids which return to the ground surface on directionally drilled land-to-land crossings has become a common feature of such projects. However, the regulatory agencies governing the crossing of Dauphin Island accepted that such a requirement was not practical when drilling was to be undertaken with entry and exit points in water depths of 4.5 m and 6.7 m. Nevertheless, the composition of the drilling fluids was limited to bentonite (a sodium montmorillonite clay) with no additives.

In view of the stringent environmental regulations governing work in Mobile Bay, the usually accepted method of mixing powdered drilling fluids with water on location was considered inappropriate. The usual method of breaking open sacks of bentonite over mixing tank hoppers causes a considerable amount of dust which in an offshore application not only produces an environmental hazard but also jeopardizes safety. The volume of fluid to be mixed on location could be minimized by using premixed bentonite transported to site in tanks on a 55 m x 16 m flat barge.

The preferred method of handling the storage of drilling fluid on location was to use seventeen 79,500 l capacity tanks on a flat barge which could hold either premixed bentonite or fresh water. The drilling fluid would be mixed at a batch plant on the mainland (some 40 km from the drilling location) and pumped into the 79,500 l tanks on the barge. The barge would then be towed to the location and the drilling fluid pumped to the rig's mixing tank when needed.

Since it would be necessary to vary the viscosity of the drilling fluid during the operations, two of the holding tanks were used to store fresh water. Also, three pneumatic tanks would be used to store dry, powdered, bentonite on location. The pneumatic tanks were connected to enable transferring the dry bentonite to a hopper over the rig's mixing tank. This arrangement provided an essentially closed system and avoided any of the environmental or safety problems associated with breaking open individual sacks on location.

The volume of drilling fluid that could be stored on location was less than half that estimated for the completion of the project. Additional premixed fluid and bulk dry gel would be transported to replenish the stores at location by way of smaller (318,000 l and 35 cu.m) delivery barges.

PIPELINE FABRICATION AND TESTING

Land & Marine subcontracted SubSea International for the pipeline fabrication, testing and anode installation. These operations were supported by SubSea's Lay Barge 278 and a class 130 SEWOP. By utilizing these vessels, the pipe laying operations could be carried out on the Gulf of Mexico side of the island where water depths varied between 6 m and 9 m. It was considered that this arrangement, with the drilling rig being positioned on the leeward side of the island, was the least weather sensitive and would minimize lost time due to weather conditions.

The SubSea class 130 SEWOP was initially positioned some 60 m from the designed drilled exit point and the lay barge anchored an additional 400 m further away. A pull cable was strung from the SEWOP's winch to the leading section of the first pipeline to be laid. As joints of 305 mm diameter pipe were welded on the lay barge, the pipeline was pulled towards the SEWOP. Having laid approximately 200 m of pipeline in this fashion, the lay barge pulled on her anchors to lay the remaining 990 m. The pipeline was then filled with water. The lay barge was repositioned 400 m from the SEWOP, the pull cable retrieved and the second 305 mm diameter pipeline laid by the same procedure along a line 3 m away from the first pipeline. The procedure was repeated for the third pipeline.With all three pipelines flooded, gauging pigs were run and pre-pull hydrostatic tests conducted.

The welded field joints were sandblasted, preheated and coated with fusion bonded epoxy. Two of the pipelines were provided with external neoprene spacers to prevent contact between the pipelines as they were bundled and pulled into the drilled hole.

DRILLING OPERATIONS

Before any equipment was mobilized to the offshore location, a survey was carried out to locate and mark the designed drilling entry and exit points. A 35 m x 12 m spud barge with a 130 ton deck crane installed the three 914 mm diameter fender piles and a 300 mm diameter casing pipe supported on two "goal post" assemblies.

Since the distance between the seabed and the drilling rig was approximately 60 m, a casing pipe was required during the initial pilot hole drilling to provide some rigidity to the drill pipe as it was pushed ahead by the rig. The 300 mm diameter casing pipe was supported on two sets of "goal posts" which each comprised of two 12 m long, 300 mm diameter piles with cross members set at predetermined elevations.

While the fender piles and casing pipe were being installed, the drilling rig and its associated equipment were being loaded onto the SEWOP's and flat barges at a nearby dock facility. Each item of equipment was tied down to the decks in readiness for mobilization to the offshore location.

On 16th November 1991 the equipment sailed from the dock and arrived at the drilling site. Hydraulic and drilling fluid hose connections were made, and the following day pilot drilling commenced.

Pilot hole was drilled in the sand and clay formations with a 73 mm jet. As the pilot hole drilling progressed, 125 mm diameter wash pipe was installed over the 73 mm diameter pilot tubing. The drilled profile used an entry angle of 6 degrees and a minimum radius of curvature of 390 m. Land & Marine had carried out stress calculations in the pre-construction phase of the project which verified that the pipelines would be within allowable stress limits.

As the pilot drilling proceeded under the island, a "TruTracker" was used to provide an additional correlation on the magnetometer and accelerometer signals being received at the drilling rig's control console. The TruTracker system was used only on the island (over a distance of approximately 100 m). To use this system over water was neither practical nor necessary.

On 20th November the pilot hole was completed as the drill bit exited on the Gulf of Mexico side of the island, some 1100 m from the drilling rig. Air was pumped through the drill string and the exit point located by divers. A survey of the actual exit point showed it to be only 0.5 m off target. Additional lengths of 125 mm diameter wash pipe were added to the string and then the pilot tubing was tripped out of the hole.

Immediately prior to the completion of pilot hole drilling, a class 205 SEWOP was mobilized and positioned at the exit point. The 125 mm diameter wash pipe was lifted to its deck and a 1 m diameter "flycutter" connected. As the flycutter was then lowered to the sea bed, joints of heavy walled 125 mm diameter drill pipe was stalked on behind it. Drilling fluid was pumped to the flycutter as it was rotated and pulled into the seabed by the drilling rig. As joints of wash pipe were removed at the drilling rig, joints of heavy walled drill pipe were attached to the drill string behind the flycutter.

After 35 hours of continuous pre-reaming the flycutter was lifted to the deck of the drilling rig's SEWOP.

On completion of the pre-pullback hydrostatic tests, each of the product pipelines was de-watered. The lay barge was repositioned at the leading end of the pipelines and the pipelines were lifted in order to attach a single pulling head to the bundle. The pulling head was then connected by way of shackles and a swivel assembly to a 950 mm diameter barrel reamer. As the assembly was lowered to the seabed, the drilling rig applied tension to the drill string.

During the fabrication and testing operations, some silting-in of the pipelines had occurred. However, the drilling rig experienced little trouble providing the force necessary to start the pullback. With drilling fluid being pumped to the reamer, and rotation and pull being applied by the drilling rig, the pipeline bundle entered the drilled hole.

The pullback operation progressed steadily; no excessive torques or pulling forces were experienced or required. After 36 hours the pullback was completed. The pulling head was disconnected from the swivel and reamer at the drilling rig before being lowered to the seabed. This was only thirteen days after the equipment had arrived at the location.

POST-PULLBACK OPERATIONS

With the pullback operation having been successfully completed, the SEWOP's and flat barges with the drilling rig and its equipment were returned to dock, unloaded and demobilized.

Each of the pipelines in the bundle was gauged and hydrostatically tested. The exposed ends of the pipelines were lowered, by jetting, to provide a transition from the entry and exit angles of the drilled crossing to the conventional pipe laying which was to follow. Anodes were attached to the ends of the pipelines.

The piles and "goal posts" were extracted and then all remaining equipment was demobilized.

CONCLUSION

The directionally drilled crossing of Dauphin Island was designed and constructed within the schedule and budget limits that had been specified by BP Exploration. Much of the success of this project can be accredited to the attention which was placed on pre-construction engineering and planning. A team spirit of co-operation and shared goals was formed and maintained throughout the project between BP Exploration, Costain Land & Marine and the subcontractors. By combining the efforts of all the parties involved, the risks associated with applying horizontal directional drilling to an offshore environment were minimized and a successful outcome ensued.

No Trenches in Town, Henry & Mermet (eds) © 1992 Balkema, Rotterdam. ISBN 90 5410 085 0

La traversée de l'île Dauphin, baie de Mobile, Alabama, par forage directionnel

Neil Smith
Costain Land & Marine, Inc., Houston, Tex., Etats-Unis

RESUME : la conception et la construction de trois gazoducs en faisceau de 305 mm de diamètre et de 1 190 mètres de long furent achevées en décembre 1991. La construction fut réalisée par la technique du forage directionnel horizontal, avec des points d'entrée et de sortie dans l'eau, à des profondeurs respectives de 4,5 mètres et de 6,7 mètres. La sélection des structures de travail pour la construction (barges de forage à élévation automatique), la logistique de l'alimentation en fluide de forage, les opérations de pose des tuyaux et les procédures d'installation sont décrites de manière à donner une idée des problèmes et des solutions concernant la construction de traversées par forage dans cet environnement extraordinaire.

INTRODUCTION

La première installation d'un gazoduc par forage directionnel fut réalisée aux Etats-Unis il y a plus de vingt ans. Depuis lors, les avantages de la technique du forage directionnel ont été reconnus par l'industrie des gazoducs. Cette technologie a été appliquée comme étant la méthode la plus économique et la plus favorable à la préservation de l'environnement pour traverser des fleuves dans le cadre de nombreux projets dans le monde entier.

L'application de la technologie à l'industrie des gazoducs offshore a été essentiellement limitée aux approches sur terre ferme. Toutefois, quelques traversées d'eau à eau ont été réalisées dans le cadre de projets dans lesquels ses avantages économiques ont été reconnus ou dans lesquels les techniques conventionnelles du dragage ou de la mise sous terre après la pose ont été jugées peu pratiques.

En 1991, Costain Land & Marine a conçu et construit deux traversées du canal maritime de la baie de Mobile (avec des longueurs de forage de 480 m et de 590 m) depuis une barge de travail marine à des profondeurs d'eau dépassant celles de n'importe quelle traversée d'eau à eau ayant été construite précédemment. Après avoir réalisé ces traversées avec succès, Costain a reçu un contrat de BP Exploration, Inc. pour installer la traversée de l'île Dauphin par forage directionnel.

Le développement par BP Exploration du Bloc Mobile 821 pour la production et le transport du gaz incluait la construction d'un gazoduc de 305 mm de diamètre au-dessous de l'île Dauphin. Ce gazoduc transporte du gaz depuis sa plateforme de production située offshore, au large de l'Alabama, dans le golfe du Mexique, jusqu'à des installations sur la terre ferme. Afin de tenir compte de l'augmentation des besoins dans l'avenir, deux gazoducs de 305 mm de diamètre supplémentaires furent installés dans la traversée de l'île (le faisceau consistait en deux gazoducs de 12,75 pouces de diamètre extérieur par 0,500 pouce d'épaisseur étanche et un gazoduc de 12,75 pouces de diamètre extérieur par 0,562 pouce d'épaisseur étanche, tous étant conformes à la norme API 5L Grade B).

En sélectionnant le forage directionnel comme méthode de construction, BP Exploration a réalisé les avantages économiques et environnementaux de cette méthode sur les procédés d'installation conventionnels. Le forage directionnel a atténué l'impact de la construction sur cette île dont l'environnement est très fragile, maintenu la stabilité de l'île, permis d'atteindre des profondeurs supérieures d'enterrement des gazoducs (tout particulièrement dans les zones de brisants), accéléré la réalisation des travaux en réduisant les retards imprévisibles et minimisé les risques associés aux conditions hivernales.

BARGES DE FORAGE

Comme la construction de la traversée a commencé en hiver, la sensibilité des équipements à utiliser aux conditions météorologiques était un facteur extrêmement important. L'impact du temps a fait l'objet de réflexions approfondies pendant la phase de planification préalable à la construction pour essayer de minimiser les risques résultant d'un temps inclément.

Une plateforme de travail stable était nécessaire pour réaliser les opérations de forage. La tour de forage n'a qu'une capacité de résistance limitée au tangage, au roulis et au lacet des navires de mer. Même de faibles mouvements de la barge de forage risquent de fatiguer excessivement la colonne de forage et les éléments mécaniques de la tour de forage. Bien que des traversées qui avaient été construites dans la baie de Mobile au début de l'année aient été forées depuis une barge de forage marine de 120 m x 30 m, il n'était pas considéré économiquement faisable d'utiliser une aussi grande barge pour ce projet. Le risque d'interruptions du travail en conséquence de tempêtes hivernales, même si le forage devait se faire depuis une plateforme flottante relativement stable, était également considéré prohibitif. Par conséquent, d'autres options furent envisagées.

La possibilité de placer la tour de forage sur une barge plate de dimensions moyennes (55 m x 16 m) lestée au fond de la mer fut prise en considération. Cependant, cette option fut considérée comme peu pratique en raison du risque d'endommagement de la barge, de la difficulté d'obtention de l'assurance requise et de la probabilité selon laquelle l'eau dépasserait le niveau du pont même lorsque la mer est modérément agitée.

La troisième option, celle qui a été sélectionnée en fin de compte, consistait à placer la tour de forage sur une barge de forage à élévation automatique. Ces bateaux élévateurs automoteurs sont capables de se soulever sur des pieds intégrés de façon à constituer une plateforme de travail stable. Les activités de forage pourraient être exécutées depuis de telles barges par tous les temps à l'exception des tempêtes.

La sélection de la classe de barge de forage à élévation automatique dépendait de la profondeur de l'eau dans la vicinité du site proposé et de la disponibilité des barges. Etant donné que seules les petites barges de forage à élévation automatique (classes 150 et 105) seraient capables de manoeuvrer dans des profondeurs d'eau variant entre 3 m et 4,5 m, il faudrait au moins deux de ces barges pour y installer la tour de forage et son module de contrôle.

Des analyses de stabilité d'une barge de forage à élévation automatique de la classe 150 furent réalisées pour déterminer les facteurs de sécurité contre les risques de retournement et de glissement, et pour calculer les contraintes prévues qui pourraient résulter des effets combinés de la capacité de tirage maximum de la tour de forage (250 tonnes) et des charges environnementales (vent et vagues). Ces analyses ont démontré que deux barges de forage à élévation automatique de la classe 150 attachées en tandem constitueraient une plateforme convenable pour installer la tour de forage.

Afin de donner au foreur une vue adéquate de la tour de forage, il a fallu utiliser une troisième barge de forage à élévation automatique de plus petite taille (classe 105) pour y installer le module de contrôle. Cette barge fut placée le long de la barge soutenant la tour de forage, mais perpendiculairement à celle-ci.

Les autres équipements associés à la tour de forage (colonne de forage, pompes à boue et cuve à boue) furent placés sur une barge plate de 55 m x 16 m qui était amarrée du côté des barges de forage à élévation automatique se trouvant à l'abri du vent. Pour protéger ces dernières contre tout risque d'endommagement par la barge plate, trois piliers constitués de tuyaux de 914 mm de diamètre furent installés pour servir d'amortisseurs entre les barges. Une deuxième barge plate de 55 m x 16 m fut également utilisée pour y placer les réservoirs de fluide de forage mélangé à l'avance (bentonite) devant être utilisé pendant l'opération. Cette deuxième barge plate fut amarrée contre la barge contenant la colonne de forage, les pompes et la cuve à boue.

FLUIDES DE FORAGE

Du fluide de forage fut pompé pendant toutes les opérations de forage (forage du trou pilote, pré-alésage et retrait) afin de stabiliser le trou foré, de transporter les débris le plus loin possible de la face de forage et de réduire le frottement sur la colonne de forage, ainsi que, en fin de compte, sur le faisceau de tubes lorsqu'ils sont attirés à l'intérieur du trou foré.

La collecte, le confinement et l'élimination ultérieure des fluides de forage qui remontent à la surface du sol lors de traversées par forage directionnel de la terre à la terre sont devenus des caractéristiques courantes de tels projets. Cependant, les autorités publiques régissant la traversée de l'île Dauphin ont admis qu'une telle exigence n'était pas réaliste lorsque le forage devait être entrepris avec des points d'entrée et de sortie à des profondeurs d'eau de 4,5 m et 6,7 m. Néanmoins, la composition des fluides de forage fut limitée à la bentonite (une argile à base de montmorillonite de sodium) sans aucun adjuvant.

En raison des règlements très stricts qui régissent les travaux dans la baie de Mobile pour des raisons de protection de l'environnement, la méthode généralement acceptée qui consiste à mélanger des fluides de forage en poudre avec de l'eau sur place a été considérée inappropriée. La méthode habituelle consistant à ouvrir des sacs de bentonite en les déchirant au-dessus de trémies de mélange entraîne l'apparition de grandes quantités de poussière - dans une application offshore, cela non seulement crée un risque pour l'environnement mais aussi peut affecter la sécurité des travailleurs. Le volume de fluide à mélanger sur place pourrait être minimisé en mélangeant de la bentonite à l'avance et en la transportant sur le site dans des réservoirs sur une barge plate de 55 m x 16 m.

La méthode préférée de manipulation du fluide de forage sur place était d'utiliser dix-sept réservoirs d'une capacité de 79 500 l sur une barge plate pouvant contenir soit de la bentonite mélangée à l'avance, soit de l'eau douce. Le fluide de mélange serait mélangé dans une usine de traitement en vrac sur la terre ferme (à quelques 40 km du site de forage) et pompé dans les réservoirs de 79 500 l sur la barge. La barge serait alors remorquée jusqu'au site de forage, et le fluide de forage serait pompé dans la cuve de mélange de la plateforme de forage au moment choisi.

Comme il serait nécessaire de faire varier la viscosité du fluide de forage pendant les opérations, deux des cuves furent utilisées pour contenir de l'eau douce. En outre, trois réservoirs pneumatiques allaient être utilisés pour contenir de la bentonite sèche en poudre sur le lieu de forage. Les réservoirs pneumatiques furent reliés les uns aux autres pour permettre le transfert de la bentonite sèche dans une trémie au-dessus de la cuve de mélange de la plateforme. Cette configuration constituait essentiellement un système clos et éliminait tous les risques pour l'environnement ou pour la sécurité qui étaient associés au déchirement de sacs individuels sur le lieu du forage.

Le volume du fluide de forage qui pourrait être entreposé sur place fut estimé à la moitié des estimations de ce qui serait nécessaire pour compléter l'ensemble du projet. Des quantités supplémentaires de fluide mélangé à l'avance et de gel sec en vrac seraient donc transportées de façon à réapprovisionner les réserves sur place en utilisant des barges de livraison de capacité inférieure (318 000 l et 35 mètres cubes).

FABRICATION ET ESSAIS DES GAZODUCS

Land & Marine engagea SubSea International comme sous-traitant pour fabriquer, tester et installer les anodes. Ces opérations furent réalisées depuis une barge de pose de SubSea de classe 278 et une barge de forage à élévation automatique de classe 130. Trois embarcations furent utilisées pour poser les gazoducs du côté de l'île qui fait face au golfe du Mexique, où la profondeur de l'eau varie entre 6 m et 9 m. Il avait été déterminé que cette configuration, avec le positionnement de la plateforme de forage du côté à l'abri du vent de l'île, était celle qui protégerait le mieux la plateforme contre les intempéries et minimiserait les pertes de temps causées par ces intempéries.

La barge de forage à élévation automatique de la classe 130 de SubSea fut positionnée initialement à quelques soixante mètres du point de sortie désigné pour le forage, et la barge de pose fut ancrée quatre cents mètres plus loin. Un câble de tirage fut déroulé depuis le treuil de la barge de forage jusqu'à la section d'attaque du premier gazoduc à poser. Des joints de gazoducs de 305 mm de diamètre furent soudés sur la barge de pose, et le gazoduc fut tiré vers la barge de forage à élévation automatique. Une fois qu'environ 200 m de gazoduc fut posé de cette façon, la barge de pose tira sur ses ancres pour poser les 990 m restants. Le gazoduc fut alors rempli d'eau. La barge de pose fut repositionnée à 400 mètres de la barge de forage, le câble de tirage fut récupéré et le deuxième gazoduc de 305 mm de diamètre fut posé en suivant la même procédure le long d'une ligne située à trois mètres du premier gazoduc. Cette procédure fut répétée pour le troisième gazoduc. Une fois que les trois gazoducs furent remplis d'eau, des furets de jaugeage furent introduits et des essais hydrostatiques de prétirage furent effectués.

Les joints soudés sur place furent décapés au jet de sable, préchauffés et enduits d'époxy lié par fusion. Deux des gazoducs furent munis d'entretoises externes en néoprène pour empêcher tout contact entre les gazoducs pendant leur installation en faisceau et leur pénétration dans le trou foré.

OPERATIONS DE FORAGE

Avant que les premiers équipements ne soient transportés sur le site au large, un levé de terrain fut effectué de façon à localiser et à identifier les points d'entrée et de sortie désignés pour le forage. Une barge à percussion de 35 m x 12 m avec une grue de 130 tonnes sur le pont installa les trois pieux de 914 mm de diamètre et un tube de sondage de 300 mm de diamètre supporté par deux ensembles de « poteaux de but ».

Etant donné que la distance entre le lit de la mer et la plateforme de forage était d'environ 60 m, il a fallu utiliser un tube de sondage pendant le percement du trou de forage pilote initial afin d'assurer une certaine rigidité au gazoduc pendant qu'il allait être poussé par la colonne de forage. Le tube de sondage de 300 mm de diamètre fut soutenu par deux jeux de « poteaux de but » comprenant chacun deux poteaux de 12 m de long et de 300 mm de diamètre, avec des barres transversales fixées à des hauteurs prédéterminées.

Pendant l'installation des pieux et du tube de sondage, la tour de forage et ses équipements annexes furent chargés dans les barges de forage et dans les barges plates dans un port voisin. Chaque équipement fut attaché aux ponts de façon à être prêt pour la mise en service dès l'arrivée sur le lieu de forage.

Le 16 novembre 1991, les équipements furent transportés du port au lieu de forage. Des raccordements furent faits pour les tuyaux de fluide hydraulique et de fluide de forage, et le forage pilote commença le lendemain.

Un trou pilote fut percé dans les formations de sable et d'argile avec un jet de 73 mm. Lorsque le percement du trou pilote a suffisamment progressé, un tube d'usure fut installé au-dessus du tube pilote de 73 mm de diamètre. Le profil du forage utilisa un angle d'entrée de six degrés et un rayon de courbure de 390 m. Land & Marine avait effectué des calculs de contrainte avant le début de la construction pour ce projet, et ces calculs avaient confirmé que la contrainte affectant ces gazoducs serait dans des limites d'effort acceptables.

A un stade ultérieur de la progression du forage pilote au-dessous de l'île, un « TruTracker » fut utilisé pour fournir une corrélation supplémentaire entre les signaux provenant du magnétomètre et de l'accéléromètre à la console de commande de la tour de forage. Le système TruTracker ne fut utilisé que sur l'île (sur une distance d'à peu près 100 m). L'utilisation de ce système sur l'eau n'aurait été ni pratique, ni réellement nécessaire.

Le 20 novembre, le trou pilote fut achevé - la mèche de forage ressortant du côté de l'île faisant face au Golfe du Mexique, à environ 1 100 mètres de la tour de forage. De l'air fut pompé à travers la rame et le point de sortie fut repéré par des plongeurs. Un levé de terrain réalisé sur le point de sortie réel indiqua qu'il n'était qu'à 50 cm du point ciblé. Des longueurs supplémentaires de tubes d'usure de 125 mm de diamètre furent ajoutées à la rame, et le tube pilote fut retiré du trou.

Immédiatement avant l'achèvement du forage du trou pilote, une barge de forage à élévation automatique de la classe 205 fut amenée et positionnée au point de sortie. Le tube d'usure de 125 mm de diamètre fut soulevé jusqu'à son pont et un couteau frappeur d'un mètre de diamètre fut connecté. Le couteau frappeur fut ensuite abaissé au fond de la mer et des joints de tube de forage de 125 mm aux parois épaisses furent placés à l'arrière à des fins de renforcement. Du fluide de forage fut pompé en direction du couteau frappeur pendant sa rotation et son abaissement dans le lit de la mer par la tour de forage. Lorsque les joints des tubes d'usure furent retirés à la tour de forage, des joints de tubes de forage aux parois épaisses furent attachés à la trame de forage derrière le couteau frappeur.

Au bout de 35 heures de pré-alésage, le couteau frappeur fut remonté sur le pont de la barge sur laquelle se trouvait la tour de forage.

A l'issue des essais hydrostatiques de pré-retrait, tous les tubes destinés à l'installation finale furent vidés. La barge de pose fut repositionnée à l'extrémité d'attaque des gazoducs et les gazoducs furent soulevés afin d'attacher une tête de tirage unique au faisceau. La tête de tirage fut alors reliée à un aléseur à tambour de 950 mm de diamètre au moyen de manilles et d'un dispositif pivotant. Lorsque l'ensemble fut abaissé au fond de la mer, la tour de forage appliqua de la tension à la trame de forage.

Pendant la fabrication et la conduite des essais, un certain envasement des gazoducs a été constaté. Cependant, la tour de forage n'a guère eu de difficultés à fournir la force nécessaire pour commencer l'opération de retrait. Le fluide de forage fut pompé dans l'aléseur, des mouvements de rotation et de tirage furent appliqués à la tour de forage et le faisceau de tubes entra dans le trou foré.

L'opération de retrait se poursuivit de façon régulière ; on ne constata pas de forces de tirage ou de couple excessif. L'opération de retrait fut achevée au bout de 36 heures. La tête de tirage fut déconnectée du dispositif pivotant et de l'aléseur sur la tour de forage avant d'être abaissée au fond de la mer. Seulement treize jours s'étaient écoulés depuis l'arrivée des équipements sur les lieux.

OPERATIONS POSTERIEURES AU RETRAIT

A l'issue de l'opération de retrait qui avait été couronnée de succès, les barges de forage et les barges plates retournèrent au port avec la tour de forage et tous les équipements accessoires, et elles y furent déchargées et mises hors service.

Tous les tubes du faisceau furent jaugés et soumis à des essais hydrostatiques. Les extrémités exposées des gazoducs furent abaissées au jet de façon à fournir une transition depuis les angles d'entrée et de sortie de la traversée forée jusqu'à la pose des tubes suivant un procédé conventionnel, qui allait constituer l'étape suivante. Des anodes furent attachées aux extrémités des gazoducs.

Les pieux et les « poteaux de but » furent extraits et tous les autres équipements furent mis hors service.

CONCLUSION

La traversée par forage directionnel de l'île Dauphin a été conçue et construite en respectant le calendrier et le budget qui avaient été spécifiés par BP Exploration. Une grande partie du succès de ce projet doit être créditée à l'attention qui a été prêtée aux travaux de planification et d'ingénierie préalables à la construction. Un esprit d'équipe et de coopération et des buts communs furent développés et entretenus pendant tout ce projet entre BP Exploration, Costain Land & Marine et les sous-traitants. En combinant les efforts de tous les participants, les risques associés à l'application d'un forage directionnel horizontal pour un environnement marin furent minimisés et le résultat de ces travaux s'est avéré très positif.

No Trenches in Town, Henry & Mermet (eds) © 1992 Balkema, Rotterdam. ISBN 90 5410 085 0

Geological aspects due to a horizontal drilling

W.de Groot
Public Works Rotterdam, Netherlands

ABSTRACT: During a drilling according to the Horizontal-Directional-Drilling method great mud-outbursts have occurred. In relation to this a motorway has been lifted.
Out of results of measuring the waterpressure, the way in which bentonite is liberated and the fact any return current fails to appear. We may conclude that a drilling-hole is hardly created. Above the layer of clay in which has been drilled, there are very weak layers of clay, sand and peat-bog. Due to this geological composition in connection to the former mentioned phenomena, a favourable judged layer of clay is "blown up". This geotechnical phenomenon could occur in any comparable circumstance in which one has to deal with the outburst of great quantities of bentonite.

1. INTRODUCTION

In the western part of the Netherlands the Horizontal-Directional-Drilling method, (HDD see fig. 1), is used comparitively often to place conduit-pipes. It is an attractive mthod because the space needed as a whole is inconsiderable and because great lengths can be abridged by only one drilling. Enough space should be available to take in the conduit-pipe. The method has proved to be very appropriate for conditions of soil which can be compared to the Delta-area in Holland.
Therefore the HDD-method has been chosen to place conduit-pipes to the "Distripark Eemhaven", situated in the harbour area of Rotterdam. The entry-point of this approximately 350 meters long drilling is at about N.A.P. + 4 meters and ends at about N.A.P. - 0.5 meters. At one drilling a main-dike, a railroad and a motorway, the A-15, have been crossed (see fig. 2)

Owners and managers of these infrastructural works have made certain conditions to the execution of this project. The main conditions are:
- The depth should be as such that a lift of surface will not be expected. (aspect of instability).
- The "pleistocene" sand must not be drilled because a so-called "salt-ooze" should be avoided (aspect of environment).
- One has a preference in drilling into the layers of clay (which should be easier according to the contractor).

During every drill some casings have been installed in which cables and conduit-pipes have been placed.

2. GEO-TECHNICAL ASPECTS

Before the choice is made to use a certain method an extensive geo-technical research is being carried out by the geo-technical department of Public Works Rotterdam. From the probes and the ground-drills one compiles the geological profile of the ground-plan which has to be followed during drilling (see fig. 3)
The most important geo-technical data are: The pleistocene layer of sand begins at approximately N.A.P. - 16 meters and is covered by a zone of clay of about 3 meters. The soil above this layer is a composition of clay, sand and much peat-bog. The groundwaterlevel is high, about 0,5 meters below the surface while the bottom has a very low bearing-power.
The maximum depth to be kept for drilling has been determined at about N.A.P. - 15 meters, meaning the layer of clay above the pleistocene sand. Figure 3 also shows the trace to be followed in the vertical direction.

Although as well in as outside Holland many HDD-drillings have been converted, still many questions remain unanswered about the geo-technical behaviour of the

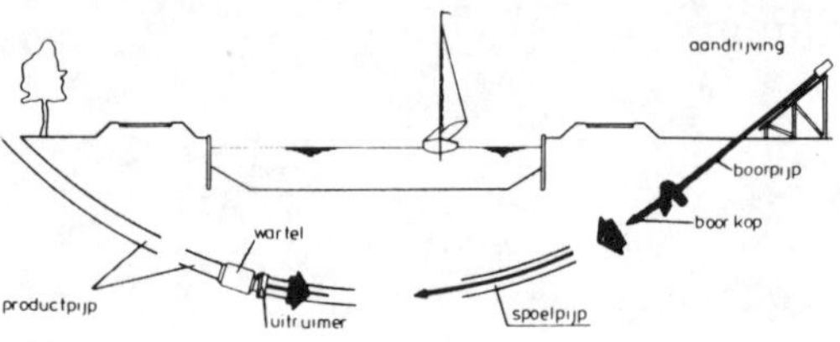

fig. 1

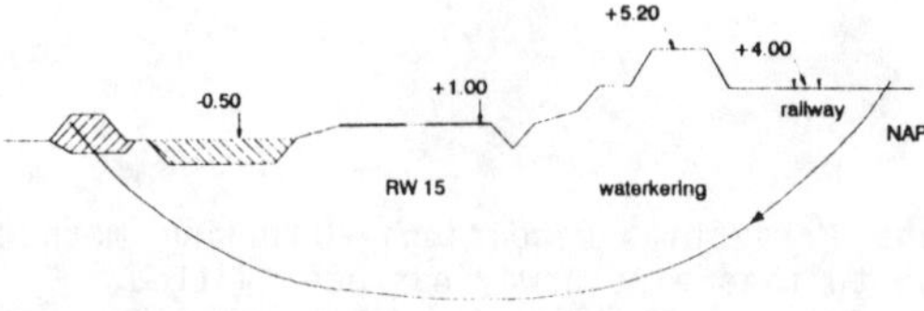

fig. 2

soil during drilling. In an attempt to gain more insight in this matter it has been decided to install waterpressure-meters (wpm's) along the drilling ground-plan.

In consultation with the managers of the dike three places for installing these wpm's have been indicated. These locations are at a distance of approximately 30, 90 and 150 meters from the entry-point of the drilling. Three wpm's have been placed at the first location, measured from the heart of the ground-plan at 1.5, 2.5 and 3.5 meters. At the second location we find two wpm's at 2.5 and 3.5 meters from the conduit-pipe, while at 150 meters one wpm is installed at 3.5 meters from the conduit-pipe (see fig. 4)

3. STARTING THE DRILLING-PROCEEDINGS

The drilling has been set in at an angle of approximately 15 degrees. First mud-outbursts, which are inadmissable in this area, have occurred beneath the railroad at about 85 meters from the entry-point. In addition to this a large quantity of mud-liquid has appeared at the surface. A next, even more serious, outburst has occured in the middle of the dike, approximately 90 meters from the entry-point, close to the second series of wpm's. The managers of the dike as well as the principal want an explanation for this outburst and, if possible, a solution to prevent any further outbursts.
By means of the installed drilling-pipes the length of the drilled track has already been determined. By staking out this length on the track that still needs to be drilled it is possible to locate the drill-head. By locating this drill-head it has been concluded that it has just reached the layer of clay in which the drilling will take place. At this location exactly at the depth of the drilling-track, wpm's have been placed. These wpm's show a slight fluctuation in pressure. However, these fluctuations are so little that a mud-outburst is hardly to be expected.
In order to reduce the outburst of mud-liquid the contractor has rapidly fitted the wash-over-pipe.

Nevertheless, against all odds, the phenomenon increases. Information given afterwards by the contractor show that also the return-currents were too slight.

After a wide discussion the idea has come up that return-currents, which have to come through the wash-over-pipe, do not appear because there is hardly any or perhaps an insufficient forming of so-called "hollow-space". By that, the draining of soil is also omitted, so that the bore-hole is not clean!

By request the contractor has retracted the wash-over-pipe to show the die-head. Retracting the wash-over-pipe has been carried out at great speed and in relation to this a serious quantity of mud-liquid has been pressed to the surface. The die-head of the wash-over-pipe, coming out of the bore-hole is completely wrapped with clay, so that it is impossible to create "hollow space" by means of this die-head.
In order to remain the bore-hole intact one has not withdrawn the die-head of the drilling-pipe and in relation to this it is not checked. According to this it has not been possible to see whether the die-head of the pilot-pipe is capable of creating a "hollow space" in the clay.

While withdrawing the die-head the wpm's again show a little fluctuation in water-pressure. Supposing that the drill-head only just reached the layer of clay the mudoutbursts were thought to be starting-symptoms.

The idea, which now comes to mind, is that neither the heads of the pilot-pipe nor those of the wash-over-pipe can create a bore-hole while wrapped so much in clay. It could be concluded that the die-head is being pressed through the pliable and changable zones, while after this the soil, which is a mixture of clay and peat-bog, closes in again around the pipe.

Forcing up mud during the time the wash-over-pipe was retracted could be caused by the fact that the wash-over-head, just like a piston in a cilinder, puts the mud under very high pressure. As a result of this the layers break open and through that the mud-liquid flows to the surface.

4. CONTINUATION OF THE DRILLING-PROCEEDINGS

At the location of the next important mudoutburst (in the middle of the dike) the soil coverage on the drilling is much bigger than during former outbursts. This coverage is the biggest of the whole trace.
Besides this, the drilling-pipe is fully down in the layer of clay, so that, according to the contractor, no outbursts could have been expected.

Before passing the motorway new outbursts occur, though they remain limited to liquid coming up along the road and they go on until the farthest point is reached. Greater outbursts take place while the bore-hole is being cleared. While the cable protection pipes are retracted at the time of the last "clearing", a large quantity of mud comes out.

At first it was thought that only in a combination of clay- and peat-bog layers no bore-hole, or a one which is too small, was formed. At second sight it showed that this phenomenon also occurred, and even at a higher rate, in the layer of clay itself. In that case, while "cleaning", the bentonite-draff has been put under pressure by the clearer! The only possibility for the bentonite to escape is to come out of the bore-hole through the thin layer of clay. By means of the high pressure the bentonite has gone under the callous road so that flowing out only becomes possible in a horizontal direction. In realtion to this the motorway was raised approximately 0.25 meters.
An examination afterwards of the measurements done by the wpm's show that, during the mudoutbursts, the waterpressure was very low. These pressures were even so low that they cannot give an explanation for the cracking of the soil.
Measurements, taken of the instrumenst at the drilling-equipment, give even less hope for straight conclusions. They only show the pressures of pressing at the ingine and the pressures in the drilling-front.
(So measurements at the most interesting locations are not being taken yet.)

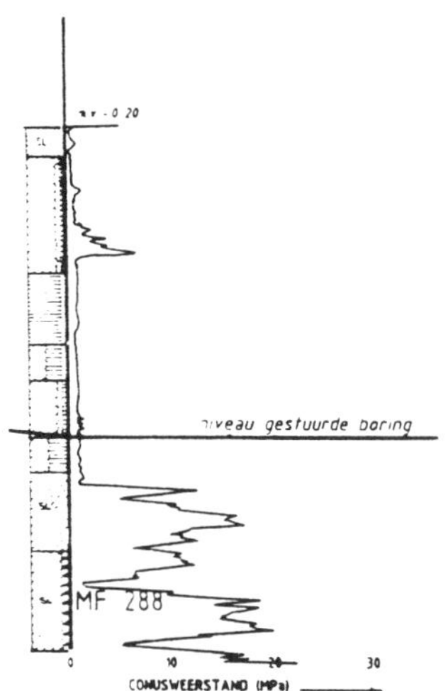

fig. 3

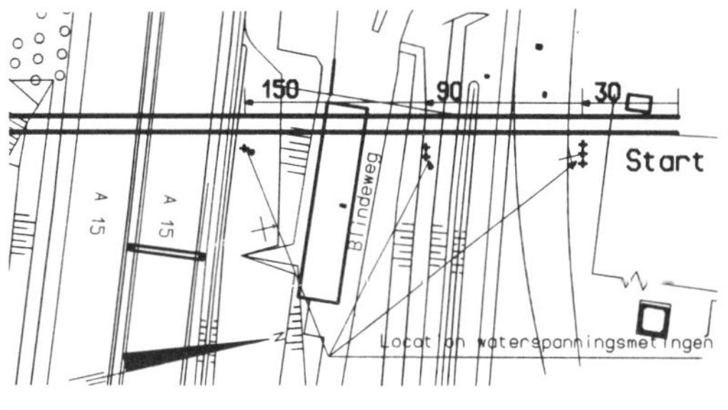

fig. 4

5. GEO-TECHNICAL PHENOMENON

While the project progresses, a hypothetis, based upon the former mentioned idea, is being framed, so that the outbursts of mud can be explained while registered waterpressures still indicate a low value.

A reasonable explanation has to be found in the construction of the ground. There is a relatively thin elastic layer of clay and above that a block of weak/plastic layers of ground which are abounding in water.

The hypothetis is based on the fact that the thin layer of clay is stretched out locally, specially in height. This is made possible by the fact that there is hardly any resistance against this transformation by the weak clay- and peat-bog layers lying above (see fig. 5)

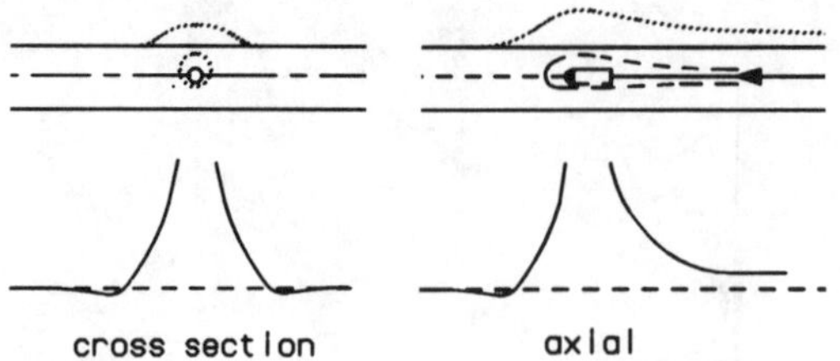

fig. 5

Because of the cohesion in the layer of clay, the adjecent clay is also being stretched. But, as only water passes through clay, the pressure of the ground-water still lags for a long time.
From this, one could conclude that the layer of clay, put under pressure near the bore-hole and in relation to this being stretched out locally, shows a lower waterpressure at a short distance from the drilling-hole.
As being established before there are hardly any return-currents in relation to the quantity of bentonite which is used.
Besides, as a result of the hampered removal, the pressure at the drilling-front is high, much higher than 1200 kPa. According to the hypothetis, it is assumed that, considering the circumstances, the layer of clay is being "blown-up".

An important diagnosis at the time of the drilling-proceeding is that during this proceeding few return-currents a few any soil-draff came out. In relation to this it is not to be expected that a bore-hole has been formed, although this is one of the functions of the wash-up-pipe and the return-currents.

6. WATERPRESSURES

The remaining question is whether the hypothesis can be supported by means of the waterpressures. One has to take into consideration that a slight deviation from the track, what is acceptable over longer distances, has a plain influence on the information given by the wpm's. During examining the results this has to be taken into consideration.

* Waterpressures at 30 meters form the entry-point
The meters were placed at a distance of respectively 1.5 meters, 2.5 meters and 3.5 meters and read approximately 2.5 meters, 2.0 meters and 2.0 meters. Just before the drill-head approaches the wpm's, waterpressures decrease. When the drill-head has passed all pressures increase about 3 kPa. After the wash-over-pipe has passed too it shows that at a distance of 1.5 meters pressure has increased till above 30 kPa. At a distance of 2.5 meters and 3.5 meters pressure decreases till below 20 kPa. At 2.5 meters we find the biggest decrease of pressure.

After the last cleaning when the conduit-pipes are withdrawn, at 1.5 meters pressure increases 40 kPa. At 2.5 meters and at 3.5 meters there is a very low increase. This distinctive behaviour remains noticable during all proceedings, although readings are very low. We find a clear rise nearby the drilling and a slight rise or even a decrease further along. The rise as well as the decline reduce when the distance becomes greater (see figure 6). A clear rise is noticed while the conduit-pipes are retracted.

Because of the lag, pressures have not regained their basic level again even two days after ending the complete proceeding.

* Waterpressures at 90 meters from the entry-point.
At the start of the activities pressure of both wpm's is almost equal. At the first passage of the drilling-front and the wash-over-head pressure is decreasing. Then we find a rise again, specially at the wpm placed a distance of 2.5 meters. Latter passages (drill-head as well as wash-over-head) show an increase of pressure at 2.5 meters, so it can be concluded that the coil-head has a very clear influence on this phenomenon.

After the first "clearer" is passed, pressure at 2.5 meters decreases while pressure at 3.5 meters rises. To withdraw the conduit-pipes the biggest "clearer" is being used. After the clearer has passed, pressure at 2.5 meters starts to fall rather spectacular, while after that, during the passage of the conduit-pipes, pressure increases approximately 32 kPa.

* Waterpressures at 150 meters from the

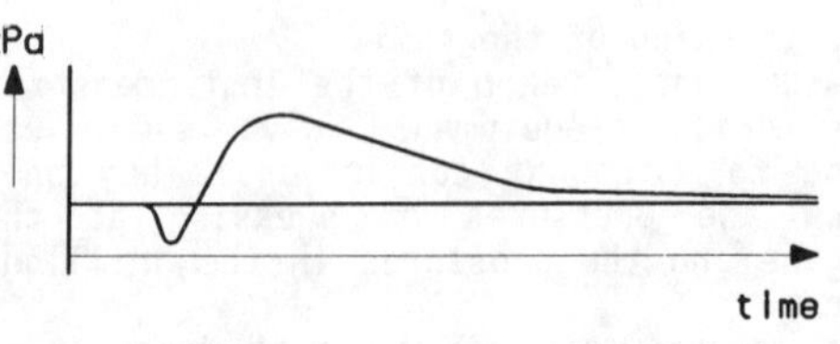

fig. 6

entry-point
In this case we can refer to the wpm at 90 meters, because the wpm's almost show the same result.
Again the wash-over-pipe exerts a clear influence on the pressure.

* Outcome of compared waterpressures
In the introduction of this paragraph we have already mentioned the possibility that the distance between drilling-front and wpm's could deviate in relation to what is presumed to be the distance. Measurements done by gyroscope do not give an unequivocal result. But, apart of the fact that a slight deviation of the trace occurs at 90 meters, both wpm's, placed at 30 and 90 meters show an almost equal result.

So the hypothetis seems to be confirmed, considering the results shown by the wpm's at different locations, during, as well as after several proceedings have been carried out.
As stated before we find a clear rise in pressure during the retraction, higher than in any other part of the proceedings.

After the withdrawal of the casings bentonite came up under the road. Therefore the retraction has been carried out at high speed to get away from this vulnerable spot. Because no "clean" bore-hole was made it is supposed that the bentonite and the clay/bentonite mixture was put under high pressure. Probably the layer of clay had already been broken so this mixture, being put under pressure, comes to the surface. In this case, under the callous road. The outcome of that has already been mentioned.

7. CALCULATION OF THE HYPOTHETIS.

After analysing measurement-results and having indicated that the hypothetis about the geo-technical has been right, it might be interesting to find out whether we could compute the same result of water-pressure in a comparable layer of clay.

To carry out the calculations we use the finite elements programme called PLAXIS. We have taken a cross-section equal to the drilling done at approximately 90 meters from the entry-point. The conduit-pipe is situated at about 16.5 meters below the surface (approximately N.A.P. - 12.5 meters) and just beneath a layer of clay we find the pleistocene sand. In the cross-section we find a bore-hole which is kept into balance by forces pressed against the wall (see figure 7). Thereupon forces in the bore-hole are being increased (raising the mud-pressure).

Calculation-results show that the transformation of the bore-hole is not equal in all directions. It shows that the vertical transformation is about twice the horizontal transformation (see figure 8). It is also noticed that transformation increases rapidly at pressures bigger than 200 kPa (see figure 9).

As indicated it could be possible for mud-pressures to reach 1200 kPa or even more during drilling.

This is a consequence of hardly present return-currents. Also during clearing-proceedings high pressures could have appeared, because the bentonite cannot flow away sufficiently.

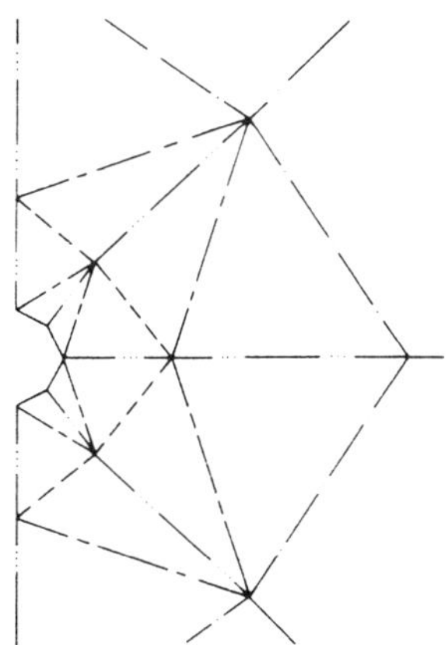

fig. 7

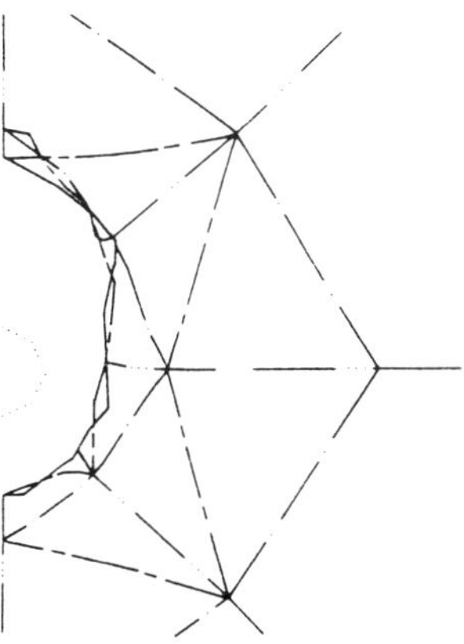

fig. 8

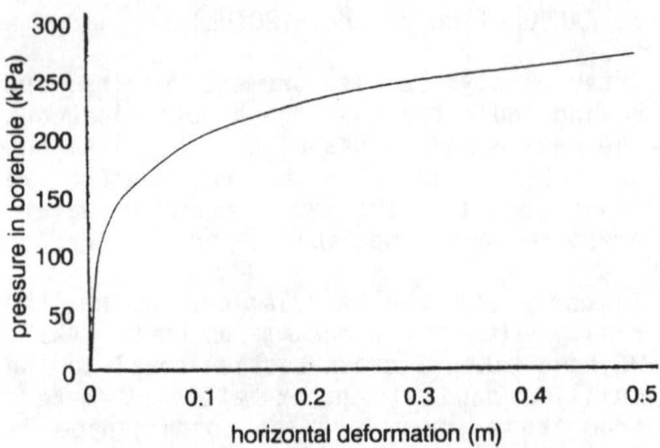

fig. 9

Therefore one could assume that the calculated transformations have taken place indeed.

A comparison of waterpressures shows that the computed waterpressures are higher than measured pressures. The explanation for this lies in the fact that the lead-in information deviates from the real data about soil-conditions.

Within the means of this calculation-programme it is impossible for cracks to occur in the model. At a computed horizontal removal of 0.3 meters and more it is to be expected that the layer of clay has cracked at the upper end. In this sort of situation the bentonite will simply disappear from the bore-hole, even at a relatively slight effective pressure.

8. CONCLUSIONS AND RECOMMENDATIONS

The measurement-data as well as the calculation-results support the hypothetis describing the behaviour of the layer of clay, influenced by pressures in the bore-hole caused by drilling. Because the die-heads have not been able to create a "clean" hole, return-currents could not take place. In relation to that the pressure in the bore-hole became so high that the layer of clay cracked and the bentonite-draff could disappear relatively easy. Pressure, getting too high, could be caused by the so-called scavenging-pressure. One has to add to this that it could also be caused by the clearing-proceedings and the retraction of the conduit-pipe because in all these circumstances bentonite-draff can be removed.

If the die-heads are not adjusted to the layers of soil and clay that need to be parted, such mudoutbursts will always occur in equal geo-technical circumstances.

So besides technical improvements it is important to check return-currents extensively, regarding off-take and composition.

From the drilling it has to be concluded that placing wpm's is hardly of any use in these clayish grounds, because data hardly indicate what is going on.

Another reason lies in the fact that the actual waterpressures still lag in the measurements.

So, by taking the HDD-method into practise in vulnerable areas, it might be useful to consider the following criteria:
- Adjust the die-heads to the soil that needs to be parted, specially to clay;
- Pressuremeasurements of the bentonite at the drilling-front should be made possible;
- The return-currents must be checked very carefully considering composition (is there any removed soil in it?) and take-off (supply and bore-hole volume);
- Having a unambiguous deviation in the return off-takes there has to be an immediate consultation of experienced geo-technical experts;
- Knowledge and experience of the drilling-crew should quarantee a solid result. (Do not push it.)

No Trenches in Town, Henry & Mermet (eds) © 1992 Balkema, Rotterdam. ISBN 90 5410 085 0

Aspects géologiques d'un forage directionnel horizontal

W.de Groot
Service des Travaux Publics de Rotterdam, Pays-Bas

RESUME

Le forage effectué selon la méthode directionnelle horizontale déclenche de (trop) grandes éruptions de boue. Cela a entraîné le soulèvement d'une route nationale. Les résultats des relevés de la pression de l'eau, la façon dont la bentonite n'arrive pas à se dégager et l'absence de tout flux de retour nous ont amené a conclure qu'il se forme à peine un trou de forage. Sur la couche d'argile dans laquelle le forage est effectué se trouvent des couches molles d'argile mêlée à du sable et de la tourbe. Cette composition géologique et l'absence des phénomènes précités entraîne le soulèvement d'une couche d'argile jugée relativement bonne. Ce phénomène géotechnique peut avoir lieu partout dans des conditions similaires. Il faut alors tenir compte de grandes éruptions de bentonite.

1. INTRODUCTION

La méthode de forage directionnelle horizontale est relativement souvent employée dans l'Ouest des Pays-Bas (voir figure 1.) pour le placement de conduites. Cette méthode est intéressante car l'espace nécessité, sur le tracé des conduites à placer, est réduit et parce qu'un seul forage permet le percement de grandes distances. Il faut suffisamment de place dans le prolongement du forage pour étendre la conduite qui doit être introduite. La méthode convient parfaitement aux sols dont la composition ressemble à ceux du domaine des deltas des Pays-Bas.
C'est, entre autres, pour cette raison que la méthode a été choisie pour le placement des conduites en direction du Distripark Eemhaven, dans le secteur portuaire de Rotterdam. Le point de départ de ce forage de presque 350 m de long se trouve environ à NAP [1] - 0,5 m. Une digue principale, une voie de chemin de fer et la route nationale A-15 ont été croisées lors d'une même opération de forage (voir figure 2.).
Des conditions ont été posées par les propriétaires et gestionnaires des infrastructures à croiser quant à l'exécution du forage. Les plus importantes sont les suivantes :

- la profondeur doit être telle qu'elle doit éviter que la pression de la boue ne crevasse le sol (aspect d'instabilité),
- le sable pléistocène ne doit pas être foré afin d'éviter des infiltrations d'eau salée (protection de l'environnement),
- le forage doit se faire de préférence dans les couches d'argile (l'entrepreneur considère que c'est plus simple).

Le diamètre du forage définitif est d'environ 1 m. A chaque forage est introduit un certain nombre de gaines dans lesquelles sont maintenant placés des cables et des conduites.

[1] NAP = Normaal Amsterdams Peil, niveau d'eau fixé à Amsterdam

2. ASPECTS GEOTECHNIQUES

Le département géotechnique du Service des Travaux publics de Rotterdam a effectué une analyse géotechnique détaillée avant de choisir la méthode de forage. Des échantillons et des forages ont permis de composer le profil géologique du tracé à suivre lors des

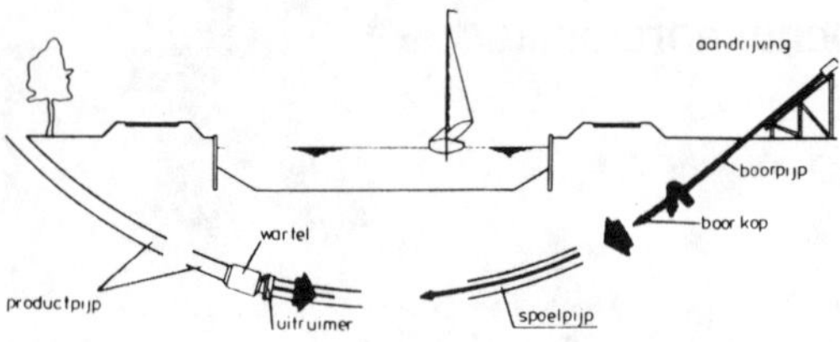

fig. 1

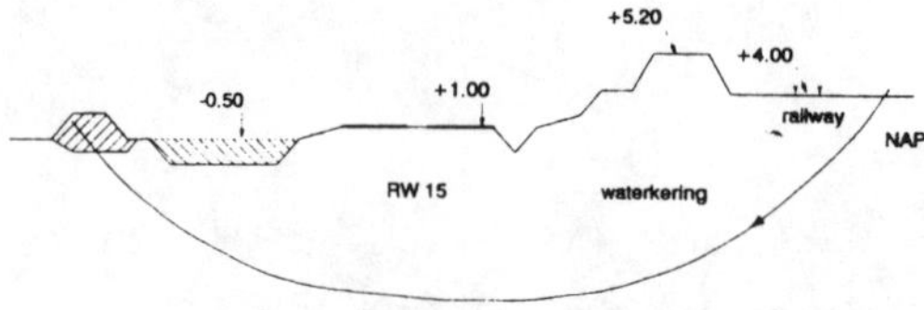

fig. 2

travaux de forage (voir figure 3.)
Les principales données géotechniques sont les suivantes :

- La couche de sable pléistocène commence à une profondeur d'environ NAP - 16 m sur laquelle se trouve une couche d'argile d'une épaisseur d'environ 3 m.
- Le sol qui recouvre cette couche d'argile est constitué de couches de composition différente d'argile, de sable et de (beaucoup de) tourbe.
- Le niveau de l'eau souterraine est relativement élevé - environ à 0,5 m. sous le niveau du sol - et la force portante du sol est faible.
- La profondeur maximale à maintenir pour le forage est fixée à environ NAP - 15 m dans la couche d'argile située sur le sable pléistocène. La figure 3. indique également le tracé à suivre, dans la verticale.

Bien que de nombreux forages faits selon la méthode directionnelle horizontale aient été exécutés aux Pays-Bas et à l'étranger, de nombreuses questions se posent en ce qui concerne le comportement géotechnique du sol au cours des travaux de forage. Pour avoir une idée plus précise de ce comportement, il a été décidé de placer des tensiomètres d'eau (TE) le long du tracé à forer.
Trois endroits ont été fixés en accord avec les responsables des digues pour le placement de TE.
Ces endroits se trouvent à environ 30, 90 et 150 m du point de pénétration du forage. Sur le premier sont placés trois TE, à environ 1,5 m, 2,5 m et 3,5 m à partir du centre du tracé prévu (voir figure 4.). Sur le deuxième endroit sont placés deux TE à 2,5 m et un à 3,5 m de la conduite. Un TE a été placé à 150 m du point de pénétration et à 3,5 m de la conduite.

3. LANCEMENT DU PROCESSUS DE FORAGE

Le forage a été commencé sous un angle de 15 degrés. Les premières éruptions de boue, inadmissibles dans un tel terrain, ont eu lieu sous la voie de chemin de fer, à environ 85 m du point de pénétration. Du liquide boueux (en assez grande quantité) est alors remonter à la surface. Une autre éruption a eu lieu au milieu de la digue, à environ 90 m du point de pénétration, à hauteur de la deuxième série de TE.
Les gestionnaires ainsi que le maître d'ouvrage veulent avoir une explication de cette éruption et, si possible, une solution afin d'éviter d'autres éruptions.

Les tuyaux de forage placés déterminent la longueur du tracé déjà foré. En reportant cette longueur sur le tracé à forer, on peut déterminer où se trouve la tête de forage. Ceci permet de conclure que la tête de forage vient d'atteindre la couche d'argile dans laquelle le forage doit avoir lieu. A cet endroit, des TE ont été placés à la profondeur du tracé à forer. Ces TE enregistrant continuellement la pression de l'eau ne montrent que peu de variations de la pression.

Ces variations sont si minimes qu'on ne prévoit pas d'éruptions de boue.
Pour limiter le dégagement de liquide boueux, l'entrepreneur a placé plus rapidement que prévu la conduite de curage. Mais, contrairement à ce qu'il prévoyait, cela n'a fait que renforcer le phénomène. Il s'est ensuite avéré, d'après les informations fournies plus tard par l'entrepreneur, que les flux de retour étaient (trop) faibles.

Après d'amples discussions est née l'idée que les flux de retour, pour lesquels sont entre autres prévus les conduites de curage, ne peuvent avoir lieu pour la bonne raison qu'il n'y a pas du tout ou pas suffisamment d'espace vide.
Ce qui fait que l'écoulement du sol ne se fait pas non plus et que le trou de forage reste encombré. L'entrepreneur a, sur demande, retiré la conduite de curage afin de montrer la tête de coupe. Le retrait s'étant effectué avec grande rapidité, une forte quantité de liquide

boueux est remonté à la surface.
A la sortie de la tête de coupe du trou de forage, on s'aperçoit qu'elle est entièrement entourée d'argile. Il est donc impossible de ménager un espace vide avec cette tête dans l'argile.
Pour maintenir le trou de forage, la tête de coupe de la conduite de forage n'a pas été retirée ni contrôlée. Il n'a donc pas été possible de voir si la tête de la conduite pilote est capable de créer un espace vide dans l'argile.

Au cours du retrait, les TE n'enregistrent que peu de variations dans la pression de l'eau.
Comme la tête de forage, selon les estimations, ne vient que de pénétrer dans la couche d'argile, on a tendance à penser que ces éruptions de boue ne sont qu'un phénomène de rodage.

L'idée se crée que ni les têtes de la conduite pilote ni celle de la conduite de curage ne peuvent, entourées d'argile comme elles le sont, pratiquer un trou de forage. On pense que la tête de coupe est poussée dans les couches modelables et déformables et qu'ensuite le sol - un mélange d'argile et de tourbe - se referme derrière la tête autour de la conduite. La tension exercée par la boue lors du retrait de la conduite de curage serait due au fait que la tête de curage, telle "un piston dans un cylindre", met la boue sous haute pression. La boue fait alors éclater les couches de sol (argileux) et remonte à la surface.

4. SUITE DU PROCESSUS DE FORAGE

A l'endroit de l'éruption de boue suivante - au milieu de la digue - la profondeur du sol au-dessus du forage est (beaucoup) plus importante que lors des éruptions précédentes. A cet endroit, la profondeur du sol au-dessus du forage est la plus importante du tracé. De plus, le tuyau de forage se trouvant maintenant entièrement dans la couche d'argile, il ne devrait plus y avoir d'éruptions.

Mais de nouvelles éruptions ont lieu avant le passage de la route nationale. Elles se limitent à du liquide qui se répand le long de la route jusqu'à ce qu'on soit arrivé au bout de ce passage. Des éruptions plus importantes ont lieu lors du déblayage du trou de forage. Une grande vague de boue s'échappe lors du retrait des gaines, au moment de l'ultime déblayage.

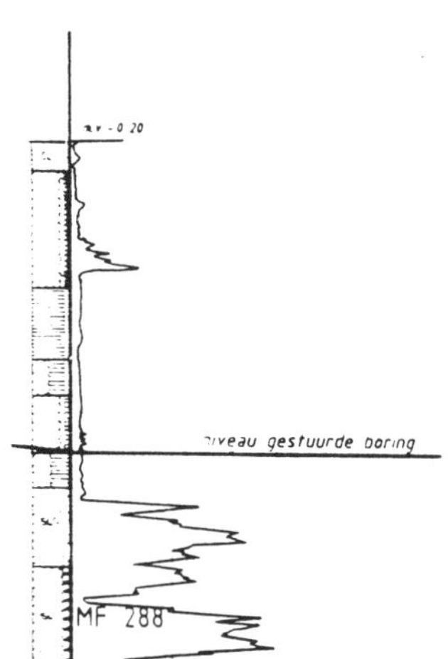

fig. 3

On a d'abord pensé qu'il ne se formait qu'un trou de forage quasi inexistant ou (trop) petit uniquement dans les couches d'argile et de tourbe.
Mais on pense maintenant que ce phénomène a également eu lieu, de façon beaucoup plus prononcée, dans la couche d'argile. Dans ce cas, le curage de la bentonite a été mis sous pression par le déblayeur pendant les activités de déblayage. Et la seule échappée possible pour la bentonite mise sous pression est le long du trou de forage à travers la fine couche d'argile. La forte pression a envoyé la bentonite sous le revêtement de la route et elle ne peut donc s'écouler que dans une direction horizontale. Ce phénomène a soulevé la route nationale d'environ 0,25 m.

Lors de l'étude effectuée plus tard sur les relevés des TE, il s'est avéré que les pressions d'eau mesurées lors des éruptions de boue n'étaient que relativement faibles. Elles sont si faibles qu'on ne peut leur attribuer la formation de crevasses dans le sol. Les relevés des instruments de forage ne nous en disent pas plus. Ils n'indiquent que les pressions près de la pompe : les pressions au front de forage (l'endroit le plus intéressant) ne sont pas encore mesurées.

5. PHENOMENE GEOTECHNIQUE

Au cours du projet, une hypothèse finit

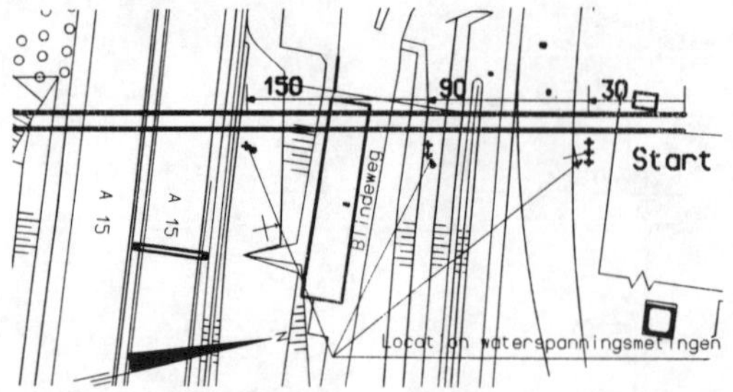

fig. 4

par se former, à partir de l'idée précitée, afin d'expliquer les éruptions de boue alors que les pressions d'eau mesurées sont aussi faibles.

L'explication est que tout serait lié à l'agencement du sol. D'abord une couche relativement fine et élastique d'argile sur laquelle reposent des couches de sol très molles, modelables et riches en eau. Il y a en outre le fait qu'il se forme à peine de flux de retour pour le mélange de sol et de bentonite.
L'hypothèse serait (voir figure 5.) que la couche fine d'argile s'étire à certains endroits sous la pression du liquide boueux et notamment vers le haut. Ceci est possible car la résistance - de la tourbe qui se trouve au-dessus - contre cette déformation augmente à peine. Comme la couche d'argile a une certaine cohérence, la couche qui s'étend à côté s'étire également. L'argile ne laissant pas facilement passer l'eau, l'effet de la pression de l'eau se répercute (très) longtemps. Cette hypothèse sous-entend que la couche d'argile mise sous pression à l'endroit du trou de forage et qui de ce fait s'étire localement indique, à une petite distance du trou de forage, une pression d'eau plus faible.

On a déjà constaté qu'il n'y presque pas de flux de retour par rapport à la quantité de bentonite introduite. De plus l'écoulement bouché augmente la pression au front de forage, on prévoit qu'elle dépasse les 1200 kPa. On suppose que, conformément à l'hypothèse et compte tenu des circonstances, la couche d'argile est soulevée à l'endroit du front de forage.

Une constatation importante a été faite lors du processus de forage, c'est qu'il n'y a guère eu de flux de retour ni de terre de curage renvoyée vers l'extérieur au cours du forage. Un trou de forage a donc à peine été formé alors que c'était là l'une des fonctions de la conduite de curage et des flux de retour.

6. PRESSIONS DE L'EAU

La question est de savoir si l'hypothèse émise peut être étayée par les pressions d'eau relevées.
Il faut dans ce contexte penser qu'un petit écart dans le tracé, chose tout à fait possible sur de telles distances, a des influences sensibles sur les données indiquées par les TE. On en tient compte lors des évaluations.

Pressions de l'eau à 30 m du point de pénétration
Les valeurs initiales des tensiomètres placés à environ 1,5 m, 2,5 m et 3,5 m de distance sont respectivement à environ 2,5 m, 2,0 m et 2,0 m. Les pressions d'eau baissent juste avant l'arrivée de la tête de forage auprès de TE. Une fois qu'elle est passée, toutes les valeurs augmentent d'environ 0,3 m de colonne d'eau (3kPa).

Après le passage de la conduite de curage, la pression dépasse les 3 m de colonne d'eau à une distance de 1,5 m. A 2,5 m et 3,5 m, la pression descend sous les 2 m de colonne d'eau et à 2,5 m, la pression baisse le plus (voir figure 6.)

Lors du retrait des conduites, précédé du dernier déblayage, la pression augmente à

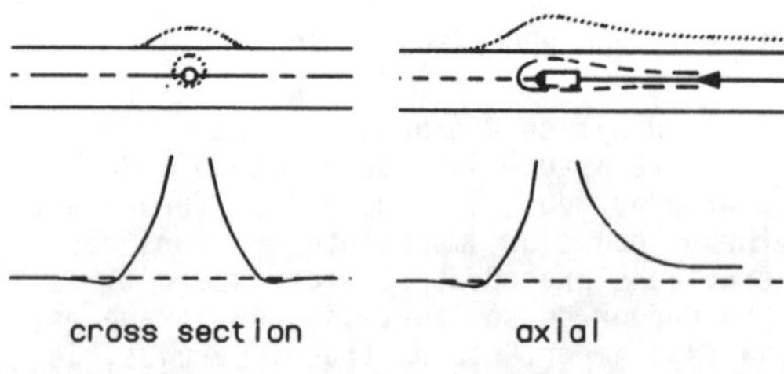

fig. 5

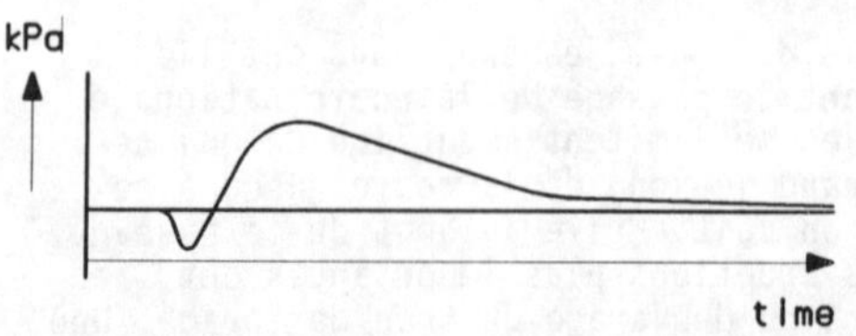

fig. 6

1,5 m d'environ 4 m de colonne d'eau. A 2,5 m et à 3,5 m, les pressions n'augmentent que très peu.

Ce comportement significatif, malgré les déviations pourtant faibles, reste présent pendant toutes les opérations de forage : on constate une nette augmentation tout près du forage et une faible augmentation ou une baisse un peu plus loin. Augmentation et baisse qui diminuent plus la distance augmente (voir figure 6.). Puis on constate une nette augmentation lors du retrait des conduites.

Les effets se répercutant assez longtemps, les pressions ne sont pas encore revenues à leur niveau d'origine deux jours après le processus de forage.

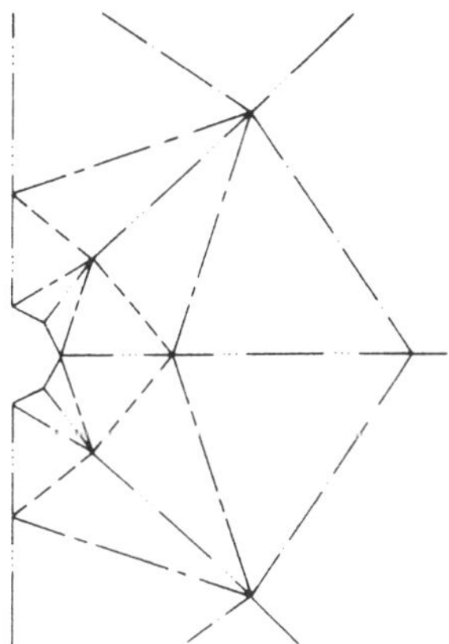

fig. 7

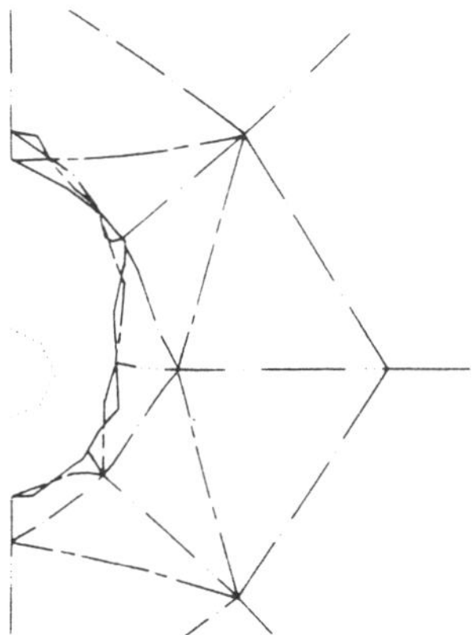

fig. 8

Pressions de l'eau à 90 m du point de pénétration
Au moment du début des travaux, les pressions des deux TE sont presque identiques. Au premier passage du front de forage et de la tête de curage, ces pressions commencent par baisser, puis elles augmentent surtout au TE situé à 2,5 m.
D'autres passages avec la tête de forage et la tête de curage indiquent une augmentation de la pression à 2,5 m. C'est dans ce cas la tête de curage qui influence nettement le phénomène.

Une fois que le premier déblayeur est passé, la pression baisse à 2,5 m et augmente à 3,5 m.
Le plus grand déblayeur est utilisé lors du retrait des conduites.
Après son passage, la pression baisse d'abord à 2,5 m, de façon relativement spectaculaire, ensuite - lors du passage des conduites - la pression augmente d'environ 3,2 m.

Pressions d'eau à 150 m du point de pénétration.
Ce tensiomètre d'eau donne à peu près le même résultat que les TE à 90 m.
L'influence de la conduite de curage sur les pressions est également très nette.
Résultat de la comparaison des pressions d'eau.
On a déjà suggéré dans l'introduction de cette section que les distances entre les TE et le front de forage ne correspondraient pas avec l'hypothèse. Un relevé de contrôle fait avec un gyroscope ne donne pas non plus de réponse univoque. Mais auprès d'un faible écart du tracé à 90 m, les données indiquées par les TE à 30 m et 90 m sont à peu près identiques.

L'image donnée par les pressions d'eau à différents endroits, au passage et après le passage des différentes activités de forage, semble confirmer l'hypothèse. Les pressions relevées au cours du retrait des conduites sont nettement plus élevées qu'au cours des autres activités de forage.

Lors du retrait des gaines, de la bentonite est remontée à l'endroit de la route. Le retrait a été rapidement fait afin de pouvoir passer au plus vite cette zone sensible. Comme il n'a pas été pratiqué de trou de forage "propre", la bentonite et le mélange d'argile et de bentonite a été mis sous (haute) pression, comme prévu. La couche d'argile

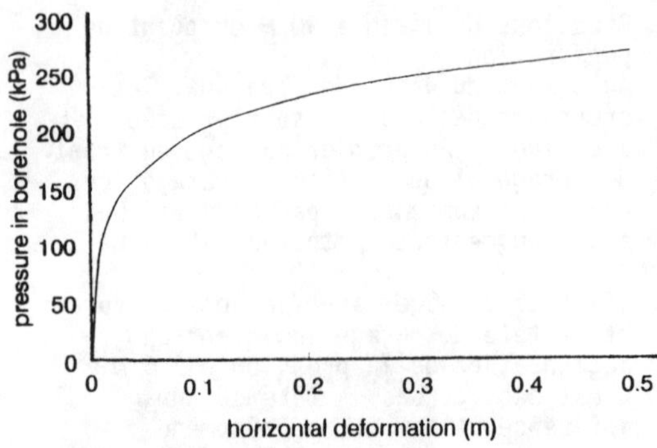

ayant déjà éclaté, la pression a fait remonté ce mélange à la surface. Cela a eu lieu sous le revêtement de la chaussée, avec les conséquences déjà citées.

7. CALCUL DE L'HYPOTHESE

Après l'analyse des résultats des mesures relevées par les tensiomètres et l'indication que l'hypothèse du comportement géotechnique est juste, il est intéressant de voir si un même schéma des pressions de l'eau peut être calculé dans une couche d'argile semblable.

Le programme des éléments finis PLAXIS a été utilisé lors des calculs.
On a considéré une coupe, correspondant au forage à environ 90 m du point de pénétration. La conduite y est placée à environ 16,5 m de profondeur (environ NAP -12,5 m), juste sous une couche de 4,5 m d'épaisseur contenant beaucoup de tourbe. Sous la couche d'argile se trouve le sable pléistocène.
Dans cette coupe, le trou de forage considéré est tenu en équilibre par les forces exercées sur la paroi (voir figure 7.)

Puis les forces dans le trou de forage sont augmentées (augmentation de la pression de la boue).

Les résultats des calculs indiquent que l'importance de la déformation du trou de forage varie selon les directions. Les calculs montrent que la déformation verticale est deux fois plus importante que la déformation horizontale (voir figure 8.). Il semble en outre que la déformation augmente rapidement à des pressions supérieures à 200 kPa (voir figure 9.).

On a déjà indiqué que les pressions de la boue au cours du forage ont pu atteindre une valeur de 1200 kPa ou plus en raison de la présence très faible de flux de retour. Des pressions élevées ont également pu se produire lors du processus de déblayage parce la bentonite ne peut s'écouler suffisamment. On peut donc s'attendre à ce que les déformations calculées aient réellement eu lieu.

Lors de la comparaison des pressions d'eau, les pressions d'eau calculées semblent supérieures aux pressions relevées. Ceci s'explique par le fait que les valeurs introduites diffèrent des données réelles.

Dans le programme de calcul, le modèle ne peut se crevasser. Dans le cas d'un déplacement horizontal calculé de 0,3 m et plus, on peut s'attendre à ce que la couche d'argile de la partie supérieure se soit crevassée. Dans une telle situation, avec une surpression proportionnellement faible, la bentonite disparaîtra simplement hors du trou de forage.

8. CONCLUSION ET RECOMMANDATIONS

Tant les relevés que les résultats des calculs étayent l'hypothèse décrivant le comportement de la couche d'argile sous l'influence des pressions causées par le forage dans le trou de forage. Comme les têtes de coupe n'ont pas été en mesure de creuser un trou de forage "propre" les flux de retour n'ont pas pu se faire. Cela a fait augmenter si fortement la pression dans le trou de forage que la couche d'argile s'est déchirée et que la bentonite a facilement pu disparaître. Une pression trop élevée peut déjà être causée par la "pression de curage".
Elle peut également être causée par les processus de déblayage et de retrait de la conduite car, dans toutes ces situations le mélange de bentonite ne peut être évacué.

De telles éruptions de boue continueront à se présenter dans des conditions géotechniques identiques si les têtes de coupe ne sont pas adaptées aux couches de sol/d'argile.

Il est important, outre les améliorations techniques, de contrôler soigneusement les flux de retour par rapport au débit et à la composition.

On peut conclure de ces activités de forage que le placement de TE dans les sols argileux ne sert presque à rien puisque les relevés donnent à peine une indication de ce qui se passe et parce que les pressions d'eau réelles se répercutent dans les relevés.

Lors de l'application de la méthode directionnelle horizontale dans des terrains vulnérables, il est judicieux de tenir compte des points suivants :

- Adapter les têtes de coupe au type de sol à forer, notamment à l'argile.
- Permettre le relevé de la pression de la bentonite dans le front de coupe.
- Contrôler scrupuleusement les flux de retour, par rapport à la composition (contiennent-ils du sol refoulé?) et au débit (apport + volume du trou de forage).
- En cas d'anomalie importante dans les débits des flux de retour, consulter immédiatement des experts géotechniques connaissant la question.
- Les connaissances et l'expérience de l'équipe de forage doivent pouvoir garantir un résultat sérieux (ne pas forcer).

1.3 Pipes

No Trenches in Town, Henry & Mermet (eds) © 1992 Balkema, Rotterdam. ISBN 90 5410 085 0

Pipes for microtunnels: Suitability of use

Ph. Henri
Société des Tuyaux Bonna, Paris, France

SUMMARY: In addition to the functional requirements of the systems of which they form part, jacked pipes are designed to absorb substantial longitudinal thrust, with the junction system between the pipes acting as an efficient link. In microtunnels, the automatic guidance system and the small diameter of the structures generate additional particular stresses which must be allowed for in the pipe design. This paper reviews the work carried out under the auspices of the FSTT on the harmonisation of the products normally used for lining microtunnels as regards both dimensions and functional technical performance. In support of the latter, it also sets out a design basis for microtunnel pipes based upon models designed from laboratory work which it is hoped will be confirmed by site experiments now being planned.

1 GENERAL INTRODUCTION

The development of microtunnelling as a "no-dig" technique, particularly in urban areas, has been made possible by a combination of new technological advances:

- Improved knowledge of the geology of sub-soils through the use of more reliable measurement techniques with a degree of complexity adapted both to the type of project (with regard for example to its extent, the risks encountered and the technical difficulties expected) and the degree of uncertainty of available data (as concerns geological and related information, and previous site operations in the neighbourhood).

This approach is of particular importance both in the choice of the microtunnelling machine (especially the cutting tools and the method of muck removal) and the design and choice of pipes for lining the microtunnel.

- The development of precision-guided tunnelling machines, particularly for the small diameters in the microtunnel range, not accessible from the inside and hence fully controlled from the outside.
- The related development of prefabricated pipes for lining microtunnels, specially designed for the purpose.

As regards large diameter pipes easily accessible from the inside (in practice, this means above a nominal diameter of 1000 mm), the jacking technique has developed considerably since it began some 30 years ago, with the result that the particular characteristics expected of jacked pipes can be specified.

After reviewing these characteristics, we shall show how the microtunnelling approach is more than a mere extrapolation to small diameters of a concept already well proven with bigger pipes. We shall review a number of practical solutions to the special problems of microtunnelling, in the design and manufacture of suitable products.

Finally we shall review the work being carried out in France by those involved in microtunnel construction, particularly that under the auspices of the FSTT with a view to obtaining a better idea of the stresses generated by this particular method of laying pipe.

2. JACKED PIPING

This process has developed considerably in

France over the last 30 years, in parallel with drilling techniques in general.

Pipe design has steadily evolved and is now better able to accommodate the particular stresses involved in jacking, for both pressurised and non-pressurised applications.

2.1 Forces acting on jacked pipe.

a) First there are the forces allowed for in the design of any buried piping :

- internal pressure if any
- external pressure from the earth and from groundwater.

The vertical pressure of the earth on top of the buried pipe can be worked out from the density and volumes of the overlying strata, corrected for the internal friction in the earth, the arch effect and the cohesion of the soil itself.

These pressures are non-uniform and were for a long time poorly understood; they are quantified using geological parameters which are occasionally difficult to assess, and they also depend on the size of the annular space left around the pipes and on the grip of the surrounding earth. They have been the subject of a great deal of research by authors such as Terzaghi, Leonards, Marston, Spangler, Protodiakonov, Szechy, Bugaïeva, who have proposed calculational models more or less adapted to project conditions (depth of cover relative to the size of the pipe, whether soils are granular or cohesive, etc.) but whose principles and results are still fairly comparable. It is often recommended that minor effects such as cohesion or the arch effect, should be simply neglected when they are poorly understood or variable.

The reactions underneath the pipe balancing these vertical forces are distributed according to the nature of the soil and the type of pipe/earth contact, that depends in particular on the excavation method used (especially the clearance the tunnelling machine leaves around the outside of the pipe) and the stiffness of the pipe itself.

In a similar way, the horizontal pressures are distributed, depending on the nature of the earth and the stiffness of the pipe. The problem can be simplified by assuming that these are constant over the height of the pipe and that there is a factor of proportionality (depending on the internal angle of friction

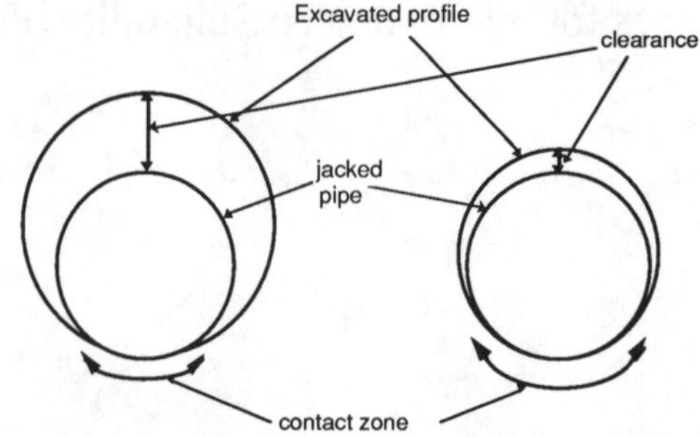

Fig.1: Effect of clearance on earth-pipe contact

of the earth) with regard to the vertical pressures.

- effect of surface live loads transmitted (and diffused) through the earth (these are very low once the buried depth of the pipe reaches 3 m, frequently the case in trenchless works).
- overall shear or bending forces induced by any movements of the surrounding earth (consolidation, settling, etc.).

The latter effects tend to be low compared with the conventional situation of pipes laid in trenches, since the jacking technique causes little redistribution of the earth and hence has little effect on its internal equilibrium.

b) There are also the forces arising from the jacking technique:

- longitudinal thrust inducing compressive forces that are to some degree deviated and off-centre in the pipe wall.

The misalignment and angular deviation made inevitable by the non-uniformities of the earth and the contingencies of the guidance system introduce an unbalanced and eccentric aspect to the thrust transmitted to the pipe column from the jacking pit and passing through all the pipe junctions.

- additional external pressure from the lubricant or consolidation fluid injected into the annular space the excavation process leaves around the pipe.

This pressure can reach several bar.

2.2 Design of pipes for jacking

The particular forces described in section 2.1 result in pipes being designed:

- to have good resistance to axial thrust:

utilisation of materials with a high compressive strength and thicknesses appropriate to the forces involved.

- provided with junctions designed to transmit these forces in the most evenly distributed manner possible, while maintaining a seal not only to keep in the effluent which will flow inside the buried pipe but also to keep out the water or lubricant circulating outside it.

The way in which these junctions respond to moderate but frequent angular deviations is a decisive factor both in the guidance of the pipe column and in its capability to withstand the thrust.

a) Determination of thrust forces :

The thrust driving the pipe column must be such as to overcome the combination of the head force at the front of the tunnel and the embrace of the surrounding earth over the whole external surface of the pipe. During actual jacking, the thrust balances the sum of:

• the head force: reaction of the tunnelling machine cutting action on the column of pipes, reaction of the cutting front, and penetration of the cutting head. The penetration is of course a function of the geometry of the cutting head, its mode of working (orientation) and of the configuration of the cutting tool inside the head; it also depends on the characteristics of the earth and may be very closely correlated with its resistance to penetration; finally this penetration force seems to depend only slightly on the depth of overlying earth above the jacked pipe.

The cutting front reaction is essentially a function of the excavation method (use of a mud shield creates a chamber at the head of the tunnel, in which the pressure is maintained at a level such as to balance out the thrust from the earth and surrounding water) and on the type of earth: it is calculated from the geological characteristics of the soils, whose active and passive thrust phenomena it involves.

• the longitudinal friction along the pipe column: this is characterised by a series of static and dynamic coefficients of friction (the angle of friction between earth and pipe depends on the roughness of the outer surface of the pipe and generally lies between two-thirds and three-quarters of the internal angle of friction of the earth), acting as multiplying coefficients of the lateral embrace of the earth on the pipe, and through a value of cohesion specific to the surrounding earth.

The value of the friction does of course also depend very directly on the weight of the pipes, the presence of water in the earth and on the quality of the contact between the outer surface of the pipes and the surrounding ground; this contact depends in turn both on the nature of the earth and on the size of the annular clearance the tunnelling machine leaves around the outer surface of the pipe.

The friction depends on the shape of the pipes (a circular shape limits its intensity) and their stiffness (a rather flexible pipe generates an additional lateral embrace as it deforms, thus leading to greater friction between earth and pipe).

Finally, friction is considerably reduced by the injection of lubricants into the annular space between the earth and the jacked pipe, whenever the nature of the terrain so permits. In practice, the lubricant is injected through openings made for the purpose in the outer skirt of the tunnelling machine. When the diameter of the jacked pipe allows, it is possible to inject lubricant from the pipe itself and thus improve penetration; in any event, the viscosity of the lubricant is a compromise between lubricating power (which improves as the lubricant is thicker) and its injectability (a lubricant of low viscosity will be distributed more uniformly).

The injection of an appropriate amount of lubricant under controlled pressure also limits the convergence of the earth on the pipe and, by helping reduce the earth's grasp, reduces still further the friction at the pipe-earth interface.

Generally speaking, the thrust forces are considerably increased by the angular deviations made necessary by the guidance corrections of the tunnelling machine. In fact these increase the total area of contact between the earth and the pipe and reorientate the forces of reaction of the earth on the pipes.

Moreover, the thrust forces may be increased by any prolonged stoppage in the driving operations which allow the earth to move inwards and embrace the pipes.

b) Designing pipes for jacking :

The two basic choices concern the length of

the pipes and the type of system used to join them together.

• The length of the jacked pipe should not exceed that of the systems making up the tunnelling machine, and particularly the length of the first follower tube. Otherwise, the tunnelling machine would have to be extended by one or more articulated sections of increasing length starting from the cutting front, so that the length of the machine tube connected to the column of pipes is not less than that of the first pipe in the column.

This precaution is essential in order to limit the number of contact points and hence the friction between the earth and the pipe as the pipe column moves forward in the chamber made by the tunnelling machine as it advances, the straightness of which is governed by a series of successive angular deviations.

The occurrence of such contact points is also aggravated by the convergence of the earth on the pipe, whatever clearance is left by the machine around the outside of the pipe.

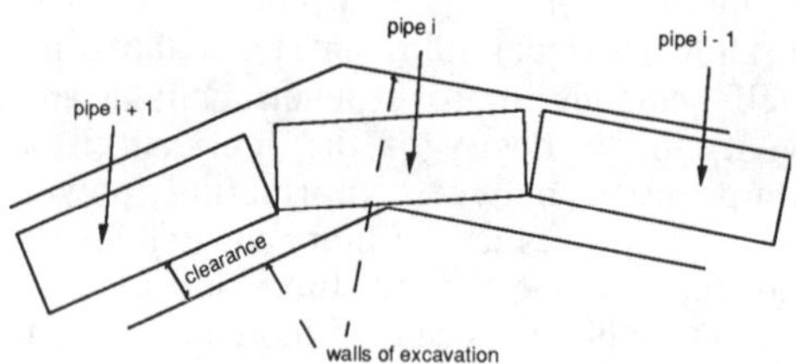

Fig. 2 : Relationship between clearance/angle of deviation/length of pipe.

It is clear therefore that for a given clearance and a given length of pipe there is (assuming a certain mode of convergence of the earth on the pipe) a maximum permissible value for the angular deviation for each excavation pass. If this value is exceeded as the tunnelling machine proceeds, friction is considerably increased and so is the thrust needed to overcome it: this effect results precisely from the angular deviation and is a dual penalty for the pipe since this increased axial force is also badly distributed in the pipe wall and hence more likely to cause damage.

It is not easy to give a precise definition of the value of such a limit, which is closely related to the assumption made about the convergence of the earth and to the semi-elastic response of the pipe junction assemblies to the deviated forces causing it; the former results from the geological approach with all its well-known uncertainties, while the second depends as much on the geometry of the junction as on the performance of the sealing system in compression.

• The junctions should be designed to remain leak-tight under moderate deviations and misalignments, while still transmitting the axial forces uniformly between two adjoining pipes.

Consequently special attention will be paid in designing the junction to the surfaces in contact, and an appropriate material can be inserted between the two contiguous pipes in order to reduce the undesirable stress concentrations that the geometrical imperfections of the surfaces in contact will certainly generate.

This material to be inserted between the pipes is therefore characterised by:

- mechanical strength of the same order of magnitude as that of the material of the jacked pipes; if this material is damaged during the driving process the required distributing effect is lost, resulting in intolerable stress concentrations in the pipes themselves.

- controlled flexibility, whereby it can deform under stress in a uniform and moderate manner so as first to avoid causing undue dimensional changes in the column of jacked pipes as a function of the thrust exerted, and secondly to accommodate a series of intermittent stresses; the latter do of course correspond to the successive thrust phases, but also to the inevitable intermittent angular deviations, applying compressive stress first to one part and then to another of the section facing the junction; hence the more elastically the distribution material behaves the more continuous will be the contact between adjoining pipes during the jacking process.

The positioning and possible attachment of this thrust distribution material in line with the junction systems will be designed in such a way that its crushing does not interfere with the junction seals.

The junction must also be stiff enough to maintain adequate compression at the seal during the driving phase, taking account of the misalignments to which the junction is subjected. The capability of the junction to accommodate misalignment is conditioned by the pipe nesting geometry. It is obviously essential that there should be no projections on the outside of a pipe which would increase the forces of friction at the earth-pipe interface.

When the female fitting consists of a metal ring attached to the body of the pipe, this must be designed to avoid damaging the seal during fitting (use of a chamfer and appropriate thickness).

In any event, care will be taken to restrict the external clearance after fitting, which might create additional friction at each junction.

3 PIPES FOR MICROTUNNELLING

3.1 Special features of microtunnelling

The first of these is naturally the diameter of the pipes, which is generally kept between 200 to 1200 mm ID. This characteristic leads to a number of basic and limiting features:

• a higher stress level in the pipe material during driving.

In fact the friction opposing the movement of the pipe column through the earth is fairly closely proportional to the external surface area of the pipe, and hence to its diameter.

On the other hand, the available thrust is proportional to the annular cross-section of the pipe and hence to the product of the diameter and the wall thickness.

In normal production pipes, regardless of the constituent material, wall thickness is roughly proportional to diameter. For this range therefore, the available thrust is proportional to the square of the diameter.

In conclusion, at stresses that are also acceptable in the material, the driven length will be shorter the smaller the diameter. Unless friction is reduced by some technical advance, jacking pipes using the usual lengths for this type of project requires the pipes to have thicker walls, which makes them more expensive and increases the cost of the work in terms of volume excavated, pipe weight, etc., or for the permissible stress to be raised to a higher level.

A palliative is of course systematically to inject lubricant.

In any event, the stress level in the materials under the effect of the axial compression of the pipes can be limited by greater precision in the dimensions of the products, especially the squaring of the end surfaces.

• the difficulty (or even the virtual impossibility in the range of the smallest diameters) of effectively transmitting the thrust inside tunnel by means of intermediate stations, a method commonly used in traditional jacking and with which there is no theoretical limit on the length of a pipe section capable of being jacked.

• the significant occurrence of leaks in the system during jacking: when the tunnelling process involves the use of a mud (bentonite for example) or takes place below the level of the water table, leaks can result in difficulties inside the microtunnel, in view of the practical difficulty of satisfactorily extracting incoming water or other effluent in a pipe of small diameter.

3.2 Design of pipes of microtunnelling

We are interested basically in microtunnels produced in a single stage, with the definitive pipes driven into the ground as the excavation proceeds.

There is also a 2-stage microtunnelling technique in which temporary metal tubes are first introduced and then replaced in the second stage by the definitive pipes; in this approach the definitive pipes are not subjected to the same stresses and as a result do not require the same characteristics. Pipes for microtunnelling are characterised by the following factors:

3.21 Dimensional requirements :

- inside diameter: to allow - especially in small sizes - the passage of muck removal tubes, cables and laser sighting lines, compatible with the control of the microtunnelling machine.
- outside diameter: ideally some 10 to 30 mm less than the outside diameter of the machine tubes (according to the pipe diameter); too small an outside diameter of the jacked pipes in fact makes the control of

the microtunnelling machine somewhat random.

- squaring off: in order to distribute the stresses uniformly at the junction.

- straightness: in order to limit any additional spurious friction.

- length: in order to accommodate fixed lengths of elements (for example, muck removal tubes) inserted in each pipe before jacking, and to be correctly inserted in the thrust rig in the jacking pit.

In general, the preferred manufacturing methods are those which can guarantee tight tolerances for these functional dimensions.

3.22 High level of mechanical performance: satisfactory material, appropriate thicknesses

For considerations of resistance to axial thrust, checks will be made to ensure that pipes with a high length/diameter ratio are unlikely to buckle.

3.23 Strictly leak-tight: with regard both to the inside and the outside; this is done by compressing an elastomeric bead between an outer ring fitted to the outside diameter of the pipes and the profile of the male end.

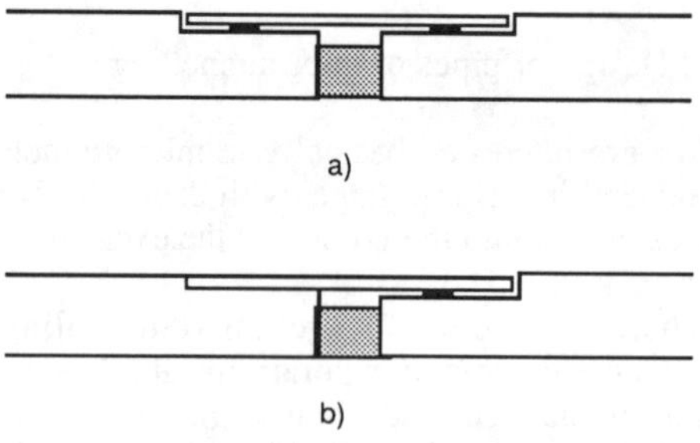

Fig. 3 : outer ring : two examples among possible configurations

The outer ring is made from a rigid material, and may float or be attached to one of the pipes; in any event it should adequately compress the seal throughout the jacking phase, having regard to the misalignments to which the junction is inevitably subjected; its length determines the dimensions of the nesting arrangement and hence the junction's capability to tolerate misalignment.

3.24 Transmission of loads in line with the junction : it is useful to introduce a thrust distribution material with the characteristics specified above (see § 3.1) in the seal plane between pipes (possibly retained by the outer ring or attached to one of the pipes).

In addition to these special requirements, pipes for microtunnelling must of course satisfy the functional requirements of the pipework systems of which they form part: sewage system operating under gravity, pressurised system, other circuits and so on.

3.3 Special test for microtunnelling pipes

It is particularly important to have a practical means of verifying the capability of absorbing the axial thrust forces together with angular misalignment at a junction.

A procedure for such a test was studied in the "Piping" workshop of the FSTT, which is one of the ten working groups established under the auspices of the Scientific Council of the FSTT to develop theoretical, practical and experimental knowledge of ten topics selected in the general area of "Trenchless Works".

The procedure involves subjecting a pair of assembled pipes to an axial compressive force developed on an appropriate test bed by a set of hydraulic jacks, and verifying performance according to the following two standards:

- no damage likely to interfere with the operation of the pipework caused by the application of the maximum thrust force P_S for which the pipe was designed, together with a misalignment of 1°.

- under a load of 2 P_S, the pipe should not reach an ultimate limiting state incompatible with the use of the product (for example a failure in the body or at the junction).

These test procedures can be used to determine the operating range possible on a plot of maximum thrust/misalignment angle and for evaluating the safety margins available as regards both load and misalignment angle, by varying these two parameters separately until failure ensues, as appropriate.

4. CURRENT WORK

a) Standardisation of sizes

The principal problem dealt with by the "Piping" workshop of the FSTT has been harmonisation of the products used in microtunnelling as regards both dimensions and functional performance.

The functional performances meet the specific requirements described in chapter 3 above and can be assessed using procedures we have already described.

The standardisation of dimensions has required allowance to be made for the current state of technology, especially the existing equipment in use on the microtunnelling market. In this way the stresses dealt with above resulted in the determination of a series of preferred outside diameters for pipes capable of being jacked by most of the microtunnelling systems in service in Europe at present.

Since the designers and operators of pipeline systems nevertheless identify pipes in terms of the inside diameter, a range of nominal diameters have been selected in parallel, sufficiently wide to cover the majority of products in any material. Depending on the material used, the actual inside diameter may differ slightly from the declared nominal diameter, which will be used by default as the value closest to the actual inside diameter.

It is clear that although this range does offer dimensional harmonisation of the different materials between one another and with the microtunnelling machines, it does not claim to completely cover the range of products so far developed and actually manufactured.

b) Design of pipes for microtunnelling

• Theoretical approach: the "Piping" workshop of the FSTT has undertaken the dual task of:

- incorporating the design of jacked pipes in service into the general methods used for designing pipes laid in trenches (calculation of pipe cross-sections under ovalizing loads) ; in fact, the microtunnel installation induces environmental conditions in the final phase which must be specified in the general framework of the design of buried pipes.

- proposing design rules for all materials for the specific effects of jacking (deviated axial thrust).

• Planned experiments: it is planned to carry out instrumented tests under axial thrust on a test rig and on site, with a view to obtaining a better idea of the stress distributions in the body of the pipes and in the junctions during the jacking phases. In parallel with the on-site instrumentation (using strain gauges), an analysis of site and machine parameters will be developed (paths actually followed, rotational torques, thrust forces, etc.) in order to correlate the values of these parameters with the measured stresses.

This approach should make it possible first to validate the chosen methods of calculation and secondly to identify the laws of behaviour of microtunnelling pipes in situ.

5 CONCLUSION

In the medium term, the theoretical and practical work undertaken on the topic of pipes for microtunnelling will meet the dual objective of making the use of these products much more reliable in a controlled technical context and of optimising their size and design according to the specific features of each project.

Table 1: Nominal diameters ND and outside diameters D_e

D_e =	360 / 405 / 535 / 640 / 750 / 850
	960 / 1090 / 1180 / 1275 / 1400
ND =	200 / 250 / 300 / 350 / 400 / 450
	500 / 550 / 600 / 650 / 700 / 750
	800 / 850 / 900 / 950 / 1000
	1050 / 1100 / 1150 / 1200

In this way it will be possible, in the field of microtunnelling, to take full advantage of the substantial recent developments implemented both in the technologies of pipe manufacture and in the intrinsic performance of the materials used.

No Trenches in Town, Henry & Mermet (eds) © 1992 Balkema, Rotterdam. ISBN 90 5410 085 0

Aptitude à l'emploi des tuyaux pour microtunnels

Ph. Henri
Société des Tuyaux Bonna, Paris, France

RESUME : Outre les prescriptions fonctionnelles des réseaux dans lesquels ils sont intégrés, les tuyaux de fonçage sont conçus pour reprendre d'importantes poussées longitudinales vis-à-vis desquelles l'assemblage des tuyaux doit assurer le rôle d'un relais efficace. Dans le cas des micro-tunnels, le guidage automatique et le petit diamètre des ouvrages installés génèrent des contraintes spécifiques additionnelles vis-à-vis desquelles les tuyaux doivent être dimensionnés. La présente communication dresse le bilan des travaux menés sous l'égide de la FSTT quant à l'harmonisation des produits couramment utilisés pour le revêtement des micro-tunnels, tant sur le plan dimensionnel que sur celui des performances techniques fonctionnelles. Elle présente parallèlement, en support de ces dernières, les bases du dimensionnement des tuyaux pour micro-tunnels, issues de modèles conçus à partir de travaux menés en laboratoire, et que des expérimentations in-situ projetées devraient utilement conforter.

1 INTRODUCTION - PRESENTATION GENERALE

Le développement, dans le cadre des techniques sans tranchée, du microtunnelage, et particulièrement en milieu urbain, est rendu possible par le faisceau convergent de nouvelles possibilités technologiques :

• L'amélioration de la connaissance géotechnique des sous-sols, par le biais de la mise en oeuvre avec une fiabilité accrue de techniques d'expertise d'un niveau de complexité adapté tant à la physionomie du projet (importance, risques encourus, difficultés techniques d'exécution supposées) qu'au degré d'imprécision attaché aux données effectivement disponibles (informations géologiques et géotechniques disponibles, chantiers précédemment réalisés dans un environnement voisin).

Cette approche revêt une importance primordiale tant pour le choix du microtunnelier (et en particulier des outils de coupe et du mode de marinage) que pour le dimensionnement et le choix des tuyaux qui devront assurer le revêtement du microtunnel.

• La mise au point de tunneliers de foration guidés avec précision, en particulier dans les petits diamètres de la gamme microtunnels, non accessibles par l'intérieur et par conséquent entièrement pilotés depuis l'extérieur.

• Le développement connexe de tuyaux préfabriqués constituant le revêtement des microtunnels, spécialement conçus et calculés à cet effet.

Dans la gamme des ouvrages de grands diamètres commodément accessibles par l'intérieur (en pratique, à partir de DN 1000), la technique du fonçage s'est considérablement développée depuis ses premières réalisations effectives il y a 30 ans, ce qui a permis de préciser les caractéristiques particulières attendues des tuyaux pour fonçage.

Après un rappel de celles-ci, nous montrerons dans quelle mesure l'approche microtunnel est davantage qu'une simple extrapolation vers les petits diamètres d'une conception déjà bien éprouvée dans la gamme supérieure du fonçage. Nous ferons le bilan des solutions pratiques apportées aux problèmes particuliers soulevés par le microtunnelage, dans les domaines tant de la conception que de la réalisation de produits adaptés.

Enfin, nous présenterons l'état des travaux engagés en France par les partenaires de la construction par microtunnels, en particulier dans le cadre de travail mis en place par la FSTT dans le but de mieux cerner les sollicitations générées par ce mode spécifique d'installation des conduites.

2 LES TUYAUX DE FONÇAGE

Depuis maintenant 30 ans en France, ils ont connu un important développement, en parallèle de l'ensemble des techniques de foration.

La conception de ces tuyaux a continuellement évolué dans le sens d'une meilleure intégration des sollicitations particulières du fonçage, tant pour les applications avec que sans pression intérieure.

2.1 Actions sollicitant les tuyaux de fonçage

a) Ce sont d'abord celles qui interviennent dans la conception de tout ouvrage enterré :

• pression intérieure éventuelle.

• pression extérieure des terres et de l'eau phréatique.

Les pressions verticales des terres en toit d'un ouvrage enterré résultent de l'intégration des poids volumiques et volumes des couches de sol le surplombant, pondérées par la prise en compte des frottements internes dans les sols, des effets de voûte et des cohésions propres des sols.

Hétérogènes et longtemps mal connues, quantifiées à l'aide de paramètres géotechniques dont l'appréciation est parfois difficile, ces pressions qui dépendent de surcroît de l'importance du vide annulaire autour des tuyaux et du mode de convergence des terres, ont fait l'objet de nombreuses études par des auteurs tels que Terzaghi, Leonards, Marston, Spangler, Protodiakonov, Szechy, Bugaïeva, qui ont proposé des modèles de calcul plus ou moins bien adaptés en fonction des conditions des projets (hauteur de couverture relativement à la taille du tuyau, sols granulaires ou cohérents, ...), mais dont les principes comme les résultats demeurent à peu près comparables. Il est souvent

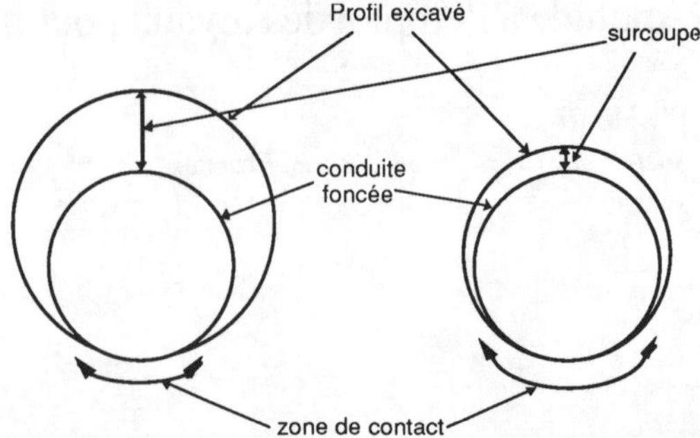

Fig.1 Incidence de la surcoupe sur le contact sol-tuyau.

recommandé de négliger prudemment les effets minorateurs (cohésion, voûte) lorsqu'ils sont mal connus ou variables.

Les réactions équilibrant en sous-face de la conduite ces actions verticales se distribuent en fonction de la nature des terres et de celle du contact entre la conduite et le sol, qui est fonction notamment du mode de creusement (et en particulier de la surcoupe ménagée par le tunnelier vis-à-vis du diamètre extérieur du tuyau) et de la rigidité propre du tuyau.

Les pressions horizontales se distribuent de la même manière en fonction de la nature du sol et de la rigidité de la conduite. Un souci de simplification conduit à les supposer constantes sur la hauteur de la conduite et dans un facteur de proportionnalité (fonction de l'angle de frottement interne des terres) vis-à-vis des pressions verticales.

• incidence des charges d'exploitation en surface, diffusées dans le sol (très faible dès que la profondeur d'enfouissement du tuyau atteint 3 m, ce qui est fréquemment le cas en matière de travaux sans tranchée).

• actions de cisaillement ou de flexion d'ensemble induites par les mouvements éventuels du sol environnant (consolidations, tassements, ...).

Ces derniers effets sont plutôt atténués par rapport aux conditions classiques de la pose en tranchée, puisque le remaniement des sols, minimal en fonçage, perturbe peu l'équilibre interne des terres.

b) Ce sont aussi les actions propres à la mise en oeuvre par fonçage :

• poussée longitudinale induisant des efforts de compression plus ou moins déviée et plus ou moins excentrée dans la paroi du tuyau.

Le désalignement et l'angulation rendues inévitables par les hétérogénéités de sols et les aléas du guidage déséquilibrent et excentrent en effet la poussée transmise au train de tuyaux depuis le puits de fonçage et transitant par tous les assemblages entre les tuyaux.

• pression extérieure additionnelle du lubrifiant ou du coulis de consolidation, injecté dans l'espace annulaire ménagé par l'excavation autour du tuyau.

Cette pression peut atteindre plusieurs bars.

2.2 Conception des tuyaux pour fonçage

Les efforts particuliers décrits au paragraphe 2.1 conduisent à concevoir des tuyaux :

- dont le corps possède une bonne résistance vis-à-vis de la poussée axiale : utilisation de matériaux caractérisés par une résistance à la compression élevée et choix d'épaisseurs adaptées aux efforts à mettre en oeuvre.
- pourvus d'assemblages étudiés pour transmettre ces efforts avec la meilleure répartition possible, tout en maintenant l'étanchéité tant vis-à-vis des effluents destinés à circuler dans l'ouvrage foncé que vis-à-vis des eaux ou lubrifiants circulant à l'extérieur de la conduite.

La réponse de ces assemblages à des déviations angulaires modérées mais fréquemment répétées est un facteur primordial tant du guidage du train de tuyaux que de la capacité résistante de cette colonne à la poussée.

a) Détermination des efforts de poussage :

La poussée destinée à faire progresser le train de tuyaux doit être telle qu'elle permette de vaincre la conjugaison de l'effort de pointe en tête de tunnel et l'intégration sur la totalité de l'enveloppe extérieure de la conduite de l'étreinte des terres au repos. En phase dynamique de fonçage, la poussée balance la somme :

• de l'effort en tête : réaction de coupe du tunnelier sur la colonne de tuyaux, réaction du front de taille, poinçonnement de la trousse coupante. Le poinçonnement est bien sûr fonction de la géométrie de la trousse coupante, de son mode de travail (orientation) ainsi que de la configuration de l'outil de coupe à l'intérieur de la trousse ; il dépend aussi des caractéristiques du sol et peut être corrélé très directement avec la résistance pénétrométrique de celui-ci ; enfin il semble que cet effort de poinçonnement soit assez peu dépendant de la hauteur de couverture au-dessus de la conduite foncée.

Par ailleurs, la réaction du front de taille est essentiellement fonction du mode de creusement (l'utilisation d'un bouclier à boue ménage une chambre en tête du tunnel, où la pression est maintenue au niveau où elle équilibre les poussées des terres et de l'eau) et de la nature des sols : elle se calcule à partir des caractéristiques géotechniques des sols, dont elle mobilise les phénomènes de poussées active ou passive.

• du frottement longitudinal de long de la colonne de tuyaux : il est caractérisé par un jeu de coefficients de frottement statique et dynamique (l'angle de frottement sol-tuyau dépend de la rugosité du fût extérieur de la conduite et il est compris généralement entre les deux-tiers et les trois-quarts de l'angle de frottement interne dans le sol), agissant en coefficients multiplicateurs de l'étreinte latérale des terres sur la conduite, et par une valeur de cohésion propre au sol environnant.

La valeur du frottement dépend aussi bien entendu très directement du poids des tuyaux, de la présence d'une nappe dans le sol et de la qualité du contact entre le fût extérieur des tuyaux et la terrain environnant ; ce contact dépend à son tour tant de la nature du sol que de l'importance de la surcoupe réalisée par le tunnelier par rapport au diamètre extérieur du tuyau.

Le frottement dépend de la forme des tuyaux (la forme circulaire en limite l'intensité) et de leur rigidité (un tuyau déformable sollicite une étreinte latérale supplémentaire de par sa déformation et par suite un frottement sol-tuyau accru).

Enfin, le frottement est largement diminué par l'usage, lorsque la nature des terrains le permet, de lubrifiants injectés dans l'espace annulaire entre terrain et conduite foncée. En pratique, cette injection est réalisée à partir d'orifices ménagés à cet effet dans la jupe externe du tunnelier. Lorsque le diamètre de la conduite foncée le permet, une injection en partie courante de la conduite est possible et améliore d'autant la glissance sol-conduite ; dans tous les cas, la viscosité du lubrifiant est l'objet d'un compromis entre pouvoir lubrifiant (d'autant meilleur que le lubrifiant est épais) et injectabilité (un lubrifiant peu visqueux se répartit de façon plus homogène).

L'injection d'un volume adéquat de lubrifiant sous pression contrôlée limite de surcroît la convergence des terres sur la conduite et, contribuant à diminuer l'étreinte du sol, réduit encore davantage le frottement à l'interface sol-tuyau.

D'une manière générale, les efforts de poussage sont considérablement accrus par les déviations angulaires rendues nécessaires par les corrections de guidage du tunnelier. En effet, celles-ci augmentent les surfaces de contact entre le sol et la conduite et réorientent les efforts de réaction des terres sur les tuyaux.

En outre, les efforts de poussage peuvent être augmentés par un arrêt prolongé des opérations de poussage, permettant aux terres de converger en étreinte radiale autour des tuyaux.

b) Définition du tuyau pour fonçage :

Les deux choix fondamentaux sont ceux de la longueur et du type d'assemblage des tuyaux entre eux.

• La longueur du tuyau foncé ne doit pas excéder celle des corps monolithes composant le tunnelier, et en particulier celle du premier tube suiveur. Si tel était le cas, il faudrait prolonger le tunnelier par une ou plusieurs parties articulées entre elles, de longueur croissante en partant du front de taille, de manière que la longueur du tube machine raccordé au train de tuyaux ne soit pas inférieure à celle du premier tuyau de la colonne.

Cette précaution s'impose si l'on veut limiter les zones de contact et par conséquent les frottements entre le sol et la conduite lors de la progression du train de tuyaux dans le boyau ménagé à l'avancement par le tunnelier et dont la rectitude est gouvernée par les déviations angulaires successives.

L'occurrence de tels points de contact est de surcroît aggravée de par la convergence des terres sur l'ouvrage, quel que soit par ailleurs la valeur de la surcoupe ménagée par la machine vis-à-vis du diamètre extérieur de la conduite.

Ainsi, il apparaît clairement que, pour une valeur de la surcoupe donnée, une longueur de tuyau donnée, il existe (une hypothèse étant faite sur le mode de convergence des sols sur l'ouvrage) une valeur de la déviation angulaire maximale admissible pour chaque passe de terrassement. Lorsque cette valeur est excédée dans la conduite du tunnelier, les frottements sont considérablement augmentés et par suite la poussée nécessaire pour les vaincre : cet effet est, de par la déviation angulaire précisément, doublement pénalisant pour le tuyau puisque cet effort axial accru se trouve de surcroît mal distribué dans la paroi du tuyau et donc d'autant plus susceptible d'y occasionner un désordre.

Il n'est pas aisé de définir précisément la valeur d'une telle limite, étroitement liée à l'hypothèse faite sur la convergence des sols et à la réponse semi-élastique des assemblages entre tuyaux aux efforts déviés qui les sollicitent ; la première relève de l'approche géotechnique avec toutes les incertitudes qu'on lui connait, et la seconde dépend tant de la géométrie de l'assemblage que du comportement vis-à-vis de la compression de la garniture qui lui confère son étanchéité.

• Les assemblages doivent être conçus pour demeurer étanches pour des déviations et désalignements modérés, tout en régularisant la transmission des efforts axiaux entre deux tuyaux adjacents.

On privilégiera par conséquent l'importance des surfaces en contact dans la conception de l'assemblage et on pourra interposer entre les deux tuyaux en contact un matériau répartiteur dans le but de diminuer les concentrations nuisibles de contraintes que les imperfections géométriques des produits en contact direct ne manqueraient pas de créer.

Ce matériau d'interposition entre tuyaux est donc caractérisé par :

- un niveau de résistance mécanique du même ordre de grandeur que celui du matériau composant les tuyaux foncés ; la ruine de ce matériau au cours du poussage entraînerait en effet la perte de l'effet répartiteur cherché, et par suite des concentrations de contraintes insupportables pour les tuyaux eux-mêmes.

- une souplesse contrôlée lui permettant de se déformer régulièrement et modérément sous contrainte afin d'une part de ne pas occasionner de variations dimensionnelles exagérées du train de tuyaux poussé en fonction de la valeur des poussées exercées, d'autre part d'accepter une série de sollicitations alternées ; ces dernières correspondent aux phases successives de poussée bien sûr, mais aussi aux inévitables déviations angulaires alternées, sollicitant en compression tantôt une partie, tantôt une autre partie de la section au droit de l'assemblage ; le contact entre les tuyaux adjacents sera donc d'autant plus continûment assuré au cours du poussage que le matériau répartiteur démontrera un bon comportement élastique.

Le positionnement et la fixation éventuelle au droit des assemblages de ce matériau répartiteur de poussée seront étudiés afin que son écrasement n'obère pas le bon fonctionnement de la garniture d'étanchéité de ces assemblages.

Par ailleurs, la rigidité de l'assemblage doit être suffisante pour conserver à la garniture d'étanchéité un niveau de compression suffisant au cours de la phase de poussage, en tenant compte des désaxements auxquels l'assemblage est soumis. La géométrie des emboîtements entre tuyaux conditionne la capacité de désaxement effective de l'assemblage. Il est bien sûr impératif qu'aucun point saillant ne soit créé par rapport au fût extérieur du tuyau, afin de ne pas augmenter les forces de frottement à l'interface sol-tuyau.

Lorsque l'emboîture femelle est constituée par une bague métallique ancrée dans le corps du tuyau, elle devra être conçue pour ne pas blesser la garniture d'étanchéité au cours de la mise à joint (chanfrein, épaisseur).

On aura soin dans tous les cas de limiter le jeu extérieur entre les abouts après emboîtement, susceptible de créer un frottement additionnel singulier au droit de chaque assemblage.

3 LES TUYAUX POUR MICROTUNNELAGE

3.1 Spécificités du microtunnelage

La première particularité est bien sûr le diamètre de ces tuyaux dont on borne généralement la gamme aux diamètres intérieurs compris entre 200 et 1200 mm. De cette particularité découlent plusieurs spécificités fondamentales et contraignantes :

• un niveau de contraintes plus élevé qu'en fonçage dans le matériau composant les

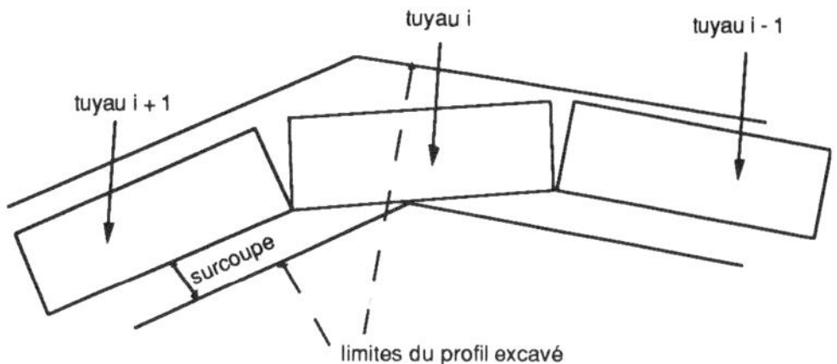

Fig.2 Relation Surcoupe/Angle de déviation/ Longueur du tuyau

tuyaux.

En effet, à très peu près, le frottement s'opposant à la progression dans le terrain du train de tuyaux est proportionnel à la surface extérieure du fût des tuyaux et par conséquent au diamètre de ceux-ci.

En revanche, la poussée disponible est proportionnelle à la section annulaire droite de la conduite et par conséquent au produit du diamètre par l'épaisseur du tuyau.

Dans la gamme courante des fabrications, et quel que soit le matériau les composant, l'épaisseur des tuyaux est pratiquement proportionnelle à leur diamètre. Il en ressort que, dans cette gamme, la poussée disponible est proportionnelle au carré du diamètre.

En conclusion, à contraintes admissibles égales dans le matériau, la longueur foncée sera d'autant plus limitée que le diamètre sera petit. A défaut de réduire par un progrès technologique la valeur des frottements, la conduite de fonçages dans les longueurs habituelles de ce type de projets nécessite que les tuyaux soient épaissis (ce qui enchérit leur coût aussi bien que celui des travaux : volume excavé, poids des tuyaux, ...) ou que la contrainte admissible soit portée à un niveau supérieur.

Un palliatif consiste bien sûr au recours systématique à l'injection de lubrifiant.

Dans tous les cas, une précision accrue sur les côtes fonctionnelles des produits, et en particulier vis-à-vis de l'équerrage des faces d'abouts, permet de limiter le taux de contrainte dans les matériaux sous l'effet de la compression axiale des tuyaux.

• la difficulté (pour ne pas dire la quasi-impossibilité dans les plus petits diamètres) de relayer efficacement la poussée à l'aide de stations intermédaires, solution communément utilisée en fonçage traditionnel et grace à laquelle il n'existe pas de limite théorique à la longueur d'un tronçon susceptible d'être foncé.

• l'incidence notable d'une perte d'étanchéité du réseau en cours de fonçage : lorsque le tunnelage utilise une boue (bentonitique par exemple) ou bien est conduit sous le niveau de la nappe, une perte de l'étanchéité peut se traduire par des incidents contraignants à l'intérieur du microtunnel, compte tenu de la difficulté matérielle à procéder dans un petit diamètre à un épuisement satisfaisant des venues d'eau ou d'un autre effluent.

3.2 Conception des tuyaux pour microtunnelage

Nous nous intéresserons essentiellement au microtunnels réalisés en une seule phase, dans lesquels les tuyaux destinés à assurer le revêtement définitif sont d'emblée poussés dans le terrain au fur et à mesure de l'excavation.

Il existe parallèlement une technique de microtunnelage en deux phases mettant en oeuvre dans un premier temps des tubes métalliques provisoires destinés à être remplacés dans une seconde phase par des tuyaux définitifs ; une telle pratique ne soumet pas les tuyaux définitifs aux mêmes contraintes et n'exige par conséquent pas d'eux les mêmes performances. Les tuyaux pour microtunnelage sont caractérisés par les éléments suivants :

3.21 respect du dimensionnel :

- diamètre intérieur : pour permettre, en particulier dans les petites dimensions, le passage des tubes de marinage, câbles et rayons de visée laser, dans des conditions compatibles avec le pilotage du microtunnelier.

- diamètre extérieur : idéalement inférieur de 10 mm à 30 mm au diamètre extérieur des tubes machines (en fonction du diamètre du tuyau) ; un diamètre extérieur des tuyaux foncés trop petit rend en effet aléatoire le pilotage du microtunnelier.

- équerrage : afin d'homogénéiser la répartition des contraintes au droit de l'assemblage.

- rectitude : dans le but de limiter les frottements parasites additionnels.

- longueur : afin de s'accommoder des longueurs fixes des éléments (tubes de marinage par exemple) introduits dans chaque

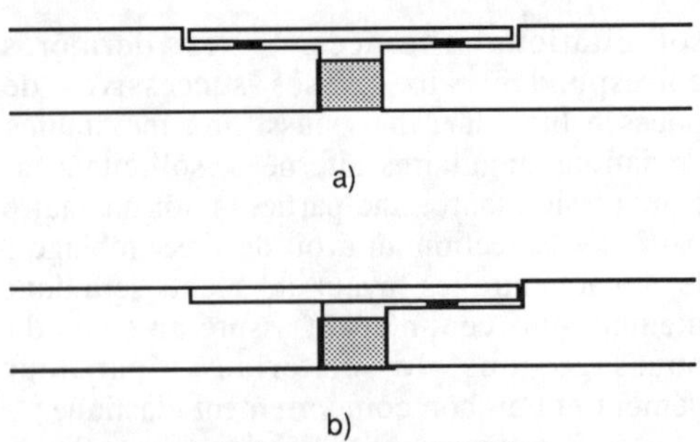

Fig.3 Bague extérieure : deux exemples parmi les configurations possibles

tuyau avant fonçage, et de manière à s'inscrire correctement dans le bâti de poussée en puits de fonçage.

On privilégiera de façon générale les moyens de fabrication permettant de garantir des tolérances serrées sur toutes ces cotes fonctionnelles.

3.22 performances mécaniques élevées :

matériau performant, épaisseurs adaptées. On vérifiera en outre pour les tuyaux présentant sur le plan de la résistance à la poussée axiale un élancement important, que la sécurité vis-à-vis du flambement est assurée.

3.23 étanchéité rigoureuse :

tant vis-à-vis de l'intérieur que de l'extérieur ; elle est obtenue par la compression d'une corde élastomère entre une bague extérieure ajustée dimensionnellement au diamètre extérieur des tuyaux et le profil de l'about mâle.

La bague extérieure, confectionnée dans un matériau rigide peut être flottante ou ancrée à l'un des tuyaux ; dans tous les cas, elle doit conserver à la garniture d'étanchéité un niveau de compression suffisant au cours de la phase de poussage, en tenant compte des désaxements auxquels l'assemblage est inévitablement soumis ; sa longueur conditionne la dimension de l'emboîture et par conséquent la capacité de désaxement effective de l'assemblage.

3.24 Transmission des charges au droit de l'assemblage :

un matériau répartiteur répondant aux besoins spécifiés précédemment (voir § 3.1) pourra être utilement interposé dans le plan de joint entre tuyaux (voire éventuellement assujetti à la bague extérieure ou à l'un des tuyaux).

Outre ces prescriptions particulières, les

tuyaux pour microtunnelage doivent bien sûr satisfaire aux exigences fonctionnelles des réseaux dans lesquels ils s'inscrivent : réseau d'assainissement gravitaire, réseau en pression, gaine technique, ...

3.3 Essai spécifique au tuyaux pour micro-tunnelage

Il est particulièrement important de disposer du moyen pratique de vérifier l'aptitude à reprendre des efforts de poussée axiale concomitants avec un désaxement angulaire au niveau d'un assemblage.

La procédure de mise en oeuvre d'un essai de ce type a été étudiée au sein de l'atelier "Tuyaux" de la FSTT, qui est l'un des dix groupes de travail mis sur pied dans le but de développer, sous l'égide du Conseil Scientifique de la FSTT, l'état des connaissances tant théoriques que pratiques et expérimentales, sur dix thèmes retenus dans le champ général des Travaux Sans Tranchée.

Cette procédure consiste à soumettre à un effort de compression axiale développé sur un banc approprié par un jeu de vérins hydrauliques, un couple de tuyaux assemblés, et à vérifier les deux niveaux de performance suivants :

- aucun désordre susceptible de nuire à l'exploitation du réseau occasionné par l'application de l'effort de poussée maximale P_S pour lequel le tuyau a été dimensionné, en concomitance avec un désaxement de 1°.
- sous une charge 2 P_S, le tuyau ne doit pas atteindre un Etat Limite Ultime incompatible avec la mise en oeuvre ou l'utilisation du produit (par exemple rupture dans le corps ou au droit de l'assemblage).

Ces modalités d'épreuve peuvent être utilisées pour déterminer la plage d'utilisation effectivement possible sur un diagramme Poussée maximale/Angle de désaxement et d'apprécier les marges de sécurité disponibles tant sur la charge que sur l'angle de désaxement, en faisant varier indépendamment ces deux paramètres jusqu'à la ruine le cas échéant.

Tableau 1. Diamètres nominaux DN et extérieurs D_e

D_e =	360 / 405 / 535 / 640 / 750 / 850 / 960 / 1090 / 1180 / 1275 / 1400
DN =	200 / 250 / 300 / 350 / 400 / 450 / 500 / 550 / 600 / 650 / 700 / 750 / 800 / 850 / 900 / 950 / 1000 / 1050 / 1100 / 1150 / 1200

4 LES TRAVAUX EN COURS

a) Harmonisation dimensionnelle

Le problème primordial auquel s'est attaché l'atelier "Tuyaux" de la FSTT est l'harmonisation tant sur le plan dimensionnel que sur le plan des performances fonctionnelles, des produits utilisés en matière de microtunnelage.

Les performances fonctionnelles répondent aux spécificités développées dans le chapitre 3 précédent et peuvent être évaluées à l'aide des procédures que nous avons précédemment mentionnées.

L'harmonisation dimensionnelle a, de son côté, nécessité que soient pris en compte l'état actuel des technologies et en particulier des matériels existants et opérant sur le marché des microtunnels.

Ainsi, les contraintes préalablement évoquées ont conduit à déterminer une série de diamètres extérieurs préférentiels pour les tuyaux susceptibles d'être poussés derrière la plus grande partie des microtunneliers en service en Europe à l'heure actuelle.

Le repérage des tuyaux se faisant malgré tout à partir du diamètre intérieur pour les concepteurs et les exploitants des réseaux, une gamme de diamètres nominaux a été retenue parallèlement, suffisamment étendue pour couvrir la plus grande partie de la production tous matériaux confondus. Selon le matériau utilisé le diamètre intérieur réel pourra légèrement différer du diamètre nominal annoncé, qui sera retenu comme valeur approchée par défaut du diamètre intérieur réel.

Il est bien clair que cette gamme, qui a le mérite de proposer une harmonisation dimensionnelle des différents matériaux entre eux et avec les microtunneliers, n'a pas la prétention de couvrir complètement la gamme des produits développés jusqu'à présent et effectivement fabriqués aujourd'hui.

b) Calcul des tuyaux pour microtunnelage

• Approche théorique : l'atelier "tuyaux" de la FSTT a entrepris la double tâche :

- d'intégrer le calcul des tuyaux en phase de service dans les méthodes générales utilisées pour calculer les réseaux installés en tranchée (calcul des sections transversales sous l'effet des charges ovalisantes) ; l'installation en microtunnel induit en effet des conditions d'environnement en phase définitive qui devront être précisées dans le cadre général du calcul des réseaux enfouis.

- de proposer des règles de dimensionnement tous matériaux vis-à-vis des effets spécifiques du fonçage (poussée axiale déviée).

• Expérimentations projetées : il est prévu de conduire des instrumentations à la poussée axiale sur banc d'essai et sur site, dans le but de mieux apprécier les distributions de contraintes dans le corps des tuyaux et au niveau des assemblages au cours des phases de poussée. En parallèle de l'instrumentation sur site (par jauges extensométriques), une analyse des paramètres de terrain et de machines (trajectoires effectivement suivies, couples de rotation, forces de poussée, ...) sera développée afin de corréler les valeurs de ces paramètres aux contraintes mesurées.

Cette démarche devrait permettre d'une part de valider les méthodes de calcul retenues, d'autre part de mieux cerner les lois de comportement en site des tuyaux pour microtunnelage.

5 CONCLUSION

L'ensemble des réflexions et des travaux engagés sur le thème des tuyaux pour microtunnelage satisfera à moyen terme au double objectif de fiabiliser l'utilisation de ces produits dans un contexte technique maîtrisé de façon plus approfondie, et d'optimiser le dimensionnement et la conception de ces tuyaux en fonction des caractéristiques propres à chaque projet.

Ceci permettra de profiter pleinement dans le domaine du microtunnelage des importants développements réalisés récemment tant dans les technologies de fabrication des tuyaux que sur le plan des performances intrinsèques des matériaux qui les composent.

No Trenches in Town, Henry & Mermet (eds) © 1992 Balkema, Rotterdam. ISBN 90 5410 085 0

Ductile iron pipe for trenchless techniques

W.W. Stevens
Stanton PLC, UK

A. Girka
Pont-à-Mousson S.A., France

ABSTRACT : In the water and sewage industries the market for trenchless techniques is still growing due to continuing development of the technology and an increasing awareness of its applications. The paper is in two parts : firstly, the market for microtunnelling is examined and the development of a one-pass ductile microtunnelling pipe is described in detail ; secondly, the techniques for pipeline renovation/ rehabilitation are reviewed with particular reference to developments in pipeline replacement with ductile iron.

1. INTRODUCTION

Ductile iron pipe with push-in joints is used extensively for water distribution pressure pipelines and, to a lesser degree, for gravity sewers.

In Europe ductile pipe has not featured in trenchless technology, whereas in Japan it has been installed directly by pipe jacking since the seventies and more recently by microtunnelling.

A description of recent European developments in trenchless technology with ductile pipe is given together with comments on the potential market.

2. DEVELOPMENT OF A DUCTILE MICRO-TUNNELLING PIPE

2.1 Microtunnelling market in Europe

Microtunnelling has shown spectacular growth, the number of machines in the U.K. has nearly doubled over the last 3 years. In 1992 more than 20 machines are likely to install between 15 and 20 km of sewer pipe. In Germany, where the method was first developed in Europe, more than 100 machines are in operation installing 60 to 70 km of sewer pipe per year. In France, the method was first introduced in 1989 and in 1992 with 12 machines the total length of pipe installed is expected to be between 5 and 7 km. Nevertheless, to put the current market size in perspective, the meterage in Europe is < 1 % on average of that installed by traditional open cut.

The technique is almost universally employed in gravity sewage schemes and is only occasionally used to install pressure pipelines for two reasons :

1. There are no manufacturers of micro-tunnelling pressure pipe in Europe. (When microtunnelling is used it is to provide casings for housing standard pressure pipe, discussed later).

2. Pressure pipelines are usually laid with only 1 m depth of cover and the level essentially follows the ground surface profile (unlike gravity sewers) making tunnelling techniques uncompetitive with traditional open cut.

A small market exists for tunnelling techniques at critical crossings in pressure pipelines where it is impossible to open cut without causing severe disruption eg under motorways, railways, etc. The continuing growth in the number of micro-tunnelling machines will make the technique more competitive. Further impetus will be given if effective legislation is introduced to account for social disruption. In Japan, where such legislation exists, microtunnelling is used far more extensively for sewers (about 10 % of all installations) and also, to a limited degree, for pressure pipelines.

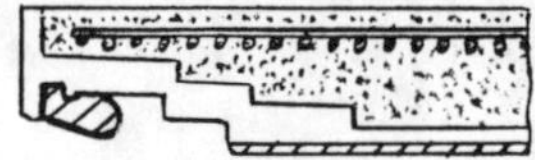

Figure 1 : D.I. microtunnelling pipe

2.2 Current practice with ductile iron

In Europe, the current practice for laying ductile iron pressure pipe at critical crossings is to install a pipe casing to act as a housing. The ductile iron pipe is usually located in the housing with a metal 'spider' next to each flexible joint to prevent damage to the pipe surface during insertion and to restrict joint deflection when the pipeline is pressurised. In some circumstances the void between the pipeline and casing is filled with grout or pea gravel.

In good ground conditions auger boring with steel pipe is the main method used for installing casings with internal diameters < 900 mm. For sizes ≥ 900 mm, pipe jacking with concrete pipe is usually preferred. In poor ground, especially with adverse ground water, microtunnelling can be competitive.

2.3 Objectives

The aim of the development was to design and manufacture a one-pass DN 600 ductile microtunnelling pressure pipe for trial use with a view to marketing a range of sizes for the water and sewage industries.

2.4 Pipe specification

The pipe was made to suit the Decon/Soltau RVS 250 A microtunnelling machine and after consultations with contractors the following specification was defined :

1. size : ID : 600 mm, OD : 787 mm and laying length of 1.2 m.
2. pressure rating : max. working of 25 bar and max. test of 30 bar.
3. joint deflection : recommended max. of 1° during jacking and max. of 3° in service.
4. jacking force : max. of 250 tonnes.

2.5 Pipe design

A longitudinal cross section of the pipe is given in Fig. 1. It consists of a cast ductile iron socket welded to a centrifugally cast ductile iron barrel and surrounded with a reinforced concrete coat. The jacking force at the joint is transferred through the socket and concrete faces via a chipboard packing piece. In this design the concrete is subjected to the jacking force and the iron barrel is virtually free of direct compressive stress. The reinforced concrete also provides :

1. a uniform external diameter compatible with the cut diameter of the tunnelling machine.
2. protection to the ductile surface against scouring during jacking.

The push-in joint is the Tyton design which is used extensively in the water industry and is well proven. Current push-in joints produced by Stanton have been designed to have a leak-tight service life well in excess of 100 years under the most arduous conditions (Greatorex 1991).

2.6 Laboratory tests

Two pipes were manufactured and assembled with packing pieces. The assembly with the joint straight was cyclically loaded 20 times to 100 tonnes. The joint was then deflected to 1.2° and further load cycles applied : 20 cycles at 100 tonnes, 10 c - 200 t and finally, 3 c - 243 t.

The pipes were restrained against straightening under load but monitoring of the joint angle revealed that the deflection reduced from the initial 1.2° to 0.83° at the load of 243 t. Inspection of the pipes after load testing revealed only minor hair-line cracks in the concrete which were of no significance.

2.7 Installations

The pipe has been used in two installations with the RVS 250 A machine in 1991 :

1. Ware in Hertfordshire (England) underneath a canal and railway for a distance of 60 m (50 pipes) for the Lee Valley Water Company. The installation was the first trial and to minimise the contractor's (M.J. Clancy & Sons Ltd, Harefield) risk the pipe was used as a casing for a DN 300 ductile iron pipe. The ground was dense, silty sand with well graded gravel and the maximum jacking force attained was 250 t. Inspection of the finished pipeline by CCTV revealed that the lining was undamaged and successful pressure testing with air at 1 bar confirmed the security of the joints.
2. The BP Ethylene Plant, Grangemouth (Scotland) for a distance of 53 m (44 pipes) beneath a road and buried plant pipework (Contractor : Sillars (B & CE) Ltd, Hartlepool). Although the pipeline

Figure 2 : D.I. microtunnelling pipes with auger flights

Figure 3 : D.I. pipe in the RVS 250 A jacking frame

would be working at gravity pressures (< 1 bar) it was a strategic fire main and ductile iron was specified for proven long-term joint security and inherent strength. The ground was wet, sandy, silty clay and the maximum jacking force was 72 t. The method of installation and pipeline integrity were verified by successful pressure testing at 15.5 bar. The contractors on both projects were impressed by the performance of the pipes and are keen to promote their use.

2.8 Economic viability

An analysis of the installed cost for ductile iron cased crossings indicated that a one-pass ductile iron microtunnelling pipe may reduce the overall installation cost particularly where poor ground conditions prevail.

2.9 Conclusions

A one-pass ductile iron microtunnelling pipe :

1. has been succesfully installed in two schemes and it has been demonstrated that the method of installation is reliable,
2. may reduce the overall costs of ductile iron crossings in poor ground but further market research is in progress to quantify the savings,
3. offers long term joint security and inherent strengh.

3. RENOVATION/REHABILITATION OF PIPELINES AND THE POTENTIAL FOR DUCTILE IRON PIPE

3.1 Classification of the techniques

The aim of renovation is to restore the performance of a pipeline. The techniques available are replacement and structural lining.

Rehabilitation is applicable to a main still performing satisfactorily mechanically. The objectives are restoration of leak-tightness at certain points, improvement in hydraulic performance, or achievement of distributed water quality in accordance with EC regulations.

The principal trenchless techniques are summarized below :

Renovation		Rehabilitation
Replacement	Structural renovation	
Bursting	Siplining	Joint sealing
Extraction	Modified slip lining	Mortar lining
		Epoxy lining
	Cured in place lining	Thin wall sliplining
		Modified sliplining
		C.I.P. lining
		Loose lining

3.2 Analysis of the techniques

All the techniques have a number of aspects in common :

1. knowledge of the pipeline's history : location of discontinuities (repairs, changes in diameter or materials),
2. preparatory scraping operations and dimensional checks (excepting replacements), removal of accessories incorporated in the main,
3. straight sections, or with bends up to 11 or 22° maximum,
4. replacement, sliplining and C.I.P. lining necessitate subsequent service connections via localised excavations.

3.2.1. In-situ cement mortar or epoxy lining

The purpose of relining is to restore the hydraulic capacity and improve the water quality. Substrate preparation is important, particularly for the thinner epoxy lining. These techniques are unsuitable if there are structural or leak-tightness problems.

3.2.2 Conventional sliplining

Depending on the circumstances, the inserted tubing can be self-sufficient structurally. Connections are made externally. Adaptation pieces are used for jointing and accessory installation. Grouting of the annular gap is recommended. The hydraulic capacity is reduced by 20 to 50 % and the excavation necessary for the insertion can be substantial, together with significant surface disruption.

3.2.3 Modified (close fit) sliplining

The advantage compared with conventional sliplining lies in keeping the loss in hydraulic performance to a minimum. There is no annular gap. Installation is more specialised and preparatory scraping and internal dimension checking is vital.

3.2.4 Cured in place (C.I.P.) lining

These procedures are still little used in Europe in the potable water field. Depending on the lining composition, they can provide structural renovation with good mechanical and leak-tightness performance.
Making of external connections is still an aspect under development. Techniques are more expensive than sliplining.

3.2.5 On-line replacement by pipe-bursting

This process has the advantage of allowing the installation of a larger diameter main than the one being replaced. However, the practice of leaving the fragments of the old main in the ground may present the risk of damage to the new one. These fragments are sometimes regarded as contaminants in environmental terms (prevents the use of this method in Berlin).

3.2.6 On-line replacement by extraction

The practice of extracting the old main removes the uncertainty of leaving a harmful bed for the new one. As in pipe-bursting, services are disconnected before the replacement is installed.
It is also possible to install a larger diameter main.

3.3 Renovation/rehabilitation in Europe

1. Germany

The old länder (provinces) contain 360.000 km of principal mains. This system is in good condition (5 to 10 % leakage) and is of recent origin (45 % of the length is under 25 years old). About 100.000 km consists of unlined grey iron pipes. Every year, 1 % of the length is systematically replaced by conventional open cut excavation (grey iron more than 50 years old) and a further 1.500 km is rehabilitated by cement mortar lining. On-line replacement by bursting and extraction (Berlin method) is also used but only represents 1 to 2 % of the total replacement.

2. U.K.

The mains system of 360.000 km is old, with 45 % of the length more than 45 years old.
Leakage rates vary between 15 to 40 %, depending on the system. Close to 150.000 km of the mains consist of the previously used grey iron. By 1991, 1.500 km were being replaced annually.
Renovation requirements are estimated as 100.000 km over a 10 years period. Open cut replacement should deal with 3.000 km/annum and the remaining 7.000 km to be dealt with each year would be essentially by in-situ lining with cement mortar or epoxy.
Bursting techniques are well established in the U.K. and already amount to

50 km/annum. Close fit sliplining and C.I.P. lining, originally developed in the U.K., are finding increasing application.

3. France

The principal system of 560.000 km is relatively young (more than 50 % of the length is under 25 years old) and is in good condition (average leakage rate is between 10 to 15 %). Nearly 250.000 km of main consists of grey iron. The renovation required has been estimated to be 30 % of the length, i.e. about 170.000 km. The principal technique used is conventional open cut replacement. Bursting is rarely used (only 3 or 4 contractors are equipped for it). A replacement technique by extraction is being developed for installing ductile iron pipes.

Polyethylene sliplining represents a few dozen km each year. C.I.P. lining tends to be applied for a few km/annum of potable water pipes.

The lengths involved in trenchless renovation/rehabilitation techniques in France are very small, well below those in Germany and the U.K.

3.4 Potential for ductile iron

On line replacement by bursting or extraction allows advantage to be taken of the characteristics of ductile iron pipeline systems :

- the intrinsic mechanical properties of the material,
- the modular design, with push-in joints.

3.4.1 On-line replacement by pipe-bursting

A short pipe (DN 200 x 2.5 m) has been designed (Robins et al 1990) and used in trials (Robins et al 1991) in the U.K. : the standard integral socket was replaced by a double socket collar as illustrated in the Fig. 4.

The external diameter of the collar was only 60 % of that of the conventional socket. The trials examined :

1. ground disturbance during passage of the new pipe,
2. influence of the socket projection on the thrust forces,
3. influence of old pipe fragments on various types of external coating.

The technique is very promising for pipes in the size range DN 150 to 450.

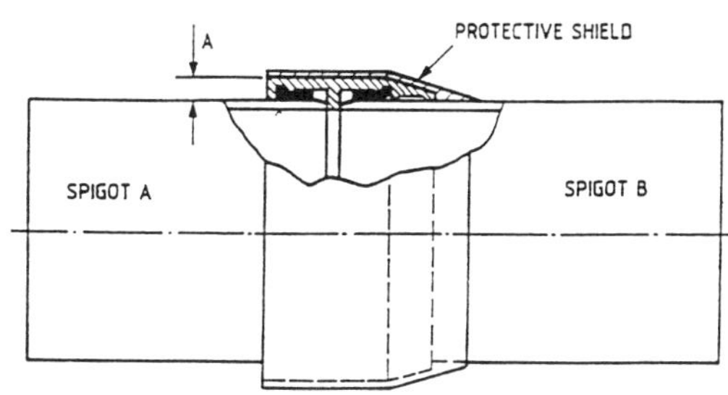

Figure 4 : D.I. design for pipe-bursting

3.4.2 On-line replacement by pipe - extraction

A problem which is often encountered with this technique is partial break up of the old main in the ground. The presence of any fragments left in the installation area justifies (similarly to the bursting technique), the use of ductile iron pipes for the new main. The functional requirements are similar to those needed for pipe-bursting.

3.5 Conclusions

The rehabilitation of potable water mains in urban areas can be carried out by a variety of the trenchless techniques. However, all require extensive preliminary preparation of the main concerned, and subsequent service connections.

Furthermore, the nature of the problems affecting the old main is a determining factor in making the right choice of method to be used. The majority of mains in small and medium diameters consist of grey iron, having an average age in excess of 50 years. Their on-line replacement by bursting, and particularly by extraction, associated with the installation of a new ductile iron main is a promising technique with the following advantages :

1. smaller launch pits with less surface disruption,
2. simple and quick push-in joints which can be assembled in the launch pit,
3. not susceptible to structural damage by the presence of any sharp fragments left in the ground.

References :

Greatorex, C.B. 1991. Stanton ductile iron flexible joints for the water industry. Stanton PLC internal report

Robins, P.J, Rogers, C.D.F. & Scott,

A.M. 1990. Design and development of ductile iron pipes for use with pipe-bursting. Pipeline management 1990
Robins, P.J., Rogers, C.D.F. & Scott, A.M. 1991. Field performance of a new ductile iron pipe for use with pipe-bursting. No-Dig International conference 1991 Hambourg.

Acknowledgement :

We wish to thank Mrs. M. HAVIOTTE for assisting in the preparation of the paper.

No Trenches in Town, Henry & Mermet (eds) © 1992 Balkema, Rotterdam. ISBN 90 5410 085 0

Numerical analysis of stress distribution around shallow buried pipes

Y.G. Diab
Dune S.A., Lyon, France

ABSTRACT : In this paper the results of numerical analysis by using a Finite Element model to evaluate loads distribution around shallow buried rigid circular pipe due to static superimposed loads are presented. In addition a comparison is realized between the computed results and an available analytical method. This class of structure is encountered frequently in urban underground constructions

1 INTRODUCTION

The amazing load carrying capacity of buried cylinders has been recognized for a long time and has been extensively used in the civil engineering practice for buried pipes of all kinds.

The most commonly used method for computing backfill loads on rigid pipes is that developed by Marston and his associates in the early part of this century (Marston and Anderson 1913; Marston 1930; Schlick 1931; Spangler 1969).

Over the last two decades, the Marston theory had been criticized as overly conservative, and some efforts have been made by many researchers to develop alternate procedures for design and to understand the phenomenon of soil-pipe interaction by employing the exact and approximate solution techniques (Burns and Richard 1964; Watkins 1966; Young and Trott 1984; Gerbault 1985).

The Finite Element method (Zienkiewicz 1971) has been used to determine pressure distribution around buried rigid pipes (Parmelee 1976; Heger 1988) and many others. Results obtained from these studies have been compared with the site measurements, and have been found to correlate well with the latter.

2 SCOPE AND OBJECTIVES OF WORK

The aim of the analysis presented in this paper is to evaluate the load distributions around shallow buried circular rigid pipes due to static external loads by using the Finite Element method. This kind of structures is encountered frequently in the urban underground constructions.

The objectives of the present study are to compute normal and tangential loads on shallow buried rigid pipes due to various loading configurations. In addition comparisons between

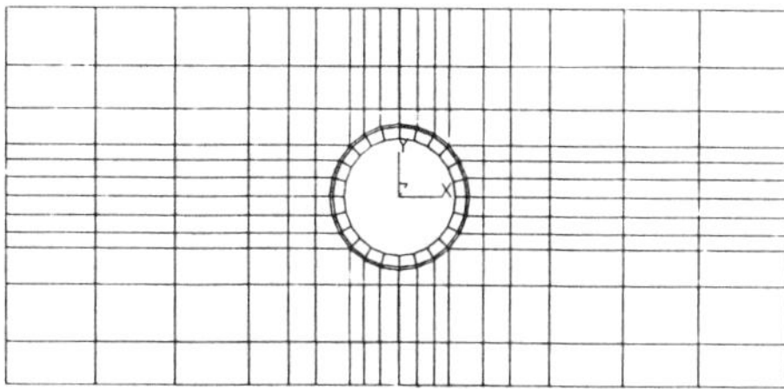

Figure (1) Finite Element mesh

the computed values and the available analytical solution developed by Burns and Richard (1964) is realized.

3 ANALYSIS PROCEDURE

Rigid pipes are defined as those having small diameters to wall thickness ratios and are accordingly not susceptible to buckling under pressure. Many authors define ratios to distinguish flexible culverts from rigid pipes (A.T.V. 1984; C.C.T.G. 1991). Pipes analyzed in this study fall in the category of rigid pipes. The depth of burial considered in the analysis is equal to one pipe diameter to satisfy the criterion of a shallow buried pipe.

It is assumed that the pipe with the loading system is long enough to permit the use of plane strain analysis. A depth of one diameter below the pipe is chosen to minimize the boundary effects and to realize a realistic comparison of the numerical results with the available analytical solution. The lateral boundary conditions are assumed at a distance of four diameters away from the center line of the pipe. In addition, a fixed condition is assumed at the bottom and a freedom in the vertical direction at the lateral boundaries of the medium surrounding the pipe.

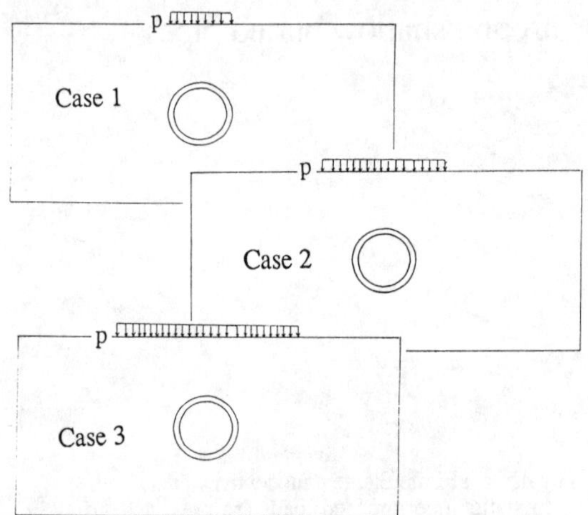

Fig. (2) Configurations of the different loading cases

Figure (1) shows the region and the Finite Element mesh used in this analysis.

In the analyses that follow, the pipe's law mechanical behavior is assumed to be elasto-plastic with a parabolic failure criteria.The mechanical properties of the pipe's material are the followings:

Initial Young modulus:	20 000 MPa
Poisson ratio :	0.2
Compression strength :	25 MPa
Tensile strength :	3 MPa

The soil surrounding the pipe is considered to behave elastically because the objective of this study is to analyze the variation of the pipe response submitted to various loadings configurations of low intensity, but we mention that our Finite Element program is able to consider different kinds of non linear laws to analyze the soil with sufficient precision (modified Cam-Clay model and Duncan law).The adopted values for the soil are:

Volumic weight :	18 KN/m3
Elastic modulus:	2 MPa
Poisson ratio :	0,35

The loading configurations analyzed in this study are listed in table (1) :

Table (1) Loading cases used in analysis

Loading case	Kind	Length	Intensity	Total magnitude
1	Distributed	d	p	pd
2	Distributed	2d	p	2pd
3	Distributed	3d	p	3pd
4	Concentrated	--	pd	--

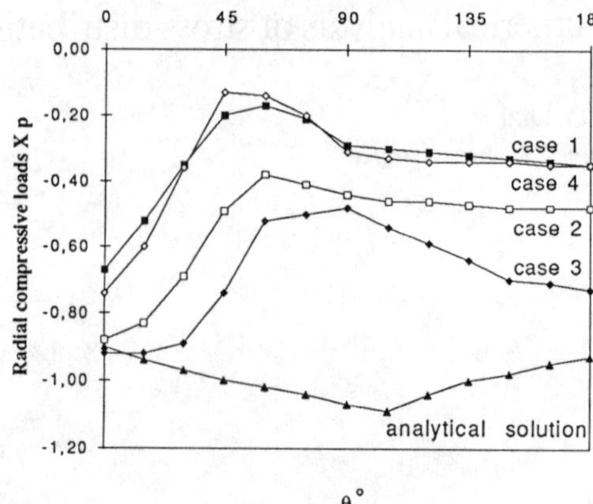

Fig 3: Radial compressive loads distribution

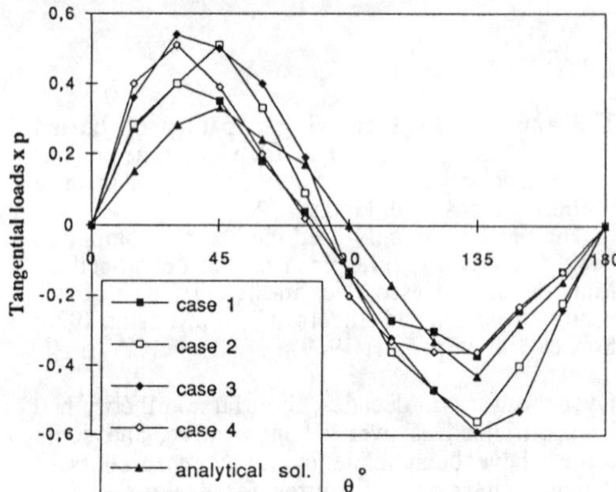

Fig. 4: Tangential loads distribution

In case 1-3, the length (b) of the symmetrical,uniformly distributed load (p) is the only variable with values ranging from d to 3d. A concentrated load of magnitude pd is applied symmetrically at the surface in case 4 (figure 2).

4 RESULTS

The magnitudes of the radial compressive and tangential loads on the pipe due to various lengths of distributed load (cases 1 to 3) and to a concentrated load (case 4) are plotted in a nondimensional form in figures 3 and 4. We observe that the radial loads applied on the pipe increase with the increasing of the load length, though this increase tends to slow down as the load length becomes equal to 3d.

Concerning the tangential loading (figure 4), there is no substantial increase in the tangential loads from loading case 1 to case 2 and from case 2 to case 3: The load distribution remains the same.

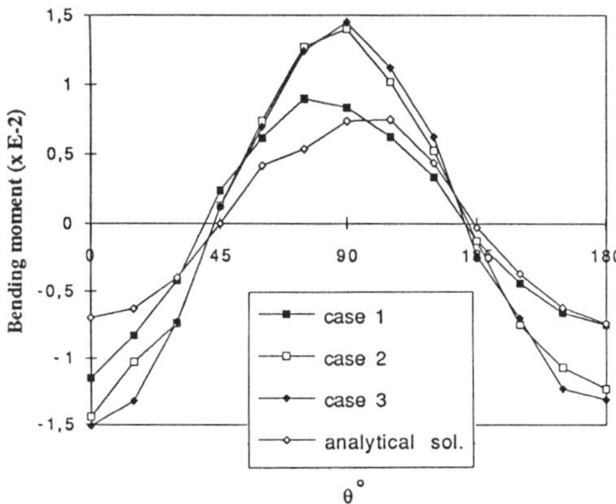

Fig 5: Distribution of bending moments

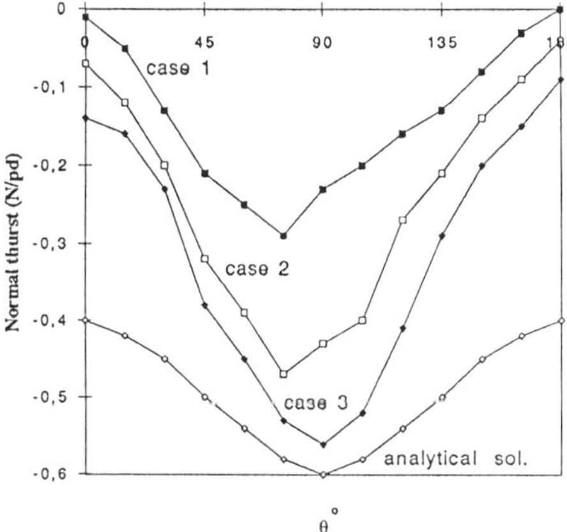

Fig. 6: Distribution of normal thrusts

Loads are also computed for our problem with the theory developed by Burns and Richard, for an elastic circular shell buried in a elastic homogeneous half space and subject to surface over-pressure. This theory is only valid for deeply buried pipes (Diab 1992). As in the Finite Element analysis, no slippage is considered between the pipe and the surrounding soil and the resulting load distribution is showed on figures (3) and (4). A comparison between case 3 and the analytical solution shows that the boundary effects, due to shallow depth of burial (d) for the pipe and a thickness of only (d) for the soil below the pipe, can not be ignored. The proposed analytical solution proposed by Burns and Richard gives a rather uniform normal load distribution around the pipe and a tangential load distribution with a maximum value of 0.35p. The values calculated by the Finite Element method range between 0.5p and 0.9p for the normal loading intensity and 0.55p for the tangential loads.

The bending moment and normal thrust in a pipe due to cases 1-3 are shown in figures 5 and 6. We observe that moment and thrust both increase as the load length is increased from (d) to (2d). An increase in the load length from (2d) to (3d) does not change the moment but increase the thrust. A comparison of case 3 and the analytical solution demonstrates that the moments given by the latter are smaller and thrusts are larger than those calculated in case (3). This confirm the results obtained previously: more radial loads and lower tangential loads predicted by the analytical solution.

Concerning the concentrated load (case 4), the resulting distribution of thrust and moment around the pipe due to elastic analysis do not show significant difference comparing to case (1). This case of loading demonstrates the inadequacy of the elastic hypothesis concerning the soil behavior. Naturally, there is no justification in the assumption that the soil below the point loading behaves elastically and in our opinion, a concentrated load undergoes a plastic flow in the soil below the load.

5 CONCLUSION

The results obtained from the analytical solution and the Finite Element method leads to the following conclusions:

The loads predicted on a pipe by the analytical solution of Burns and Richard, for no slippage at the pipe- soil interface, are not realistic.
The existence of free surface at a distance of one diameter above the pipe has a significant influence on the load distribution around a pipe which is not reflected by the analytical analyses. The loads calculated by the Finite Element analysis show that the bending moment which must be carried by the pipe is higher than the moment predicted by theoretical solution. On the other hand, the normal thrust is much higher in the analytical solution. A much more severe design condition occurs in a pipe due to high bending moment than to large normal thrusts.

The assumption concerning the elastic law behavior of the soil does not give satisfactory results in the case of concentrated loads (soil above the loading point). But for a rigid pipe with a depth of cover (d) or more, the load distribution around it, due a concentrated load acting symmetrically at the surface, is approximately the same as that obtained by assuming the total load to be uniformly distributed over a length of (d). In other words, structures loads transferred at a distance (d) or more above a pipe may be assumed to act as distributed over a length (d).

Although considerable insight into the problem of load distribution around shallow buried rigid pipe has been obtained in this study. The analyses of the parameters controling the pipe response require more information concerning:

a- The elastoplastic behavior of the pipe's material, The plasticity and the fracture criteria under a multi-axial configuration of loading.

b- The mechanical behavior of the soil, and the relation between the physical properties of the soil and its mechanical law behavior, for this reason specific laws have been developed and introduced in our Finite Element model to improve the modeling of the soil surrounding the pipe.

c- The quality of the interface between the pipe and the surrounding soil (perfect contact, friction, slippage). A special element is introduced in our Finite Element model, and the analysis of the importance of a good modeling of the interface will be presented in a next communication.

REFERENCES

A.T.V. 1984. Specifications for the structural design of wastewater drains and sewers, worksheet A127, *Abwassetechnischen Vereinigung in junction with the Verband Komunular Stadterinigungsbetriebe*, Hamburg: 182 pp.

Burns, J.Q.; Richard, R.M. 1964. Attenuation of stresses for buried cylinders. *Proceedings of the symposium soil-structure interaction*,.Tuscon, Arizona: 505-536.

C.C.T.G. Annexe 4 1991. Ouvrages d'assainissement, Fascicule 70 modifié, Paris: 184 pp.

Diab, Y. 1992. Comportement des conduites rigides enterrées.*Thèse de Doctorat, Université Claude Bernard.* Lyon: 400pp.

Gerbault, M. 1988. Calcul des canalisations circulaires semi-rigides. *Annales de l'I.T.B.T.P.* 439: 70-95.

Heger, F.J. 1988. Earth load and pressure distribution and new installation criteria for concrete buried pipes. *Proceedings of the conference Pipeline Infrastructure*,A.S.C.E, Boston, Massachusetts: 117-136..

Marston, A.; Anderson, A.O. 1913. The theory of loads and pipes in ditches and tests of cement and clay drain tile and sewer pipe. *Iowa State college Bulletin 31*: 104 pp.

Marston, A. 1930. The theory of external loads on closed conduits. *Records of the Highway Research Board 9*: 138-170.

Parmelee, R.A. 1976. A study of soil- structure interaction of buried concrete pipe. *Proceedings of the symposium : Concrete pipe and the soil-structure system.*A.S.T.M., Chicago,Illinois: 66-76.

Schlick, W.J. 1931. Structural design of rigid pipe sewers and drains. *Iowa Engineering Society 6*: 56-73.

Spangler, M.G. 1969. Factors of safety in the design of buried pipelines. *Highway Research Record 269*: 9-16

Watkins, R.K. 1966. Structural design of buried circular conduits. *Highway Research Record 145* :1-16.

Young, O.C.; Trott, J.J. 1984. Buried rigid pipes, structural design of pipelines. London,*Elsevier*: 227pp.

Zienkiewicz,O.C. 1971. The Finite Element in engineering science. London, *McGraw Hill*: 520pp.

No Trenches in Town, Henry & Mermet (eds) © 1992 Balkema, Rotterdam. ISBN 90 5410 085 0

Frictional resistance of jacked concrete pipes at full scale

Paul Norris & G.W.E. Milligan
Department of Engineering Science, University of Oxford, UK

ABSTRACT: For the past 6 years a programme of research on the pipe jacking technique has been in progress at Oxford University to clarify how pipes interact with each other and the ground. This paper is concerned with pipe-soil interface data obtained from the incorporation of an instrumented pipe on five construction sites. The results show that large localised radial and frictional stresses can be generated during jacking, which appear to be a function of pipeline misalignment.

1 INTRODUCTION

An important consideration for pipe jacking is the amount of friction generated when the pipe is pushed into the ground. This friction contributes to the jacking resistance and is a major factor in determining the required capacity of the main thrust rams and whether intermediate jacking stations will be required. The magnitude of the pipe friction depends on the pipe size and material, type of soil, its moisture content and grading, depth of cover and the details of the construction equipment and procedures employed. Factors such as the amount of overcut by the shield, misalignment of the pipes, excavation methods, duration of work stoppages, and whether or not a bentonite injection system is used will also affect the amount of friction.

This paper presents some of the results obtained from instrumentation and monitoring of pipejack tunnels under construction on five sites. The instrumentation system is shown in Figure 1. The instruments of particular significance to the paper include:-

1. Contact stress cells which measure both radial and shear interface stresses between the pipe and the ground;
2. Pore pressure probes, close to the stress cells;
3. Jack load cells attached to the rams in the jacking pit;
4. A displacement transducer to measure the forward movement of the pipe string.

Details of the five schemes are given in Table 1, where it will be seen that the instrumented pipe was located close behind the shield in the first scheme, and positioned progressively further back in the pipe string in subsequent schemes. Data has been selected from schemes 1, 3 and 4 to investigate global and local friction values in the various soil types and the effects of construction factors on their magnitude.

Ground conditions at these three sites are summarised in Table 1.

2 TOTAL JACKING RESISTANCE

The provision of sufficient jacking capacity is largely based on previous experience from schemes in similar ground conditions. The total jacking load depends upon both the force required to push the shield into the excavation, referred to as face resistance and the frictional resistance along the pipe length, Table 2. The face resistance varies between 6-9% of the total jacking load for the two cohesive drives. The high face resistance for scheme 2 is a result of the shield trimming mudstone in the invert. Scheme 4 provided an opportunity to deliberately

Table 1. Details of the instrumented schemes

	1	2	3	4	5
Pipe i.d (mm)	1200	1350	1800	1500	1200
Cover (m)	1.5-1.7	7-11	11-21	5-7	4-7
Length (m)	65	110	80	160	350
Test Pipe	No.3	No.10	No.15	No.15	85m
Lubrication	No	No	No	No/Yes	Yes
Excavation	Hand	Hand	Hand	Hand	Slurry TBM
Ground Type	Stiff Glacial Clay	Weathered Mudstone	London Clay	Dense Silty Sand	Loose Sand and Gravel
Particle Size Dist. (clay, silt, sand)	Not determined		36, 62, 2	15, 55, 30 Laminated sandy silt 5, 10, 85 Silty sand	
Soil density (Mg/m^3)	2.1-2.3	Mudstone properties not available	2.0-2.1	1.7-1.9	Scheme not considered in this paper.
Index Tests w PL % LL	12-15 17 22-29		24-29 23-30 67-80	17 Analysis NP for 22 LSS	
Triaxial s_u (kPa)	150-300		225-600	-	
Insitu s_u (kPa)	130-260		Too Stiff	N/A	
Consolidated undrained ϕ' deg	5 33		50 31	-	

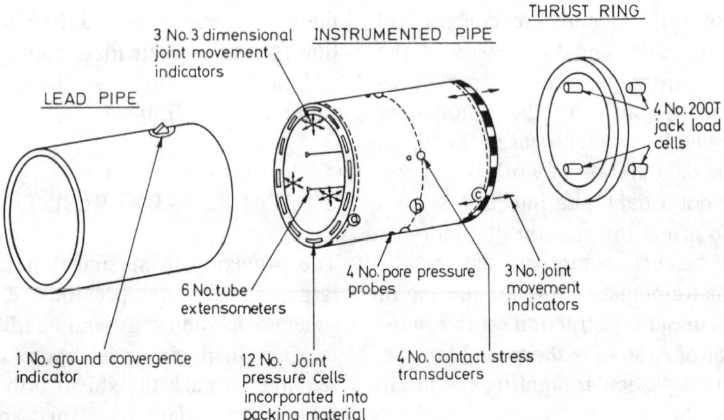

Fig.1. Schematic of instruments

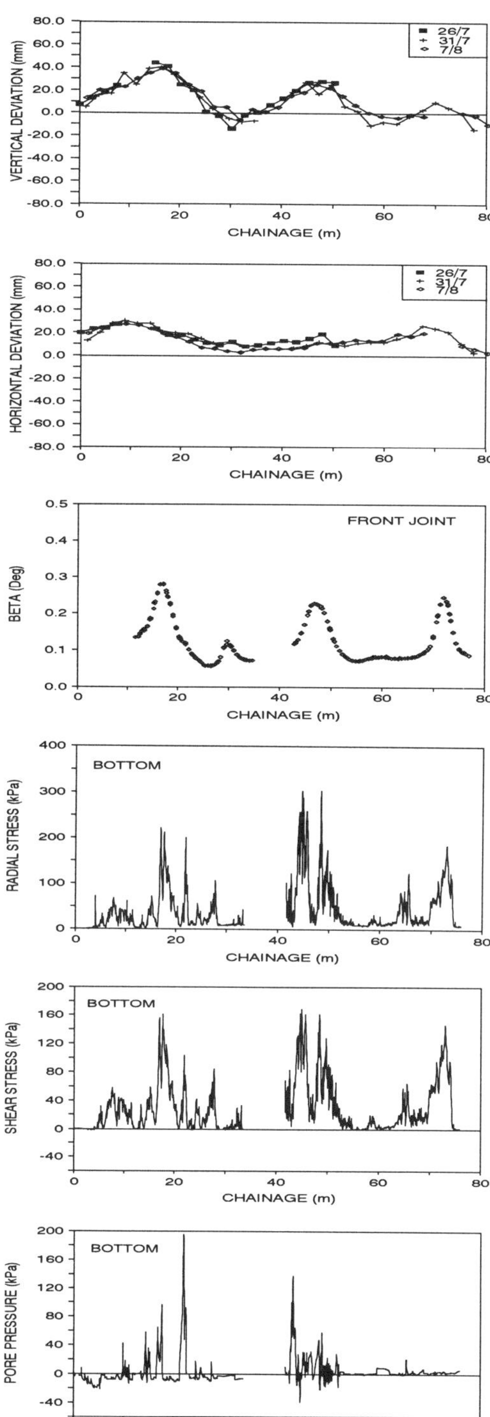

Fig.2 Local contact stresses from scheme 4.

monitor the effects of changes in shield trimming and hence face resistance. The large variation of 3% to 25% is a result of changing the extent of excavation from outside the shield to 20mm or so inside with extra trimming in the invert. The frictional resistances are generally at the lower end of the Craig limits in Table 2, which is probably a function of the competent nature of the ground and the well controlled alignment of the drives.

3 LOCAL INTERFACE STRESSES

Local contact stresses have been recorded using contact stress cells and pore pressure probes positioned at crown, invert and pipe axis. The major contact stresses were mobilised at the bottom of the pipes, Table 3, apart from scheme 3 through heavily overconsolidated London clay in which lateral ground stresses up to 650kPa were sufficient to damage the instruments on the tunnel axis. A typical set of responses along the bottom interface of scheme 4 are presented in Figure 2. The plots clearly show that peak values of radial, frictional and pore water pressures are obtained over short lengths of the drive with a lower overall average value for the total length. The local peak skin friction values are nearly two orders of magnitude larger than the average friction values in Table 2. The radial stresses exhibit peak values up to 300kPa over short lengths of 1-2m.

4. GROUND RELATED FACTORS

4.1 Soil type

To establish whether there is a fundamental difference in interface behaviour between cohesive and non cohesive soils plots of shear stress against total and effective radial stress have been produced for schemes 1 and 4. The data from the bottom probes of both schemes are presented in Figure 3. A number of salient features are evident. Best line fits to the plots indicate only small differences between the total and effective stress behaviour. This suggests that a partially drained state exists at the pipe-soil interface with

Table 2. Average face resistance and pipeline friction values

Scheme		Measured face resistance		Measured average friction		Craig (1983) limits (kPa)
		kN	% Total	(kN/m)	(kPa)	
1	Dry Wet	120	9	7.2 29.8	1.5 6.2	5-18
2		950	43*	18.0	1.5	2-3
3		300	6-8	54.4	7.6	5-20
4	unlub lub	100-800	3-25	23.1 9.4	4.2 1.7	5-20

* Value based on 40m monitored length. Total drive length 100m

Table 3. Maximum local interface stresses

Scheme	Full overburden (γh)	Radial (kPa)			Shear (kPa)			Pore Pressure (kPa)		
	(kPa)	Top	axis	bottom	Top	axis	bottom	Top	axis	bottom
1	37	5	10	550	5	10	250	-5	NI	400
2	242	NI	NI	550	NI	NI	160	NI	NI	700
3	300	450	650	IF	150	150	IF	IF	250	250
4	126	100	100	300	60	60	60	10	10	190

NI - No instrument IF - Instrument failure

Table 4. Predicted and measured coefficents of skin friction

Scheme	Internal angle of friction ϕ' (deg)			Skin friction coeff.	Predicted δ' (deg)			Field
	Peak	Critical State	Residual		Peak	Critical state	Residual	δ (deg)
1	33av	30 Bolton (1979)	25.3 (Lupini et al 1981)	0.68 (Potyondy 1961)	22.4	20.4	17.2	19
4	47 Bolton (1979)	32 Bolton (1979)	-	0.87	40.9	27.8	-	37.7

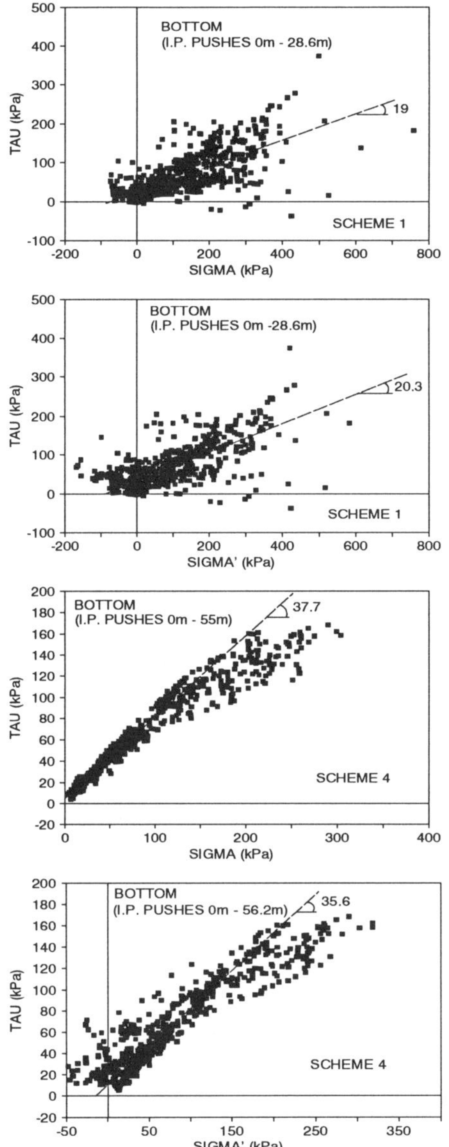

Fig.3 Shear stress/radial stress relationships.

the concrete pipe acting as a local drainage path. Both materials exhibit frictional responses with coefficients of skin friction between the pipe and soil of 19° for the glacial clay and 37.7° for the dense silty sand. These values fall within the range of typical peak, critical and residual friction angles in Table 4.

Given the degree of scatter of the field results, there is some evidence to suggest that the repeated passage of pipes subjects the clay to large shear displacements which reduces the friction angle below the critical state value while the silty sand material illustrates a post peak response which has not reached the critical state.

During the later stages of scheme 1 the moisture content of the soil was increased following a period of heavy rain and temporary inundation of the tunnel due to a fractured water main. As a result, the measured angle of skin friction was reduced to 13.5°; this is consistent with values of skin friction between smooth concrete and cohesive granular soils at varying moisture content measured in laboratory tests by Potyondy (1961). However the overall resistance to jacking increased considerably (see Table 2) due to larger radial effective stresses as the soil swelled, closed the overbreak and made contact with the pipe around its full perimeter.

4.2 Pipe self weight friction

Comparison of the pipeline self weight friction with measured average frictional resistance is presented in Table 5.
Reasonable agreement is obtained for schemes 1, 2 and 4 which are through relatively stable bores. The larger recorded values are possibly a function of limited ground closure onto the pipe and the effects of pipeline misalignment. A greater understanding of the imposed radial loading from the highly plastic overconsolidated clay is required to obtain closer agreement for scheme 3.

5. CONSTRUCTION RELATED FACTORS

5.1 Misalignment

Comparison of tunnel alignment data and local interface stress profiles, Figure 2, illustrates reasonable agreement between the positions of peak stress values and maximum angular deviation (β). The relationship is reproduced in Figure 4 as a scatter plot. Although there is considerable variation in the relationship, larger stresses generally occur at larger angular deviations. The scatter is probably a function of the precise profile of the ground in relation to the locus of pipe movement, the necessary ground

Table 5 Pipe self weight friction

Scheme	Field skin friction δ	W tan δ (kN/m)	Av friction (kN/m)
1	19°	6.1	7.2 wet 29.8 dry
2	16°	6.6	8
3	14°	8.8	54.4
4	37.7°	18.7	23.1 unlub 9.4 lub

support varying along the pipe length.

5.2 Time factor

Increases in pipejacking forces in clay soils following a lapse in jacking are routinely recorded. The effect was evident in the jacking records for the low plasticity drive of scheme 1 although a more dramatic effect was observed in the high plasticity London Clay of scheme 3 which forms the basis for discussion. The jacking record for a short section of the drive and the associated total radial stresses acting on the pipe at the start and end of each stoppage are presented in Figure 5. The rapid and repeatable increase in restart jacking forces is a pronounced feature. It has previously been suggested that the time factor allows consolidation of the ground to take place around the pipe increasing the radial pressure and hence the related frictional resistance. The total radial stress plot however indicates reductions during the rest periods which are not a function of changes in jack ram loading during the stoppage. Unfortunately there is no reliable pore pressure data from the drive to establish whether the reduction is a result of large pore pressure dissipation.

The occurrence of large restart forces after short pauses have been observed during laboratory controlled shear interface tests between steel and London clay, Tika (1989).
The tests were conducted under constant applied normal stresses and subject to fast rates of shearing of 110mm/min and 1100mm/min. The responses are included as Figure 6 and highlight a peak resistance which drops rapidly to a minimum value with fast displacement; the magnitude of frictional resistance increasing with increasing rate of shear. Typical jacking rates fall between these two values, scheme 3 was 335mm/min (2 rams) and 185mm/min (4 rams). Tika suggests that the fast rate of shearing may

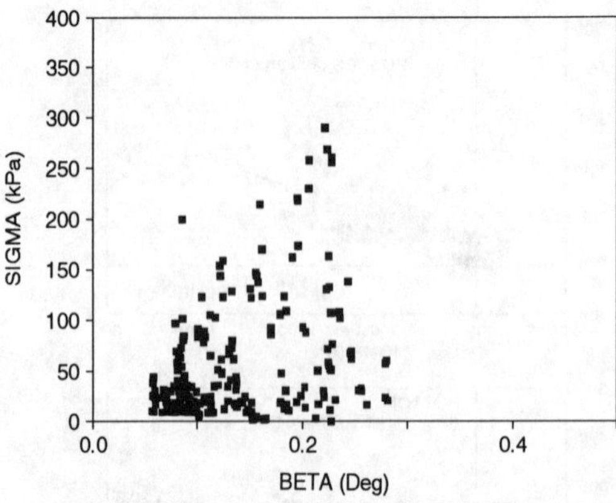

Fig.4 Relationship between local radial stress and Beta; Scheme 4.

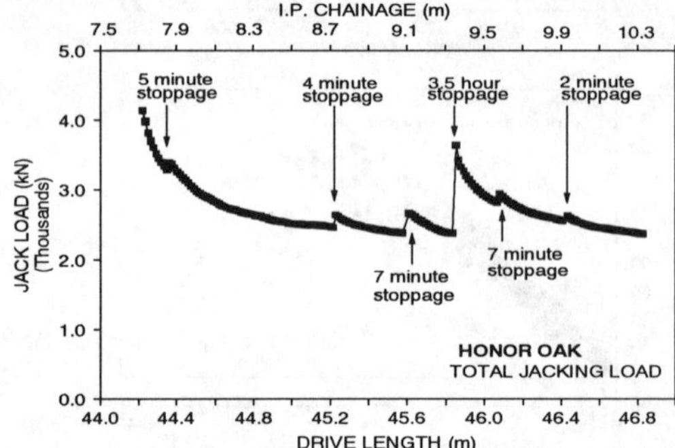

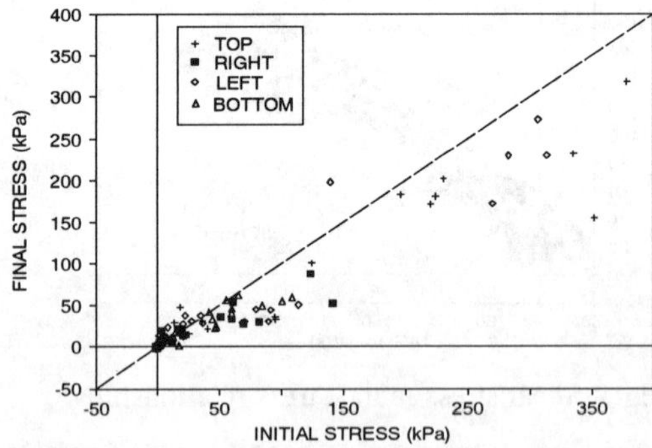

Fig.5 Time dependent changes in jacking load and local interface stresses.

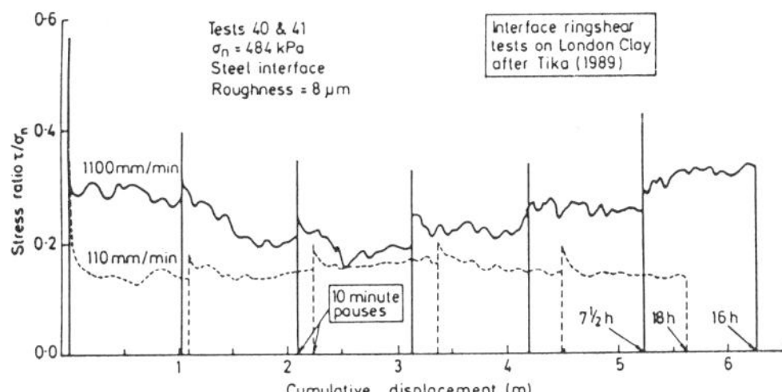

Fig.6 Interface ring shear tests on London clay.

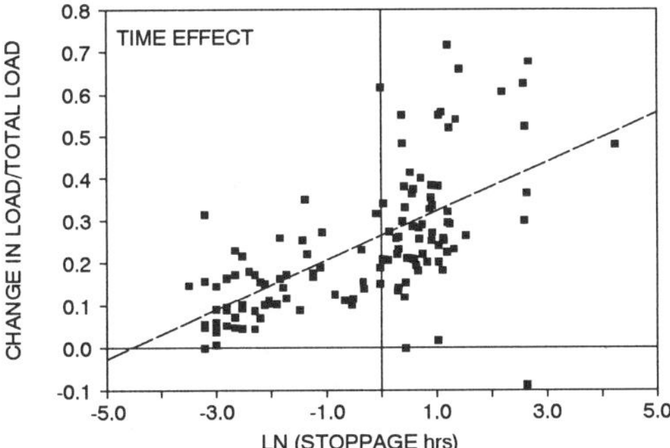

Fig.7 Time factor for scheme 3.

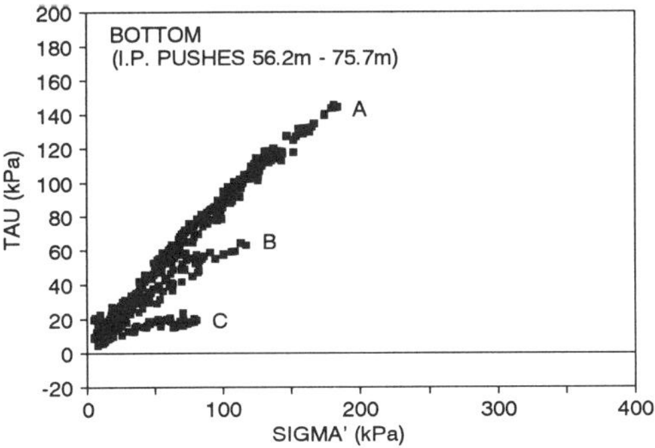

Fig.8 Effective skin friction during lubrication of scheme 4.

change the basic shearing mechanism of the soil from pure sliding to sliding with some turbulence of the platy clay particles combined with a possible transient (viscous) effect. It will be noted that although the global response of the pipejack of scheme 3 seems to match the laboratory trends, the shear mechanism may be further complicated by both the radial and shear stresses varying during pushes.

The data from all the stoppages have been compiled in Figure 7. Although there is some variation which is probably a function of changing face resistance, the general trend fits the relationship:

$$\frac{\delta P}{P} = 0.259 + 0.058\ln(t)$$

and demonstrates that a major part of the increase occurs within minutes of stopping.

5.3 Lubrication

Use of lubricants was avoided in the first three drives, introduced during the later stages of scheme 4 and used throughout the machine drive of scheme 5. The effectiveness of lubrication is dependent upon many factors, including minor changes in ground conditions and correct selection of equipment and procedures for the injection system. Figure 8 provides some indication of the local variability of its effectiveness. The data is taken from scheme 4 over a section of tunnel through silty sand which varied in silt content from 70% down to 15%. At the time of monitoring it was unlikely that the full annulus around the pipe was filled with lubricant. Line B and C correspond to areas with a high silt content while line A corresponds to the coarser grained material. The apparent coefficient of skin friction given by line C is one third of that given by line A (which is similar to the unlubricated value). Closer scrutiny of the site log of activities reveals that the difference in B and C is probably because injection had taken place overnight in the area of the instrumented pipe prior to the pushes along line C whereas B was recorded later in the shift when lubrication was being concentrated in a different section of the drive. These observations suggest that an insufficient quantity of lubricant was introduced into the drive causing localisation of its effectiveness. Inspection of the ground surrounding the instrumented pipe upon removal of the contact stress cells confirmed that the annulus was not completely filled although a layer of soil-lubricant mixture, typically 10mm thick, had formed adjacent to the pipe over the bottom half of the pipe

In contrast data from scheme 5 suggests that the lubricant around the pipe annulus was pressurised, supporting the surrounding ground and causing pipeline buoyancy and low interface frictional stresses.

6. CONCLUSIONS

The field data clearly demonstrates that large localised radial, shear and pore pressures can be generated during jacking. Peak values typically occur over short 1m to 2m lengths and generally correspond to positions of maximum angular misalignment. Critical pressure distributions, which are highly non symmetrical around the pipe circumference, interact with non-uniform pipe end load transfer, Norris & Milligan (1992), requiring consideration in the pipe design process.

A partially drained frictional material response is exhibited at the pipe interface in both cohesive and non-cohesive ground. There is limited evidence that the material within the rupture zone approaches a critical state condition. Use of peak friction angles from drained shear box tests should provide conservative estimates of the appropriate coefficients of friction. The main area of uncertainty remains the prediction of imposed radial loading during jacking. Further field measurements are required to establish the suitability of using Marston trench loading conditions which are the basis of UK design.

If stoppages occur in cohesive ground then a greater force is needed to restart the process than would have been required to maintain continuity. The "time factor" effect is pronounced in the high plasticity clay and may be a function of jacking speed rather than increases in the radial ground pressure during the stoppage.

Lubrication is an effective method of reducing frictional resistance but care is needed in matching lubricant properties to existing ground conditions, providing sufficient pumping capacity and adopting suitable injection sequences.

ACKNOWLEDGEMENTS

This research has been made possible by the generous financial support of the Science and Engineering Research Council, the Pipe Jacking Association, and five water service companies: Northumbrian, North West, Severn Trent, Thames and Yorkshire.

REFERENCES

Bolton, M. 1979. A guide to soil mechanics. London : MacMillan.

Craig, R.N. 1983. Pipejacking. A state of the art review. CIRIA. Technical Note 112. London.

Lupini, J.F., Skinner, A.E. and Vaughan, P.R. 1981. The drained residual strength of cohesive soils. Geotechnique 31: 181-213.

Norris, P. and Milligan, G.W.E. 1992. Pipe end load transfer mechanisms during pipe jacking. Proc. 8th Int. Conf. on Trench less Construction, No Dig 92, Washington.

Potyondy, J.G. 1961. Skin friction between various soils and construction materials. Geotechnique 11: 339-353.

Tika, T.M. 1989. The effect of rate of shear on the residual strength of soil. PhD Thesis, Univ. of London (Imperial College).

No Trenches in Town, Henry & Mermet (eds) © 1992 Balkema, Rotterdam. ISBN 90 5410 085 0

Résistance de frottement des canalisations en bétons poussées

Paul Norris & G.W.E. Milligan
Department of Engineering Science, University of Oxford, Royaume-Uni

RÉSUMÉ: Au cours des 6 ans passés un programme de recherche s'avancait à l'Université d'Oxford au sujet de la téchnique de soulevation des canalisations pour clarifier l'interaction des canalisations entre eux et entre les pipes et le sol. Cet article s'interesse à l'information de l'interaction entre les pipes et le sol obtenu de l'incorporation d'un canalisation instrumenté à cinq chantiers de construction. Les resultats montrent que des grandes efforts radiales et de frottement localisés peuvent être produits pendant le soulèvement, ceux qui paraissent être une fonction du défaut d'alignement des conduites.

1 INTRODUCTION

Une considération importante pour le soulèvement des pipes au cric est le quantité de frission géneré quand on pousse le pipe à travers le sol. La frisson contribue à la résistance de soulevation et c'est un facteur majeur pour décider la capacité necessaire des premiers pilons de pousse et si les points de levage intermédiare seront necessaire.La grandeur de la frisson des tuyaus se dépend au grosseur des tuyaus et leur matériaux, le genre de sol,son état hygrométrique et calibration, le profondeur de la couverture et les détails de l'équipement de construction et les procédures employées. Les facteurs comme la quantité de déraillage par le bouclier, le mésalignement des tuyaus, les méthodes d'excavation,la durée des interruptions, et si ou non on utilise une système d'injection de bentonite affecteront tous l'extente de la frisson.

Cet article présente quelque des résultats obtenus par l'instrumentation et le contrôle des tunnels de soulevation des conduits sous construction en cinq endroits. La système d'instrumentation se montre en Figure 1. Les instruments particulièrement significants au sujet de cet article s'inclus:

1. Les cellules de tensions contactes qui mesurent les tensions radiales même que les césaillages interfaces entre le tuyau et le sol;
2. Les sondes de poussée des pores, près des cellules de tension;
3. Les cellules de charges au cric attachés aux pilons aux fosses au soulèvement;
4. Une transducteur déplacement pour mésurer le mouvement en avant du cordon de conduits.

Les détails de ces cinq schémas sont montrées au Table 1, où on verra que le conduit instrumenté se situait près derrière du bouclier au premier schèma, et positioné progressivement plus loin en arrière en le cordon de conduits en les schèmas suivants. On a pris les renseignements des schèmas 1, 3 et 4 pour investiger les valeurs globale et locales de frisson en les genres de sol et les effets des facteurs de constructions sur leur magnitude.

Les conditions de terre à ces trois endroits sont summarise en Table 1.

2. RÉSISTANCE TOTALE POUR LE SOULÈVEMENT AU CRIN

La provision d'une capacité aussi grand pour le soulèvement en crin est basée largement a l'experience antérieure des schémas au conditions de sol semblable. Le charge de soulèvement totale dépends au force necessaire pour pousser le bouclier dans l'excavation, referré comme résistance de face et la résistance de frission le long du tuyau, Table 2. La résistance de face varies entre 6-9% de charges au cric pour les deux entraînement de cohesion. La résistance de face en haut pour la schema 2 est un résultat de l'ecavement du bouclier par le schiste argileux

Table 1. Details des projets à instrument

	1	2	3	4	5
Pipe i.d (mm)	1200	1350	1800	1500	1200
Coavercle	1.5-1.7	7-11	11-21	5-7	4-7
Longueur	65	110	80	160	350
Pipe d'Essai	No.3	No.10	No.15	No.15	85m
Lubrification	Non	Non	Non	Non/Oui	Oui
Excavaction	A Main	A Main	A Main	A Main	Boue de rivière TBM
Type à sol	Argile rigide glaciale	Pierre de Boue Climatisée	Argile de Londres	Sable à dépôt vaseux	Sable et gravier libres
Grandeur de Particule Dist (Argile, Dépôt Vaseux, Sable)	Non déterminé		36, 62, 2	15, 55, 30 Dépôt Vaseux à sable Laminé 5, 10, 85 Sable Laminé	
Densite du Sol (Mg/m^3)	2.1-2.3	Proprietes de Perre de Boue ne sont pas disponibles	2.0-2.1	1.7-1.9	Projet non-consideré dans cetle feuille
Essais indexes W PL % LL	12-15 17 22-29		24-29 25-30 67-80	17 Analysis NP for 22 LSS	
Triaxial S_u (kPa)	150-300		225-600	-	
Insitu S_u (kPa)	130-260		Trop rigide	N/A	
Consolidé nonépuisé deg	5 33		50 31	-	

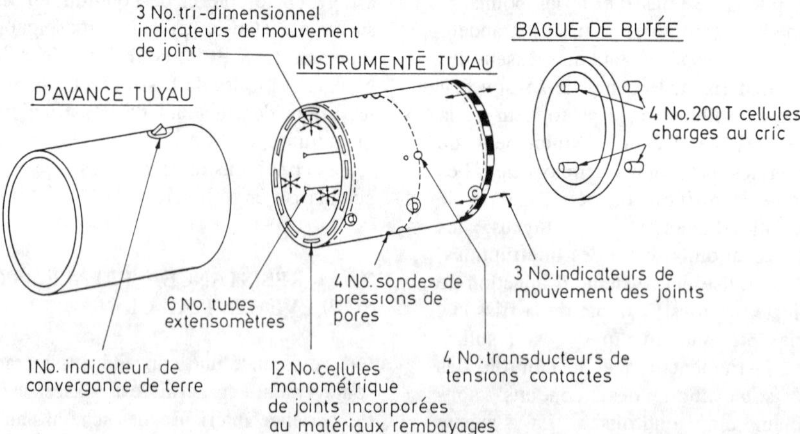

Fig 1. Schéma des instruments.

dans l'inverte. Le schema 4 a donné l'occasion pour monitoire les effects des changes a l'enclavement du bouclier et puis la résistance en face. La grande variation de 3% à 25% est le résultat de la changement de l'étendue de l'excavation de ydehors du bouclier à 20mm ou environs en dedans avec l'enclavement de plus dans l'inverte. Les résistances frissonnelles sont generallement au limites inférieures Craig en Table 2, qui est probablement une fonctionne de la nature competente de la terre et l'alignement bien controllées des transmissions.

3. TENSIONS DE L'INTERFACE LOCALE

On a enregistré les tensions contactes locales utilisant des cellules de tensions de contactes et des probes de pressions de pores situés à la couronne, l'inverte et l'axis des tuyaux. Les contactes tensions majeures étaient mobilisés en bas des conduits, Table 3, au dehors de Schema 3 par l'argile de Londres bien surcompactée dans qui les tensions lateralles de terre jusqu'à 650 kPa étaient assez pour faire dommage aux instruments sur l'axe tunnel. Une série de réponses typique le long de l'interface d'en bas de schema 4 se présente en Figure 2. Les lignes montrent clairement que les valeurs de crête des pressions radiaux, frissionelles et de l'eau interstitielle sont obtenues sur des longueurs courts du transmission avec un valeur moitié moins surtout pour le longueur total. Les valeurs pour le frottement superficiel maximum locale sont presque deux ordres de magnitude plus grands que les valeurs de frissons moyennes en Table 2. Les efforts radiales montrent les valeurs maximums jusqu'à 300kPa sur les durées courtes de 1-2m.

4. LES FACTEURS RELATIVES AU SOL

4.1. Les types de sol

Pour établir s'il y a une différence entre l'action d'interface entre les sols cohésives et non-cohésives, on a produits les plans des cisaillages contre l'effort radial total et effective pour les schemas 1 et 4. Les renseignements des sondes inférieures des deux schemas se présentent en Figure 3. Quelques characteristiques saillents sont évidents. Les mieux accords de lignes aux levés indiquent seulement des petits différences entre les fonctionnements de tensions totales et effectives. Cela suggère qu'un état partiellement drainé existe à l'interface conduit-sol où le canalisation en béton sert d'une route locale de drainage. Tous les deux matériaux exhibitent les réponses frisonnelles avec les coéfficients de frottements superficiels entre le conduit et le sol de 19° pour l'argile glaciaire et 37.7° pour le sable dense limoneux. Ces valeurs tombent dans le rang d'angles maximums, critiques et residuel typiques en Table 4.

Donné le degrès de dispersion des résultats de champs, il est évident de suggérer que le passage

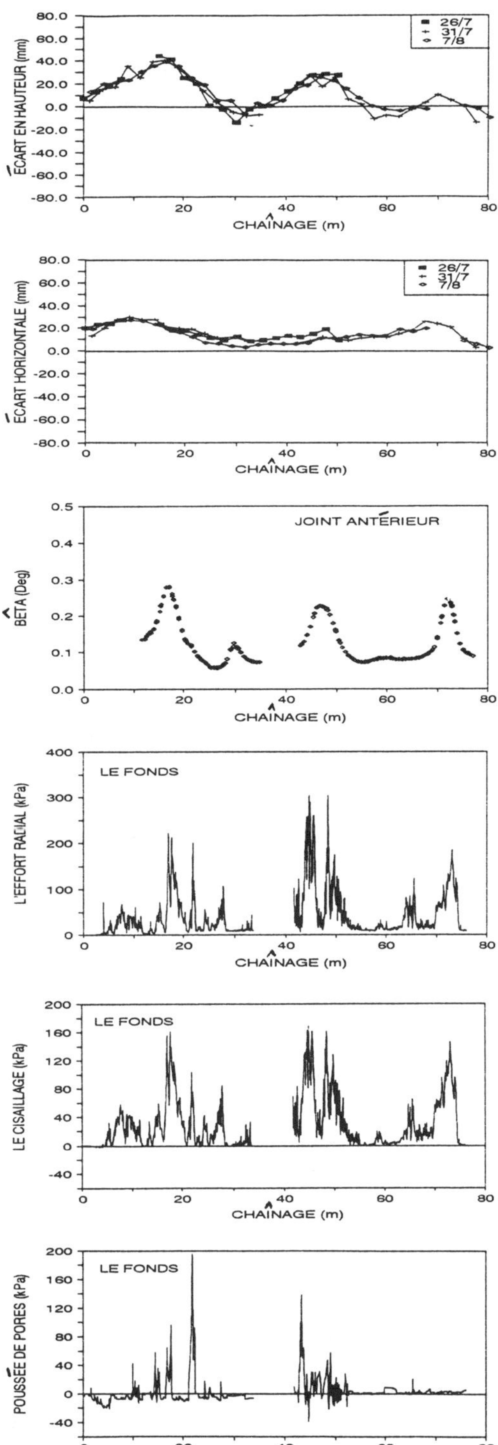

Fig.2 Efforts de contactes locales de schéma 4.

Table 2. Résistance de face moyenne et valeurs de frottement de la Tyauterie

Projet		Résistance de face mesurée		Frotlement moyen mesurée		Limites Craig (1983) (kPa)
		kN	% Totale	(kN/m)	(kPa)	
1	Sec Moite	120	9	7.2 29.8	1.5 6.2	5-18
2		950	43*	18.0	1.5	2-3
3		300	6-8	54.4	7.6	5-20
4	Non-lubrifié Lubrifié	100-800	3-25	23.1 9.4	4.2 1.7	5-20

* Valeur basée sur longueur de 40m suivie. Longueur totale de chassage

Table 3. Efforts d'entre face locale maximums

Projet	Surcharge pleine ( )	Radiale (kPa)	Cisaillement (kPa)	Pression de pore
	(kPa)	Bout d'axe supérieur	Bout d'axe supérieur	Bout d'axe supérieur
1	37	5 10 550	5 10 250	-5 NI 400
2	242	NI NI 550	NI NI 160	NI NI 700
3	300	450 650 IF	150 150 IF	IF 250 250
4	126	100 100 300	60 60 60	10 10 190

NI - Pas d'instrument IF - De'faut d'instrument

Table 4. Coefficients de Frottement de peau prédits et mesurés

	Angle de frottement intern (deg)				Prédit (deg)			Champ
Projet	À Pic	État critique	Résiduel	Coefficient de frotlement de peau	À Pic	État critique	Résiduel	(deg)
1	33av	30 Bolton (1979)	25.3 (Lupini et al 1981)	0.68 (Potyondy 1961)	22.4	20.4	17.2	19
4	47 Bolton (1979)	32 Bolton (1979)	-	0.87	40.9	27.8	-	37.7

répété des conduits soumis l'argile à de grands displacements de cisaillements qui réduirent l'angle de frisson en desous du valeur de l'état critique pendant que le matériel de sable limoneux illustre une réponse après maximum qui n'a encore atteigni l'état critique.

Pendant les derniers étages de schéma 1 l'etat hygrométrique du sol avait aggrandi après une période de pluie forte et l'inundation temperaire du tunnel grâce à une conduite d'eau cassée. Comme résultat, l'angle mesuré de frottement superficiel s'était réduit à 13.5°;cela est consistent avec les valeurs de frottement superficiel entre le concret lisse et les sols granulés cohésives à des états hygrométriques variés mesurés en essais de laboratoire par Potyondy (1961). Néanmoins la résistance surtout à soulèvement au crin est considérablement augmentée (voir Table 2) à cause des plus grands efforts radials effectifs

Table 5. Frottement d'auto-poids de Pipe

Projet	Frottement de Peau de Champ	W tan (kN/m)	Frottement Moyen
1	19	6.1	7.2 moite 29.8 sec
2	16	6.6	8
3	14	8.8	54.4
4	37.7	18.7	23.1 non-lubrifié 9.4 lubrifié

comme le sol gonflait, fermait le surbris et faissait contacte avec le conduit autour de sa pleine périmètre.

4.2 Le frisson conduit pression

La comparison du frisson conduit pression avec la résistance frissionnelle moyenne mesurée se présente en Table 5.

L'accord raisonnable est obtenus pour les schémas 1, 2 et 4 qui sont par des forages relativement stables. Les plus grands valeurs notées sont peut-être une fonction de clôture de terrain limitée sur le conduit et les effects de mésalignement de canalisation. Unc compréhension plus grande du chargement radial imposé de la très plastique surconsolidee argile est demandée pour obtenir l'agreement plus près pour le schéma 3.

5. LES FACTUERS RELATIVES AUX CONSTRUCTION

5.1. Mésalignement

La comparison des renseignements de l'alignement tunnel et les profiles de tensions d'interface locales, Figure 2, illustrée l'agrément raisonnable entre les positions de valeurs maximum tensions et la déviation maximum angulaire (ß). Ce relation est reproduit en Figure 4 comme un levé dispersion. Malgré qu'il y a une variation considérable en le relation, des plus grands tensions se trouvent géneralleme nt à des deviations angulaires plus grandes. La dispersion est probablement une fonction de la profile précise du terrain en relation à la situation de mouvement des conduits, le soutien de terrain nécessaire varié le long du conduit.

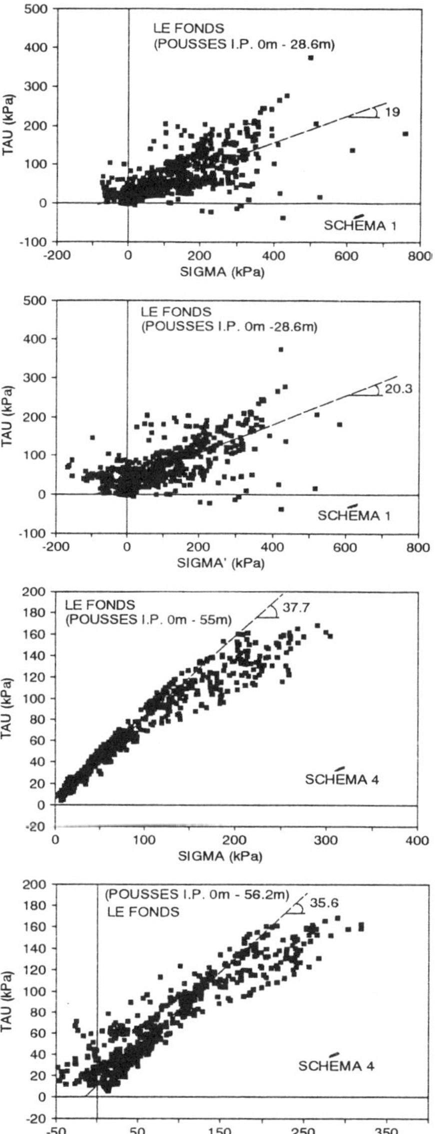

Fig.3 Relations entre le cisaillage et l'effort radial.

5.2 Coefficient de temps

Des accroissements dans les efforts du soulèvement au cric des pipes dans les glaises suite à un laps de temps dans le soulèvement au cric sont enregistrés à la routine. L'effet à été patent dans les enregistrements du soulèvement au cric pour le chassage à basses plasticité du projet 1, bien qu'on aurait eu noté dans l'argile de

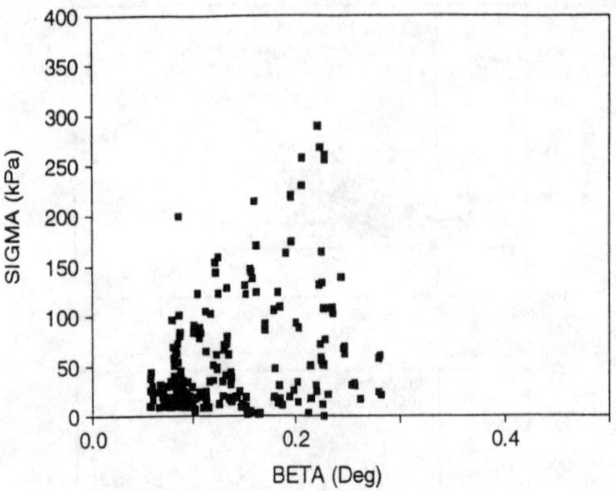

Fig.4 Relations entre l'effort radial locale et Bêta; Schéma 4.

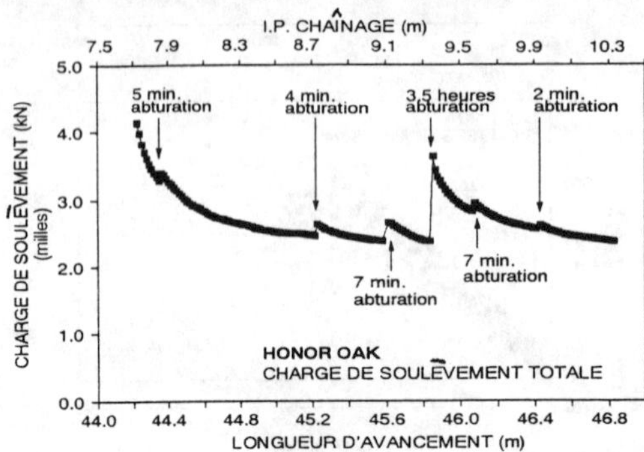

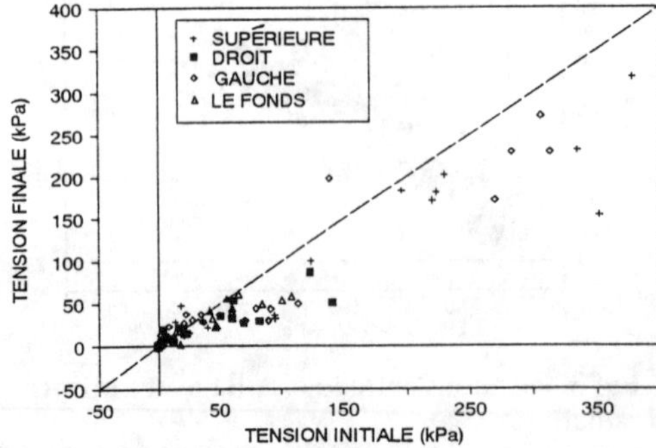

Fig.5 Changés dé pendent au durée au charge de soulèvement et au tensions localisés d'interaction.

Londres de haute plasticité du projet 3 un effet plus dramatique, ce qui forme la base pour discussions. L'enregistrement du soulévement au cric pour une section courte du chassage et les efforts radiaux totaux qui agissent sur la pipe au commencement et à la fin de chaque bouchage se trouvent montrés dans la Figure 5. L'accroissement rapide et qu'on peut répéter dans les efforts du soulévement au cric à reamorçage est un caractère bien marqué. On a suggéré dans le passé. que le facteur du temps permet à la terre de se consolider autour de la pipe en augmentant la pression radiale et par cela la résistance à frottment relative. Cependant le diagramme entier de l'effort radial indique des réductions dans les périodes de repos qui cric pendant le bouchage. Malheureusement il n'y existe aucunes données sûres de la pression de pore du chassage pour établir si la réduction est-elle un résultat de la grande dissipation de la pression de pore.

L'occurrence des grands efforts to reamorcage après des arrêts brefs a été noté au cours des essais contrôlés en laboratoire du glissement entre l'acier et l'Argile de Londres, Tika (1989).

Les essais ont été condiit sous des efforts normaux appliqués constamment et sujets aux grands taux de vitesse de cisaillement de 110 mm./min. et 1100 mm./min. Les réponses se trouvent montrées dans la Figure 6 et elles soulignent une résistance à pic qui tombe rapidement jusqui'à une valeur minimume avec déplacement vita; la grandeur de la résistance à frottment augmente comme le taux de cisaillement augmente aussi. Les taux de soulèvement au cric typiques tombent entre ces deux valeurs, dans le projet 3 - 335 mm./min. (2 plongeurs) et 185 mm/min. (4 plongeurs). Tika suggère, que le grand taux de cisaillement rapide peut faire changer le mécanisme de cisaillement basique de la terre à glissement pur jusqu'à glissement avec quelque turbulence des particules d'argile en forme de nappe jointe à un effort transitoire (visqueux) possible. On notera que, bien que la réponse globale du cric à pipes du projet 3 paraitrait apparier aux tendances du laboratoire, le mécanisme de cisaillement peut être compliqué plus loin par les effets radiaux et les effets de cisaillement tous les deux qui varient pendant les poussées.

Les données de tous les bouchages ont été receuillit dans la Figure 7. Bien qu'il aurait quelques variation ce qui, probablement, est le résultat de la résistance de face changeante, les tendances générales s'accomodent au rapport:

$$\frac{\partial P}{P} = 0.259+0.581n(t)$$

et démontre qu'une partie majeure de l'accroissement a lieu dans les minutes de bouchage.

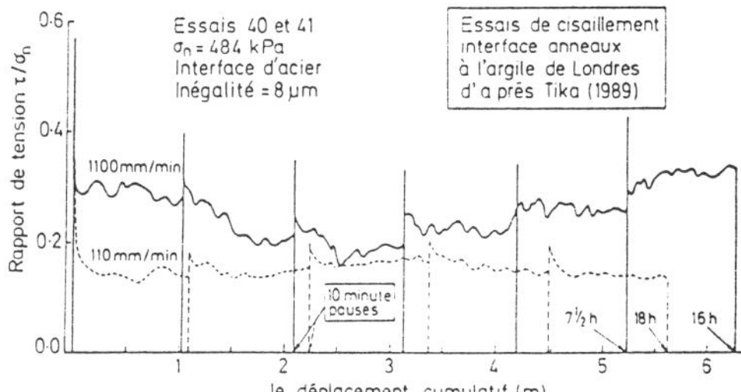

Figure 6: Essais de Glissement à Bague Entre Deux Faces.

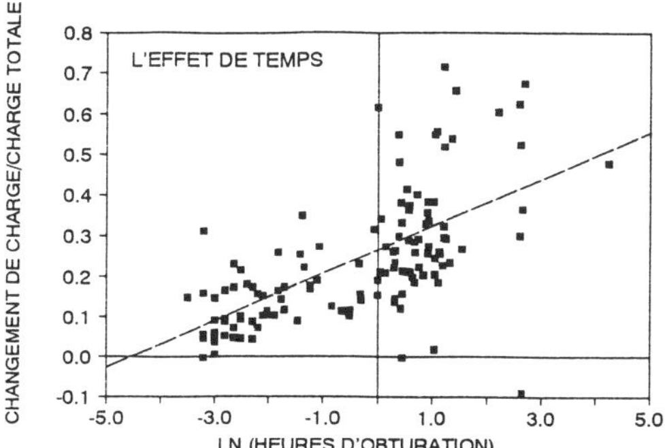

Fig.7 Coefficient de temps pour schéma 3.

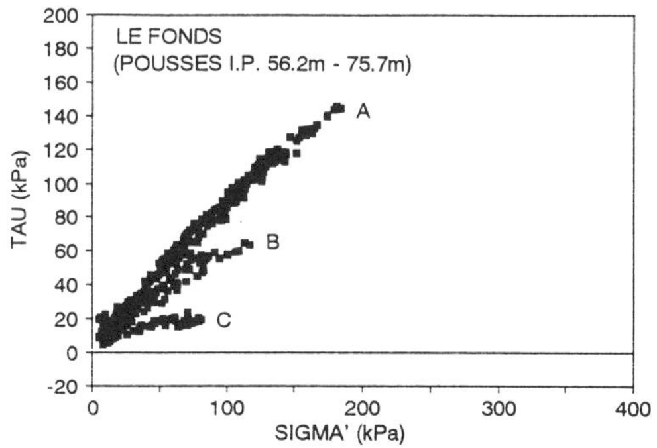

Fig.8 Le frottement superficiel effective pendant la lubrication de schéma 4.

5.3 Lubrification

Dans les premiers trois chassages on s'était empêché de se servir des lubrifiants, introduits au cours des étapes dernières du projet 4 et employés par toute la transmission de machine du projet 5. L'efficacité de la lubrification est dépendente de beaucoup de facteurs, y inclus des petits changements dans les conditions de la terre et un choix correct de l'appareillage et les procédés pour le système d'injection. La Figure 8 donne quelques détails sur la variabilité locale de son efficacité. Les données sont mises du projet 4 sur un section de tunnel à travers la sable à dépôt vaseux qui varie en teneur de dépôt vaseux de 70% jusqu'à 15%. Au moment de contrôler il était peu probable que la couronne pleine autour de la pipe était remplie de lubrifiant. La ligne B et C se conforme au materiau à grain plus grossier. Le coefficient apparent de frottement de croûte donné par la ligne C est une tierce de celle-là donné par la ligne A (ce qui est semblable à la valeur non-lubrifiée). Un examen plus minitieuse du journal du pied d'oeuvre fait connaître que la différence en B et C est causée par une injection ayant lieu par la nuit dans l'aire de la pipe à instruments à l'enlèvement des cellules d'effort de contact a confirmé que la couronne n'était pas remplie complètement bien qu'une couche de mélange à lubrifier la terre, typiquement d'une épaisseur de 10 mm., aurait se former avoisinante à la pipe sur la moitié inférièure de la pipe.

Selon les données faisant contraste du projet 5, celles-ci suggèrent que le lubrifiant autour de la couronne de la pipe a été mis sous pression en soutenant la terre entourante et en causant une poussée de la tuyauterie et des bas efforts de frottement entre les faces.

6. CONCLUSIONS

Les données de l'enquête sur les lieux démontrent avec netteté que des grandes pressions radiales, de cisaillement et de pore peuvent être générées pendant le soulèvement au cric. Les valeurs à pic ont lieu typiquement sur les longueurs courtes de 1 m. jusqu'à 2 m., et généralement elles se conforment aux positions de mauvais alignement angulaire maximum. Les distributions de pression critiques, qui sont hautement non-smétriques autour de la circon férence de la pipe, réagissent avec transfert non-uniforme de la charge de bout de la pipe, Norris et Milligan (1992), en demandant d'une considération du processus du dessin de la pipe.

Une réponse d'un materiau de frottement partialement épuisé se montre entre les deux faces de la pipe, dans une terre cohésive. Il y a de la preuve limitée, que le matériau en dedsns de la zone de rupture s'approche a une condition d'état critique. L'emploi des angles à frottement à pic des essais de boîte de cisaillement épuisé devrait fournir des évaluations prudentes des coefficients de frottement appropriés. L'aire principale d'incertitude reste la prédition de chsrge radiale imposée pendant le soulèvement au cric. Il faut des mesures additionnelles afin de pouvoir établir la convenance d'employer les conditions de chsrge de fossé Marston. qui forment la base du dessin anglais.

S'il y auraient des bouchages dans une terre cohésive, puis il faudrait une plus grande force pour réamorcer le processes qu'il aurait nécessaire pour maintenir de la continuité. L'effort "Facteur du Temps" est marqué dans la haute argile plastique, et il pourrait être une fonction de la vitesse de soulèvement au cric plutôt que des augmentations dans la pression radiale pendant le bouchage.

La lubrification est une méthode efficace pour une réduction de la résistance à frottement, mais il faut prendre tous les soins en assortissant les propriétés lubrifiantes aux conditions de terre actuelles, pourvu une possibilité de pompage suffisante et d'établissement des séquences d'injection convenables.

RECONNAISSANCES

Cette recherche a été possible par le soutien financier généreux du Conseil des Sciences et de la Recherche du Génie, l'Association du Soulèvement au cric, et cinq compagnies de services dex eaux; Northumbrian, North West, Severn Trent, Thames et Yorkshire.

REFERENCES

Bolton, M. 1979. Livret-guide au Mécanisme de la Terre. Londres: McMillan.

Craig, R.N. 1983. Soulèvement Au Cric des Pipes. Critique de la Dernière position. CIRIA. Note Technique 112, Londres.

Lupini, J.F., Skinner, A.E. et Vaughan, P.R. 1981. La Force Residuelle des Terres Cohésives. Géotechnique 31: 181-213.

Norris, P. et Milligan, G.W.E. 1992. Mécanismes de Transfert de la Charge Finale pendant le Soulévement au cric de Pipes. Proc. 8ieme Conf. Int. sur la Construction sans Fossé. No. Dig. 91, Washington.

Potyondy, J.G. 1961. Frottement de Peau entre Terres Varieés et Matériaux de Construction. Géotechnique 11: 339-353.

Tika, T.M. 1989. L'Effet du taux de Cisaillement sur la Force Residuelle de la Terre. Thesis PhD, Univ. de Londres (Imperial College).

No Trenches in Town, Henry & Mermet (eds) © 1992 Balkema, Rotterdam. ISBN 90 5410 085 0

Basic analysis for modelling the dynamic pipe bursting

C. Falk & D. Stein
Ruhr-Universität Bochum, Germany

ABSRTACT: In recent years, there had been realized extensive theoretical investigations and large-format model tests at the Ruhr-Universität Bochum for soilmechanical modelling the dynamic pipe bursting process. There had been developed basics to calculate the effects of the dynamic pipe bursting on the surroundings and to determine the load related to the new pipe run.

1. INTRODUCTION

Dynamic pipe bursting for the renewal of defective nonman-size supply and sewage lines has proved its general suitability in numerous pipe renewals (Stein et. al., 1989). It is an economical and environmental-conserving method. Up to now, however, there has been only insufficient knowledge regarding the effects of dynamic pipe bursting on the surroundings and the load related to the new pipe run.

During the past few years, due to this research deficit, the Ruhr Universität in Bochum, Germany, has made extensive theoretical investigations and large-format model tests for soil mechanical modelling the dynamic pipe bursting. These tests and investigations aimed at the development of basics which were to be used as a standard for statical calculations of the load related to the new pipe run and of the effects of dynamic pipe bursting on the surroundings.

2. TESTS

The emphasis was put on large-format dynamic bursting tests to evaluate the deformation process of the model soil, the secondary state of stress and the load related to the pipe. On this occasion a pipe run DN 100 ($\varnothing_i$ = 100 mm, $\varnothing_a$ = 130 mm) made of stoneware was renewed by dynamic pipe bursting. The installed new pipe was a PVC pipe DN 100 ($\varnothing_i$ = 104 mm, $\varnothing_a$ = 110 mm). During the dynamic bursting operation the settlements of the ground line and the deformation of the model soil were indicated by extensometers installed in the soil. The model soil consisted of sand with a natural water content (w = 4%, U = 3.12).

The load related to the installed PVC pipe was determined by strain measurements on the inner pipe surface, the characteristic values of the soil by triaxial tests used as compression tests.

The result of these tests is as follows: With regard to soil mechanics dynamic pipe bursting can be described in two successive phases; cavity expansion and displacement of surrounding soil and of fragments of the destroyed pipe which had to be renewed (phase I) and a radial redeformation of the soil owing to the difference in diameter between the displacement body and the new pipe (phase II). The displacement of soil which had been measured during the bursting pipe operation occurred in a radial direction with a distance of 1 $\varnothing_v$

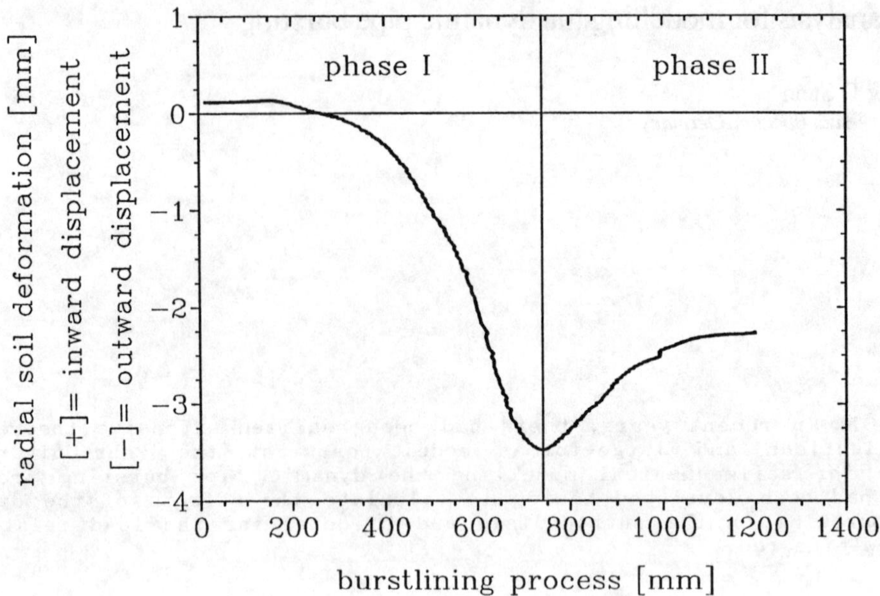

Picture 1. Description of the two phases of dynamic pipe bursting with the help of an extensometer displacement (example).

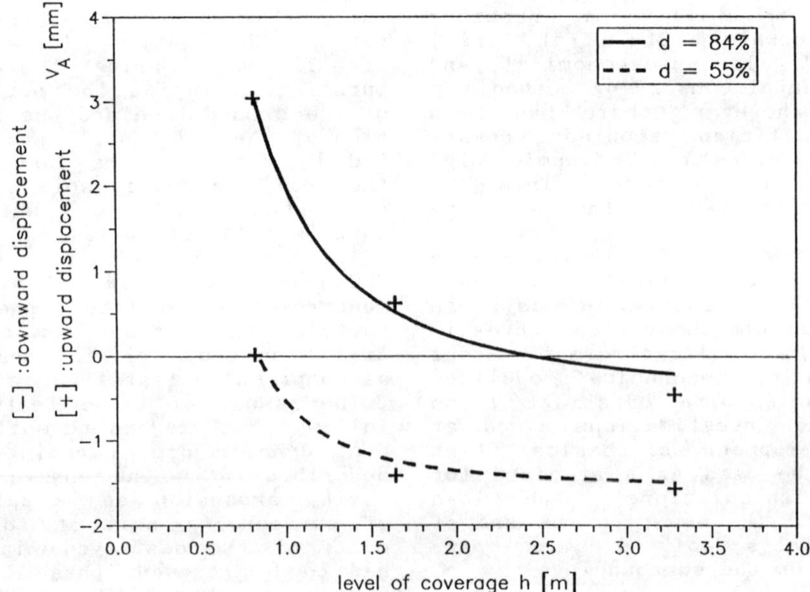

Picture 2. Vertical displacement of pipe axis

to the displacement body ($\varnothing_v$ = $\varnothing$ of the displacement body, $\varnothing_v$ = 130 mm). This is shown as an example in picture one.

It could be derived from the measured displacements of soil and the strain mesurements at the installed pipe that the radial soil redeformations at the cavity cover were only of a few mm and that the ring fissure of 10 mm between the displacement body and the successive new pipe remained unchanged in all essential parts. Thus the new installed pipe run lay virtually unloaded in a cavity created by the dynamic bursting operation.

Apart from radial soil deformations there occurred deformations due to a displacement of the bursting unit axis with regard to the pipe axis of the pipe to be renewed.

The reason for this could be found in vertical, downward directed soil displacements owing to a compaction of soil caused by percussions of the soil displacement hammer. The hammer had been modified as a bursting unit. This effect was overlapped by an upward directed vertical yield of the bursting unit, which had also been observed in tests by Leach et. al. (1989), Poole et. al., (1985) and Robins et. al., (1990). The more the primary stress level respectively the level of coverage decrease and the initial compactness of soil increases, the more the effect of the upward directed yield of the bursting unit increases and the compaction becomes less and also, as a consequence, the displacement of soil particles directed downward. This can be clearly seen in picture 2.

3. THEORETICAL INVESTIGATIONS

Basics for modelling dynamic pipe bursting have been elaborated to determine the state of stress/strain in the soil, caused by a dynamic bursting operation, and to estimate the load related to the installed pipe run.

The soil mechanic modelling of the above described phase I, that is to say the cavity expansion, can be derived from pressuremetertheories (Baguelin, 1978). In order to calculate the deformed state of a cavity with an initial inner pressure of P_o, due to an extension from P_o to P (picture 3), the Lamè solution is necessary which can be obtained by the equilibrum conditions in radial and tangential direction with

$$\frac{d\sigma_r}{dR} + \frac{\sigma_r - \sigma_\varphi}{R} = 0$$

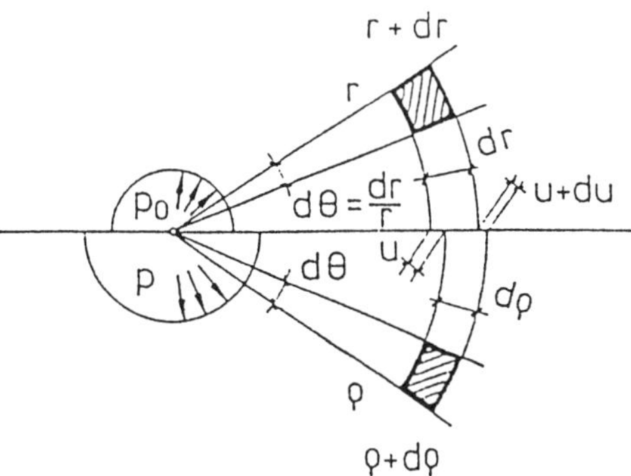

Picture 3: Geometrical change of a soil particle during the deformation of a cavity (Baguelin, 1978)

After having inserted the known twodimensional stress/strain relation into the Lamè equation the differential equation is obtained

$$R_2 \frac{d^2U}{dR^2} + R \frac{dU}{dR} - U = 0$$

The solution of this equation provides the stress distribution in the elastic zone.

$$\sigma_r = P_0 + 2\,G\,(\epsilon_0 R_0^2)/(R^2)$$

$$\sigma_\varphi = P_0 - 2\,G\,(\epsilon_0 R_0^2)/(R^2)$$

$$G = E\,/\,(2(1+\nu))$$

The stress distribution in the plastic zone can be derived from the condition that radial and tangential stresses of both, the known Mohr-Coulomb failure criterion and the Lamè equation have to be sufficient.

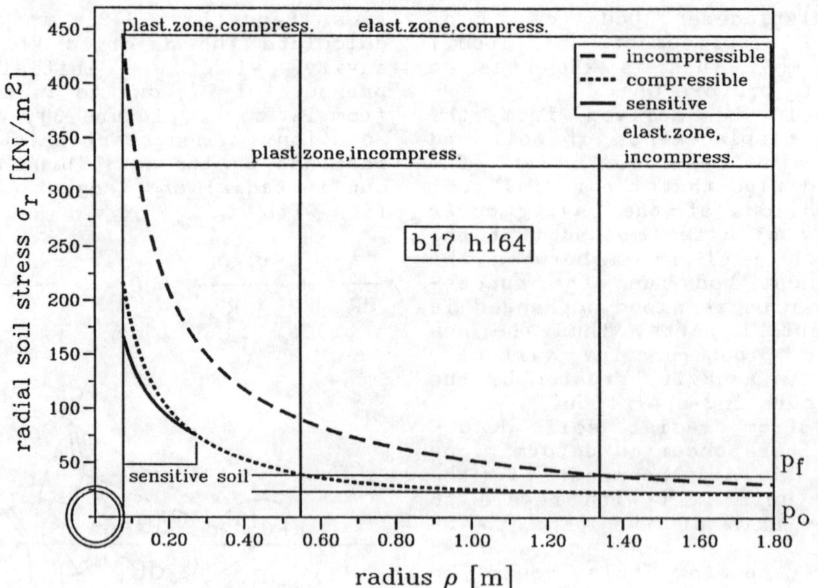

Picture 4. Radial stresses caused by dynamic expansion - compression and dynamic exciting of soil are considered - and comparison to stresses in an incompressible, "insensitive" soil.

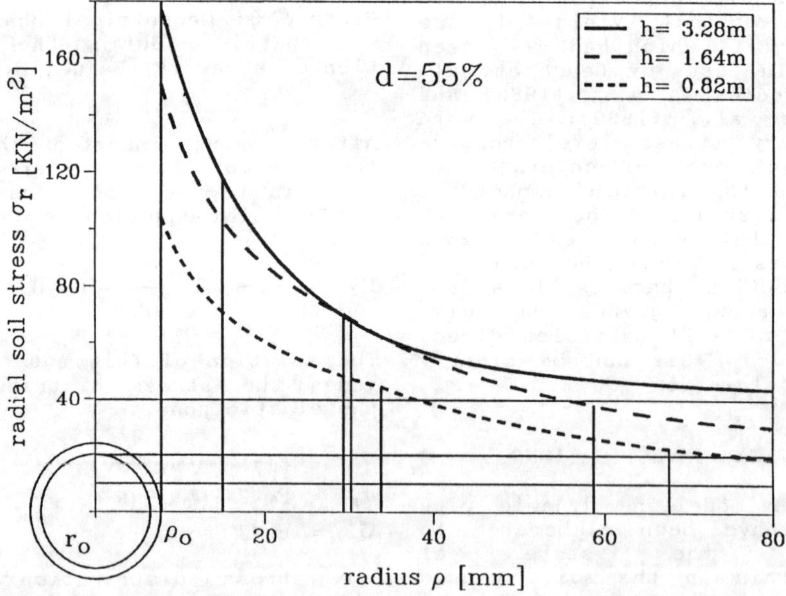

Picture 5. Radial stresses at different levels of coverage directly after pipe bursting operation and after expansion of cavity caused by bursting unit.

$$\sigma_r = (P_F + c \cot \varphi) \left[\frac{\rho_F^2}{\rho^2} \right]^{\frac{1-K_a}{2}} - c \cot \varphi$$

$$\sigma_\varphi = K_a \, (P_F + c \cot \varphi) \left[\frac{\rho_F^2}{\rho^2} \right]^{\frac{1-K_a}{2}} - c \cot \varphi$$

$$\rho_F = \sqrt{(a^2/2\alpha_F)}$$

$$\alpha_F = (P_F - P_0)/2G$$

$$P_F = P_0 (1 + \sin \varphi) + c \cos \varphi$$

'F' indicates transition from the elastic to the plastic zone.

The compaction of soil in the line zone can be considered by using the value of a_o for the radial cavitiy expansion a:

$a_o = \sqrt{[a^2/(m_v-(m_v-1)/2\alpha_F)]}$,

a: radius of a circular section with equal surface in relation to displaced cross-section surface,

α_F Almansi strain at failure.

Following Youd (1970), Barkan (1962) and Smoltczyk (1975), this model considers the dynamic exciting of soil ('sensitive soil') in a circular zone near the displacement body by supposing a diminuished shear strength of soil in this area.

The stresses in the surroundings of the cavity, thus calculated, are shown as examples in pictures Nos. 4 and 5. Picture 4 compares stresses for an incompressible soil.

It is apparent that the secondary stress state and the size of the influence on the surroundings depend considerably from the initial stress state, that is to say from the level of coverage H. The cause lies in an increased compression, respectively in a decreased dilatation of the surrounding soil with a raising level of stress as well as a bigger modulus of elasticity and a bigger elastic strain at failure α_F with an increasing initial stress.

The strain/stress state in the second phase of the dynamic pipe bursting can approximately be determined with the help of elasto-plastic plain-strain modes around a circular hole, which is used in tunnel construction. Herewith it was possible to confirm the stability of the expanded cavity in the used soil, which had also been observed in tests.

4. SUMMARY AND PROSPECT

The above described investigations and tests show that dynamic pipe bursting causes extremely complexe soilmechanical effects on the surrounding line zone. The load related to those pipes which are relocated by means of dynamic pipe bursting differs considerably from previous initial loads for pipe runs that are built with open cut method or trenchless construction method. Contrary to these methods the new installed pipe in the used soil can be discharged considerably by dynamic pipe bursting. Leach/Reed (1989) and Rogers et.al. (1991) had derived from their tests a redeformation of the surrounding soil after the deforming process until the soil had lain close to the new installed pipe. This could not be confirmed with respect to the used model soil. In relation to traditional methods for pipe renewal, dynamic pipe bursting should have another rank in future times. If the results were confirmed under different boundary conditions and during the installed state, the consequenze would be a lesser wall thickness and a lower potential of damage on those pipes renewed by dynamic pipe bursting.

However, the described investigations and tests do not yet allow a definitive derivation as to how the load of a new installed pipe in an installed state developes in the long term. Particularly dynamic loads in traffic and a changement of the ground water level might lead to a "wetting-induced-collaps" of the pipe-surrounding cavity, which was created by pipe bursting operation, and thus they might cause a considerable increase in load. In this case an existing ground pressure might be transmitted to single and line loads by those fragments surrounding the new pipe run. Nanegrungsunk's (1988) tests show that this, as a consequence, could lead to a four times higher load on the new installed pipe run - compared with the pure load on ground pressure.

Further tests should aim at clarifying the long-term development of pipe/soil interaction under changeable boundary conditions as well as preventing, in the long term, possibly arising point loads and line loads due to fragments of the old pipe.

The effects of dynamic pipe bursting on the surroundings, for example on neigbouring pipe runs, can approximately be estimated by the help of an evolved analytic method which is based on used pressuremetertheories. They depend decisively on the compressibility of the sourrounding soil, the size of expansion and the level of coverage.

BIBLIOGRAPHIE

Baguelin, F.; Jèzèquel, J.F.; Shields, D.H.: The Pressuremeter and Foundation Engineering. Series on Rock and Soil Mechanics, Trans Tech Publications, Clausthal, 1978.

Barkan, D.D.: Dynamics of Bases and Foundations, translated from Russian by L. Drashevska, McGraw-Hill Book Co., Inc., New York, 1962.

Leach, G.; Reed, K.: Observation and Assessment of the Disturbance Caused by Displacement Methods of Trenchless Construction. Proc. NO-DIG 89, London April 1989, p. 2.4.1 - 2.4.12.

Nanegrungsunk, B.: Belastung der im Berstverfahren verlegten Rohre. Disseration Ruhr-Universität Bochum, 1988.

Poole, A.D.; Rosbrook, R.B.; Reynolds, J.H.: Replacement of small diameter pipes by pipe bursting. Paper 4.1 "NO-DIG 85". First International Conference and Exhibition on Trenchless Construction for Utilities. 16.-18.4.1985. Kensington Exhibition Centre, London. Veranstalter: Institution of Public Health Engineers, London.

Robins, P.J.; Rogers, C.D.F.; Scott, A.M.: Design and development of ductile iron pipes for Pipe Bursting. Proceedings of Pipeline Management '90 Conference, London, Juni 1990, p. 4.1 - 4.13.

Rogers, C.D.F.; Robins,P.J.; Scott,A.M.: Field Performance of a new ductile iron pipe for use with pipe bursting. No-Dig '91 und 3 rd International Conference On Pipeline Construction. Hamburg, Oktober 91.

Smoltczyk, U.: Graving Dock Foundation on Deep Fill. Proc. 1. Baltic Conf. Soil Mech. Found. Engng. III, 1975, S. 213-222.

Stein, D.; Möllers, K.; Bielecki, R.: Microtunneling, Verlag Ernst & Sohn, Berlin, 1989.

Youd, T.L.: Densification and shear of sand during vibration. Journal of the Soil Mechanics and Foundation Division, May 1970, p. 863-880.

No Trenches in Town, Henry & Mermet (eds) © 1992 Balkema, Rotterdam. ISBN 90 5410 085 0

Recherches fondamentales pour le modelage du procédé d'éclatement dynamique

C.Falk & D.Stein
Ruhr-Universität Bochum, Allemagne

ABSTRACT: La Ruhr-Universität Bochum a fait pendant ces dernières années des recherches théoriques ainsi qu'une série d'expériences modèles concernant le modelage mécanique du sol par le procédé d'éclatement dynamique. On a trouvé la base d'une norme pour le calcul statique de lignes installées au moyen du procédé d'éclatement dynamique et les effets que ce procédé peut avoir sur les zones avoisinantes.

1. INTRODUCTION

Le procédé d'éclatement dynamique a prouvé lors de nombreux mises en oeuvre ses qualités de procédé à la fois économique et favorable à la protection de la nature qui sert au renouvellement des lignes du réseau de distribution que l'on ne peut pas inspecter à pied (Stein et. al., 1989). Mais jusqu'ici on n'a eu que des constatations insuffisantes quant aux conséquences du procédé pour la zone de lignes avoisinante et quant à la sollicitation de lignes installées au moyen du procédé d'éclatement dynamique.

Initié par ce déficit de recherche on a fait pendant ces dernières années des recherches théoriques ainsi qu'une série d'expériences modèles concernant le modelage mécanique du sol par le procédé d'éclatement dynamique. La recherche avait pour but de trouver la base d'une norme pour le calcul statique de lignes installées au moyen du procédé d'éclatement dynamique et les effets que ce procédé peut avoir sur les zones avoisinantes.

2. RECHERCHES EXPERIMENTALES

Au centre des recherches expérimentales ont eu lieu des expériences de grand format concernant l'éclatement dynamique destinées à la constatation du comportement de déformation du sol modèle, de l'état secondaire de contrainte et de la sollicitation des tuyaux. Pour cette expérience on renouvelait une ligne DN 100 ($ø_i$ = 100 mm, $ø_a$ = 130 mm) de grès qu'il fallait restaurer à l'aide du procédé d'éclatement dynamique. On installait un tuyau de CPV (chlorure de polyvinyle) DN 100 ($ø_i$ = 104 mm, $ø_a$ = 110 mm) en tant que tuyau nouveau. Au cours du procédé d'éclatement dynamique on fixait les abaissements du bord supérieur du terrain et la déformation du sol expérimental par des extensomètres installés dans le sol. On choisissait du sable avec une teneur d'eau naturelle en tant que sol expérimentel (w = 4%, U = 3.12).

La sollicitation de la nouvelle ligne de CPV était fixée à l'aide de mesurages d'extension à l'intérieur du tuyau. On trouvait les paramètres du sol à l'aide d'expériences triaxiales qui étaient faites en tant qu'expériences de compression.

Comme les expériences l'ont démontré, on peut décrire le procédé d'éclatement dynamique au niveau de la mécanique du sol à

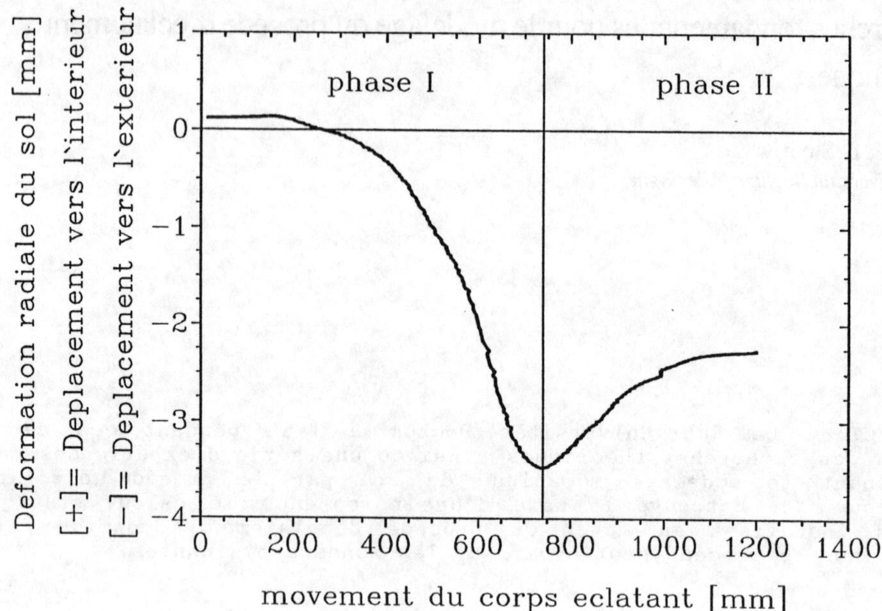

Illustration 1. Exemple de démonstration des deux phases du procédé d'éclatement dynamique à l'aide d'un déplacement d'extensomètre.

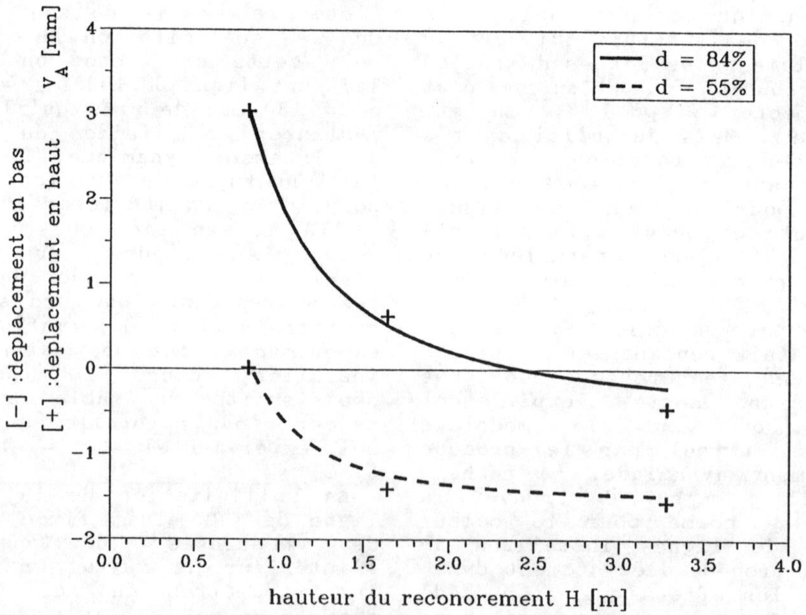

Illustration 2. Déplacement vertical de l'axe du tuyau.

l'aide de deux phases qui se poursuivent: a) une extension d'une cavité et un déplacement du terrain avoisinant ainsi que des morceaux du tuyau détruit à renouveler (phase I) et b) une redéformation radiale du sol à cause de la différence transversale du corps déplaçant par rapport à la nouvelle ligne (phase II). L'illustration no.1 montre de façon exemplaire les déplacements du sol pendant le procédé d'éclatement dynamique en direction radiale avec une distance de 1 øv par rapport au corps déplaçant (øv = ø du corps déplaçant, øv = 130mm).

On pouvait conclure des déplacements du sol mesurés et des mesures de l'extension dans le tuyau installé que les redéformations radiales du sol à la paroi de la cavité ne faisaient que quelques millimètres et qu'il n'y avait pas de déformations importantes de la faille causées par la décyclisation entre le corps déplaçant et le nouvau tuyau suivant. La nouvelle ligne se trouvait donc quasiment sans pression dans la cavité qui s'était produite au cours du prodédé d'éclatement dynamique.

A côté des déformations radiales du sol on constatait des déformations à cause d'un déplacement de l'axe du corps éclatant jusqu'à l'axe du tuyau de la ligne à renouveler (cf. illustration 2).

Ce déplacement était causé par des déplacements verticaux qui se dirigeaient vers le bas à cause d'une compression du sol par les coups de la torpille modifiée en tant que corps éclatant. A cet effet on superposait un déplacement vertical du corps éclatant vers le haut ce qui avait déjà été constaté également par Leach et.al. (1989), Poole et.al. (1989) et Robins et.al. (1990) dans leurs recherches. Avec une tension primaire diminuant voire une hauteur de recouvrement et avec une densité croissante du gisement de départ l'effet du déplacement vers le haut du corps éclatant se renforce; en même temps la compression diminue ainsi que le déplacement vers le bas des parcelles du sol ce qui est montré dans l'illustration no. 2.

3. RECHERCHES THEORIQUES

Pour fixer l'état de tension du sol causé par un procédé d'éclatement dynamique et pour pouvoir juger la sollicitation de la ligne installée on élaborait les bases pour un modelage du procédé.

Le modelage mécanique du sol par la phase I décrite ci-dessus, c'est-à-dire l'extension de la cavité se laisse dériver de théories de pressiomètres [5]. Pour l'état d'expansion de la cavité avec la pression intérieure Po causée par l'élargissement de Po à P (cf. illustration 3) il résulte des conditions d'équilibre en direction radiale et tangentielle la solution Lamé avec la

$$\frac{d\sigma_r}{dR} + \frac{\sigma_r - \sigma_\varphi}{R} = 0$$

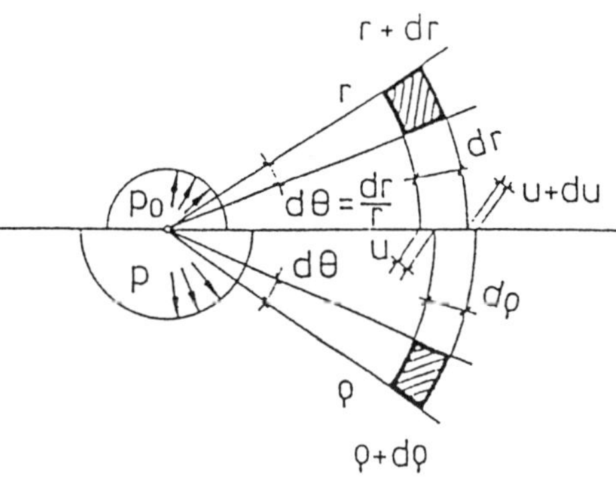

Illustration 3. Changement géométrique d'un élément du sol pendant l'élargissement d'une cavité. (Baguelin et. al., 1978)

Après l'insertion de la relation de tension et extension à deux dimensions il résulte de l'équation Lamé la équation

$$R_2 \frac{d^2U}{dR^2} + R \frac{dU}{dR} - U = 0$$

dont le résultat donne la distribution des contraintes dans le domaine élastique.

$$\sigma_r = P_0 + 2\,G\,(\epsilon_0 R_0^2)/(R^2)$$

$$\sigma_\varphi = P_0 - 2\,G\,(\epsilon_0 R_0^2)/(R^2)$$

$$G = E/(2(1+\nu))$$

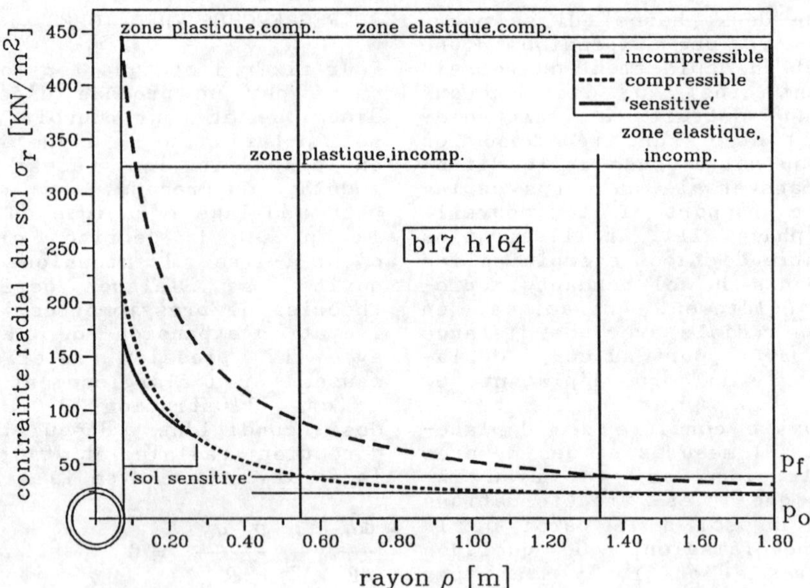

Illustration 4. Cours de la contrainte radiale causée par un procédé d'éclatement dynamique en considération de la compression et de l'excitation dynamique du sol; comparaison au cours de la contrainte pour un sol incompressible, 'insensitive'.

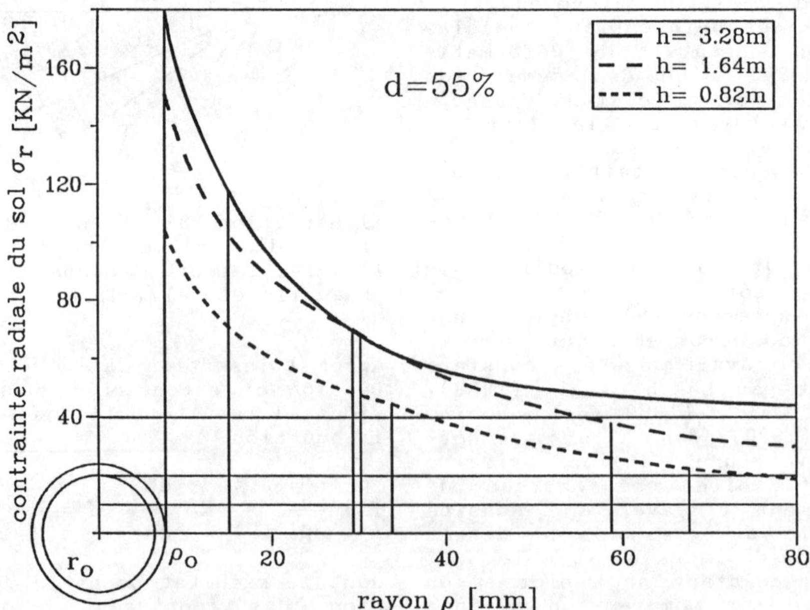

Illustration 5. Contraintes radiales avec des hauteurs de recouvrement différentes juste après le procédé d'éclatement et l'élargissement de la cavité par le corps éclatant.

La distribution des contraintes dans la zone plastique résulte de la prémisse qui prescrit que les contraintes radiales et tangentielles doivent correspondre et à la prémisse de fraction d'après Mohr qui est généralement connue et à l'équation de Lamé:

$$\sigma_r = (P_F + c \cot \varphi) \left[\frac{\rho_F^2}{\rho^2} \right]^{\frac{1-K_a}{2}} - c \cot \varphi$$

$$\sigma_\varphi = K_a \, (P_F + c \cot \varphi) \left[\frac{\rho_F^2}{\rho^2} \right]^{\frac{1-K_a}{2}} - c \cot \varphi$$

$$\rho_F = \sqrt{(a^2/2\alpha_F)}$$

$$\alpha_F = (P_F - P_0)/2G$$

$$P_F = P_0\,(1 + \sin \varphi) + c \cos \varphi$$

- la transition de la zone élastique à la zone plastique est signée par 'F'.

La compression du sol dans la zone de la ligne est prise en considération par la fixation de l'élargissement radial de la cavité à l'aide de la valeur a_o avec:

$a_o = \sqrt{[a^2/(m_v-(m_v-1)/2\alpha_F)]}$,
a: rayon d'une surface de cercle à superficie identique à celle de la superficie transversale déplacée,
α_F: dilatation Almansi en état de fraction.

L'excitation dynamique du sol ('sensitive soil') dans un terrain circulaire dans les alentours du corps déplaçant est prise en considération dans le modèle suivant l'exemple de Youd (1970), Barkan (1962), Smoltczyk (1975) par la supposition d'une résistance au cisaillement diminuée dans ce terrain.

Les contraintes dans les alentours de la cavité trouvées par un tel procédé sont décrites par les illustrations 4 et 5 qui pourraient servir d'exemple. L'illustration 4 rend possible des comparaisons en donnant le cours de la contrainte pour le sol incompressible.

Il est apparent que l'état de contrainte secondaire et la grandeur de la zone influençant les alentours dépendent surtout de l'état de contrainte primaire, c'est-à-dire de la hauteur de recouvrement H. Ceci peut être expliqué par une compression augmentée voire une dilatation diminuée du sol avoisinant avec un niveau de tension augmentant ainsi que par un module d'élasticité plus grand et une dilatation limite élastique augmentée a_F avec une tension primaire augmentante.

L'état de contrainte/d'extension dans la phase II du procédé d'éclatement dynamique se laisse fixer de façon approximative avec des modèles de disques perforés élasto-plastiques utilisés dans la construction de tunnels. De cette manière on pouvait vérifier la consistance de la cavité élargie pour le sol choisi pour l'expérience qu'on avait déjà pu observer dans les expériences décrites ci-dessus.

4. RESUME ET PERSPECTIVES

Les expériences et recherches décrites ci-dessus montrent que le procédé d'éclatement dynamique a des conséquences extrêmement complexes par rapport à la zone de lignes avoisinante. L'état de sollicitation des tuyaux installés à l'aide d'un procédé d'éclatement dynamique se distingue tout à fait des états de sollicitation observés dans les lignes installées d'après la méthode de construction ouverte ou fermée. Le procédé d'éclatement dynamique peut - contrairement aux méthodes mentionnées ci-dessus - mener à un délestage considérable de la nouvelle ligne. Une redéformation du sol avoisinant après le procédé d'élargissement jusqu'à un adjacement au nouveau tuyau installé qui était dérivée sur la base d'expériences différentes par Leach/Reed (1989) et Rogers et.al. (1991) ne pouvait pas être affirmée. Désormais il faudrait revaloriser le procédé d'éclatement dynamique par rapport à des procédés traditionnels de renouvellement de lignes. Si les résultats trouvés pouvaient être affirmés dans d'autres conditions de base et en état de service, cela pourrait engendrer une épaisseur des parois et un potentiel de dommages réduit pour des lignes renouvelées par un procédé d'éclatement dynamique.

On ne peut pas encore conclure définitivement des expériences décrites l'état de sollicitation d'une nouvelle ligne installée en état de service à long terme. Surtout des compressions causées par le trafic et un changement du niveau d'eau souterraine pourraient initier un 'wetting-inducedcollaps' de la voûte entourant le tuyau causée par le procédé d'éclatement ce qui pourrait provoquer une augmentation considérable de la sollicitation. Dans ce cas une pression de la terre déjà existante sera disposée autrement par les éclats qui entourent la nouvelle ligne en faveur de sollicitations singulières et linéaires. Selon les recherches de Nanegrungsunk (1988) ceci pourrait causer une sollicitation quatre fois plus grande de la nouvelle ligne par rapport à la sollicitation nette de la compression de la terre.

C'est pourquoi les recherches ont pour but de constater le comportement à long terme de l'interaction entre le tuyau et le sol sous des conditions marginales changeantes ainsi qu'un empêchement à longue date de sollicitations de point ou linéaires par les éclat du vieux tuyau.

Les effets du procédé d'éclatement dynamique sur les alentours, p.e. sur des lignes avoisinantes, peuvent être très bien prévus de façon approximative par un procédé de calcul analytique élaboré qui est basé sur des théories de pressiomètre connues. Ces effets dépendent surtout de la capacité de compression du sol adjacant, de l'ampleur de l'élargisssement et de la hauteur de recouvrement.

BIBLIOGRAPHIE

Baguelin, F.; Jèzèquel, J.F.; Shields, D.H.: The Pressuremeter and Foundation Engineering. Series on Rock and Soil Mechanics, Trans Tech Publications, Clausthal, 1978.

Barkan, D.D.: Dynamics of Bases and Foundations, translated from Russian by L. Drashevska, McGraw-Hill Book Co., Inc., New York, 1962.

Leach, G.; Reed, K.: Observation and Assessment of the Disturbance Caused by Displacement Methods of Trenchless Construction. Proc. NO-DIG 89, London April 1989, p. 2.4.1 - 2.4.12.

Nanegrungsunk, B.: Belastung der im Berstverfahren verlegten Rohre. Disseration Ruhr-Universität Bochum, 1988.

Poole, A.D.; Rosbrook, R.B.; Reynolds, J.H.: Replacement of small diameter pipes by pipe bursting. Paper 4.1 "NO-DIG 85". First International Conference and Exhibition on Trenchless Construction for Utilities. 16.-18.4.1985. Kensington Exhibition Centre, London. Veranstalter: Institution of Public Health Engineers, London.

Robins, P.J.; Rogers, C.D.F.; Scott, A.M.: Design and development of ductile iron pipes for Pipe Bursting. Proceedings of Pipeline Management '90 Conference, London, Juni 1990, p. 4.1 - 4.13.

Rogers, C.D.F.; Robins,P.J.; Scott,A.M.: Field Performance of a new ductile iron pipe for use with pipe bursting. No-Dig '91 und 3 rd International Conference On Pipeline Construction. Hamburg, Oktober 91.

Smoltczyk, U.: Graving Dock Foundation on Deep Fill. Proc. 1. Baltic Conf. Soil Mech. Found. Engng. III, 1975, S. 213-222.

Stein, D.; Möllers, K.; Bielecki, R.: Microtunneling, Verlag Ernst & Sohn, Berlin, 1989.

Youd, T.L.: Densification and shear of sand during vibration. Journal of the Soil Mechanics and Foundation Division, May 1970, p. 863-880.

1.4 Detection and geotechnical investigations

No Trenches in Town, Henry & Mermet (eds) © 1992 Balkema, Rotterdam. ISBN 90 5410 085 0

Synthesis of microtunneling projects in the 'Val de Marne'

Alain Guilloux
Terrasol, France

Christian Legaz
DSEA, Département du Val de Marne, France

Abstract : this paper presents a synthesis of various microtunneling projects performed in the "Val de Marne", suburn of Paris. The geotechnical conditions of these projects and the performances of the machines are presented. It evidences the major role played by the soil investigations and the driving parameters of the machine on the success or not of such a project.

1 INTRODUCTION

Hereafter is proposed a synthesis of various microtunneling projects realized in the "département du Val de Marne", in the suburns of Paris, with the aim of analysing the role played by the geotechnical conditions in such works.

These works were realized under the authority of the D.S.E.A. ("Direction des Services de l'Eau et de l'Assainissement") between 1989 and 1992. They are of great interest because there were done in various soil conditions and with different machines; the first projects can be considered as experimental ones in France.
Each one of the projects presentation includes :

-a short desciption of the site and of the machine,

-an analysis of the driving parameters,

-a description of the main difficulties when existing.

With these data, we shall present the specificities of a microtunneling project with regard to a traditional trench construction. Some indications about the geotechnical investigations for these kinds of projects will be presented, and the influence of the soil conditions on the progress of a microtunneling project (feasability, machine choice, hazards).

2 EXPERIENCES OF VARIOUS PROJECTS IN THE "VAL DE MARNE"

2.1 BOISSY I (june 1989)

The project BOISSY I, for a rainwater pipe ϕ 500 mm, was done in sandy-clayey soils, at 2 - 2.5 m depth, in two sections 26 m and 70 m long, with a microtunnelling machine "Unclemole" with slurry discharge.

The jacking thrust (fig. 1) stays low and somewhat constant (about 200 kN), which shows the efficiency of the lubrification by "gel" to reduce the friction along the pipes. On the other hand, the progress was slower (1H to 1H30 for a 2 m long pipe) for the fisrt section, in a more clayey soil, than for the second one (average 0H30).

2.2 CRETEIL I (december 1989)

The project CRETEIL I, for a rainwater pipe ϕ 300 mm, was done in more or less clayey sands with flints, at 2.5 m depth, in two sections 40 m and 26 m long, with a microtunneling machine

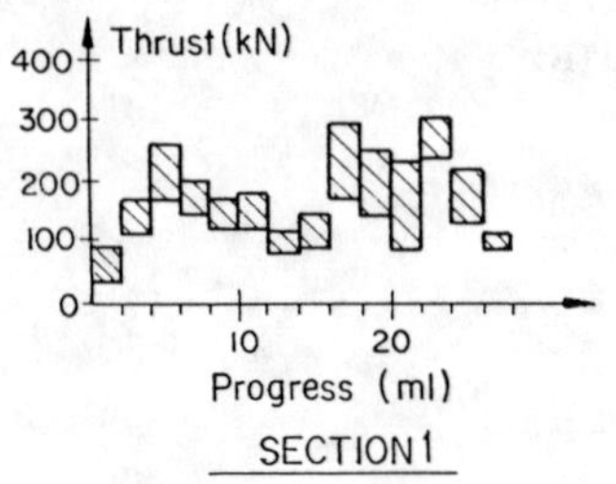

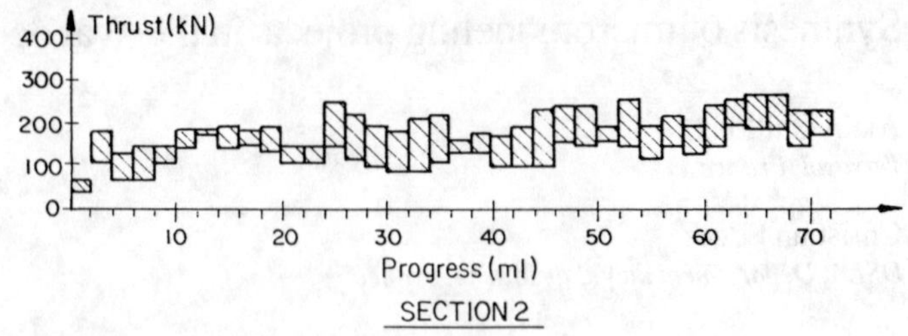

Fig 1. VARIATION OF THRUST VERSUS PROGRESS (BOISSY I)

HERRENKNECHT (slurry discharge).

The jacking thrust (fig. 2) increases from 0 to more than 600 kN all along the progress, which corresponds to an average shear stress of 20 kPa along the pipes. It seams to be in correlation with the deviations, mainly horizontal, which reached more than 40 mm on different points (with differential deviations up to 80 mm on two pipes, i. e. 2 %).

The progress was generally fast : 0H10 to 0H20 for a 2 m long pipe.

2.3 BOISSY II (february 1990)

The project BOISSY II, for a rainwater pipe ϕ 500 mm, was done in fill and sandy-clayey soils, at a depth between 2.5 and 5 m, in two sections 35 and 54 m long, with a MARKHAM "Super mini" microtunneling machine. Furthermore, it passed at 9 m below a metro line.

On the second section, where the deviations were negligible (less than 10 mm), the jacking thrust remains somewhat constant and low (between 200 and 250 kN - fig. 3).

On the other hand, for the first section, large deviations (up to 80 mm both horizontally and vertically, and 120 mm of differential deviations on 3 m, i. e. 4 %) seem to have induced an increase of the jacking force with the progress (fig. 3); it reached 1300 kN (which corresponds to a shear stress of 20 kPa in average). For a longer drive, the maximum thrust of 2000 kN could have been approached.

The progress rate was almost constant (0H30 average for a 2 m long pipe). An hard block could be passed thanks to the possibility of drawing back the temporary steel pipes. The measured settlements under the metro line did not exceed 3 mm.

2.4 CHAMPIGNY I (april 1990)

The project CHAMPIGNY I, for a sewer line ϕ 400 mm, was done with a MARKHAM machine with slurry discharge. It included two sections :

-the first section, 200 m long, was abandonned after 30 m driving due to "sticking" of clay in the cutting head, in spite of different tests of fluidifying products and of increase of the pumps power;

-the second section, 40 m long, had to pass under a railways line at 2 - 2.5 m depth. It was decided to perform a horizontal borehole in order to investigate the soils before to take the decision of driving. This borehole revealed alluvial soils (clayey gravels) and debris of plastic marls over 5 m.

In spite of these unfavourable conditions (clayey material) and of the presence of the railway line, which forbidded to dig any shaft in case of great difficulties to drive the pipes, it was decided to do the project as foreseen.

The jacking thrust revealed to increase slowly, but far below the allowable forces (100 to 400 kN, compared to 2000 kN). It appears to be much dependant upon the horizontal deviations. The progress rate was generally almost constant (about 0H40 for driving a 2 m long pipe), except in the most clayey soils, where it reached 2 to 3 H (up to 6 H for the tube drived in the plastic marls zone).

2.5 LA QUEUE EN BRIE (1991)

The project of LA QUEUE EN BRIE, for a sewer line ϕ 400 mm, was much more than an experimental work as it reached 1450 m long. It was to be drived in very heterogeneous soils (fig. 5) :

-over about 250 m, the top of the "Brie limestone", including very hard silicified beds : this section was done by using "conventionnal trench techniques" due to the foreseen difficulties to pass the hard subhorizontal beds : stop of the driving or large deviations;

-over about 300 m clayey sands, where a DECON-SOLTAU RVS 100 machine, with screw conveyor, was foreseen. After two 3 and 4 m long driving tests, where the screw was deteriorated on hard blocks, this section was ended by trench technology;

-the remaining sections, about 900 m long, went through plastic clays including "meulières" (very hard silicificied millstones). Therefore some difficulties were foressen due to sticking clays and hard blocks.

The work was done with a HERRENKNECHT AVN 500 machine, with slurry discharge, which drived about 800 m, and had to face the main following diificulties :

- a large block, about 0.5 m³, could be pushed over 2 to 3 m to the neighbouring shaft, but a larger block (1 m³) required to dig a shaft in order to get out the machine;

-the discharge was difficult, due to the sticking clays, when the machine progressed quickly. The numerous mechanical breakdowns seem to be partially due to this problem;

-when going through a soft clay zone,

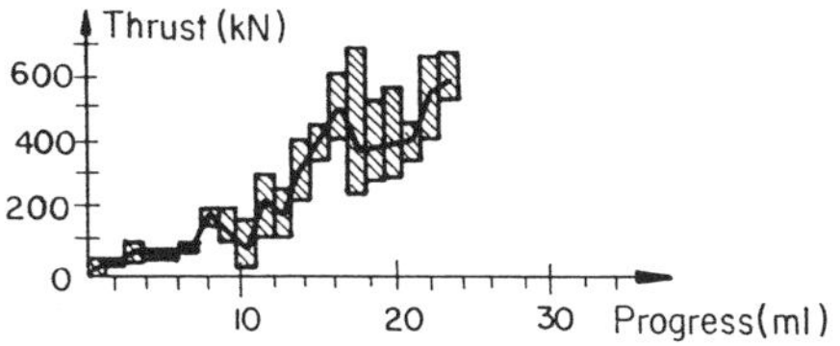

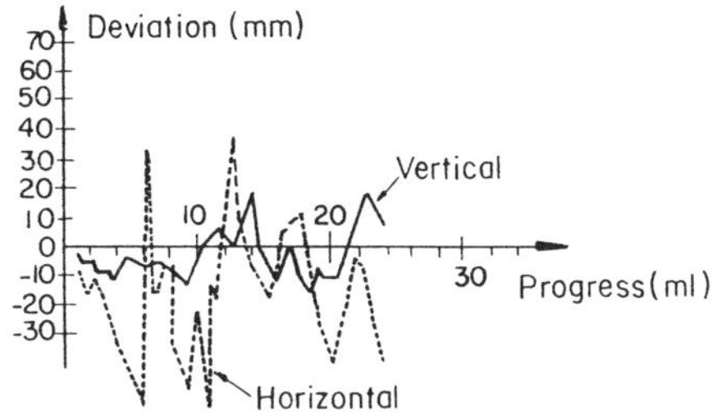

Fig2. CORRELATION BETWEEN THRUST AND DEVIATIONS.(CRETEILI. SECTION 1)

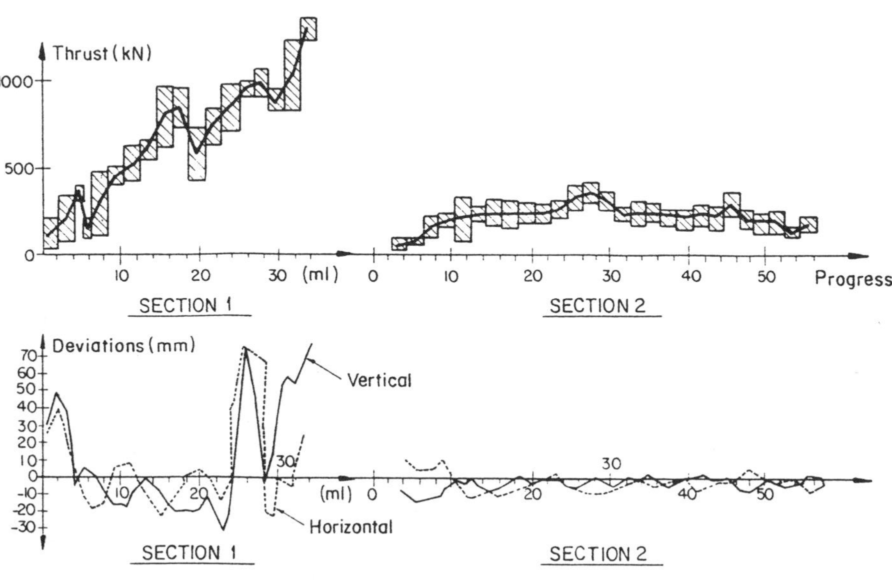

Fig 3. CORRELATION BETWEEN THRUST AND DEVITIONS (BOISSY II)

the machine sunk about 50 cm, which required once more to dig a shaft and to end the drive in trench;

-the cutting head was worn excessively, and required its repair after each drive.

The jacking force was very variable, in average 300 to 400 kN, but reached up to 2000 kN when passing the hard blocks.

The progress rate, with two shift a day, revealed rather slow : for 60 to 94 m long drives, the average rate was 3.2 m per day (1.4 m for the minimum, and a record of 16 m).

2.6 CHEVILLY-LARUE (1991)

The CHEVILLY-LARUE project, for a water pipe ϕ 600 mm, had to be drived through the "Brie marly limestones", with hard blocks ("meulières" and silicified limestones). When digging the shafts, it appeared that these blocks were more numerous and harder than previously imagined

It was therefore decided to lower the project level of about 4 m, in order to place it in the "green clays", which are very plastic clays not very favourable for microtunneling, but with the main advantage to be much homogeneous and without blocks.

The drive, was done with a HERRENKNECHT AVN 600 with "widener" machine, and could be ended in reasonnably good conditions, with a slow progress, but without any redhibitory stop. This microtunneling was the opportunity for testing various additives (polymeres) in the slurry, which could reduce the friction along the pipes and "coat" the clay lumps, which reduced the "sticking" problems in the cutter head.

2.7 Abstract of the main observations

These various works, the experience of which is corroborated by many others, show that the main difficulties to be faced were the following ones :

-presence of hard blocks or beds, the dimensions of which are larger than the third of the diameter, which cannot be easily crushed, and may in some cases stop the machine. Of course man-made obstacles (olds sewers, foundations) may also induce the same problems;

-presence of plastic clays or marls, which plug the cutting head, create plugs in the slurry pipes, and therefore reduce significantly the progress rate;

-control of the deviations during driving; when they exceed 1 or 2 %, the jacking thrust may increase significantly, and the stresses in the pipes may be very large (due to the unaxed thrust, and therefore to bending moments), which appears to be unfavourable for their behaviour.

Moreover, it may be noticed that the progress rates are very different from one site to another : for a 2 m long pipe, the driving time generally varies between 20 and 60 min (which corresponds to an instantaneous speed of 30 to 100 mm/min), and the total time for a whole cycle (driving + placement of the new pipe) between 35 and 100 min/pipe. This leads to progress of about 8 to 20 m per shift, but which may be much lower in clayey soils (down to 2 or 3 m).

3 SPECIFICITIES OF A MICROTUNNELING PROJECT

If a classical trench technology for pipes installation is generally not very sensitive to geotechnical conditions, a microtunneling project has to be considered as any underground project, with all the usual hazards.

This specificity of microtunnels with respect to traditionnal tunnels is emphasized by the following items :

-due to the small dimensions, it is very difficult to identify the problem which may be encountered at the face and to intervene directly;

-most of the projects are done at shallow depth, which means in very heterogeneous soil layers : top soils, alluvium, colluvium, with numerous horizontal variations in the soil types, various fills with buried structures.

This means that a microtunnel is very

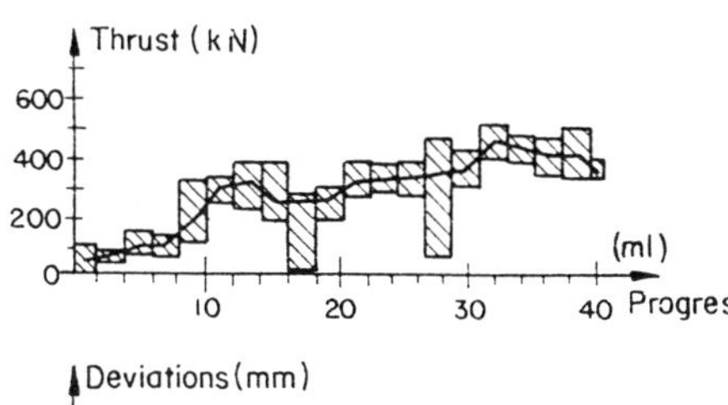

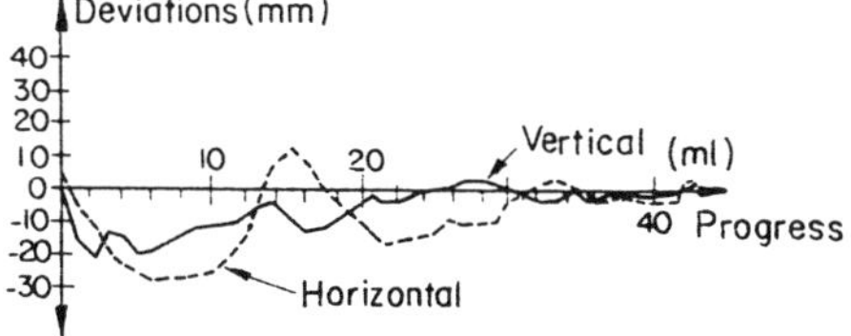

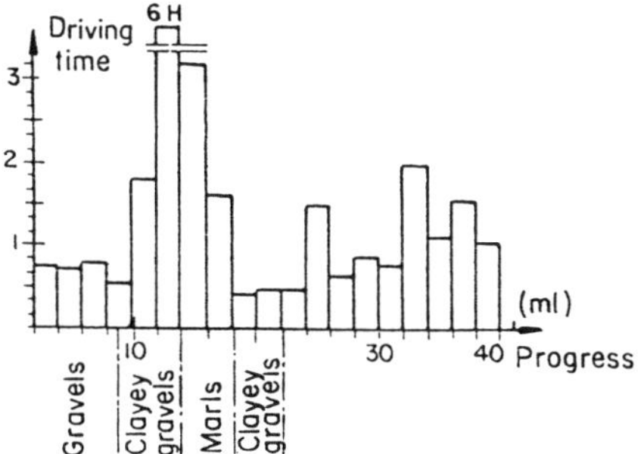

Fig 4. CORRELATIONS THRUST - DEVIATION- DRIVING TIME (CHAMPIGNY 1. SECTION 2)

sensitive to ground heterogeneities, because first they are more numerous, and secondly the small dimensions of the excavation make it much more sensitive to variations in the kind and hardness of the soils.

Furthermore, when it is reminded that a microtunneling projects cost is rather low (generally 5,000 to 10,000 FF/m), the cost of a geotechnical investigation which could avoid the main aleas may appear too much expensive when regarding the price of the project. Nevertheless the geotechnical parameters have a large influence on the selection of an adapted machine and therefore on the success of the work.

4 SPECIFIC SOIL INVESTIGATIONS

A microtunneling soil investigation has to answer to the main following questions :

-shafts and reaction masses : type and mechanical characteristics of the ground, in order to choose the methods and to design the retaining structures. These are classical questions, which can be answered by usual methods;

-excavation and mucking : the machine choice has to be done with a good knowledge of the ground to be drived through, and mainly of its heterogeneity. If the local experience of the geotehnical engineers can bring much informations, these data have to be completed by boreholes, augers, and with mechanical showel, which allow a visual evaluation of the soils and sampling for laboratory tests.

It must be pointed out that the in-situ tests (pressuremeter and cone penetration tests) have to be used cautiously : they can be useful to define a stratigraphy between boreholes, but their main backdraw (mainly pressuremeter) is to "average" the strength characteristics and therefore to "hide" decimetrical heterogeneities, the effect of which may be deciding for a microtunneling project. This kind of in-situ testing must be preferingly used for the design of shafts and pipes.

It is much useful to prefer the laboratory tests, which give a good evaluation of the main characteristics useful for a microtunnel :

-grain-size distribution and water content;

-Atterberg limits, for "sticking" risks evaluation : it may be considered that this risk is noticeable when the plasticity index exceeds 20 or 30;

-unconfined compression strength on hard blocks (eventually specific hardness and abrasavity tests) in order to evaluate the possibility for the machine to crush these blocks. One can consider that it is difficult to pass large blocks when their compression strength exceeds 20 to 50 MPa.

Moreover, the main difficulty is to evaluate the dimensions and frequencies of these blocks, as classical investigation methods by boreholes can only give partial answers. Only large diameters boreholes or shafts (0.5 to 1 m) can give reliable informations, but

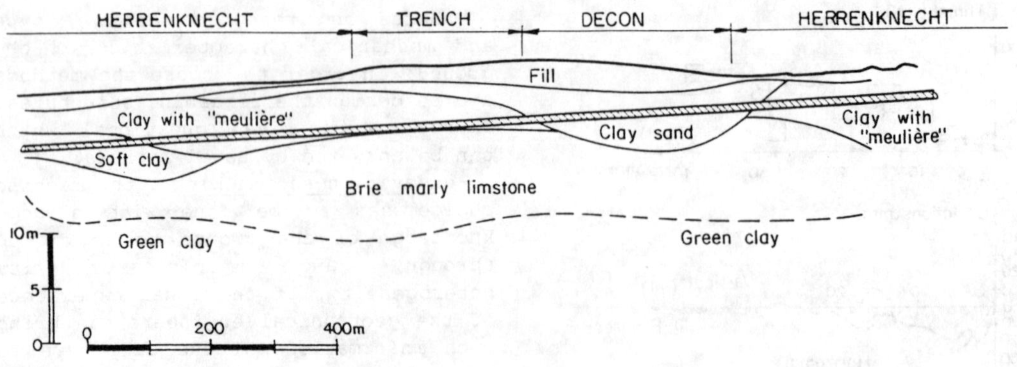

Fig 5. LA QUEUE EN BRIE : GEOLOGICAL CROSS SECTION

they are evidently much more expensive. Driving and reception shafts may be useful in this mind, but often at a stage when it is difficult to change the project.

Endly the knowledge of the water table level is necessary to choose an eventual confining pressure at the face.

-<u>obstacles detection</u> : the major problem of large dimensions obstacles (buried works or large blocks - > 1 m³ -) seems to be only partially solved by the geophysical methods.

Among the various avalaible techniques, the "radar" appears to be presently the one which can give many informations, with an easy use and interpretation, and therefore low costs on most of the projects, mainly because it is not very sensitive to the environmental conditions (vibrations, topography).

But its investigation depth remains rather limited (often 3 to 5 m, sometimes 10 m or more in favourable conditions), and may be insufficient pour some projects. When it is inside of its application range, it can detect buried obstacles, but scarcely garanty expressly their absence.

5 INFLUENCE OF GEOTECHNICAL CONDITIONS ON THE CHOICE OF THE MACHINE AND ON THE WORK PROGRESS

5.1 Adequation soil- machine

A good knowledge of the ground is essential in order :

1. <u>first to optimize the project, mainly the profile</u> : when possible, it is always preferible to select a profile which stays in an homogeneous geological layer, and to avoid the zones with blocks concentration;

2. <u>secondly to optimize the choice of the machine</u> :

-microtunneling machines with open head (without any confining pressure) and with screw have to be reserved for soils with some cohesion and above the water table; they appear to be well adapted to clayey soils, but pass with some difficulties the hard blocks; moreover in stiff ground (compressive strength larger than 5 or 10 MPa), the progress may be reduced and the machine suffer excessively;

-microtunneling machines with slurry discharge are well adapted to silty-sandy soils, even under the water table, and with a crusher, can pass blocks of compressive strength up to 20 to 50 MPa, when their dimensions do not exceed the third of the diameter. They appear to suit reasonably well to heterogeneous ground, although driving trough plastic clays or marls causes some difficulties for mucking (partially solved by use of adjuvants).

-microtunneling machine with "rock head" are allowable in rock formation (compressive strength of 20-150 MPa), but may induce some risks in "soft passes".

Moreover, the possibility of drawing back the machine must be focused, as it offers a good safety as regard the obstacles. This safety is of major impor-tance when passing under built sites (raiways lines, buildings) where the digging of a shift is excluded.

5.2 Influence of driving parameters

Even when the ground conditions of a project are well known, the driving parameters are of major influence.

Deviations have a particularly large influence on the behaviour during the driving : actually, when the guidance does not keep the deviations to reasonable values (1 to 2 % of differential deviations), the jacking force increases strongly with the progress, and can reach the limit of the equipment; moreover these high thrusts associated to unaxed efforts on the pipes, may induce excessive stresses, sometimes up to cracking.

The control of the progress speed is also very important in heterogeneous ground : when nearing a hard zone, a slow speed can allow a "sweet nibbling" of the hard point and avoid large deviations. Such a reduction causes obviously a decrease of the progress, but much less than a long driving stop! In clayey soils, a slow speed is also useful in order to limit the "sticking" problems.

6 CONCLUSIONS

Due to the recent use in France of microtunneling techniques, some questions stay raised, a large number of them are due to the ground heterogeneity existing particularly in the region of Paris : sands, compact clays and marls, sandstones and limestones, with frequent hard blocks. Therefore the problem of soils and obstacles investigation is of large concern, and not yet always well resolved.

Moreover, when the ground are well known, the choice of an adequate material is a matter of discussions and experiences, which allow to progress in the trenchless technology knowledge, even when they appear to be negative.

Finally the machine driving also plays a major role : deviations have to be restrained in strict limits, and the driving speed may be usefully reduced in some cases.

No Trenches in Town, Henry & Mermet (eds) © 1992 Balkema, Rotterdam. ISBN 90 5410 085 0

Synthèse de chantiers au microtunnelier dans le Val de Marne

Alain Guilloux
Terrasol, France

Christian Legaz
DSEA, Département du Val de Marne, France

Résumé : Cette communication présente une synthèse de différents chantiers réalisés au microtunnelier dans le département du Val de Marne. Elle analyse les conditions géotechniques de ces projets et les performances enregistrées de la machine. Elle permet de mettre en évidence l'importance des reconnaissances préalables et des paramètres de conduite de la machine sur le déroulement de ce type de chantiers.

1 INTRODUCTION

On présente ci-après une synthèse de différents chantiers réalisés au microtunnelier dans la région parisienne (Département du Val de Marne), afin d'analyser le rôle joué par les conditions géotechniques dans un projet de microtunnelier.

Ces chantiers, réalisés par la D.S.E.A. (Direction des Services de l'Eau et de l'Assainissement) entre 1989 et 1991, présentent l'intérêt d'avoir été réalisés dans des terrains variés et avec des machines différentes, les premiers d'entre eux ayant un caractère parfois expérimental.

Nous présenterons pour chacun des chantiers :

-une description du site et de la machine,

-une analyse des paramètres de foration,

-une description des principales difficultés éventuellement rencontrées.

Nous préciserons les spécificités d'un chantier au microtunnelier, par rapport à une pose traditionnelle en tranchée, puis des indications sur les reconnaissances géotechniques adaptées à de tels ouvrages, et sur l'incidence des paramètres de terrains dans l'élaboration d'un projet de microtunnelier (faisabilité, choix de la machine, aleas).

2 EXPERIENCES DE CHANTIERS DANS LE VAL DE MARNE

2.1 BOISSY I (juin 1989)

Le chantier BOISSY I, pour un collecteur EP ϕ 500 mm, a été réalisé dans des éboulis sablo-argileux, à une profondeur de 2 à 2,5 m, en deux tronçons de 26 m et 70 m, avec une machine ISEKI "Unclemole" à marinage hydraulique.

La poussée sur la machine (fig.1) est faible et peu variable (environ 200 kN) ce qui montre l'efficacité de la lubrification par le gélifiant incorporé dans la boue pour réduire le frottement sur les tubes. Par contre, l'avancement s'est avéré sensiblement plus long (1H à 1H30 de foration par tube) pour le tronçon 1 dans un terrain argileux, que pour le tronçon 2 (en moyenne 0H30).

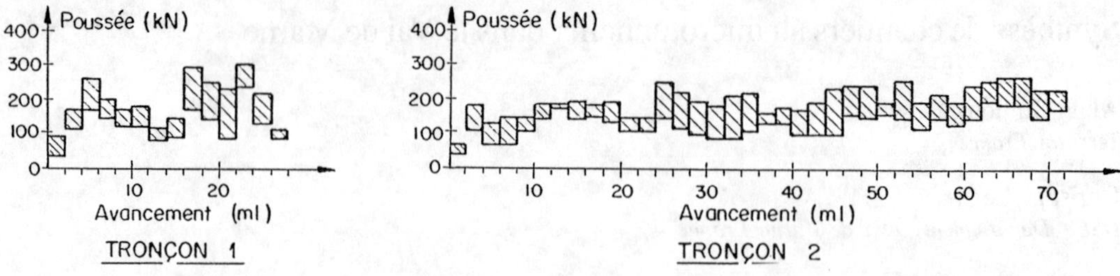

Fig 1. VARIATION DE LA POUSSEE EN FONCTION DE L'AVANCEMENT (BOISSY I)

2.2 CRETEIL I (décembre 1989)

Le chantier CRETEIL I, pour un collecteur EP ϕ 300 mm, a été réalisé dans des sables plus ou moins argileux avec silex, à une profondeur de 2,5 m environ, en deux tronçons de 40 m et 26 m, avec une machine HERRENKNECHT (marinage hydraulique).

La poussée sur la machine (Fig. 2) augmente fortement de 0 à plus de 600 kN en fonction de l'avancement (soit une contrainte de cisaillement moyenne de l'ordre de 20 kPa); elle apparait être en corrélation avec les déviations, surtout horizontales, qui ont atteint plus de 40 mm à plusieurs reprises (avec des déviations différentielles jusqu'à 80 mm sur deux tubes, soit 2 %).

Par contre, l'avancement a été en moyenne rapide (0H10 à 0H20 de foration par tube) .

2.3 BOISSY II (février 1990)

Le chantier BOISSY II, pour un collecteur EP ϕ 500 mm, a été réalisé dans des remblais et terrains sablo-argileux, à une profondeur de 2,5 à 5 m, en deux tronçons de 35 m et 54 m, avec une machine MARKHAM "super mini" (marinage hydraulique). Il passait en outre à 9m sous un remblai du RER.

Sur le tronçon 2 passant sous le RER, où les déviations sont restées tout à fait négligeables (moins de 10 mm), la poussée sur la machine est faible et peu variable (entre 200 et 250 kN - cf Fig. 3).

Par contre, sous le tronçon 1, de fortes déviations (atteignant 80 mm horizontalement et verticalement, et 120 mm en déviation différentielle sur 3 m environ, soit 4 %) semblent avoir conduit à une poussée qui augmente fortement avec le linéaire foré (Fig. 3), et qui atteint 1300 kN (soit une contrainte de cisaillement de l'ordre de 20 kPa). Pour un drive plus long, on aurait se rapprocher de la poussée maximale disponible (2000 kN).

L'avancement s'est avéré sensiblement constant (0H30 de foration par tube en moyenne). Le passage d'un bloc dur a pu être fait grâce au recul possible de la machine (tuyaux provisoires en acier). Les tassements sur les voies RER n'ont pas dépassé 3 mm.

2.4 CHAMPIGNY I (avril 1990)

Le chantier CHAMPIGNY I, pour un collecteur EU ϕ 400 mm, réalisé avec une machine MARKHAM à marinage hydraulique, comprenait deux tronçons:

-le tronçon n° 1, long de 200 m, a dû être abandonné après 30 m de foration, par suite du colmatage de la tête par de l'argile collante, malgré différents essais d'injection de produits fluidifiants et d'augmentation de puissance de la pompe,

-le tronçon n° 2, de 40 m, passant sous des voies SNCF à une profondeur de 2,5 à 5 m, a fait ensuite l'objet d'une reconnaissance par sondage horizontal avant de décider du lancement de la machine; il a rencontré des alluvions anciennes (graves argileuses) et des éboulis marneux plastiques sur 5 m environ.

Malgré ces conditions peu favorables du fait de la nature argileuse des terrains et de la présence des voies SNCF, qui interdisait de faire un puits en cas de blocage de la machine, il a été décidé de réaliser le projet comme prévu.

La poussée sur la machine est variable mais faible par rapport aux possibilités de la machine (100 à 400 kN pour 2000 kN). Elle apparait dépendre des déviations horizontales différentielles, qui ont atteint 40 mm sur 2 tubes, soit 1 % (Fig. 4).

La vitesse d'avancement s'est avérée sensiblement constante, en moyenne 0H40 de foration par tube, sauf à la traversée des terrains argileux, où elle a atteint 2 à 3 H (jusqu'à 6 H pour le tube traversant le passage d'éboulis marneux).

2.5 LA QUEUE EN BRIE (1991)

Le chantier de LA QUEUE EN BRIE, pour un collecteur EU ϕ 400 mm, long de 1450 m, dépassait le stade expérimental. Il traversait des terrains de nature et comportement très variables (Fig. 5) :

-sur 250 m environ, le toit du calcaire de Brie, avec des bancs silicifiés très durs : ce passage a été réalisé en tranchée du fait des difficultés prévisibles pour passer les bancs durs subhorizontaux : blocages, déviations excessives;

-sur 300 m environ des sables argileux, où avait été prévue une machine DECON-SOLTAU RVS 100, avec marinage à vis. Après deux essais de 3 et 4 m au cours desquels la vis a été détériorée sur des blocs, ces zones ont été faites à ciel ouvert;

-le reste du tracé sur environ 900 m traversait les argiles à meulières, constituées d'argiles plastiques avec des bancs de meulière silicifiée très durs, ce qui laissait craindre des risques de "collage" de l'argile et de blocage

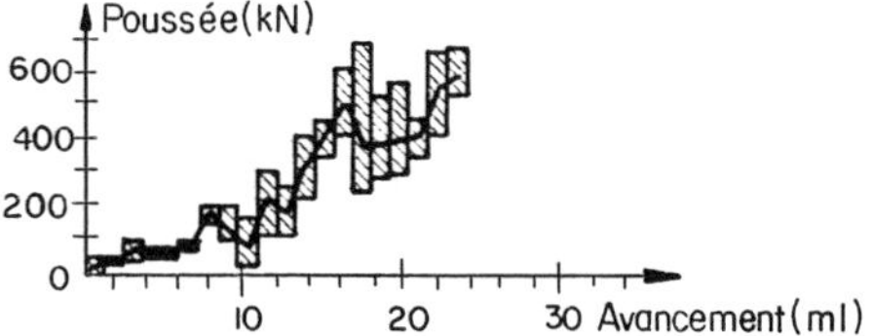

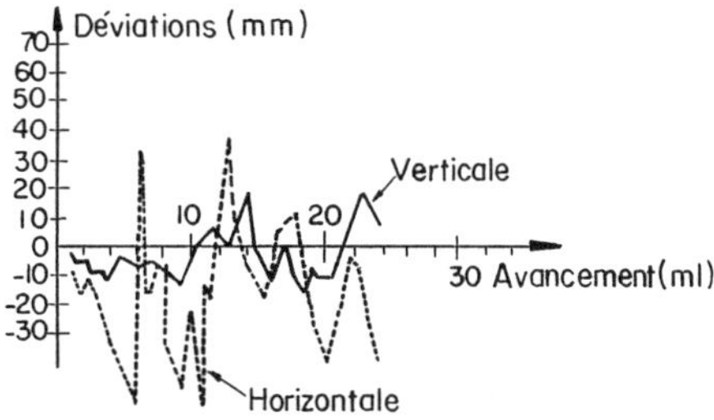

Fig 2. CORRELATION ENTRE POUSSEE ET DEVIATIONS (CRETEIL I - Tronçon 1)

ou déviation sur les bancs durs.

Le chantier a été réalisé avec une machine HERRENKNECHT AVN 500 à marinage hydraulique, qui a fait environ 800 m, et a rencontré les principales difficultés suivantes :

-un gros bloc de 0,5 m³, poussé sur 2 à 3 m jusqu'au puits voisin, et une dalle encore plus importante (1 m³), qui a imposé le creusement d'un puits pour dégager la tête;

-marinage difficile (colmatage, bourrage) lorsque la machine avançait rapidement. Les nombreuses pannes mécaniques semblent être en partie liées à ces problèmes;

-à la traversée d'une poche d'argile molle, plongement de la machine de 50 cm, ce qui a conduit à terminer le tronçon à ciel ouvert;

-usure importante de la tête, qui nécessitait sa réfection après chaque drive.

La poussée sur la machine a été très variable, en moyenne 300 à 400 kN, mais a atteint plus de 2000 kN lors des passages de blocs.

L'avancement, à deux postes par jour, s'est avéré assez faible : pour des drives longs de 60 à 94 m, l'avancement moyen a été de 3,2 m par poste (1,4 m au minimum, et un "record" de 16 m).

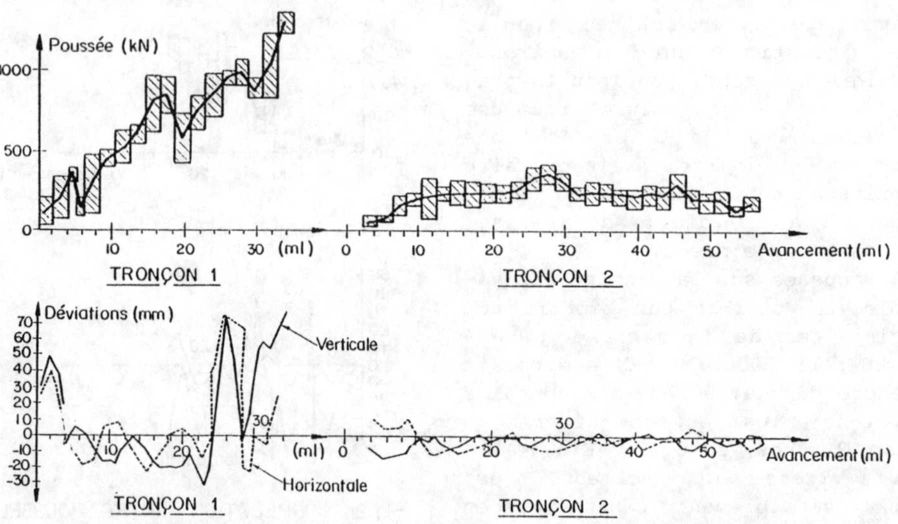

Fig 3. CORRELATION ENTRE POUSSEE ET DEVIATIONS (BOISSY II)

2.6 CHEVILLY-LARUE (1991)

Le chantier de CHEVILLY-LARUE (collecteur EP ϕ 600 mm), long de 50 m, traversait les marno-calcaires de Brie avec des calcaires siliceux et meulières. Lors du creusement des puits, les blocs se sont avérées plus nombreux et durs que prévu.

Il a été décidé de rabaisser le profil en long de 4 m, pour l'inscrire dans les argiles vertes, argiles plastiques peu favorables pour un microtunnelier, mais avec l'avantage de constituer un horizon très homogène et sans blocs.

Le chantier, réalisé avec une machine HERRENKNECHT AVN 600 équipée d'un élargisseur, a pu se terminer dans des conditions acceptables, avec un avancement lent (30 jours), mais sans blocage. Ce chantier a fait l'objet d'essais d'additifs biodégradables (polymères) dans la boue, qui ont réduit le frottement sur les tuyaux et bien enrobé les mottes d'argile, évitant de les agglomérer dans la tête.

2.7 Résumé des observations

Ces différents chantiers, dont l'expérience est corroborée par d'autres, ont montré les principales difficultés suivantes :

-présence de blocs ou de bancs durs, de dimensions supérieures au tiers environ du diamètre, que la tête ne peut broyer facilement, et qui peuvent dans certains cas bloquer la machine. Les obstacles "artificiels" (anciens collecteurs non inventoriés) peuvent bien sûr conduire aux mêmes problèmes;

-présence d'argiles ou de marnes collantes, qui colmatent la tête et créent des bouchons dans les circuits de marinage, et limitent ainsi fortement l'avancement.

-maîtrise des déviations lors du guidage, qui lorsqu'elles sont supérieures à 1 ou 2 % conduisent à une forte augmentation de la poussée, et peuvent créer des sollicitations parasites dans les tuyaux (poussée non axée, d'où des moments fléchissants) défavorables à leur bon comportement.

De plus, on peut noter que les vitesses d'avancement sont très variables : le temps de foration est souvent compris, pour un tube de 2 m, entre 20 et 60 min (soit une vitesse instantanée de 30 à 100 mm/min), et le temps global d'un cycle entre 35 et 100 min/tuyau. Ceci

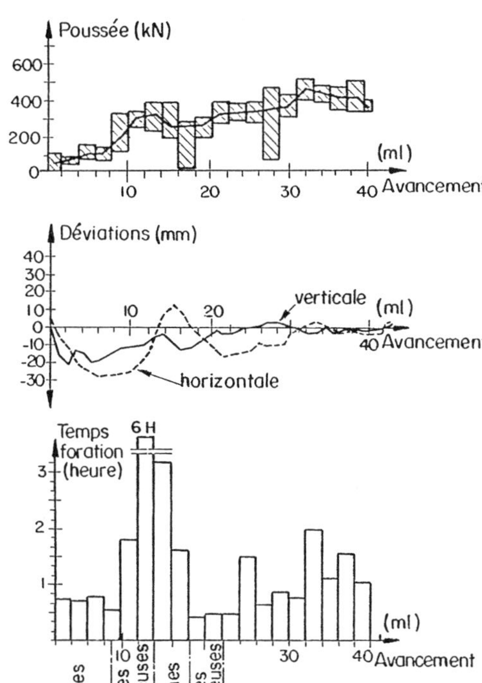

Fig 4. CORRELATIONS POUSSEE, DEVIATION, TEMPS DE FORATION (Champigny 1. Tronçon 2)

conduit à des avancements de l'ordre de l'ordre de 8 à 20 ml par poste de 8 H, mais parfois beaucoup plus faibles, de l'ordre de 2 à 3 m, dans les terrains argileux.

3 SPECIFICITES D'UN PROJET DE MICROTUNNELIER

Si la pose de canalisations en tranchée se déroule en général dans des conditions acceptables malgré des données géotechniques incomplètes ou inexistantes, un projet de microtunnelier est à considérer comme un chantier de souterrain, avec tous les aleas inhérents à ce type d'ouvrage.

Cette spécificité par rapport aux souterrains traditionnels est encore accentuée par les aspects suivants :

-les petites dimensions rendent la tâche plus délicate : il est difficile d'identifier les problèmes au front, et d'y intervenir;

-la majorité des chantiers se font à faible profondeur, c'est à dire dans des horizons très hétérogènes : formations superficielles, alluvions, colluvions, avec de fréquents passages latéraux de faciès, remblais divers avec ouvrages enterrés.

C'est dire qu'un microtunnel est particulièrement vulnérable aux hétérogénéités de terrains, parce qu'elles sont plus fréquentes, et que la petite taille de l'excavation rend le creusement beaucoup plus sensible à des variations dans la nature et la dureté des terrains.

Si on considère le fait que le coût d'un microtunnelier est relativement réduit (en général de 5000 à 10000 F/ml), le coût d'une reconnaissance géotechnique permettant de lever les principaux aleas peut sembler exhorbitante par rapport au coût de l'ouvrage. Pourtant, les facteurs géotechniques ont une incidence déterminante sur le choix d'une machine adaptée et donc sur la réussite d'un chantier.

4 RECONNAISSANCES POUR UN MICROTUNNEL

Une reconnaissance d'un projet au microtunnelier doit permettre de répondre aux questions suivantes :

-Puits et massifs de réaction : nature et propriétés mécaniques des terrains, pour choisir les techniques d'exécution et dimensionner les blindages et le massif de réaction. Ce sont là des questions faisant appel aux méthodes de reconnaissance usuelles.

-Excavation et marinage : le choix du matériel doit s'appuyer sur une bonne connaissance de la nature des terrains à traverser, et de leur hétérogénéité. Si l'expérience locale peut apporter beaucoup de renseignements, cette connaissance doit être complétée par des sondages carottés, à la tarière ou à la pelle, qui permettent d'observer les

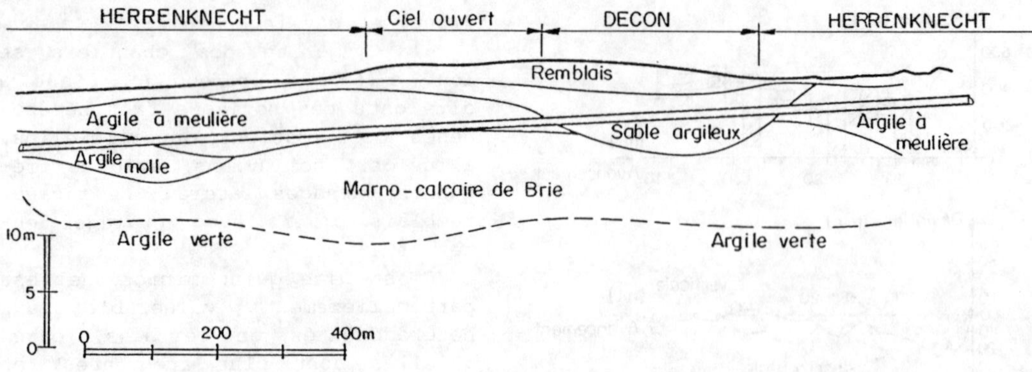

Fig5. LA QUEUE EN BRIE : PROFIL EN LONG GEOLOGIQUE

terrains et de prélever des échantillons pour essais en laboratoire.

Il convient de souligner que les essais in-situ (pressiomètre et péné-tromètre) sont à utiliser avec précaution : en effet, s'ils peuvent parfois permettre de préciser la stratigraphie par continuité entre sondages, ils présentent l'inconvénient (surtout pour le pressiomètre) de donner des caractéristiques mécaniques qui "moyennent" le comportement des terrains et masquent des hétérogénéités décimétriques, dont les conséquences peuvent être déterminantes pour un microtunnel. Ces essais in-situ sont donc à réserver au dimensionnement des puits et des tuyaux.

Il est préférable de privilégier les essais de laboratoire, qui permettent d'identifier les terrains :

-analyses granulométriques et mesure de teneur en eau;

-limites d'Atterberg pour évaluer les risques de collage; on peut considérer que ce risque devient important lorsque l'indice de plasticité est supérieur à 20 ou 30,

-résistance en compression simple sur les blocs durs, voire essais de dureté et d'abrasivité, pour évaluer la possibilité de passer ces blocs. On peut estimer qu'il est difficile de passer des blocs de dimensions importantes dont la résistance en compression dépasse 20 à 50 MPa .

De plus, la principale difficulté réside dans l'évaluation de la taille et fréquence des blocs, qu'aucun sondage classique ne permet d'apprécier correctement. Seuls des forages ou puits de grand diamètre (0,5 à 1 m) peuvent fournir des indications fiables, mais il sont bien sûr chers. Les puits de fonçage et de réception peuvent être utiles dans cet esprit, mais à un stade où il est souvent difficile de modifier le projet.

Enfin, la connaissance du niveau de la nappe par piézomètres, est également utile pour le choix d'un confinement éventuel du front.

-Détection des obstacles : le problème des obstacles de grandes dimensions (ouvrages enterrés ou gros blocs - > 1 m³ -) reste un problème majeur auquel seules les méthodes géophysiques semblent apporter une réponse partielle.

Parmi la panoplie des méthodes disponibles celle du "radar" semble être actuellement la méthode pouvant apporter le plus de données. En effet, son emploi et son interprétation relativement simples permettent de l'utiliser à un coût réduit sur la grande majorité des projets, et il est peu perturbé par l'environnement (vibrations, topographie).

Mais sa profondeur d'investigation reste limitée (souvent 3 à 5 m, parfois 10 m ou plus dans des conditions favorables), et peut être insuffisante pour certains projets.

Dans son domaine d'application, elle peut détecter des obstacles enterrés, mais peut rarement en garantir formellement l'absence.

5 INCIDENCE DES TERRAINS SUR LE CHOIX DE LA MACHINE ET LE DEROULEMENT D'UN CHANTIER

5.1 Adaptation de la machine

Une bonne connaissance des terrains est indispensable :

1. d'une part pour optimiser le tracé et en particulier le profil en long : lorsqu'on dispose d'une certaine latitude, il est préférable de choisir un profil qui reste dans un niveau géologique homogène, et d'éviter les horizons avec concentration de blocs;

2. d'autre part pour optimiser le choix de la machine :

-les microtunneliers à tête ouverte (sans pression de confinement) et à vis sont à réserver aux terrains avec une cohésion et/ou hors d'eau; ils apparaissent bien adaptés aux terrains argileux, mais passent difficilement les blocs durs; de plus dans les terrains raides (R_C > 5 à 10 MPa), l'avancement peut être ralenti et solliciter la machine de façon excessive;

-les microtunneliers à marinage hydraulique conviennent bien aux sols sablo-limoneux, même sous nappe, et permettent avec un broyeur de passer des blocs de résistance assez forte (R_C = 20 à 50 MPa) si leurs dimensions restent inférieures au 1/3 du diamètre. Ils semblent assez bien adaptés aux terrains hétérogènes, bien que la traversée d'argiles ou de marnes plastiques crée des difficultés de marinage et nécessite l'utilisation d'adjuvants dans la boue;

-enfin, les microtunneliers à "tête rocher" que permettent de forer des terrains rocheux (R_C = 20 à 150 MPa), mais présentent des risques lors des passages "mous".

De plus, il faut attirer l'attention sur la possiblité de recul de la machine, qui offre une sécurité certaine vis à vis des obstacles. Cette sécurité est importante pour la traversée de sites (voies SNCF, constructions) où il est exclu de réaliser un puits pour aller dégager une machine bloquée.

5.2 Paramètres de conduite

Même lorsque l'on a déterminé avec précision les conditions géotechniques d'un projet, la conduite de la machine lors de la foration joue encore un rôle essentiel.

En particulier les déviations ont une grande incidence sur le comportement lors du poussage : en effet, si le guidage ne maintient pas les déviations dans des limites raisonnables (1 à 2 % de déviation angulaire), la poussée augmente fortement avec le linéaire et peut atteindre les limites de la machine; de plus ces fortes poussées, associées à des efforts non axés sur les tuyaux, conduisent à des sollicitations pour lesquelles ils ne sont pas conçus, et pouvant aller jusqu'à la fissuration.

La maîtrise de la vitesse des vérins de poussée joue également un rôle important dans des terrains hétérogènes : à l'approche d'un passage dur, une faible vitesse peut permettre de "grignoter" doucement ce point dur et d'éviter des déviations importantes. Une telle réduction fait chuter les cadences, mais finalement beaucoup moins qu'un arrêt prolongé ! Dans les terrains argileux une faible vitesse permet également de limiter les problèmes de colmatage.

6 CONCLUSIONS

L'utilisation récente en France des microtunneliers laisse subsister certaines questions, dont une grande part est liée à l'hétérogénéité des terrains que l'on rencontre en particulier la région parisienne : sables, argiles et marnes compactes, grès et calcaires, avec souvent des blocs durs importants. Cela pose la question, pas toujours bien résolue, des méthodes de reconnaissance des terrains et des obstacles.

De plus, lorsqu'on peut considérer que les terrains sont "bien connus", le choix d'un matériel adapté fait encore l'objet de débats et d'expériences, qui ont le mérite de faire avancer la connaissance du procédé, même lorsqu'elles apparaissent négatives.

Enfin la conduite de la machine lors de la foration joue également un rôle déterminant : les déviations doivent être limités au mieux, et la vitesse de foration peut être dans certains cas être utilement réduite.

No Trenches in Town, Henry & Mermet (eds) © 1992 Balkema, Rotterdam. ISBN 90 5410 085 0

Survey of river floors before horizontal drillings

Richard Lagabrielle
Laboratoire Central des Ponts et Chaussées, Bouguenais, France

ABSTRACT : Horizontal bore-holes which are designed for the crossing of rivers by various utilities (gas pipes, electrical cables, etc...) should be drilled in materials as homogeneous as possible and keep clear of gravelous alluviums, especially if they might contain blocks. Indeed, this type of material very often forms the river floors and is often thicker than 10 m. The drilling of one or several survey bore-holes on each river shore makes it possible to identify the material but is not liable to correctly define the geometry of the alluviums layer. We intend to describe a geophysical method -direct current resistivity prospecting in water- which, when the measurements are performed in the river before the survey bore-holes, makes it possible to optimize their position and make the maximum profit from them in order to know the geometry of the interface between the alluviums and the homogeneous material over which they are. This method is now well known, its price is several orders of magnitude lower than the price of bore-holes drilled in a river and of course than the price of the horizontal bore-hole. It should result in significantly lower costs for the works. The method will be described from the point of view of its principles, its field implementation, the interpretation techniques. Examples of actual field surveys will be shown.

INTRODUCTION

Horizontal drillings which are designed for the crossing of rivers or canals by various utilities (gas or water pipes, electrical cables) are more and more currently used. The drilling techniques have been adapted from the accumulated experience of the drillers in the oil industry. Though very efficient, their cost is also quite high and the added costs, occuring after the works stops and the starting again of drillings due to the presence of undetected underground obstacles, may sometimes double or even triple the expected price of the horizontal borehole. It is thus the interest of the owner to make sure that not ill surprise occurs, due to uncertainties in the knowledge of the underground constitution, through the organization of a good quality geological survey before the drilling begins.

Among the survey methods, mechanical vertical soundings are nearly always performed. It is indeed the only method which makes it possible to gather soil samples on which tests can be carried out. It also allows an accurate determination of the position of the interfaces between the encountered geological layers. It has however important drawbacks which it is suitable to remember :

- It is a survey method based on the sampling of the ground and, in order to get representative results, there should be a great number of soundings. If not, the conslusions drawn cannot be trusted
- Boreholes performed in water are expensive. Their number is consequently kept as low as possible, which means that the uncertainly is increased.

Mechanical soundings are necessary. Their number is never large enough if they constitute the sole survey method used. We propose here a complementary survey method -a geophysical method- the purpose of which is to build a general image of the ground structure and constitution along the profile of the horizontal drilling.

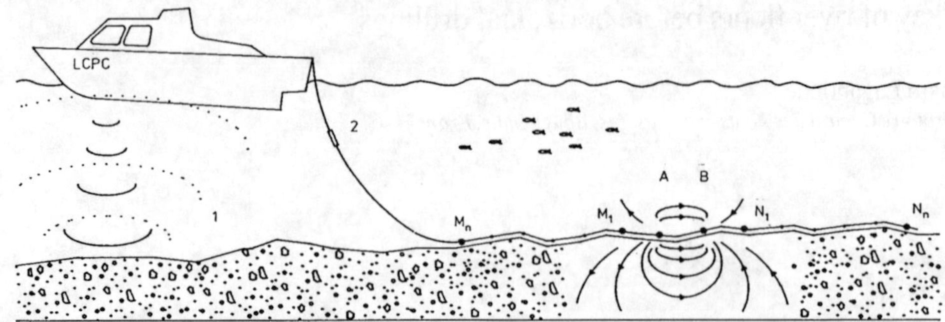

Fig. 1. Diagram showing the operation of direct current electrical prospection with electrodes on the water floor. The current flows from electrodes A and B in the water and in the underwater layers. The electrodes pairs M_1, N_1..., M_iN_i, ..., M_nN_n are used to measure apparent resistivities. Simultaneously, the water resistivity is measured with a small probe (mark 2), as well as the water depth with a sonar (mark 1). The *n + 2* data are interpreted to determine the geolectric structure of the medium. The device is moved countinuously at a speed of some km/h.

It is a technique of direct current electrical prospecting, with the electrodes along the water floor, which is easy and fast to operate and which produces information related to the nature and quality of the materials, through the repartition of their resistivity, as well as to their quantity : layers thickness, limits of the geological formations.

In this paper, we briefly describe the principles and operation of the method and we show through a few examples how the measurements results can be interpreted.

1. PRINCIPLES OF THE METHOD

The principles of the method and its theoretical justification have been described by Lagabrielle and Teilhaud (1981), a theoretical complement has been given by Lagabrielle (1983) and the adaptation of the method to offshore conditions has been discussed by Lagabrielle (1984).The basis of quantitative interpretation are briefly described by Lagabrielle (1992). Shroder et Bistrup (1987) have designed a method not very different and quoted its application to archeological prospection at sea. Whiteley (1974) and Scott et Maxwell (1989) describe a similar method for mining exploration at sea. However, in the three latter cases, the electrodes are on the water surface.

Let us recall the basic principles of our method (fig. 1) :

A multi-conductor cable, along which a pair of current electrodes A and B has been installed as well as several pairs of potential electrodes Mi and Ni for i = 1 to n, is laid on the water floor. The cable is drawn behind a boat which carries the current generator, the potential measurement device, the data acquisition system, a sonar and a water resistivity probe. With this system, in a lateraly countinuous manner, one can record n apparent resistivity profiles or a profile of electrical soundings with n array lengths.

The digital data acquisition makes it possible to plot the apparent resistivity profiles and the water resistivity (Fig. 2). The examples described bellow show what type of documents can result from the measurements interpretation.

Remarks

- We use a regulated current, alternatively positive and negative. The alternance duration is 5s. Potential is measured at the end of each alternance. The half difference of the corresponding potentials is the relevant result, already corrected from the electrodes polarization. The half sum of the two potentials measures the said polarization which can eventually be interpreted.
- Current electrodes A and B are placed in the vicinity of the centre of the quadrupolar array Mi, A, B, Ni. The reciprocity theorem is applied. These arrays are not of a classical type such

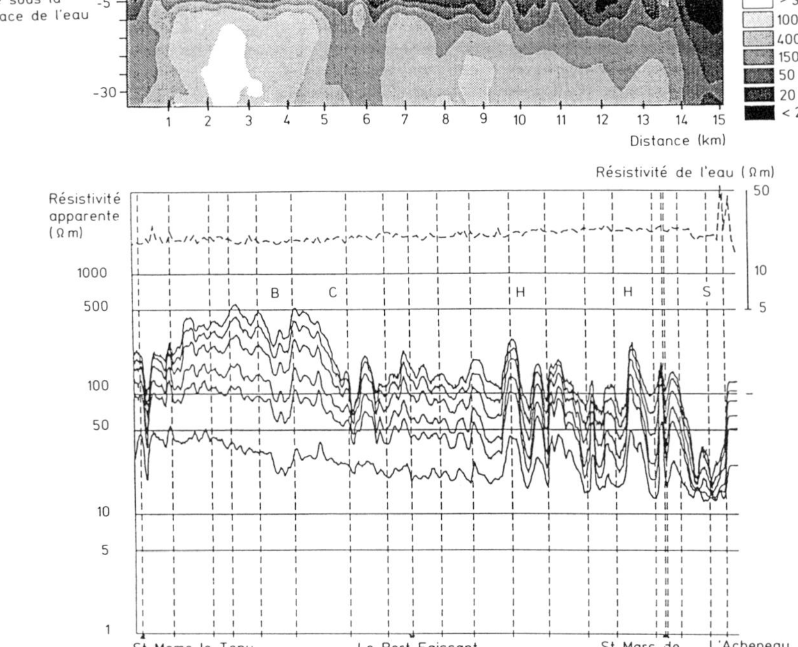

Fig. 2. Six apparent resistivity profiles in a soft water river along a distance of about 15 km. The lowest apparent resistivity values were measured with a 10 m long array ; the highest with a 100 m long array. Above : quantitative interpretation of the apparent resistivity profiles and vertical section of the ground above the profiles. The section is a ground resistivity model.

as Schlumberger or Wenner type. The interpretation of electrical soundings is done with a computer program and no master curves catalog is needed.

- In the examples bellow, the number of potential electrodes pairs is generally 7, lengths being as long as 200 m. This can appear to be a small number of data for a correct quantitative interpretation of an electrical sounding. The number of data for a given sounding is actually 9 because the water depth and its resistivity are known. In a three layer model, only three unknows are still to be determined : thickness and resistivity of the second layer (the soft alluvium) and resistivity of the bed rock. The first layer (water) being precisely known, the interpretation is strongly constrained.

2. THEORETICAL EXAMPLE

Figure 3 illustrates, on an artificial example, what can happen if one is satisfied with mechanical sounding as sole survey method.

The problem here is to cross a river with a horizontally drilled borehole. The river is about 500 m wide. It flows in its alluvium which is made of mud, sand, gravels and pebbles. They lay on a thick layer of rather soft marls. The maximum depth of the river is some 10 m, the alluvium layer is some metres thick.

To avoid the drilling difficulties in the pebbles and the blocking of the drill bit due to the presence of an undetected boulder, as well as the orientation problems which always occur is such a material, it is decided to drill entirely in the marls, which are homogeneous, resistant enough and easy to drill. In order to monitor the drilling process in such a way that the hole always remains deeper than the interface between the alluvium and marly layers, one has thus to determine its position.

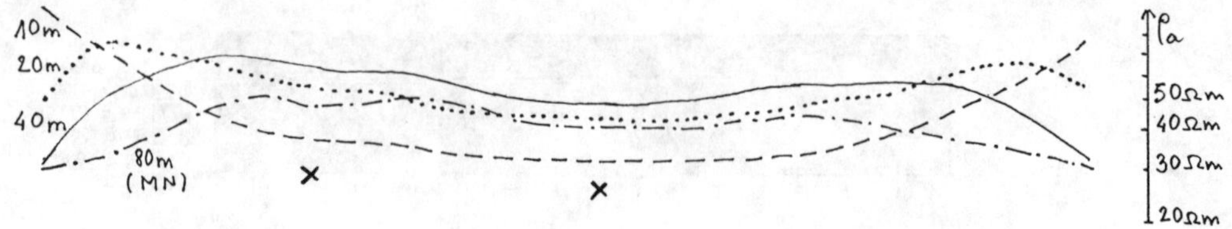

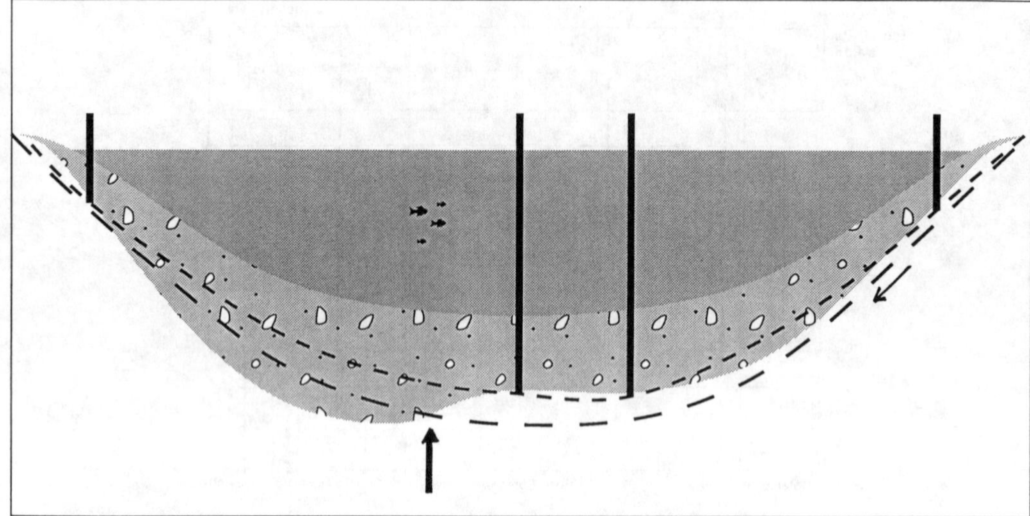

Fig. 3. Below : Geological section across a river. It flows in an alluvium layer made of mud, sands and gravels, above a thick layer of marls. The thick vertical lines are the survey boreholes made without any previous geophysical survey. Short dotted line : estimated interface between alluvium and marls. Long dotted line : planned horizontal drilling profile. Vertical arrow : blocking of the horizontal drilling.

Above : Four apparent resistivity profiles along the water floor with array lengths from 10 to 80 m. The two cross marks show the suggested positions of the survey boreholes.

Without any other previous information, the soundings have been placed as shown on the figure : two near the river shores and two in the middle of the river.

The designer knows then four points given by the boreholes and two points where the marl outcrops on the shores. They can be used to determine the interface geometry. He draws the short dotted line as its best estimate. With up to two metres of security, he can draw under it the long dotted line which is the planned horizontal drilling profile.

If the drilling starts from the shore on the right of the figure, it will carry on without any difficulty along two thirds of the profile and then it will get stuck by the gravels. After that, their will be no other solution than to start again from the beginning with a slighty deeper profile, that is with a larger curvature, without knowing to what amount this curvature should be increased. The cost of this works stop and starting again will certainly have to be paid by the owner.

It is here clear that the errors occur from a bad positionning of the survey boreholes. The previous operation of the geophysical method described here, across the river, could have helped to correctly choose the position of the survey holes.

The top of the figure exhibits indeed four apparent resistivity profiles measured along the water floor above the planned horizontal drilling profile. They show, before any quantitative interpretation, that the thickness of the alluvium layer varies

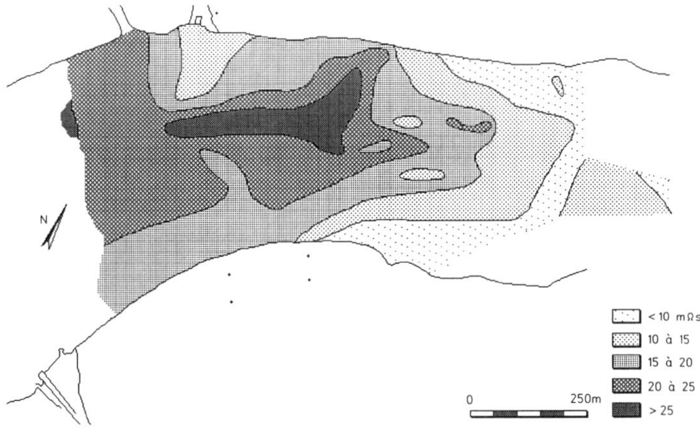

Fig. 4. Salt water river. Map of the longitudinal conductance C of the soft alluvium. This map can be read as a thickness map.

in a more complicated manner than the water depth. The two cross marks indicate the suggested positions for the survey mechanical soundings. The results obtained from the latter make it possible to quantitatively interpret the geophysical data and thus to correctly determine the alluvium marls interface.

3. SURVEY OF THE ALLUVIUM IN A BRACKISH WATER RIVER

The following example shows results obtained in a real application, which was not related to any horizontal drilling, but which was a problem of bridge construction. In fact, it is a type of survey which can be performed in the same manner, this is why we quote it.

Figure 4 shows the result of a survey aimed at the evaluation of the amount of soft alluvium under a brackish water river (water resistivity : 2 Ωm) over which a bridge construction is planned. The bridge position as well as the foundations design depend on the alluvium repartition, thickness and quality.

The river is about 500 m wide. 8 longitudinal, 2 km long, resistivity profiles have been recorded, simultaneously with 7 quadrupolar arrays 10 to 200 m in length.

The measurements results show that the electrical soundings data can be interpreted with a three layer model : water, alluvium and bedrock (of infinite resistivity). The longitudinal conductance C of the alluvium is easily calculated, since the water longitudinal conductance is known (C is the ratio of thickness to resistivity of a layer). Figure 4 is a map of C. Boreholes drilled ashore, near the zone where C exhibits the highest values, have shown that the alluvium thickness is about 30 m. This means that the map can be read as a map of alluvium thickness.

Only one day of work has been necessary for the data acquisition. Their quantitative interpretation took one other day.

CROSSING OF THE RIVER GARONNE DOWNSTREAM FROM BORDEAUX

Last, Figure 5 is a photocopy of a raw field document. The problem was to cross the Garonne downstream from the city of Bordeaux. As it is often the case, no complete previous survey had been made, but for two vertical boreholes in the middle of the river (vertical continuous arrows). They exhibited a thickness of the gravels layer of 12 and 13 m. By pure bad luck, the boreholes had been placed where the layer had the smallest thickness as it has been shown by the two successive works stops due to too small an estimate of this gravels layer thickness. We have operated after the two stops, asked by the owner.

Seven apparent resistivity profiles with array lengths from 20 to 200 m have been recorded and were placed near the survey boreholes. The latter have

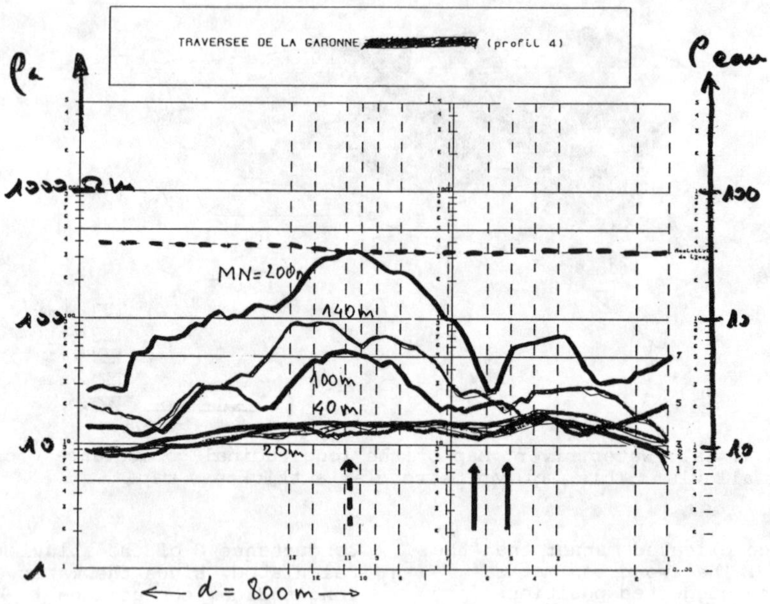

Fig. 5. Seven resistivity profiles along the water floor across the river Garonne. The shortest four arrays (from 20 to 40 m) are not influenced by the alluvium thickness (water is too deep). Both arrays, 100 and 140 m long, correctly reflect the thickness variations. The longest array (200 m) is difficult to interpret (lateral effects due to the shores). The two countinuous arrows indicate where the two survey boreholes have been performed. The discountinuous arrow indicates where one of them should have been bored.

helped us to feed the quantitative interpretation software. It is not so easy because the assumption of tabularity is not stictly valid for all the array lengths. For the longest array (which corresponds to the highest apparent resistivities) the data have not been used for quantitative interpretation for they are clearly influenced by the lateral variations of the structure. The six other profiles have been used to determine the thickness of the alluvium with the help of the drilling data for calibration. The interpreted thickness is everywhere bigger than 14 m, which readily explains the works stops. After the geophysical survey, the drilling started again, taking the survey results into account, and it came to its end without any further difficulty.

On the figure a dotted vertical arrow has been added to show where a survey borehole would have been suggested instead of one of the two which have been made (countinuously line arrows).

Half a day was necessary for the field measurements, one other day for their quantitative interpretation (which was not absolutely necessary).

CONCLUSION

In the case of the underground survey across a river, quantitative interpretation is not often possible, when no mechanical soundings results are available, for the tabularity assumption for the underground structure is rarely valid.

One should be convinced that the various survey methods should neither be considered as competitive nor as excluding one another. On the contrary, one should use a real survey methodology which harmoniously marries geology, geophysics and boreholes with in situ tests and laboratory tests on soil samples.

In any case, mechanical soundings are necessary. In water surrounding, they are very expensive. Direct current electrical prospecting is very fast and cheap. It makes it possible to choose the position of the survey boreholes in an optimal manner and to make the maximum profit of them with a negligible added cost, compared with the money saved by the suppression of unforeseen geological risks.

REFERENCES

LAGABRIELLE R., TEILHAUD S. (1981), Prospection de gisements alluvionnaires en site aquatique par profils continus de résistivité au fond de l'eau. Bull. Liaison Labo. P. et Ch. 114 : 17-24.

LAGABRIELLE R., (1983), The effect of water on direct current resistivity measurement from the sea, river or lake floor. Geoexploration, 21 : 165-170.

LAGABRIELLE R., (1984), La prospection électrique par courant continu en mer. Bull. Liaison Labo. P. et Ch. 132, juil-août : 5-11

LAGABRIELLE R., (1992), Quantitative interpretation of vertical electrical soundings underwater with any quadrupolar array. 54th meeting EAEG, Paris : 358-389

SCOTT W.J., MAXWELL F.F. (1989), Marine resistivity survey for granular materials, Beaufort Sea. Can. Jour. Expl. Geophys., 25, 2 : 104-114.

SHRODER N., BISTRUP T. (1987). Resistivity at sea : Inversion and interpretation of measurements. SEG. New Orleans expanded abstract : 94-95.

WHITELEY R.J. (1974), Design and preliminary testing of a countinuous offshore resistivity method, Bull Aust. Soc. Explor. Geophys., 5, 1 : 9-14.

No Trenches in Town, Henry & Mermet (eds) © 1992 Balkema, Rotterdam. ISBN 90 5410 085 0

Reconnaissance du fond des cours d'eau en vue de leur traversée par des forages dirigés horizontaux

Richard Lagabrielle
Laboratoire Central des Ponts et Chaussées, Bouguenais, France

RESUME : Les forages horizontaux dirigés destinés à la traversée des cours d'eau pour des installations diverses (conduites de gaz, câbles électriques, etc...) doivent être réalisés dans des matériaux aussi homogènes que possible et en tout état de cause éviter les matériaux graveleux, surtout s'ils sont susceptibles de contenir des blocs. Or ce type de matériaux forme souvent le fond des cours d'eau sur des épaisseurs qui peuvent être variables latéralement et souvent supérieures à 10 m. La réalisation d'un ou deux forages de reconnaissance dans le lit de la rivière et d'un forage sur chacune des rives permet de caractériser les matériaux mais est tout à fait insuffisante pour définir correctement la géométrie de la couche d'alluvions. Nous proposons de décrire une méthode géophysique -la prospection électrique par courant continu en site aquatique- qui, mise en œuvre dans les cours d'eau avant la réalisation des forages de reconnaissance, permet d'implanter ces derniers de manière optimale et d'en tirer le maximum de parti pour connaître la géométrie de l'interface séparant les alluvions du matériau homogène sous-jacent. Cette méthode est maintenant bien au point, le prix de sa mise en œuvre est sans commune mesure avec celui des forages en site aquatique et à plus forte raison avec celui de l'ouvrage (le forage horizontal dirigé). Elle doit conduire à des économies substantielles sur la réalisation des travaux.
La méthode est décrite du point de vue de son principe, de sa mise en œuvre, des techniques d'interprétation. Des exemples de reconnaissances effectivement réalisées sont montrés.

INTRODUCTION

Les forages horizontaux dirigés destinés à faire traverser des fleuves, des rivières ou des canaux par des canalisations (d'eau, de gaz, d'hydrocarbures...) ou par des câbles de transport d'énergie électrique sont de plus en plus souvent utilisés. Ils font appel à des techniques adaptées à partir de l'expérience acquise par les foreurs de l'industrie pétrolière. Ces techniques, très performantes, sont aussi très chères et les surcoûts entraînés par les arrêts de chantiers et les reprises de forages à cause de la rencontre d'obstacles souterrains imprévus peuvent parfois doubler et même tripler le prix prévu de l'ouvrage.
Il est donc de l'intérêt du maître d'ouvrage de se prémunir contre les mauvaises surprises dues à une connaissance insuffisante du sous-sol en s'assurant qu'une reconnaissance géologique préalable de bonne qualité soit faite.

Parmi les méthodes de reconnaissance, la réalisation de forages mécaniques est presque toujours pratiquée. C'est en effet la seule méthode qui permette le prélèvement d'échantillons de sol sur lesquels des essais peuvent être réalisés. Elle fournit aussi une détermination précise de la position des interfaces entre couches géologiques traversées. Elle souffre cependant d'inconvénients qu'il convient de rappeler :
- C'est une méthode de reconnaissance par échantillonnage et, pour que les résultats qu'on en tire soient représentatifs, il faut donc que les forages de reconnaissance soient très nombreux. Dans le cas contraire, les

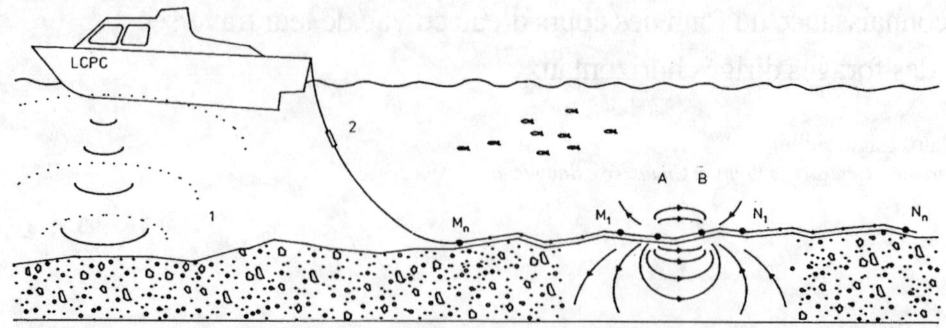

Figure 1 - Schéma illustrant la mise en oeuvre de la prospection électrique par courant continu avec électrodes au fond de l'eau. Le courant est injecté par les électrodes A et B et circule à la fois dans l'eau et dans les matériaux subaquatiques. Les paires d'électrodes M_1, N_1, …, MiNi, … MnNn, servent à mesurer n résistivités apparentes. Simultanément, on mesure la résistivité de l'eau au moyen d'une petite sonde (repère 2) et la profondeur au moyen d'un écho-sondeur (repère 1). Ces n + 2 données sont interprétées pour connaître la structure géoélectrique du milieu. L'ensemble est déplacé en continu à une vitesse de quelques km/h.

conclusions sont nécessairement incertaines.

- Les forages réalisés dans l'eau sont chers. On a donc tendance à en réduire le nombre, augmentant ainsi l'incertitude des résultats.

Les sondages mécaniques sont indispensables. Ils sont toujours en nombre insuffisant s'ils constituent la seule méthode de reconnaissance utilisée.

Nous proposons ici une méthode de reconnaissance complémentaire- une méthode géophysique -dont le but est de fournir une image générale de la structure et de la constitution du sol le long du tracé du forage horizontal dirigé.

Il s'agit d'une technique de prospection électrique par courant continu, avec électrodes au fond de l'eau, dont la mise en œuvre est très rapide et qui fournit des renseignements aussi bien sur la nature et la qualité des matériaux, à travers la répartition de leur résistivité, que sur leur qualité : épaisseur des couches, limites des formations géologiques.

Dans cet article, nous rappelons le principe de la méthode et de sa mise en œuvre et nous décrivons sur quelques exemples la manière dont on interprète les résultats des mesures.

1. PRINCIPES DE LA METHODE

Les principes de la méthode et sa justification théorique ont été décrits par Lagabrielle et Teilhaud, 1981, un complément théorique a été donné par Lagabrielle, 1983, et l'adaptation de la méthode à la reconnaissance en mer a été discutée par Lagabrielle, 1984. Les bases de l'interprétation quantitative sont évoquées par Lagabrielle, 1992. De leur côté, Shroder et Bistrup, 1987, ont mis au point une méthode à peine différente et cité son application à la prospection archéologique en mer. De même, Whiteley, 1974 et Scott et Maxwell, 1989 décrivent une méthode analogue pour la prospection minière en mer. Cependant, dans ces trois derniers cas, les électrodes sont à la surface de l'eau.

Rappelons les principes de base de notre méthode (Figure 1).

Un câble multiconducteur le long duquel on a réparti une paire d'électrodes, A et B, d'injection de courant et plusieurs paires d'électrodes, Mi et Ni i = 1 à n, de mesure de potentiel, est posé au fond de l'eau. Ce câble est traîné derrière un bateau sur lequel se trouve le générateur de courant, le système de mesure des différences de potentiel, le système d'acquisition de données ainsi qu'un échosondeur et une sonde de mesure de la résistivité de l'eau. On réalise ainsi de manière

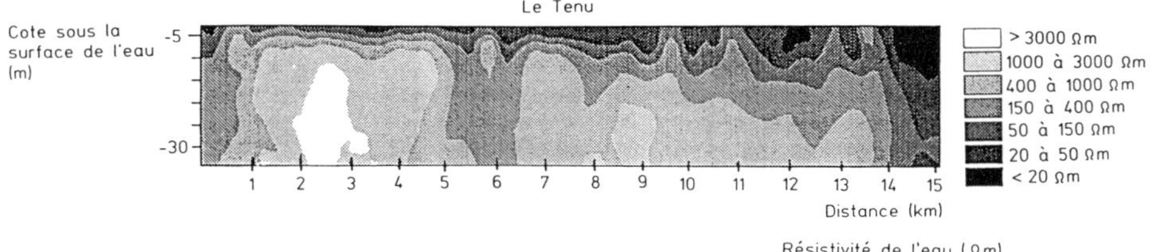

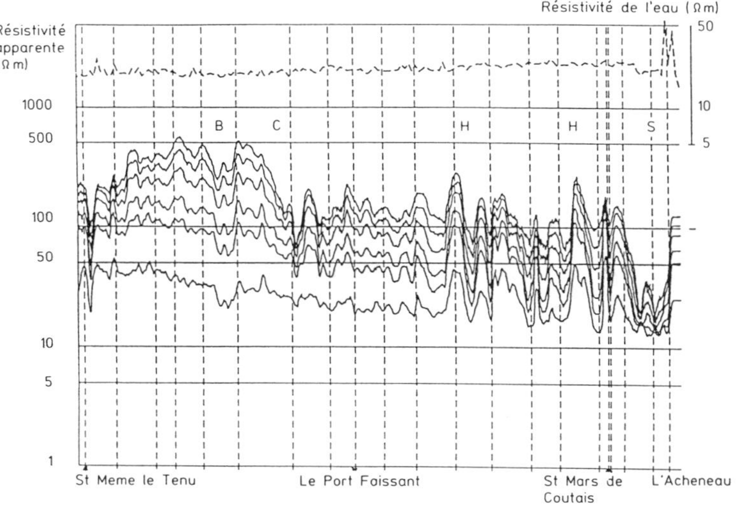

Figure 2 - Six profils continus de résistivité apparente le long d'une rivière d'eau douce sur environ 15 km de distance. Les résistivités apparentes les plus basses correspondent à une longueur de dispositif de 10 m ; les plus fortes à une longueur de dispositif de 100 m.

latéralement continue n profils de résistivité apparentes ou un profil de sondages électriques à n longueurs de lignes. L'acquisition numérique des données permet de restituer les profils des résistivités apparentes ainsi que l'enregistrement de la résistivité de l'eau (Figure 2). Les exemples d'application décrits plus loin montrent quels types de documents peuvent résulter de l'interprétation des mesures.

Remarques

1. Le courant injecté est un courant régulé alternativement positif et négatif. La durée des alternances est de 5 s. La mesure du potentiel est faite à la fin de chaque alternance. La demi différence des potentiels correspondants constitue le résultat utile, corrigé de la polarisation des électrodes, la demi somme est une mesure de cette polarisation qui peut éventuellement être interprétée.

2. Les électrodes d'injection A et B sont situées au voisinage du centre des dispositifs quadripôles Mi, A, B, Ni, le théorème de réciprocité étant appliqué, ces dispositifs ne sont pas de types classiques Schlumberger ou Wenner.L'interprétation des sondages électriques est faite sur ordinateur sans que l'utilisation d'abaques soit nécessaire.

3. Dans les exemples ci-dessous, les nombres de paires d'électrodes de potentiel était généralement de 7, les longueurs de dispositif allant jusqu'à 200 m. Ce nombre peut paraître faible pour interpréter quantitativement un sondage électrique. En réalité, le nombre de données par sondage est de 9 car on connaît l'épaisseur et la résistivité de l'eau et dans un modèle tabulaire à trois couches, il n'y a que 3 inconnues : l'épaisseur et la résistivité de la deuxième couche (le matériau meuble) et la résistivité du substratum. La première couche (l'eau) étant parfaitement déterminée, l'interprétation est fortement contrainte.

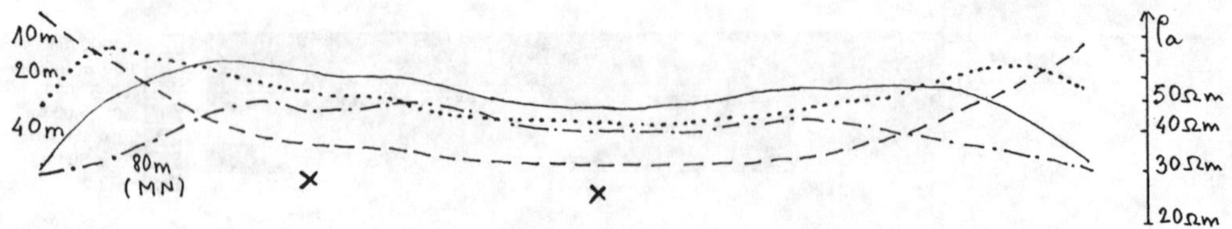

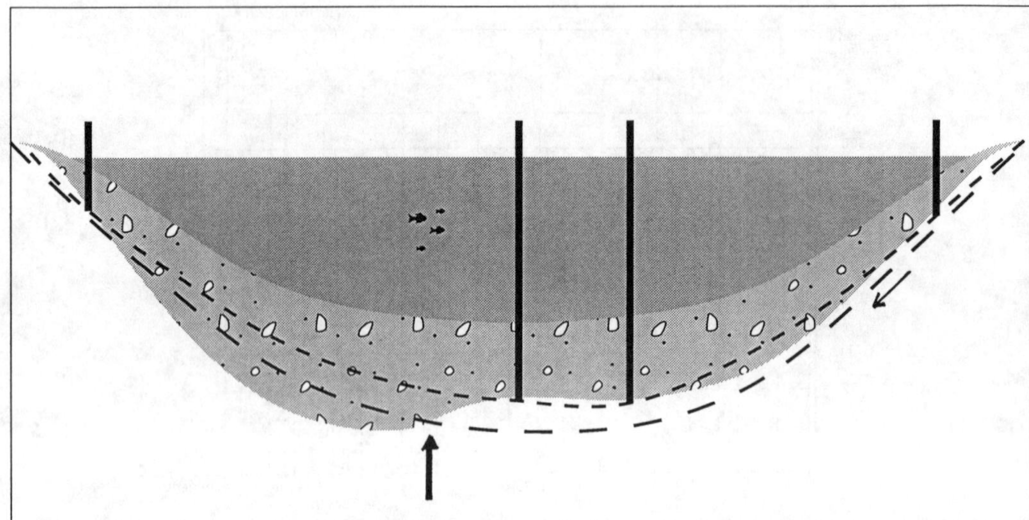

Figure 3 - En bas : coupe géologique en travers d'un fleuve. Il coule dans des alluvions de vases, sables et graves, reposant sur un substratum marneux. Les traits verticaux épais représentent des forages de reconnaissance réalisés sans étude géophysique préalable. Pointillés courts : interface estimée entre graves et marne. Pointillés longs: tracés du forage horizontal dirigé. Flèche verticale : arrêt du forage horizontal dirigé.

En haut : 4 profils de résistivité au fond de l'eau avec des longueurs de dispositif de 10 à 80 m. Les deux croix représentent la position préconisée pour les deux forages de reconnaissance.

2. EXEMPLE THEORIQUE

La figure 3 illustre, sur un exemple imaginaire, ce qui peut en coûter de se contenter de forages de reconnaissance. Le problème est ici de traverser un fleuve par un forage dirigé. Ce fleuve a une largeur d'environ 500 m. Il coule dans ses alluvions constituées de vases, sables et graves. Les graves reposent sur un substratum marneux assez tendre. La profondeur maximum est d'une dizaine de mètres, les alluvions présentant une épaisseur de quelques mètres.

Pour éviter les difficultés de foration dans les graves et le blocage de la tête de forage par la rencontre d'un bloc imprévu, ainsi que les problèmes d'orientation de l'outil dans un tel matériau, on décide de forer entièrement dans les marnes, matériau homogène, de bonne tenue et facile à forer. Il faut donc déterminer la position de l'interface qui sépare les alluvions du substratum marneux afin de diriger le forage de manière à ce qu'il soit toujours plus profond que cette interface.

Sans autre information préalable, les forages peuvent être placés comme indiqué sur la figure : deux sur les bords du fleuve et deux autres au milieu du cours.

Le projeteur dispose donc de quatre points fournis par les forages, ainsi que des affleuvements des marnes sur les rives, pour déterminer la géométrie de l'interface. Il trace donc la ligne en pointillés courts comme la meilleure estimation de l'interface. En se

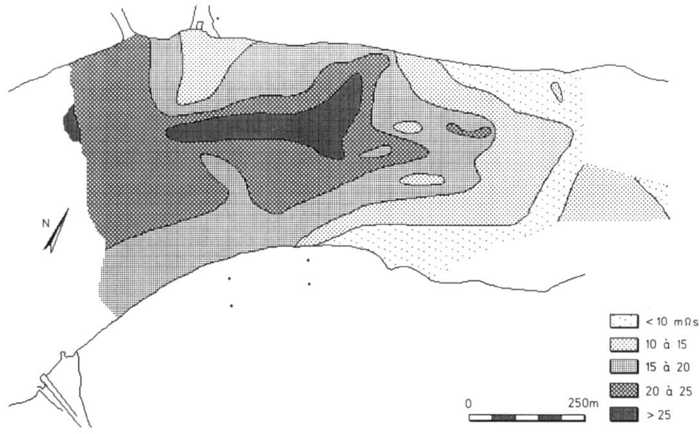

Figure 4 - Fleuve d'eau saumâtre. Carte de la conductance longitudinale C des sédiments meubles. Cette carte peut être lue comme une carte d'épaisseur.

donnant jusqu'à deux mètres de sécurité, il trace en dessous la ligne en pointillés longs comme tracé prévu de forage horizontal dirigé.

Si la foration commence par la rive qui se trouve sur la droite sur la figure, le forage se fera sans difficulté sur les deux tiers du trajet et il sera ensuite bloqué par la rencontre des graves. Il n'y aura pas alors d'autre solution pour l'entrepreneur que de reprendre le forage depuis le début avec un trajet un peu plus profond, donc une courbure plus grande, sans savoir de combien il doit augmenter cette courbure. Le coût de cet arrêt de chantier et de cette reprise sera vraisemblablement supporté par le maître d'ouvrage.

On voit que les ennuis proviennent ici d'un mauvais positionnement des forages de reconnaissance. La mise en œuvre préalable de la méthode décrite ici, en travers du fleuve, aurait permis de positionner correctement les forages de reconnaissance.

Le haut de la figure représente en effet quatre profils de résistivité apparente mesurés au fond de l'eau à l'aplomb du tracé du forage horizontal dirigé. Ces profils montrent, avant toute interprétation quantitative, que l'épaisseur des graves varie de manière plus complexe que la profondeur de l'eau. Les deux croix indiquent la position préconisée pour les forages de reconnaissance. Les résultats donnés par ceux-ci permettent à leur tour une interprétation quantitative des données de la géophysique et donc de placer correctement l'interface alluvions-marnes.

3. RECONNAISSANCE DES ALLUVIONS D'UN FLEUVE D'EAU SAUMATRE

L'exemple suivant montre des résultats obtenus dans une application réelle, qui n'était pas celle du tracé d'un forage horizontal dirigé, mais qui était un problème d'implantation de pont. Il s'agit en fait d'une reconnaissance que l'on peut traiter de la même façon, c'est pourquoi nous l'évoquons ici.

La figure 4 est le résultat d'une étude destinée à évaluer la quantité d'alluvions meubles au fond d'un fleuve d'eau saumâtre (résistivité de l'eau : 2 Ωm) sur lequel la construction d'un pont est projetée. La position du pont et le dimensionnement de ses fondations dépendent en effet de la répartition, de l'épaisseur et de la qualité des alluvions.

Le fleuve ayant une largeur de 500 m environ, 8 profils longitudinaux de 2 km ont été réalisés avec 7 dispositifs quadripôles simultanément de taille variant de 10 m à 200 m. Les mesures montrent que les sondages électriques peuvent être interprétés grâce à un modèle de terrain tabulaire à 3 couches : l'eau, les alluvions et le substratum isolant. On calcule donc aisément la conductance longitudinale C (rapport de l'épaisseur à la résistivité de la couche), de la couche d'alluvions, celle de l'eau étant connue. La figure 4 est une carte de C.

Des forages réalisés à terre au droit de la zone où C est la plus forte, ont montré que les sédiments avaient environ 30 m d'épaisseur, ce qui signifie que les sédiments ont une résistivité voisine de celle de l'eau et que la carte de C peut être interprétée pratiquement comme une carte d'épaisseur.

Une journée de travail a été suffisante à la réalisation des mesures, l'interprétation quantitative nécessitant une journée supplémentaire.

4. TRAVERSEE DE LA GARONNE EN AVAL DE BORDEAUX

Enfin, la figure 5 est la photocopie d'un document brut de terrain. Il s'agissait de traverser la Garonne en aval de la ville de Bordeaux. Comme c'est souvent le cas, aucune reconnaissance préalable complète n'avait été réalisée à l'exception de deux forages verticaux au milieu du fleuve à l'endroit repéré par les flèches. Là, l'épaisseur de la couche d'alluvions graveleuses était de 12 et 13 m. Par un hasard malheureux, les forages avaient été implantés à l'endroit où ces dernières présentaient la plus faible épaisseur comme l'ont montré les deux arrêts de chantier successifs dus à une estimation trop faible de l'épaisseur de graves. Nous sommes intervenus à la suite des ces deux arrêts de chantier à la demande du maître d'ouvrage. 7 profils de résistivité apparente avec des longueurs de ligne variant de 20 à 200 m ont été réalisés en passant à l'aplomb des forages de reconnaissance qui nous ont servi à alimenter les programmes d'interprétation quantitative. Cette dernière est délicate car nous sommes à la limite de validité de l'hypothèse de tabularité - la plus grande longueur de ligne (qui correspond aux plus fortes résistivités apparentes) n'a pas été utilisée pour l'interprétation quantitative car elle est clairement influencée par les variations latérales de la structure. Les six autres profils ont servi à déterminer l'épaisseur de graves en se calant sur les données des forages. L'épaisseur trouvée était partout

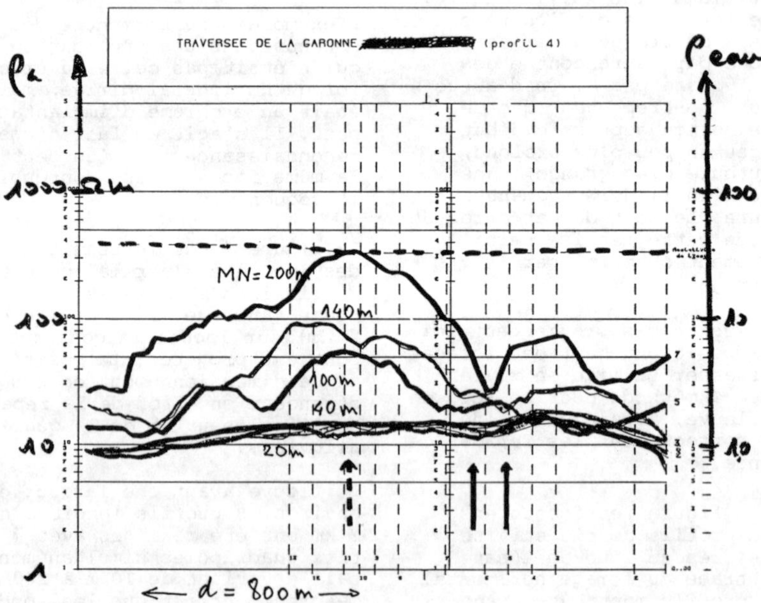

Figure 5 - Sept profils de résistivité au fond de l'eau en travers de la Garonne. Les quatre dispositifs les plus courts (de 20 à 40 m) ne sont pas influencés par l'épaisseur des alluvions (l'eau est trop profonde). Les deux dispositifs de 100 m et 140 m reflètent bien la variation de cette épaisseur. Le dispositif le plus long (200 m) est difficilement interprétable (non tabularité due aux effets de rives). Les deux flèches verticales en trait plein indiquent l'endroit où les deux sondages mécaniques de reconnaissance ont été pratiqués. La flèche en pointillé indique l'endroit où il aurait fallu reporter l'un des deux.

supérieure à 14 m, ce qui explique bien les arrêts de chantier. A la suite de cette campagne, le chantier a été repris en tenant compte des résultats et il s'est terminé sans difficulté particulière.

Sur le schéma, on a rajouté une flèche verticale en pointillé. Il s'agit de l'endroit où l'on aurait préconisé un forage de reconnaissance à la place de l'un des deux qui ont été réalisés (flèches en trait plein).

Une demi-journée a été nécessaire pour les mesures, une journée pour leur interprétation quantitative (qui n'était pas absolument indispensable pour le chantier).

CONCLUSION

Dans le cas de la reconnaissance du sous-sol en travers d'un cours d'eau, l'interprétation quantitative des résultats n'est pas souvent possible, en l'absence des résultats de sondages mécaniques, car l'hypothèse de tabularité de la structure du sous-sol est rarement réaliste. Il faut bien se convaincre que les méthodes de reconnaissance ne doivent pas être considérées comme concurrentes ni exclusives les unes des autres, mais qu'il faut faire appel à une véritable méthodologie de la reconnaissance qui marie avec harmonie la géologie, la géophysique et les forages avec essais en place et essais de laboratoire sur les échantillons prélevés.

Dans tous les cas, les sondages mécaniques sont nécessaires. Pratiqués, dans l'eau, ils sont très chers. La prospection électrique par courant continu est très rapide et bon marché. Elle permet d'implanter les forages de manière optimale et d'en tirer le meilleur profit possible pour un surcoût négligeable devant les économies entraînées grâce à la suppression des risques géologiques imprévus.

REFERENCES

Lagabrielle, R. & Teilhaud, S. 1981, Prospection de gisements alluvionnaires en site aquatique par profils continus de résistivité au fond de l'eau. Bull. Liaison Labo. P. et Ch. 114 : 17-24

Lagabrielle, R. 1983. The effect of water on direct current resistivity measurement from the sea, river or lake floor. Geoexploration. 21 : 165-170

Lagabrielle, R. 1984. La prospection électrique par courant continu en mer. Bull. Liaison Labo. P. et Ch. 132 : 5-11

Lagabrielle, R. 1992, Quantitative interpretation of vertical electrical soundings underwater with any quadrupolar array. 54th meeting EAEG, Paris : 358-389

Scott, W.J. & Maxwell F.F., 1989, Marine resistivity survey for granular materials, Beaufort Sea. Can. Jour. Expl. Geophys., 25, 2 : 104-114.

Shroder, N. & Bistrup, T. 1987. Resistivity at sea : Inversion and interpretation of measurements. SEG. New Orleans expanded abstract : 94-95

Whiteley, R.J. 1974, Design and preliminary testing of a continuous offshore resistivity method, Bull Aust. Soc. Explor. Geophys., 5, 1 : 9-14

No Trenches in Town, Henry & Mermet (eds) © 1992 Balkema, Rotterdam. ISBN 90 5410 085 0

Appraisal of electric and microgravimetric methods in underground works

J.C. Erling & F. Lantier
Techsol Company, Nivolas Vermelle, France

ABSTRACT: In underground projects, geological disturbances would be mapped to enable a better design improvements or to completing a tunnel or gallery. Two very specific geophysical techniques are described in the paper : the underground microgravity method helps to map density anomalies, results can be checked and refind by use of the electric panel method along the rock surfaces.

SURVEY BY MICROGRAVITY METHOD

1. SCOPE OF THE METHOD

Microgravity has been used for over 20 years for surface surveys of building sites, civil engineering projects and hydrogeological investigation (Lakshmanan 1963, Lakshmanan, Erling, Bichara 1977). The use of this method was extended to underground structures (Deletie, Erling, Lakshmanan, 1988).

Microgravity measures changes of earth attraction using a very high resolution microgravimeter (Lacoste and Romberg) with an accuracy of 5 μgal.

The measurement reading varies will respect to other points according to the location of the density abnormality in relation to the reading point.

Figure 1 shows an abnormality created by an empty cylinder shape at different depth Z, with soil density value 2.

2. SURFACE SURVEYS USING

Measurements used to be taken

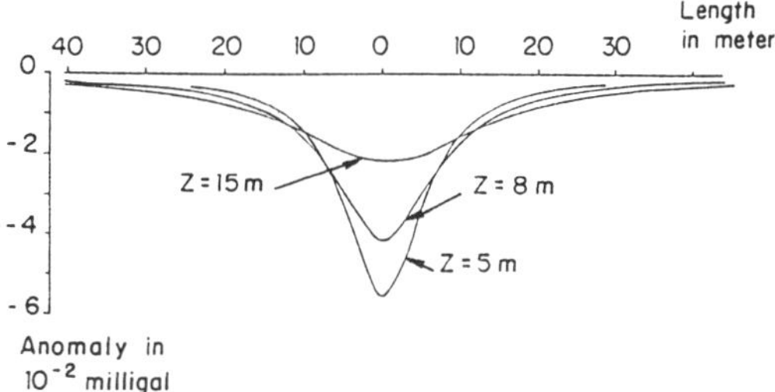

Figure 1. Microgravity anomaly of 4 meter diameter cylinder at Z depth.

on the ground surface, where the elevation is always higher than the abnormalities elevation. Different corrections are applied :

* Bouguer correction (altitude, latitude, instrument coefficient, luni-solar variations).

* Field corrections (lateral influences, topography)

When these corrections have been applied, every reading is referenced to a horizontal plan, to compile a "Bouguer abnormality" with the same density at all points.

Specific calculations allow

to determine :

* Residual abnormality by screening out deep geological influences.

* The density value of each abnormality.

3.THE METHOD APPLIED TO UNDERGROUND WORKS

The recent development of endoscopic microgravity allows the use of the microgravity method in underground works.

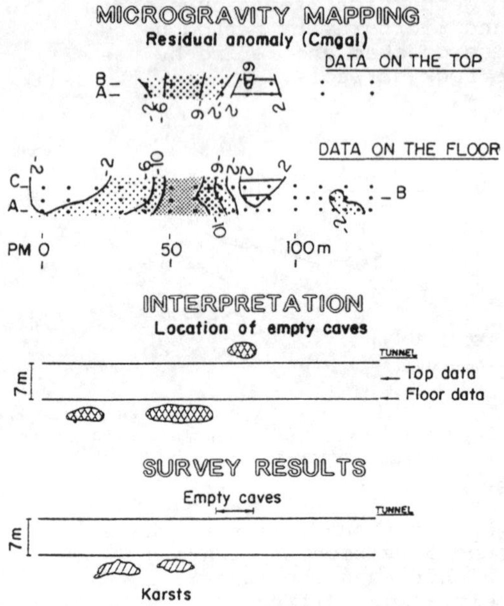

MICROGRAVITY MAPPING
Residual anomaly (Cmgal)
DATA ON THE TOP
DATA ON THE FLOOR
PM 0 50 100m
INTERPRETATION
Location of empty caves
TUNNEL
Top data
Floor data
7m
SURVEY RESULTS
Empty caves
TUNNEL
7m
Karsts

Figure 2.

3.1.Application example 1:

Figure 2 shows the survey results inside a tunnel under excavation, close to Monaco

The electric panel method was used (cf chapter 2), to investigate and refind reading on the tunnel rock surfaces.

3.2.Application example 2:

Close to Paris, a sewer under repair. Figure 3 shows 2 profiles, one at ground level, the other at the same place, but at -4m below ground level.

At PM 200, the abnormality is very large (-5 cmgal) as is the surface ground profile. This means that a density deficit occurs below the floor of the gallery.

At PM 55, the small negative abnormality underground is correlated with a positive abnormality in the ground level profile. This means the overburden is denser than the deeper layers.

At PM 140, a negative abnormality on the ground level profile is invisible on the underground profile. The overburden is probably less dense in that place.

In the same study the floor of the gallery was investigated by the electric

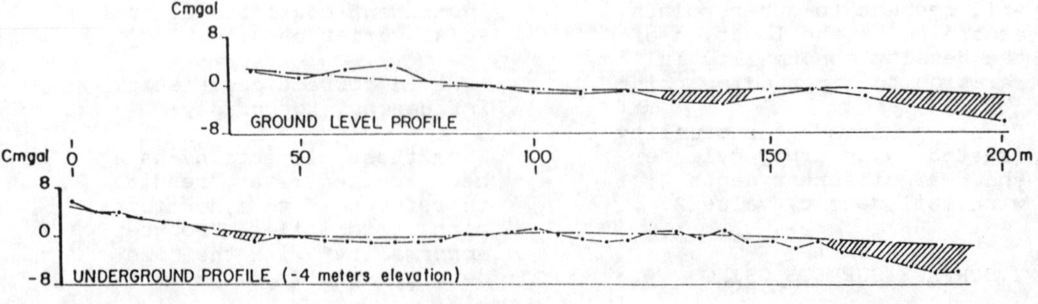

Figure 3.

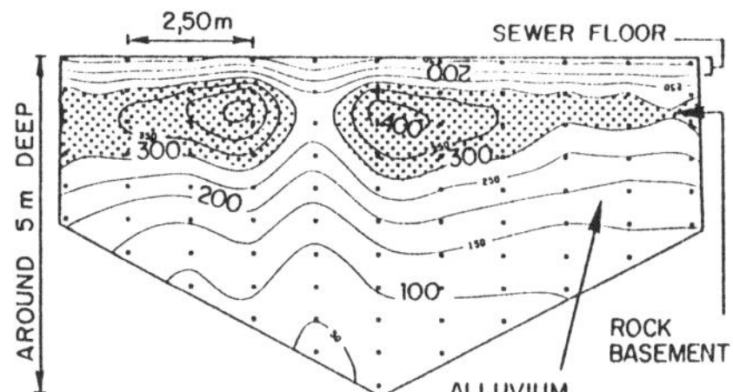

Figure 4. Electric panel investigation.

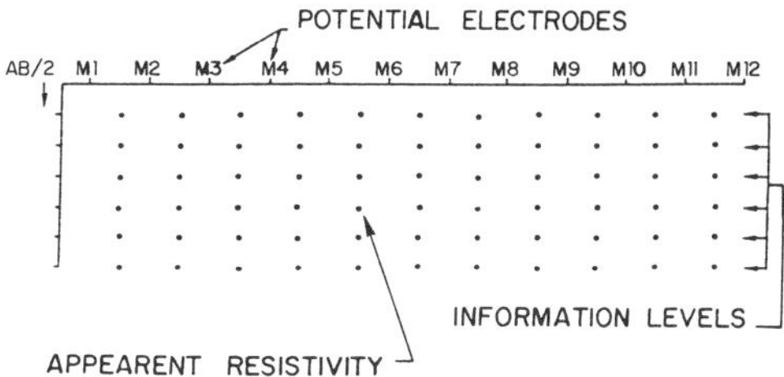

Figure 1. Results schedule.

panel method (cf next chapter).

The figure 4 shows a vertical cross section (around 5 meters deep).

The rock basement is visible (resistivity over 300 Ωm) and is below sandy clay alluvium.

ELECTRIC PANEL

1.SCOPE OF THE METHOD

This method is commonly used in the field, at ground level.

The purpose of this method is to gain information, progressively deeper, on the soils and rocks along apparent resistivity.

12 voltage electrodes M1 to M12 are equally spaced along the same line. The space between each electrode is 1, 5, 10 or 20 meters, according the required depth of investigation. To get readings for different depths, direct current is supplied by "A" electrode (left and right sides of the spread), and so that six progressively deeper apparent resistivity levels, relative to the voltage electrode level, are recorded. (cf figure 1). This information is recorded for each voltage electrode pair.

2. EQUIPMENT

Composed of a 386 Sx Computer connected to a regulated direct current box, with a digital potentiometer - Anmeter.The digital potentiometer is connected through a multi-channel connector, and reads voltage between electrodes M1-M2, M2-M3, M3-M4 ... M11-M12, for a single position of "A" electrode (direct current). The computer receives different readings, computes the appearent resistivity for each position of M1- M2 and A, and memorises it.

3. APPLICATION EXAMPLE:

A survey gallery φ 3m was excavated in a karstic limestone, before excavating a φ 12m tunnel (road tunnel).

The purpose of the study was to gain information about the rock, beyond the survey gallery the rock surfaces, prior to excavating the rock for the road tunnel. Two electric panels were installed along each rock side.12 potential electrodes were disposed evenly.The space between 2 electrodes was 5 meters. The maximum investigation was around 15 - 20 meters far. Each panel was installed independently of the other.

Results

Upper and lower positions of figure 2 .

The hatched area indicates the conductive area (less than 35Ωm). Left and right panels are indentical. The upper part relates to mudstone clay and lignite (Trias). The lower part relates to very fissured and weathered mudstone (interlayed).

The left and right side of the central portion of the figure shows karstic clay anomalies, 3 to 8 meters beyond the rock surfaces and checked by drills

4. APPLICATION EXAMPLE 2:

Checking of concrete grout beyond a gallery rock face.

The electric panel was applied to this purpose.

Voltage electrodes were equally spaced on the rock sufaces in the selected area (6 pannels).

Two steps were necessary to resolve the problem:

1. Before grouting, the resistivities of the six monitored levels were recorded, using small metallic plugs as electrodes. The results are shown in the figure 3. This is called "statement 0".

2. After concrete grouting from the gallery into the surrounding rock, the resistivities of the six monitors levels were recorded using the same plugs, one week later. In this case, concrete is a very conductive layer. Figure 4 shows "statement 1" after grouting the same area

Comparison between "statement 0" and statement 1 : resistivity "0"

resistivity "1"
We can see where concrete is and where it has never reached, figure 5 shows where resistivity as changed. The curves "20" mean the resistivity of the formation has decreased by 20 between statement 0 and statement 1.

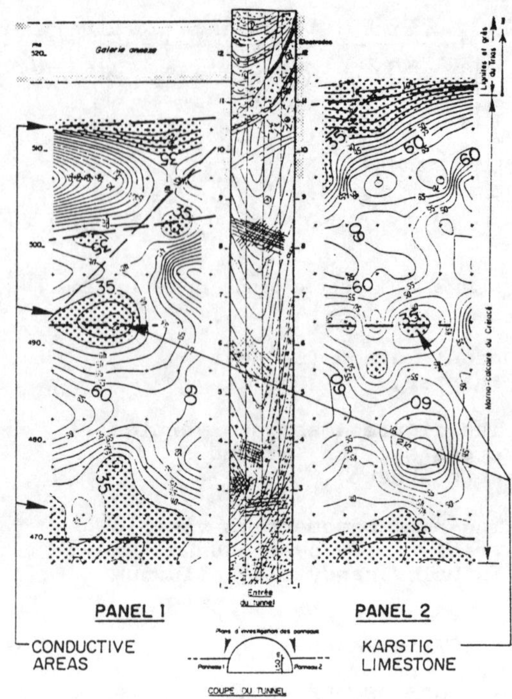

Figure 2. Horizontal plan crossing the galery (Isoresistivity curves).

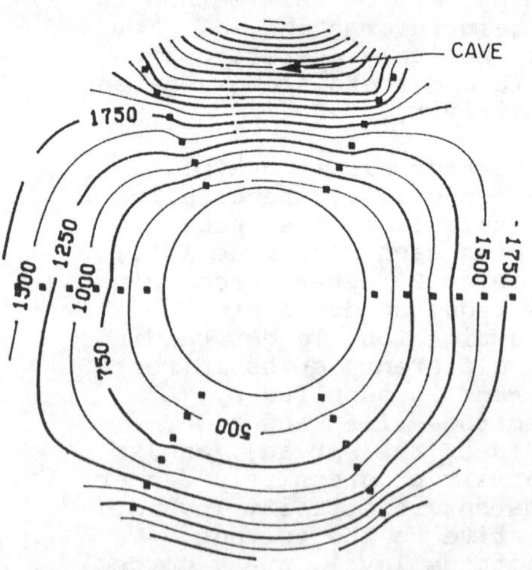

Figure 3. Isoresistivity curves Statement 0 before jetting.

5. GENERAL CONCLUSION

Specific geophysical methods (gravity and electric panel method) can be very efficient

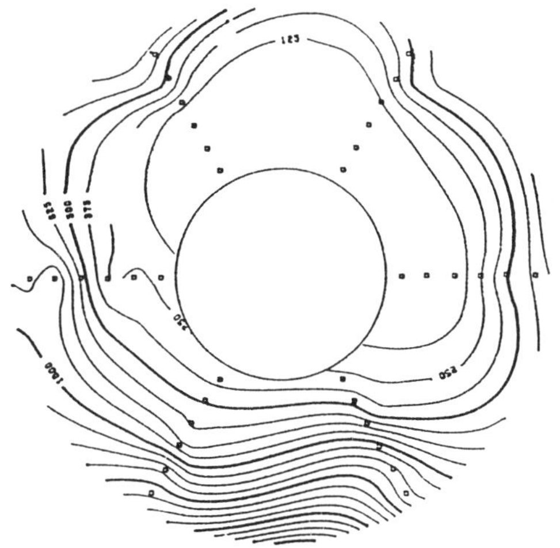

Figure 4. Isoresistivity curves Statement 1 after jetting.

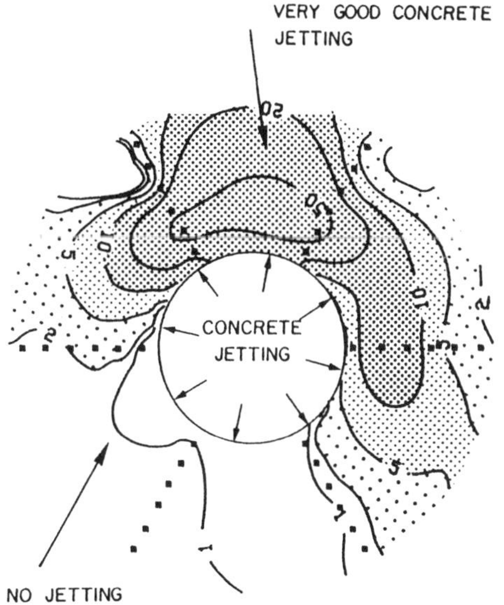

Figure 5. Mapping of concrete jetting beyond the galery.

in locating, an abnormalities in rock, beyond a rock surface in a gallery or tunnel, in any direction.

An interesting application of the electric panel method enables the integrity of concrete grout to be checked over 2 meters behind a rock surface.

References

J. Lakshmanan, (1965) "Reconnaissances de cavités dans le sous-sol par procédés électriques et gravimétriques" sol-soils.

J. Lakshmanan, JC. Erling, M. Bichara (1977) "Etude de fondation en terrain caverneux : place de la gravimétrie". Bulletin de liaison du labo des Pont et Chaussée, Paris.

JC. Erling, G. Roques (1983) "Reconnaissance et traitement de cavités naturelles et artificielles dans le domaine ferroviaire".

Assoc. Int. Géol. Ing. - Sympsosium Internationnal sur les essais in situ - Paris

JC. Erling, Y. Bertrand, Kutman (1987) "Prospection détaillée du sous-sol urbain en vue de travaux souterrains". Colloque collectivités territoriales et utilisation du sous-sol, Bordeaux.

P. Deletie, JC. Erling, J. Lakshmanan (1988) "L'endoscopie microgravimétrique adaptée à l'auscultation d'ouvrages", Journées Franco-marocaines de géotechnique, Marrakech.

No Trenches in Town, Henry & Mermet (eds) © 1992 Balkema, Rotterdam. ISBN 90 5410 085 0

Electrostatic quadrupole: A new tool for preliminary surveys to trenchless digging

Sophie Geraads
GEOMEGA, Europarc, Creteil, France

Abdo Mounir
GEOMEGA, Europarc, Creteil, Université Paris VI, France

Alain Jolivet
CNRS, Centre de Recherches Géophysiques, Garchy, France

Alain Tabbagh
Département de Géophysique appliquée, Université Paris VI & CNRS, Centre de Recherches Géophysiques, Garchy, France

ABSTRACT: Preliminary surveies are recommanded to detect obstacles (blocks, pipeworks,...) and lateral variations of geological facies.The new method which is proposed here allows to work even in urban areas, where neither classical electrical prospecting nor electromagnetic method can be easy to use : there is no need to drive electrodes into the ground, and there is no perturbation by surrounding metallic pieces.

This method is the direct extension of the electrical resistivity one, but it differs by the origin of the current, created by electrostatic influence, and by the fact that the potential difference is measured in the air.
This is why the apparatus is called "**electrostatic quadrupole**"

Apparent resistivity of the soil is determined by the mutual impedance of the quadrupole : it has been established that the known formulation for the electrical method is always valid when using low frequencies ($f < 1$ MHz).
The apparatus is made of four poles which are small metallic pieces placed inside the wheels.
The unit is completely independent and pulled by a vehicle. It allows a continuous acquisition. As for fast electrical surveying, data are recorded on a P.C.
The measurements were first compared with electrical results on test sites.
Then studies on roads or pavements, that corresponds to insulating cover, are presented.

It is also possible with this method to determine the apparent dielectric permittivity at low frequency. This will be useful for improvements of the interpretation of **georadar** data.

Fig.1: *Electrostatic quadrupole pulled by a vehicle, the poles are located inside the wheels.*

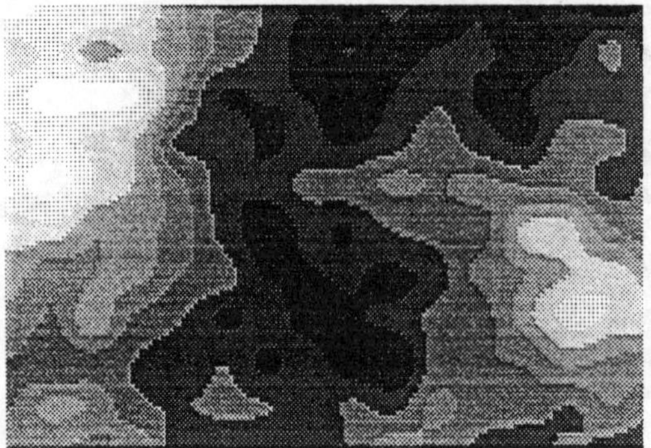

ELECTROSTATIC QUADRUPOLE 44kHz

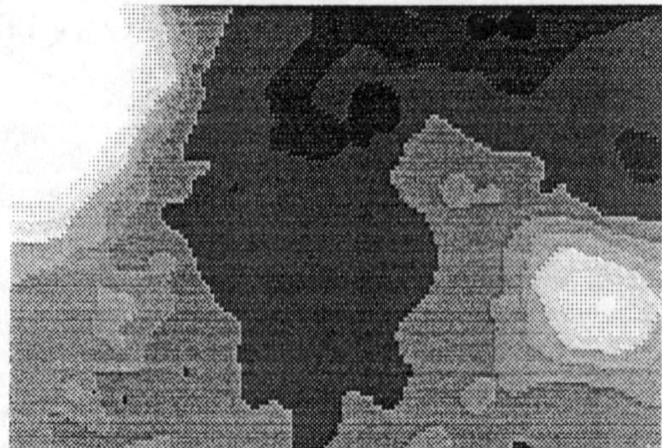

ELECTRICAL WENNER ARRAY (a=1m)

25.0 37.1 50.0 67.3 100.0 Ohm.m

0 10 20m

ELECTROSTATIC QUADRUPOLE 128 kHz

Fig.2: *Apparent resistivity on the first test site in Garchy (measurements mesh 2m x 2m),*
- electrostatic quadrupole, square array of 2 m side, 44 kHz frequency
- electrostatic quadrupole, rectangular array 1.17m x 1m, 128 kHz frequency
- Wenner array (a=2 m).

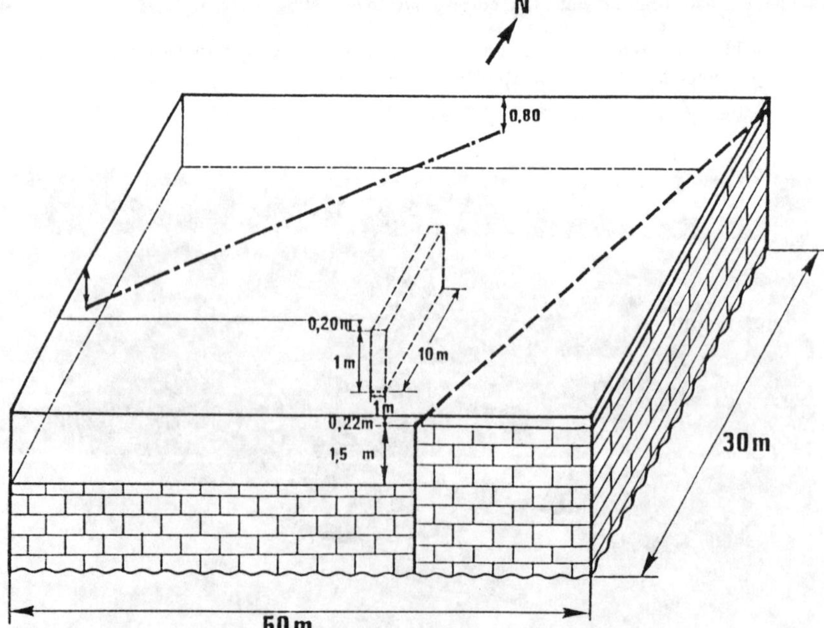

Fig. 3: *Features existing in the second test site at Garchy: buried cable, wall and fault.*

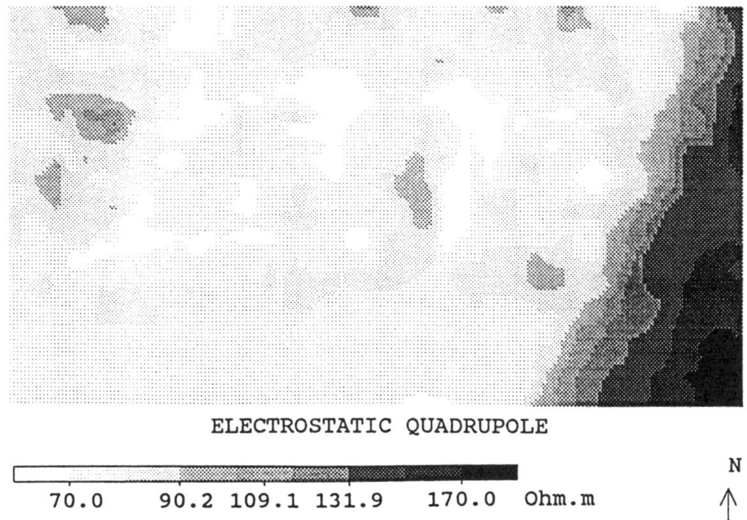

Fig.4: *Apparent resistivity at 44 kHz on the second test site (electrostatic quadrupole, square array, 1.20 m side)*

One of the first methods of geological prospecting, if not the first, applied to civil engineering or at shallow depth was the electrical method.

In this method data acquisition and interpretation are simple and robust, and the physical property involved, the electrical resistivity, is fairly correlated with major mechanical characteristics of soils, granularity and moisture.

Its use is however limited by the requirement of electrical contact between the electrodes and the ground and by the measurement slowness when using d.c.current.

Several electromagnetic methods can be used to overcome these difficulties, but they have their own limitation and a great sensitivity to metallic bodies and a low sensitivity to resistive features.

The introduction of a.c. current meters with very short measurement time allowed the application of the electrical method in continuous profiling (Dabas & Jolivet 1987) but it remains unusable on asphaltic covers wich limits its use in town for surveying prior to trenchless works.

PRINCIPLE:

The **electrostatic quadrupole** we present here is an extension of the electrical method, it derives from quadrupoles poles used for measuring the complex dielectric permittivity of the ionosphere (Storey & al 1969) and does not necessitate any electrical contact with the soil surface.

The principle of the method is the following (Grard et Tabbagh 1991): the potential in the vicinity of an electrostatic charge depends on the permittivity of the surrounding medium, if the charge or the point of measurement are approached to the soil its dielectric permittivity and its electrical resistivity arise into the measure.

But a continuous unique charge cannot be generated and it is necessary to use two alternative charges of opposite signs, they constitute a transmitting circuit locked by displacement or conduction current following the properties of the surrounding media.

The voltage generated by the two charges is measured between two poles:

We use a quadrupole and measure its transfer impedance, $Z= \delta V/I$, ratio of the voltage between the measurement poles to the current in the transmitting circuit.

At low frequencies and by locating the poles at a negligible height by comparison with their separation, the impedance is directly proportionnal to the ground resistivity (for an homogeneous ground), $Z = k.\rho$, the coefficient k depending only on the geometry of the quadrupole and being identical to that of the electrical method ($k=2\pi/(1/MA-1/MB-1/NA+1/NB)$).

If we only consider the shallow depths (0 to 15 meters), all the way of using, all the interpretation procedures and all the acquired experience are transposable to the electrostatic method.

By increasing the frequency it is also possible to determine the dielectric permittivity but this ability is

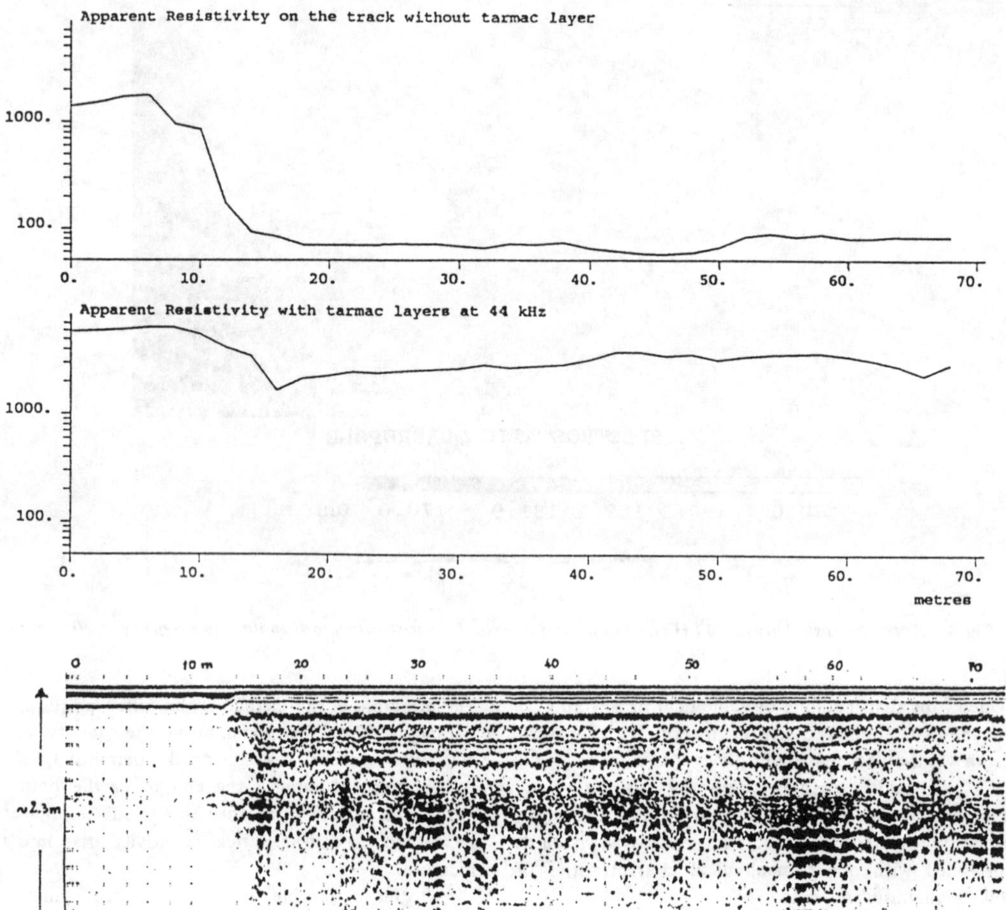

Fig.5: *Apparent resistivity profile before and after the installation of tarmac layers, and corresponding georadar profile.*

out of the scope of the present paper.

The apparatus with have developped (figure 1) is aimed for rapid continuous profiling on causeway with a metric investigation depth. the frequency used is 44 kHz, the poles are located at the corners of a square array of 1.2 m side.

The poles consist in metallic pieces located inside the wheels (SEPTA-CNRS patent).

A Doppler-radar allows the determination of the distance along the profile and data are recorded on a P.C.

We present hereafter a series of examples, the first two correspond to test sites with natural cover (grass) and allows comparison with other methods and to check the instrument responses on known features. The two other examples correspond to measurements on causeways.

EXAMPLE 1:

The first test site consists in a small area of 20 by 30 metres size located in the C.R.G. at Garchy.

The resistivity map obtained with quadrupole is presented on figure 2, the injection line was E-W orientated and the measurement grid is 2m x 2m.

This map can be compared with that obtained using Wenner array (with 1 m interpole distance) and a.c. resistivity meter operating at 122 Hz or that obtained using an **electrostatic quadrupole** prototype of rectangular array 1.17m x 1m and 128 kHz frequency.

The agreement between the three maps is very good, in spite of small differences corresponding to the geometry of the arrays and the seasonal variations of the resistivity.

This coherence stands both for the absolute

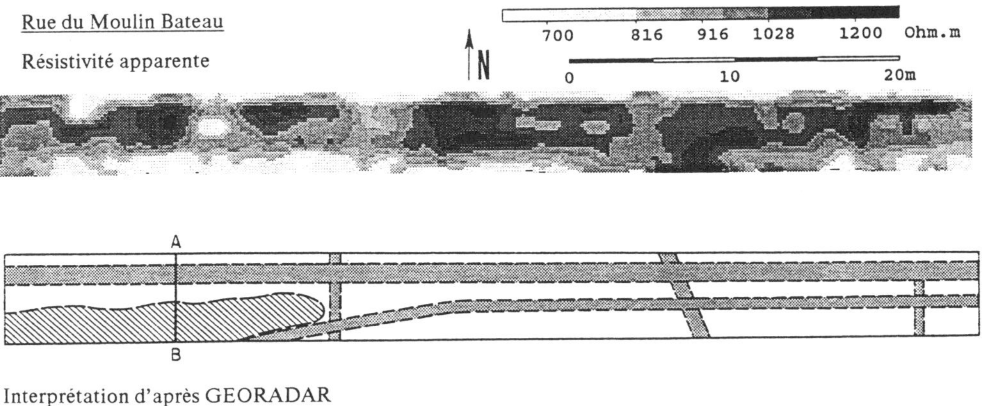

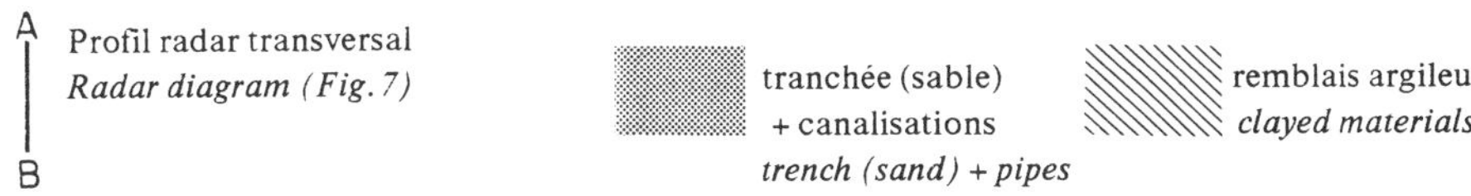

Fig.6: *Apparent resistivity on the "Moulin Bateau" street, electrostatic quadrupole, square array of 1.20 m side, 44 kHz and interpretation scheme of the georadar results.*

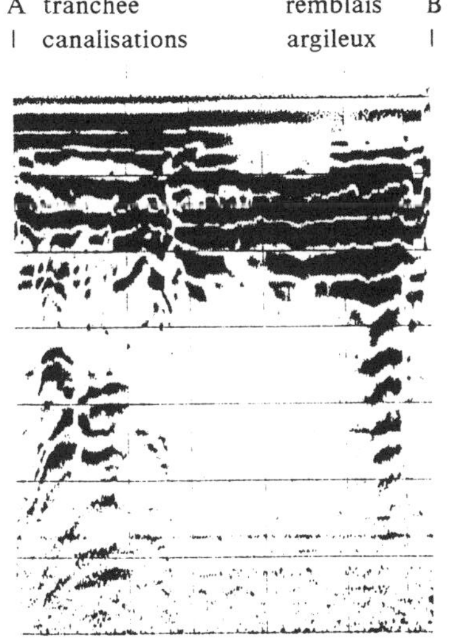

Fig.7: *Radar diagram crossing the street (500 MHz antenna, duration 50 ns, investigation depth 1.50 m)*

values of the resistivity and for the topology of the variations wich clearly exhibit clayed conductive zones existing on that area.

EXAMPLE 2:

The second test site presents, figure 3, three different features inside a small area of 30m x 50m.

In the eastern part exists a natural vertical fault wich separates a resistive medium (jurassic limestone) covered by 20 cm of topsoil from a conductive one (weathered limestone) more than 1.50 m thick.

In the center we built a wall (resistive) of 1 x 1 m section and 10 m length roughly orientated in the N-S direction.

In the western part an old electrical power line was buried at 80 cm depth.

The resistivity maps obtained with the quadrupole is presented in figure 4.

The fault and the wall are both very clearly observed.

At the line location exists a slight anomaly corresponding probably to the trench where it was located.

This result is totally coherent with a Wenner array of 2 m interpole distance (Hesse & al 1986).

EXAMPLE 3:

The third experiment took place on a road beside Mesves s/Loire (Nièvre).

A first profile was executed on the road track before it was covered with a tarmac layer. It can be seen on figure 5 that there is a very strong change in the apparent resistivity from 2000 Ωm to 80 Ωm located around the ten meter point.

The measurements achieved later, after the installation of the tarmac layers, show an increase in the absolute value of the apparent resistivity and a relative reduction of the magnitude of the change, from 5000 Ωm to 3000 Ωm.

According to the indications given by the road builder we can consider the track to be covered with two successive layers: a first concrete-graved 28 cm thick for wich a 15000 Ωm resistivity can be assumed, then a superficial asphaltic layer of zero conductivity with an equal thickness of 28 cm.

If we use a 1D layer model, the resistivity modification ranges from 80 Ωm to 3000 Ωm in the conductive portion of the profile and from 2000 Ωm to 5000 Ωm in its resistive portion.

Then there is a good agreement between the apparent resistivity before the tarmac deposit and after. The radar profile, figure 5 also, with a 500 MHz antenna shows the soil-concrete interface in good agreement with the known thickness of the layers.

EXAMPLE 4:

The last experiment took place in a street (Sucy en Brie, Val de Marne, France) where the resistivity map can be checked by **georadar** study.

The map presented here, figure 6, covers an area of 50m x 5m, the radar interpretation scheme of this area is presented on figure 6 and an example of radar profile on figure 7.

The map shows two parallel resistive features extending all along the street interrupted by small more conductive zones.

The resistive features correspond to the two main concrete pipes that exist beneath, the northern one having a diameter of 1.20 m and a top at 80 cm depth, the other the same depth but a smaller diameter.

The direct effect of the pipes on the apparent resistivity adds to the effect of the sand filling the trenches wich does not retain water.

The areas of lower resisitivity can be interpreted by the presence of more clayed filling material. This is confirm by the high reflection that exists in the S-W zone between the causeway layers and the substratum.

CONCLUSION

In conclusion we have to emphasize that the **electrostatic quadrupole** opens wide prospects for the study of the ground in urban areas.

The existing intrument gives valuable informations about the first metre and we are working now on larger arrays to take into account in measurements the first ten metres.

There exists also a possibility of measuring and mapping the dielectric permittivity, this will bring in the future important improvements for the interpretation of GEORADAR data.

REFERENCES

Dabas M. & Jolivet A., 1987, Profilage électrique de subsurface par quadripôle tracté à contact continu avec le sol; in "Pose de cables souterrains et sous-marins", SEE, 4 et 5 février 1987, A2 8p.

Grard R. et Tabbagh A., 1991, A mobile four electrodes array and its application to the electrical survey of planetary grounds at shallow depths ; J.G.R., 96-B3, 4117-4123.

Hesse A., Jolivet A. & Tabbagh A., 1986, New prospects in shallow depth electrical surveying for acheological and pedological applications; Geophysics 51-3, 585-594.

Storey L.R.O., Aubry M.P. et Meyer P., 1969, A quadrupole probe for the study of ionospheric plasma resonance ; Plasma Waves in Space and Laboratory, edited by J.O. Thomas and B.J. Landmark, 1, 303-332, Edinburgh University Press.

No Trenches in Town, Henry & Mermet (eds) © 1992 Balkema, Rotterdam. ISBN 90 5410 085 0

Le quadripôle électrostatique: Un nouvel outil pour les reconnaissances préalables aux travaux sans tranchée

Sophie Geraads
GEOMEGA, Europarc, Creteil, France

Abdo Mounir
GEOMEGA, Europarc, Creteil, Université Paris VI, France

Alain Jolivet
CNRS, Centre de Recherches Géophysiques, Garchy, France

Alain Tabbagh
Département de Géophysique appliquée, Université Paris VI & CNRS, Centre de Recherches Géophysiques, Garchy, France

RESUME: Les reconnaissances préalables ont pour but de déceler la présence d'obstacles (blocs, concessionnaires) et les variations latérales de faciès. La technique proposée permet d'intervenir en ville, dans un domaine où la méthode électrique dans sa version habituelle ne peut être employée. Elle permet aussi de s'affranchir de l'influence des perturbateurs métalliques environnants qui sont une gêne pour l'utilisation de méthodes électromagnétiques.

Cette méthode est le prolongement direct de la méthode électrique mais elle en diffère par l'origine du courant injecté, créé par influence électrostatique, et par le fait que la différence de potentiel est mesurée dans l'air.

On a donc adopté la dénomination **quadripôle électrostatique.**

La résistivité apparente du sol est déterminée par l'impédance mutuelle du quadripôle; l'étude théorique a établi que la formulation utilisée pour l'électrique en courant continu reste valable dans la plupart des cas, avec une fréquence de quelques dizaines de KHz.

L'appareillage comprend quatre pôles constitués de petites pièces métalliques placées dans des roues. L'ensemble, parfaitement autonome, tiré par un véhicule, permet une mesure en continu. Comme pour la prospection électrique rapide, l'enregistrement est fait sur P.C.

La méthode a été validée sur des sites bien connus, par comparaison avec les résultats de prospections électriques traditionnelles: le mode d'interprétation des mesures, et les résultats, sont identiques.

Deux autres exemples récents, présentés ici, ont été obtenus sur chaussée où la surface est isolante.

Enfin, en plus de la résistivité, cette technique permet de déterminer la permittivité diélectrique en basse fréquence, ce qui peut être d'un grand intérêt pour approfondir l'interprétation d'une prospection par **géoradar**.

L'une des premières méthodes, sinon la première, de prospection géophysique employée en génie civil ou plus généralement pour des études à faible profondeur a été la prospection électrique.

Son emploi et son interprétation sont en effet relativement robustes et la propriété physique identifiée, la résistivité électrique, présente une large gamme de variations et une corrélation très forte avec des caractéristiques déterminantes en mécanique des sols que sont la granularité ou l'humidité du terrain.

L'utilisation de cette méthode a toutefois été limitée par la nécessité d'un contact électrique entre le sol et l'électrode et par la lenteur des mesures si l'on utilise un courant continu.

Plusieurs méthodes électromagnétiques peuvent être employées pour pallier ces deux difficultés, mais elles ont leurs propres limites, une grande sensibilité aux perturbateurs métalliques, une faible sensibilité aux structures résistantes.

L'introduction en prospection électrique des résistivimètres à courant alternatif à mesure rapide a permis son application en profilage continu (Dabas et Jolivet, 1987) mais elle reste inutilisable sur des surfaces isolantes comme c'est le cas pour les chaussées. Son utilisation en ville pour les reconnaissances préalables aux travaux sans tranchée restera donc limitée.

PRINCIPE :

Le **quadripôle électrostatique** que nous présentons ici est une extension de la méthode électrique. Il dérive des sondes quadripolaires utilisées

Fig. 1: Quadripôle électrostatique tracté par un véhicule où les pôles sont à l'intérieur des roues.

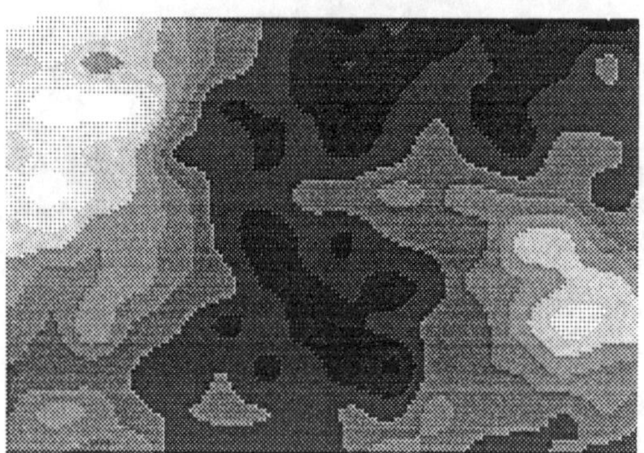

ELECTROSTATIC QUADRUPOLE 44kHz

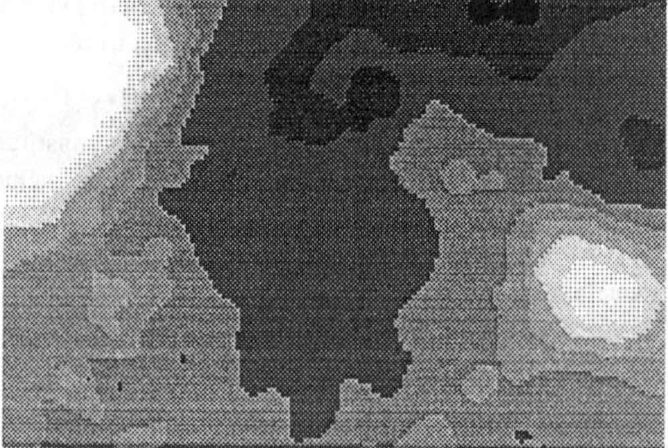

ELECTRICAL WENNER ARRAY (a=1m)

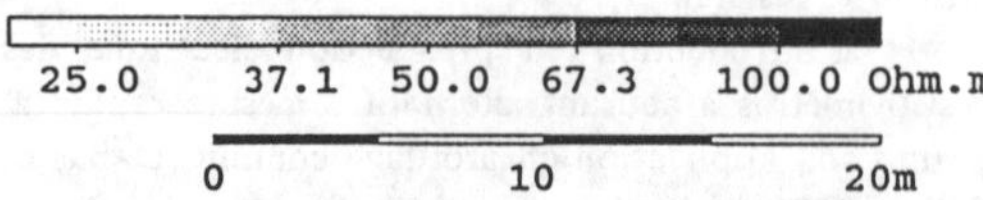

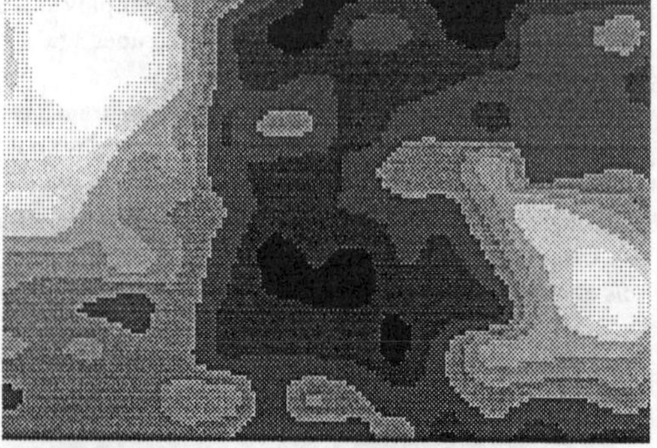

ELECTROSTATIC QUADRUPOLE 128 kHz

Fig. 2: Résistivité apparente sur la première zone test (maille de mesure 2m x 2m),
- quadripôle électrostatique carré de 1,20 m de côté fonctionnant à 44 kHz
- quadripôle électrostatique rectangulaire 1,17m x 1m, fonctionnant à 128 kHz
- quadripôle Wenner (a=1 m).

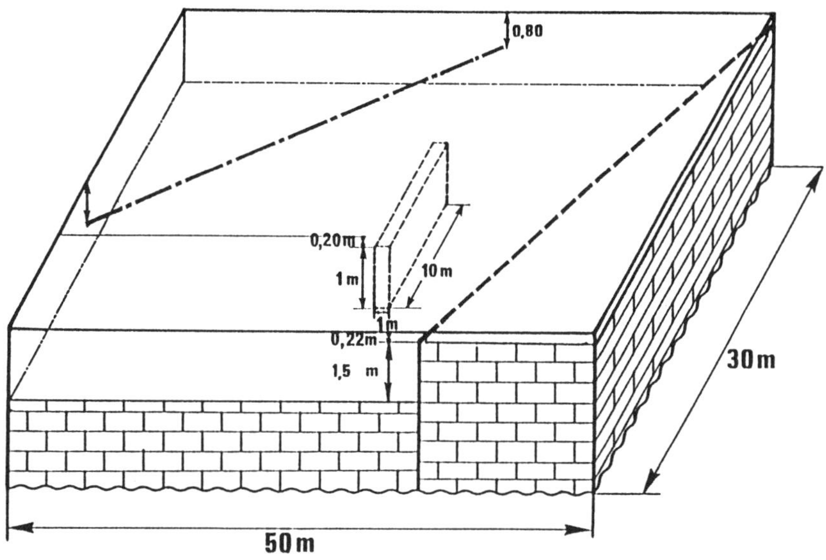

Fig. 3: Structures présentes sur le deuxième site test du C.R.G. Garchy: cable enterré, mur et faille.

pour mesurer la permittivité diélectrique complexe de l'ionosphère terrestre (Storey et al., 1969) et permet d'éviter tout contact électrique avec le sol.
Le principe en est le suivant (Grard et Tabbagh, 1991): le potentiel au voisinage d'une charge électrostatique dépend de la permittivité du milieu environnant. Si cette charge ou le point de mesure sont approchés du sol, la permittivité diélectrique et la résistivité électrique de celui-ci vont intervenir dans la mesure.

Mais on ne sait pas créer de charge unique on utilise donc deux pôles de signes opposés chargés par un courant alternatif ; ce circuit d'émission est fermé par des courants de déplacement et/ou par des courants de conduction selon les propriétés du milieu environnant. La présence de ces deux charges crée une différence de potentiel entre deux autres pôles.

On utilise donc un quadripôle dont on mesure l'impédance de transfert $Z=\delta V/I$, δV étant la différence de potentiel entre les pôles de mesures et I le courant dans le circuit d'émission.

Pour une fréquence suffisamment basse et en plaçant les pôles à une altitude négligeable devant leur écartement on a sur un sol homogène $Z=k\rho$ où k est un coefficient dépendant de la géométrie du quadripôle et identique à celui qui intervient dans la méthode électrique ($k= 2\pi/(1/MA-1/MB-1/NA+1/NB)$).

Si on se limite aux faibles profondeur (0-15 m) tous les modes d'utilisation, les méthodes d'interprétations et l'expérience acquise dans ces domaines sont transposables d'une méthode à l'autre.

Si l'on augmente la fréquence la mesure de la permittivité devient aussi possible mais le présent exposé n'abordera pas cette possibilité.

Après avoir testé la méthode avec plusieurs prototypes, l'appareil que nous avons développé (figure 1) est conçu pour des mesures en déplacement rapide sur chaussée avec une profondeur d'investigation proche du mètre.

La fréquence est de 44 kHz, le quadripôle est un carré de 1,2 m de côté, les pôles sont des pièces métalliques situées à l'intérieur des roues (Brevet SEPTA-CNRS).

La mesure du déplacement est assurée par un radar-Doppler et les données sont enregistrées sur PC.

Nous présentons ci-après une série d'exemples, en commençant par des exemples sur sol naturel de façon à pouvoir confronter les résultats avec ceux de la méthode électrique, les exemples suivants correspondent à des explorations sur chaussée.

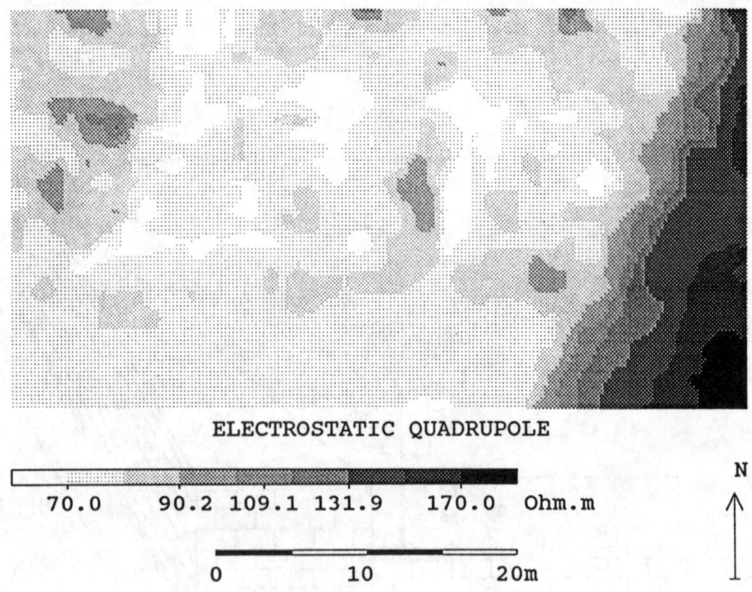

Fig.4: Résistivité apparente à 44 kHz sur le deuxième site test (quadripôle électrostatique carré, 1,20 m de côté)

EXEMPLE 1:

Le premier site d'essai a une surface de 20m x 30m. Il est situé sur le terrain du CRG de Garchy.

La carte de résistivité obtenue avec le quadripôle est présentée sur la figure 2. La maille de mesure est de 2m x 2m et la ligne d'injection était orientée dans la direction E-W.
Elle peut être comparée à la carte de résistivité obtenue avec un quadripôle électrique Wenner (l'écartement entre les électrodes était de 1 m), où à la carte obtenue précédemment avec un prototype électrostatique rectangulaire de 1,17m x 1m et de fréquence 128 KHz.

L'accord entre les trois cartes est très bon, malgré de légères différences liées à la géométrie des quadripôles et aux variations de la résistivité avec la saison.

Cet accord porte à la fois sur les valeurs absolues des résistivités observées et sur leurs variations qui montrent de façon très claire les deux poches argileuses présentes dans le calcaire.

EXEMPLE 2:

Sur le deuxième site test, il existe trois structures différentes à l'intérieur d'une surface de 30m x 50m (fig.3).

Dans la partie Est, une faille verticale met en contact un calcaire résistant couvert de 20 cm de terre arable et un calcaire altéré conducteur sur 1,5 m d'épaisseur.

Au centre, a été installé un mur résistant d'un mètre carré de section et de 10 m de long, orienté approximativement N-S.

A l'Ouest, un ancien cable électrique de puissance est enterré à 80 cm de profondeur.

La carte de résistivité obtenue avec le **quadripôle électrostatique** est présentée sur la figure 4.

La faille et le mur y apparaissent très clairement.

L'emplacement du cable est marqué par une anomalie légère correspondant à la tranchée de pose.

Le résultat obtenu est tout à fait comparable à ce qu'avait donné une prospection électrique avec un quadripôle Wenner (a = 2 m) (Hesse & al. 1986).

EXEMPLE 3:

La troisième expérimentation a été réalisée sur route à Mesve-sur-Loire (Nièvre).

Un premier profil électrique avait été exécuté sur la piste avant la pose du bitume.

Ce profil montre (figure 5) un changement très marqué, au point 10 mètre, entre une zone très résistante dont la résistivité se maintient proche de 2000 Ωm et une zone beaucoup plus conductrice, autour de 80 Ωm.

Le profil exécuté avec le **quadripôle électrostatique** après la mise en place d'une couche de grave-ciment de 28 cm et de deux couches de bitume d'épaisseur cumulée de 28 cm également montre une

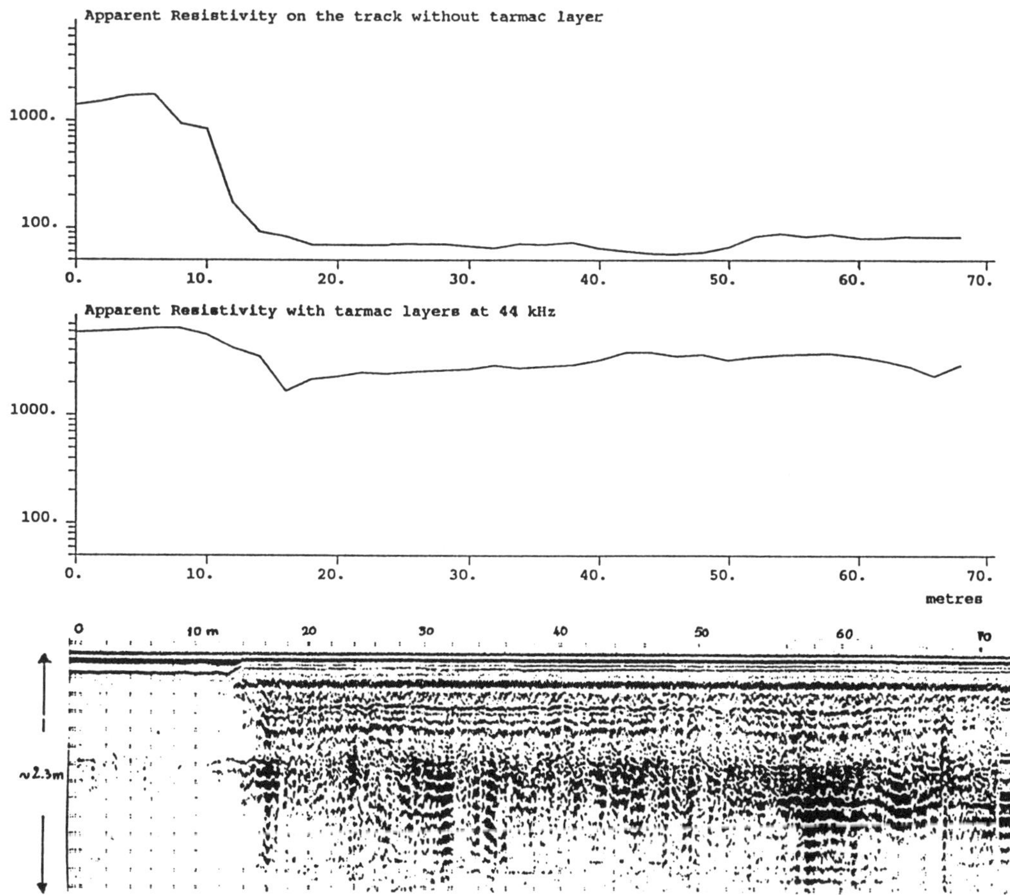

Fig.5: Profil de résistivité apparente avant et après la mise en place des couches de bitume et profil géoradar correspondant.

augmentation de la résistivité apparente, avec un maintien de l'anomalie dont l'amplitude relative est réduite.

En affectant aux couches de bitume une résistivité infinie et à la couche de grave-ciment une résistivité de 15.000 Ωm, on obtient avec un modèle 1D une modification de la résistivité apparente en parfait accord avec les observations puisque, du fait de la présence de la couverture, à un sol de 80 Ωm doit correspondre une résistivité apparente de 3.000 Ωm et à un sol de 2000 Ωm, une résistivité de 5000 Ωm.

La position de l'anomalie et les épaisseurs des différents recouvrements ont pu être contrôlées avec un profil de mesure par **géoradar** en utilisant une antenne 500 MHz (figure 5).

EXEMPLE 4:

La dernière expérience présentée a été réalisée rue du Moulin Bateau à Sucy en Brie (Val de Marne) et la prospection **électrostatique** a été contrôlée à l'aide du **géoradar.**

La carte de résistivité présentée (figure 6) couvre une surface de 50m x 5m.

Sur la même figure, on a représenté le schéma d'interprétation des données radar et sur la figure 7, un exemple de profil radar transversal à la route.

La carte montre deux structures parallèles résistantes, s'étendant tout au long de la rue, interrompues par plusieurs petites zones plus conductrices.

Les structures résistantes correspondent à deux tranchées de pose de canalisations importantes, enterrées sous la chaussée. La conduite la plus au Nord est un collecteur de 1,2 m de diamètre dont le toit est à 80 cm de profondeur et la deuxième, un ouvrage à même profondeur mais de diamètre plus faible.

A l'effet direct des canalisations s'ajoute l'effet

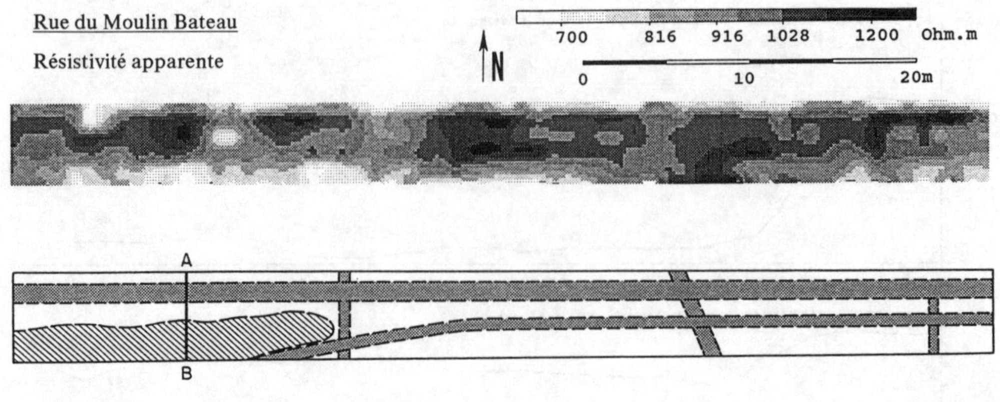

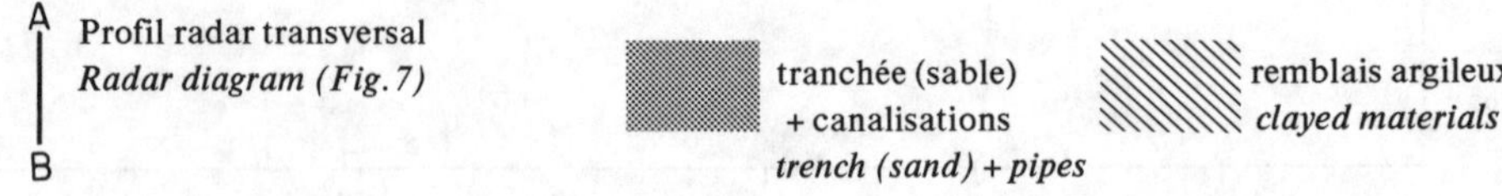

Interprétation d'après GEORADAR

A Profil radar transversal *Radar diagram (Fig.7)* B

tranchée (sable) + canalisations *trench (sand) + pipes*

remblais argileux *clayed materials*

Fig.6: Résistivité apparente sur la rue du Moulin Bateau, quadripôle carré 1,20 m de côté, 44 kHz et schéma interprétatif des résultats géoradar.

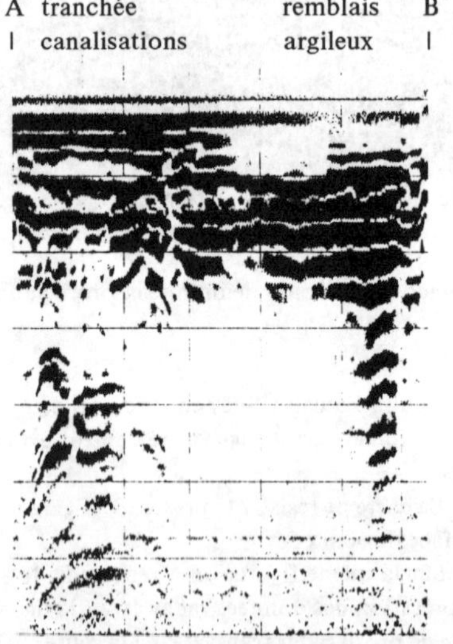

Fig.7: Coupe radar transversale à la rue (antenne de 500 MHz, durée d'écoute 50 ns, profondeur d'investigation estimée 1,50m).

du remplissage des tranchées par du sablon qui, drainant mieux l'eau, est plus résistant.

Ces structures sont nettement mises en évidence par radar.

Les zones plus conductrices doivent correspondre à des remblais plus argileux ,de moindre qualité et plus humides. Ceci est confirmé par les fortes réflexions radar observées au contact de la chaussée et de son substratum dans la zone Sud-Ouest.

CONCLUSION

En conclusion, on doit souligner que le **quadripôle électrostatique** ouvre d'importantes perspectives pour l'exploration du sous-sol en zone urbaine.

Le matériel existant est adapté à une auscultation sur une profondeur métrique et nous travaillons actuellement à la mise au point d'un appareil permettant une exploration décamétrique.

Par ailleurs, existe la possibilité de déterminer et de cartographier la permittivité électrique, ce qui sera d'un grand intérêt pour approfondir l'interprétation des résultats du GEORADAR.

REFERENCES

Dabas M. & Jolivet A., 1987, Profilage électrique de subsurface par quadripôle tracté à contact continu avec le sol; in "Pose de cables souterrains et sous-marins", SEE, 4 et 5 février 1987, A2 8p.

Grard R. et Tabbagh A., 1991, A mobile four electrodes array and its application to the electrical survey of planetary grounds at shallow depths ; J.G.R., 96-B3, 4117-4123.

Hesse A., Jolivet A. & Tabbagh A., 1986, New prospects in shallow depth electrical surveying for

acheological and pedological applications; Geophysics 51-3, 585-594.
Storey L.R.O., Aubry M.P. et Meyer P., 1969, A quadrupole probe for the study of ionospheric plasma resonance ; Plasma Waves in Space and Laboratory, edited by J.O. Thomas and B.J. Landmark, 1, 303-332, Edinburgh University Press.

No Trenches in Town, Henry & Mermet (eds) © 1992 Balkema, Rotterdam. ISBN 90 5410 085 0

Apparent resistivity profiling on a road with an electrostatic quadrupole

Georges Chevassu
Laboratoire Régional des Ponts et Chaussées, St Brieuc, France

Richard Lagabrielle
Laboratoire Central des Ponts et Chaussées, Bouguenais, France

Alain Tabbagh & Jean-Paul Decriaud
Centre de Recherches Géophysiques, CNRS, Garchy, France

ABSTRACT : Continuous apparent resistivity profiling makes it possible to survey the nature and lateral variations of the state of an existing road supporting ground. The experimental field is a 1.3 km long section of a local road which exhibits large resistivity variations and important deformations of the pavement, and where the measurements can be compared with other data : a Radio-MT profile recorded with a 711 kHz wave frequency, a pavement deflection profile along the pavement side, a series of penetrometric tests. The experiments performed show a very good correlation between the various methods based on the apparent resistivity measurements which can be used to precisely identify rock zones and fine soils zones. The electrostatic quadrupole proves, in this example to be particularly well fit for the survey of the ground under roads.

Continuous profiling, that means measurements while continuous moving of the instrument, is a very good solution to survey the underground at low cost. A great number of electrical and electromagnetic techniques can be used in such a way for resistivity recording. So far, the application of the electrical method on ground covered with an asphaltic insulating layer was not possible an E.M. magnetic field (for instance Slingram conductivity meter) and/or electrid fields (for instance Radio-MT, Guineau & Dupis, 1973, Chevassu & Lagabrielle, 1990) measured in the air can only be applied. The concept of electrostatic quadrupole derives from quadrupolar probes used for measuring the permittivity of the ionosphere. The electrical potential in the vicinity of a charge depends linearly on the permittivity of the surrounding medium. If the charge is approched to the ground the potentiel is modified and depends also on the complex permittivity of the ground. It was established by calculations and experiments (Grard & Tabbagh, 1991) that for a quadrupole of metric dimensions, at a frequency lower than about 300 kHz, the real part of the transfer impedance of the quadrupole is proportional to the ground resistivity.

If the four poles are sufficiently close to the ground surface, the quantitative formula is the same as in DC resistivity prospecting and this technique offers the ability of making resistivity measurements on any type of surfaces.

The experimental test site is a 1.3 km long section of a local road which exhibits large resistivity variations, corresponding to various types of materials in the road supporting ground from moist silt to weathered hard rock. Three types of data were available to evaluate the results : a Radio-MT profile recorded at 711 kHz frequency (which corresponds roughly to an investigation depth comprised between 1.5 and 5 m depending on resistivity (Chevassu & Lagabrielle, 1990), a pavement deflection profile along the road, a series of 15 dynamic penetrometric tests.

The prototype we used is a rectangular array 1 x 1.17 m composed of one pair of electrostatic current poles and one pair of electrostatic potential poles, the depth of investigation is then around one metre and is limited to the road itself and its immediate support. It operates at a 128 kHz frequency, the sampling interval was 5 m.

Figure 1 shows two parts of the profile. A good agreement can be observed between the quadrupole and MT resistivity profiles which both reflect the granularity and the moisture of the supporting layers. The following table sumarizes the correspondance observed between resistivity and the geotechnical quality of the subsoil :

The relations between pavement deflection and apparent resistivity is presented on figure 2. The correlation is very high, in perfect agreement with the classification proposed in the preceeding table (similar data for radio-MT -not presented here- show that, though quite good for MT, the correlation is better with the

Table 1.

Supporting soil	**Apparent resistivity (Ωm)**		**Quality of the soil**
	Radio MT	Electrostatic quadrupole	
Moist and fine	30 to 150	40 to 130	Weak supporting capacity
Sand, Gravel, pebbles	50 to 300	50 to 250	Variable supporting capacity
Rock	200 to 800	160 to 500	Good support

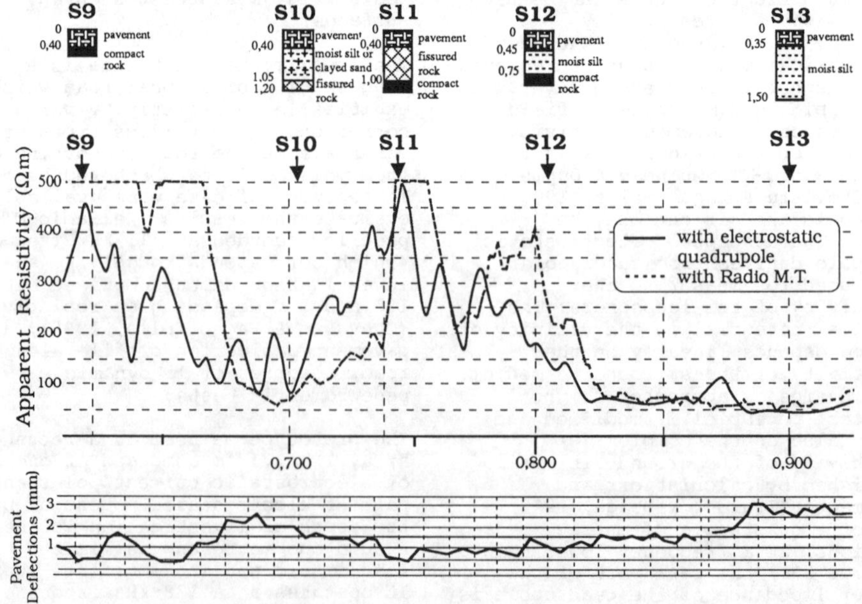

Figure 1. Profiles of apparent resistivity compared with pavement deflections and geological sections.

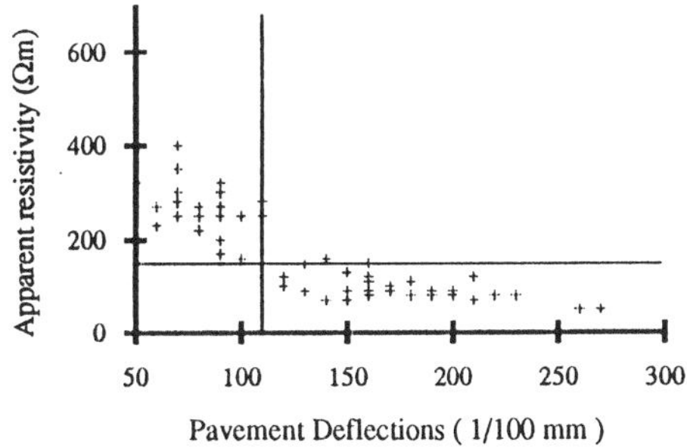

Figure 2. Comparison between pavement deflections and apparent resistivity.

electrostatic quadrupoles)

This test, not directly aimed at obstacle detection, shows that the resistivity measurement is very relevant to evaluate the mechanical properties of the soil beneath a cause way. Among the advantages of the electrostatic quadrupole one has to emphasize that the thickness of soil taken into account in the measurement can be chosen by defining the size of the array, that it can be easily extended to multiple arrays supplying a more detailed underground image of soil structure and constitution and that as well as the electric method this technique is better than any technique based on the sole magnetic field measurement for resistive target detection.

REFERENCES

Chevassu, G. et R. Lagabrielle 1990, Radio-Magnetotellurics apparent resistivity high velocity continuous profiling for shallow works. 6th International IAEG Congress, Balkema, Rotterdam : 899-905

Grard, R et A. Tabbagh 1991, A mobile four-electrode array and its application to the electrical survey of planetary grounds at shallow depths. JGR 96, B3 : 4117-4123

Guineau, B et A. Dupis, 1973, Dispositif pour la prospection magnétotellurique de subsurface. Agence Nationale pour la Valorisation de la Recherche, Paris, Brevet n° 73-11-573.

No Trenches in Town, Henry & Mermet (eds) © 1992 Balkema, Rotterdam. ISBN 90 5410 085 0

Searching for services – The role of impulse radar in the location of buried utilities

Jon Baston-Pitt
GB Geotechnics Limited, UK

ABSTRACT: In the United Kingdom, local authorities are increasingly keen to minimise the costs involved in street works, whether new developments and refurbishment projects, whilst at the same time seeking to ensure that there is minimal disruption to pedestrians and traffic in these urban areas.

The use of Impulse Radar as a non-destructive testing tool can be used on two fronts: by both local authorities as a cost-effective and extremely efficient method of locating buried utilities and by the service providers as a quality control tool to check the quality of any reinstatement.

Using relevant case studies, the role of Impulse Radar in the location of buried utilities is examined, and the potential of the method as a means of quality control, with particular reference to the new Road and Street Works Act is explored.

THE PRINCIPLES OF IMPULSE RADAR

Impulse Radar is a completely non-destructive testing technique, which is based on the interpretation of electrical and electromagnetic signals to locate and map features, such as services, within man-made structures.

Surveying by this technique is achieved by drawing the transducing radio antennae over the surface to be investigated at a slow speed; pulses of energy are transmitted into the material which are reflected from any internal surface or structural change. The returning signals from each vertical scan build, as the transducer is moved, into a continuous radio profile of the interior of the material, which can be translated into detail of thicknesss, compaction and inclusions within the subsurface.

A plane of energy, rather than a ray or beam is transmitted. This plane is normally set parallel to the direction of the survey, such that all information streams refer to material directly beneath the survey line.

Selected centre frequencies are used to provide data on different features and depths: the deeper the feature the lower the frequency that is required to identify it. The recovered signals can be recorded on graphic scanning systems, either to allow on-site interpretation of the data in real time, or a data storage for subsequent off-site analysis.

Applied to service location, this non-destructive testing (NDT) technique, apart from causing no damage to the structure of the road, allows investigation and assessment of the area to proceed without serious disruption to traffic flow, pedestrians or the normal everyday functioning of the street. Using Impulse Radar scanning techniques provides an overall and continuous assessment of what lies beneath street level and hence the location of buried utilities.

It is important to note, however, that whilst utilities can be detected to a depth of >1500mm, the actual range depending largely on the materials in which they are buried, it is not possible to identify specific

utilities. The signals from a pipe can be readily located using Impulse Radar and characterised as either metal or plastic, but, from the radar data alone, it is not possible to define whether it carries gas, water, electricity or nothing. As the case studies illustrate however, by drawing upon information from the service providers and using the visual evidence in the streets, such as manhole covers, a detailed picture of the active service runs can be built up.

CASE STUDY ONE: TESTING THE HYPOTHESIS

Over the course of the last five years or so, a great deal of time and money has been invested in the design and development sophisticated computer-based analytical systems, capable of taking the raw data from radar surveys and interpreting the information either in real time or in post processing without the need of human interpretation.

These analytic programs have generally been based on either developing specific algorithms to identify specific signals based on synthetic models, or allowing AI or Artificial Intelligence based systems to 'learn' the features which are to be found.

Such developments require considerable investment, which was justified on the assumption that although the variables were finite, they generally exceeded the ability of easy human interpretation, and necessitating uneconomic use of expert resource. Even then, the expert judgement was likely to be subjective.

The problem to be solved can be stated fairly simply: at best, profiling radar systems can, on any one survey line at one pulsed centre frequency, hope to locate between 60% and 80% of the cylindrical targets below an urban pavement. Those not located may be locally obscured due to a variety of ground conditions, and the ground conditions may locally include resolvable cylindrical targets which are not services - such as tree roots, building debris and vermin runs.

On an immediately adjacent survey line different ground conditions may obtain, and a different 60% - 80% of the targets may be resolved. This process can be repeated for a number of survey lines and iterated at a

number of different frequencies. The task for either a computer or a human interpreter is to collate and map the different streams of data until all the services have been located and all the false targets rejected.

As yet the computer based systems have not achieved better than about 60-80% success to a depth of about 1m in real conditions: these case studies examine the application of dense survey grids interpreted manually, to show that the work involves is both hard and meticulous, but produces a better than 95% confidence to a depth of around 1.5m, thus including all but the deepest services, which can often be located but with a lower confidence.

The methodology has been developed in a commercially cost conscious environment, and competes favourably against traditional trial pitting and excavation, without even taking into account the low environmental impact and minimal disruption.

As part of the programme of research and development, G B Geotechnics surveyed a short stretch of single carriageway urban pavement in Cambridgeshire, which was due for extensive excavation and remedial work, which would allow any survey results to be tested more fully than could be achieved by normal confirmatory excavation, and provide a test bed in a practical site rather than a fabricated model. The object was to test the hypothesis that Impulse Radar data relating to services could be analysed accurately and efficiently using manual interpretative methods.

The pavement under investigation was some 6m in width. It was selected as being a typical examples of a suburban street: little was known of the original pavement structure, it had been heavily reinstated and reconstructed and no plans or service details were available prior to the survey.

The research plan was formulated with a number of clear stages:

* carry out an Impulse Radar survey

* manually analyse the data and present a report

* compare the results with those from an AI-based computer system

* corroborate the findings of both surveys by excavating the pavement

For the initial Impulse Radar survey, the entire section was marked at 10m centres along the the kerb line to provide a relocation reference net. The investigation was restricted to the main service trunks, and did not include the connections to individual houses, so most survey lines were set transverse to the pavement.

A few profiles were collected from survey lines set along the length of the pavement which provided a continuous assessment of the pavement structure itself, which information could be used to assist in the understanding of the initial transmission path through the surface materials.

All surveying was relocated using a standard surveying wheel, referenced to the 10m relocation net, with spacing between survey lines typically at 1m centres.

The information recovered not only identified the structure of the pavement and the position of the underlying services, but also identified the thickness and compaction of material layers and located areas of voiding and deconsolidation in the subsurface.

A large number of longitudinal features were detected and mapped under both the footpaths and the carriageway ranging in depth from the near surface to a depth of c.1500mm. Most of the services appeared to have been placed in the footpaths, especially those carrying telephone and water services.

The manual analysis of this data has been compared both with the real evidence from excavation and that achieved from one of the more recently developed computer-based systems. Results at the time of writing are incomplete, but indicate that in the clay rich environment of this street, a real AI system achieved 25-30% detection of services, against an accuracy in excess of 95% achieved by the manual approach, relying upon intelligent observation and experience to achieve an efficient interpretation.

At the time of going to press, the County Council has yet to complete excavation of the site in order to verify the findings. These, however, will be available to be reported at the conference.

CASE STUDY TWO: STRATFORD UPON AVON

Stratford upon Avon District Council, responsible for one of the UK's most historic and high- profile tourist centres, commissioned, prior to the third phase of their town centre pedestrianisation scheme, a non-destructive investigation using Impulse Radar of the subsurface of Henley and Meer Streets.

The development plans for these streets included the planting of trees and installation of other items of street furniture. Before this could be undertaken, the Council needed to know where the ground was unobstructed, so that there was no danger of damage to the services.

They had obtained plans from the various utilities, but these appeared to provide neither a complete nor an accurate picture. It was likely that some of the recorded services had been repaired and moved, and that some of the older, and in some cases abandoned, services were not indicated at all. Finding these services quickly and accurately could only be done using a technique such as Impulse Radar: the alternative would have been to dig up the pavement at random, disrupting pedestrians, traffic, tourists and possibly causing unnecessary and unsightly damage to the street.

The Impulse Radar survey was conducted over an area of approximately 3,500 square metres and was completed in one Sunday to avoid disruption to the normal life of the city centre.

No construction details about the streets was available prior to the survey but it was evident that they had been extensively reinstated and reconstructed over the years, covering manholes and other indicators of the service layout.

The survey provided Council Engineers with a clear map of the hidden subsurface of the streets, with sections through selected points to provide depth detail, accurate to a few tens of millimetres, and identifying those areas which contained no services and in which tree planting could take place with minimal risk of damage to the pipe runs. Any services which would require relocating in order to maintain the visual impact of the pedestrianisation were identified well in advance of site works, such that the cost and inconvenience to the local shopkeepers and the tourists could be both properly planned and minimised.

The survey successfully located water, gas and other service pipelines, as well as the line of services to individual properties: by working with the various utilities, the District Council was able to cross correlate this information in order to recognise individual services, and to ensure that no services were cut accidentally or unexpectedly.

Two trial pits were opened by the Local Authority to confirm the extremely detailed plans showing both layout of services and sections through the pavement to show the relative depths of the various pipes: these excavations showed the survey to have achieved an accuracy well above the expected tolerances.

The pedestrianisation scheme could therefore proceed with a clear understanding of the arrangement of all obstructions below the surface.

ROAD AND STREETWORKS ACT 1991

Coming into force currently in the UK, is the Road and Streetworks Act 1991, which amongst other changes will radically alter the responsibility for pavement reinstatement following the installation or repair of utilities.

Previously, the responsibility for final reinstatement lay with the local Highway Authority, and the service provider through its contractor was only required to backfill and provide temporary reinstatement of any excavation. This was typically spoil from the excavation, with only minimal compaction applied, with an overlay of unspecified bound material. A fee was paid to the Highway Authority, who were then responsible for the completion of a permanent high quality reinstatement.

Management of this system was poor, both in reporting by the utilities, and execution by some authorities, and temporary works were frequently left for extended periods, during which settlement frequently took place, leading to severe rutting and hazards for road users. Permanent reinstatement at the time of backfilling is obviously more sensible and more economic, and was the management system for reinstatement preferred by both the Utilities and most Local Authorities.

Local Authorities have however retained a coordinating and inspecting function, which is intended to make sure that street works do not cause havoc and are properly done.

Even with good Quality Control however, such schemes can go wrong: there has to be confidence that the contractor providing the reinstatement has done a good job, and the contractor needs to be able to confirm and show that a good job has been done whenever a query arises.

Given the size of a typical reinstatement - a narrow trench, about the size of a trial pit, the only way to prove the quality of the job is to dig it up - and then reinstate the excavation, the quality of which may need to be checked, whereupon endless and pointless iteration is likely.

Clearly, in order to avoid this situation, it is important that both sides are able to rely on factual evidence, without any destructive probing, and the role of Impulse Radar in providing a fully non destructive investigation is obvious.

Under the terms of Section 75 (Inspection) of the 1991 Act, utilities pay an annual fee based on an average of previous years' works and the authority inspects, randomly, 30% of the works, but sectioned as 6% of the works each, in five different prescribed stages from start up to the end of the reinstatement guarantee period.

Non-destructive testing, and in particular Impulse Radar, has a major role to play: Authorities are concerned that if only "random" sites are inspected, they will be unable to pinpoint works which they consider feel are below acceptable standards. Section 72 allows the authority to inspect where it thinks it is "necessary to ensure it is undertaken to comply with standards". Thus it is perfectly feasible, and perfectly legitimate within the wording of the new Act for the authorities to use the technique of Impulse Radar, as a quality control tool for reinstatement assessment, at every site.

For their part, utilities are very conscious of the need to reinstate properly and their contractors are training up special excavation and reinstatement gangs. Not only will any poor quality work have to be redone at the utilities' expense, plus the additional burden of extra defect inspections, heavy fines can also be imposed and performance lists published - a move which has caused one major UK utility British Gas to declare that it "will come out top or know the reason why".

It is interesting to note that British Gas has already invested considerable time and money over the last two years in assessing the benefits of Impulse Radar, again in terms of its use as a quality control tool.

The utilities, in much the same way as the Local Authorities, can take the initiative by using NDT to ensure that its employees or contractors have in fact carried out the necessary reinstatement to the required standard. All work completed could be "handed over" to the local authority with, in effect, a certificate of compliance based on an indirect and non-invasive, but conclusive test by Impulse Radar.

This would avoid any unforeseen problems for the utilities at a later date and would help to build up the confidence and mutual understanding between the two parties - widely regarded as the most important criteria in the successful working of the new Act.

CONCLUSIONS

As is being actively proven in the UK, Impulse

Radar has an important role to play both as a non-destructive technique to locate services and as a tool in the Quality Control of trench reinstatement. Increased pressure on Local Authority and Contractors' finances, combined with the rigorous application of new criteria in street works, ensures that its use will become increasingly widespread and that the technology is set to become the de facto standard for all such projects in the not too distant future.

At present its use is largely tied to operation and interpretation by experienced investigators, well versed in both civil construction and radar transmission theory, and care must be taken in the selection of operatives to avoid misinformation: the advances currently being made in computer technology however are likely to change this over the next few years, and a relatively simple tool may be available for more general use.

Radar alone will however always be a mapping tool and can never be sufficiently selective to identify unquestionably the difference between a cast iron fresh water pipe and a cast iron storm water pipe: the intelligent cartographer will always be needed.

No Trenches in Town, Henry & Mermet (eds) © 1992 Balkema, Rotterdam. ISBN 90 5410 085 0

À la recherche de canalisations – Le rôle du radar à impulsion dans la localisation de réseaux enfouis

Jon Baston-Pitt
GB Geotechnics Limited, Royaume-Uni

Résumé

En Grande-Bretagne, les autorités locales montrent de plus en plus d'enthousiasme quand il s'agit de diminuer les frais qu'entraînent les travaux de la route, que ce soient de nouveaux projets de développement ou des travaux de remise à neuf, tout en recherchant l'assurance d'un minimum de perturbation pour les piétons et la circulation dans les zones habitées.

L'utilisation du radar à impulsion comme dispositif d'essai non destructif peut se faire sur deux plans: d'un côté les autorités locales peuvent s'en servir comme méthode extrêmement efficace ainsi qu'économique pour la localisation de réseaux enfouis, tandis que de l'autre ce sont les compagnies fournissant ces utilités qui peuvent l'employer comme méthode de contrôle de la qualité de toute restauration effectuée.

En se servant de l'étude de cas appropriés, le rôle du radar à impulsion dans la localisation de réseaux enfouis est examiné, et le potentiel de cette méthode en tant que dispositif de contrôle de qualité est exploré en se référant spécialement au tout récent Acte des Travaux des Chaussées et des Routes.

Principes d'utilisation du radar à impulsion

Le radar à impulsion est un dispositif d'essai entièrement non destructif, qui est basé sur l'interprétation de signaux électriques et électromagnétiques et utilisé pour localiser et tracer le plan détaillé de particularités telles que les canalisations à l'intérieur des structures artificielles.

Pour faire un relevé en se servant de ce dispositif, on fait passer à petite vitesse l'antenne radio du capteur de mesure au-dessus de la surface à étudier; les pulses d'énergie sont transmises dans le matériau et sont réfléchies par toute surface interne ou changement dans la structure. Les signaux renvoyés lors de chaque balayage vertical élaborent progressivement au cours du mouvement du capteur de mesure le profil radio continu de l'intérieur du matériau, qui peut se traduire en termes de détail de l'épaisseur, degré de compactage et détail d'inclusions des couches existant sous la surface.

C'est un plan d'énergie, plutôt qu'un rayon ou un faisceau, qui est transmis. Ce plan est généralement placé parallèlement à la direction du relevé, de telle sorte que tout le flot d'information se rapporte au matériau situé immédiatement sous la ligne du relevé.

Des fréquences de foyer bien précises sont utilisées pour fournir des données sur différentes caractéristiques et épaisseurs: plus un détail particulier est profond, plus la fréquence nécessaire à son identification doit être basse. Les signaux récupérés peuvent être enregistrés sur des réseaux de balayage graphiques, soit pour permettre l'interprétation sur place des données en temps réel, soit pour la mémorisation de ces données pour l'analyse qui se fera par après, hors du site.

Quand on l'applique à la localisation de canalisations, ce dispositif d'essai non destructif (END) non seulement ne cause aucun endommagement de la structure de la route mais encore permet l'étude et l'évaluation de la section examinée de se dérouler sans grande perturbation du flot de la circulation, des piétons, ou de l'usage courant de la route. L'emploi de la technique de balayage du radar à impulsion fournit ainsi l'évaluation globale et continue de ce qui se trouve sous le niveau de la route, et donc la localisation de réseaux enfouis.

Il est toutefois important de noter que bien que des canalisations puissent être détectées jusqu'à une profondeur d'au moins 1500mm, la gamme réelle de cette valeur dépendant des matériaux dans lesquels elles ont été enterrées, il n'est cependant pas possible d'identifier la nature spécifique des conduites. Les signaux réfléchis par une conduite permettent aisément sa localisation par le radar à impulsion ainsi que sa caractérisation métallique ou plastique, mais il n'est pas possible à partir de la seule étude au radar de définir son contenu, qu'il s'agisse de gaz, d'eau, d'électricité, ou même de rien du tout. Par contre, les études qui suivent illustrent comment, avec l'aide d'information venant des compagnies des utilités et en se servant de l'évidence visuelle qu'apporte l'inspection de la route, telle que par exemple la présence de tampons d'égout, il est tout à fait possible de former une image détaillée des canalisations actives.

Etude Numéro Un: essai de l'hypothèse

Au cours des cinq ou six dernières années on a investi énormément de temps et d'argent dans la conception et le développement de systèmes informatiques d'analyse sophistiqués, qui soient capables d'entrer les données brutes recueillies lors de l'étude au radar et de traiter l'information soit en temps réel soit en post-traitement, sans le recours à l'interprétation humaine.

Ces programmes analytiques ont généralement été basés soit sur le développement d'algorithmes spécifiques capables d'identifier des signaux spécifiques suivant des modèles de synthèse, soit sur l'extension de la capacité des systèmes d'Intelligence Artificielle (IA) à "apprendre" à reconnaître les détails particuliers que l'on veut rechercher.

De tels progrès nécessitent un investissement considérable,que l'on a justifié par la supposition que bien que les variables soient limitées, elles excédaient généralement l'aptitude d'une interprétation humaine aisée, et avaient besoin du recours peu économique aux ressources des experts. Et même alors, il était vraisemblable que l'avis de l'expert pouvait être subjectif.

Voici un énoncé assez simple du problème à résoudre: les dispositifs d'esquisse du radar peuvent, le long d'une seule ligne de relevé et fonctionnant sur une seule fréquence de foyer, espérer tout au mieux de localiser entre 60% et 80% des cibles cylindriques situées sous une chaussée urbaine. Celles qui ne sont pas localisées sont soit masquées localement à cause de diverses conditions de terrain, soit il se peut que dans le sol se trouvent localement des cibles cylindriques résolubles qui ne sont pas des conduites- telles que les racines d'un arbre, des débris de bâtiment ou des tunnels d'animaux nuisibles.

Si l'on fait un relevé sur une ligne immédiatement adjacente à la première, les conditions du terrain sont probablement légèrement différentes, et l'on peut ainsi obtenir un nouveau tracé contenant encore une fois 60-80% des cibles. Ce processus peut être répété sur de nombreuses lignes de relevé et repris en utilisant plusieurs fréquences différentes. La tâche revient alors à l'ordinateur ou bien à l'interprète humain de faire la relation entre les différents flots de données et de tracer leur esquisse jusqu'à ce que tous les réseaux soient localisés et toutes les fausses cibles identifiées et rejetées.

Jusqu'à présent les méthodes de traitement par ordinateur n'ont pas dépassé le taux de 60-80% de succès jusqu'à une profondeur d'à peu près 1m dans des conditions réelles: les études qui suivent de cas particuliers examinent l'application de grilles serrées de relevés interprétées manuellement, pour démontrer que bien que la tâche demandée soit à la fois ardue et méticuleuse, elle produit des résultats de plus de 95% de certitude jusqu'à une profondeur d'à peu près 1500mm, donc comprenant la grande majorité des conduites sauf les plus profondes qui peuvent néanmoins souvent être localisées mais avec moins de certitude.

Cette méthodologie a été développée dans un environnement commercialement conscient des frais, et fait favorablement concurrence aux méthodes traditionnelles d'excavation et de creusement de puits d'exploration, sans même prendre en considération le peu d'impact sur l'environnement et le degré minimum de perturbation.

Comme partie intégrante du programme de recherche et de développement, GB Geotechnics fit l'étude d'un court segment d'une chaussée urbaine à deux voies dans le Comté de Cambridgeshire qui allait être soumise à une excavation importante ainsi qu'à des travaux de correction, et allait donc permettre à tout résultat de l'étude d'être contrôlé de manière beaucoup plus complète que par l'excavation normale de vérification, et ainsi fournir un lit d'essai dans un site pratique plutôt que dans un modèle fabriqué. Notre objectif était de tester l'hypothèse que les données obtenues par le radar à impulsion qui se rapportent aux canalisations pouvaient être analysées de façon précise et efficace en se servant de méthodes manuelles d'interprétation.

La chaussée à étudier avait à peu près 6m de large. Elle fut sélectionnée comme étant un exemple typique d'une route de banlieue: on ne savait que peu sur la structure d'origine de la chaussée, elle avait été intensément remaniée et reconstruite et avant d'en entreprendre l'étude aucun plan ni détail des réseaux de distribution n'étaient disponibles.

Le plan de recherche fut formulé suivant un certain nombre de stades bien précis:

* entreprendre l'examen au radar à impulsion

* analyser manuellement les données et présenter un rapport

* comparer les résultats obtenus avec ceux d'un réseau informatique d'IA

* corroborer les résultats des deux études par l'excavation de la chaussée

Pour le levé initial au radar à impulsion on marqua la section complète sur tous les 10m le long de la ligne du trottoir pour fournir un filet de points de repère pour la relocalisation. L'étude fut limitée aux lignes principales des canalisations, et les connexions aux maisons individuelles n'avaient pas besoin d'être inclues, si bien que les lignes de relevé furent établies transversalement par rapport à la chaussée.

Quelques profils furent recueillis à partir de lignes de relevé placées le long de la direction de la chaussée fournissant ainsi une évaluation continue de la structure même de la chaussée, information qui elle-même pouvait venir en aide à la compréhension du trajet initial de la transmission à travers les matériaux de la surface.

Chaque levé fut relocalisé en se servant d'une roue d'arpentage standard, se référant au filet de points de repère placés tous les 10m, avec un espace typique de 1m entre les lignes de relevé.

L'information obtenue non seulement permit d'établir l'identité de la structure de la chaussée ainsi que la position des canalisations sous-jacentes, mais aussi identifia l'épaisseur et le degré de compactage des couches de matériau et localisa des zones de vide intersticiel et de déconsolidation sous la surface.

Un grand nombre de structures longitudinales fut détecté et leur plan fut tracé à la fois sous les trottoirs et la chaussée, leur situation variant en profondeur depuis proche de la surface jusqu'à une profondeur d'à peu près 1500mm. La majorité des conduites sembleraient avoir été placées dans les trottoirs, en particulier les réseaux de distribution du téléphone et ceux de l'eau.

L'analyse manuelle de ces données a été comparée à la fois avec l'évidence de la réalité apportée par l'excavation ainsi qu'avec

l'analyse accomplie par un système informatique comptant parmi les plus récemment développés. Au moment de l'édition de cet article les résultats ne sont pas encore au complet, mais indiquent que dans l'environnement riche en argile de la route, un système réel d'IA ne parvint à détecter que 25-30% des canalisations, alors que la méthode manuelle obtint une précision au-delà de 95%, se fiant à l'expérience et l'observation intelligente pour obtenir une interprétation compétente.

Au moment de la parution de cet article le Conseil Régional n'a pas encore complété l'excavation du site et donc nous ne sommes pas encore en mesure de vérifier tous les résultats. Par contre, les résultats définitifs seront disponibles pour la présentation d'un rapport qui se fera à la conférence.

Étude Numéro Deux: Stratford Upon Avon

Le Conseil Régional de Stratford Upon Avon, responsable pour l'un des centres touristiques comptant parmi les plus historiques et les mieux reconnus de Grande-Bretagne, commissionna préalablement à la 3ème phase du projet de transformation de leur centre ville en zone piétonnière un examen non destructif faisant usage du radar à impulsion des structures situées sous Henley Street et Meer Street.

Les plans de développement de ces rues comprenaient la plantation d'arbres et l'installation d'autres articles de mobilier urbain. Avant que ceci puisse être entrepris, le Conseil avait besoin de savoir où le terrain était non obstrué, de telle sorte qu'il n'y ait aucun danger de dommage aux réseaux de distribution.

Ils avaient obtenu des plans des réseaux divers, mais la description donnée par ceux-ci ne semblait être ni complète ni exacte. Il était probable que certains des réseaux enregistrés avaient été réparés et déplacés, et que certains des plus anciens réseaux, dans certains cas même abandonnés, n'étaient pas du tout indiqués. La seule méthode capable d'identifier rapidement et avec précision l'emplacement de ces canalisations était d'utiliser le radar à impulsion: la seule autre solution aurait été de creuser des trous dans la chaussée au hasard, causant ainsi des perturbations pour les piétons, la circulation, les touristes, et probablement des dégâts à la rue inutiles et peu attrayants.

Le relevé au radar à impulsion fut pris au-dessus d'une zone couvrant approximativement 3500 mètres carrés et fut complété en un seul dimanche pour éviter de perturber le niveau d'activité normal du centre ville.

Avant de mener l'étude aucun détail de construction de ces rues n'était disponible mais il était évident qu'elles avaient été considérablement remaniées et reconstruites au cours des années, recouvrant ainsi des tampons d'égout et d'autres indicateurs du plan des réseaux.

Le relevé permit de fournir aux ingénieurs du Conseil le plan détaillé du terrain caché sous la surface des rues, avec sections à des points spécifiques pour procurer le détail de la profondeur, exact à quelques dizièmes de mm près, et identifiant les zones qui ne contenaient pas de conduites et où la plantation d'arbres pouvait se faire avec un minimum de risque d'endommager des canalisations. Tous les réseaux ayant besoin d'être relocalisés afin de maintenir l'impact visuel du projet de la zone piétonnière furent ainsi identifiés bien avant le commencement des travaux au chantier, de telle sorte que cela a permis de planifier ainsi que de minimiser convenablement les frais et les dérangements pour les commerçants locaux et les touristes.

Le levé localisa avec succès les conduites d'eau, de gaz et autres, ainsi que le parcours des réseaux de distribution vers les propriétés individuelles: en travaillant avec les diverses compagnies, le Conseil Régional parvint à mettre cette information en corrélation d'une manière telle qu'ils pouvaient reconnaître la nature des canalisations individuelles et assurer qu'aucune conduite ne soit coupée accidentellement ou inopinément.

Deux puits d'exploration furent creusés par les autorités locales afin de confirmer les plans extrêmement détaillés montrant l'emplacement des canalisations ainsi que des sections à travers la chaussée indiquant la profondeur relative des conduites diverses: ces excavations démontrèrent que l'étude avait réussi un taux de précision bien au-dessus des tolérances attendues.

Ainsi donc le projet de transformation du centre ville en zone piétonnière put se dérouler avec une bonne compréhension de la disposition de toutes les obstructions existantes sous la surface.

Acte des Travaux des Chaussées et des Routes de 1991

Actuellement en train d'entrer en vigueur en Grande-Bretagne, l'Acte des Travaux des Chaussées et des Routes de 1991 va entre autres changer radicalement à qui revient la responsabilité de restaurer les chaussées après l'installation ou la réparation de canalisations.

Auparavant, la responsabilité de la restauration finale restait celle des autorités des voies publiques locales, et la compagnie responsable d'une distribution était par l'intermédiaire de son entrepreneur seulement dans l'obligation de remblayer et de fournir un rétablissement temporaire après toute excavation. Il s'agissait habituellement des déblais de l'excavation avec seulement un degré de compactage minimal appliqué et une couche de recouvrement d'un matériau lié non spécifié. Une cotisation était payée aux autorités des voies publiques qui étaient alors tenues responsables de l'achèvement d'une restauration permanente de bonne qualité.

L'administration de ce système fonctionnait mal, à la fois du point de vue du signalement donné par les compagnies, et du point de vue de l'exécution des travaux par certaines autorités, et donc des travaux temporaires étaient fréquemment abandonnés pendant des périodes étendues au cours desquelles il se produisait souvent un tassement, entraînant la formation d'ornières profondes et des risques pour les usagers de la chaussée. Une restauration permanente au moment même du remblayage était de toute évidence plus judicieux et plus économique, et fut donc le système de gestion de restauration préféré par les services des utilités ainsi que par la majorité des autorités locales.

Les autorités locales ont néanmoins retenu la fonction de coordonner et d'inspecter les travaux, dans l'intention d'assurer que les travaux de la route n'engendrent pas le chaos et soient exécutés de manière correcte.

Cependant, même à l'aide d'un bon contrôle de qualité des travaux, de tels procédés peuvent mal tourner: il doit y avoir confiance que l'entrepreneur fournissant la restauration ait fait un bon travail, et de son côté l'entrepreneur doit être capable de confirmer et de démontrer que le travail a été bien exécuté si jamais il y a une remise en question.

Considérons par exemple l'étendue d'un travail de restauration typique d'une étroite tranchée ayant à peu près la même taille qu'un puits d'exploration- le seul moyen de prouver la bonne qualité du travail est de creuser, mais alors il faut à nouveau remblayer et restaurer l'excavation, travail dont il faudrait vérifier la qualité, et il s'ensuit donc une liste sans fin et inutile de répétitions.

Il est clair qu'afin d'éviter ce genre de situation, il est important que les deux partis puisent dépendre d'une évidence basée sur les faits réels, sans nécessiter des moyens d'investigation destructifs, et donc le rôle du radar à impulsion dans son aptitude à fournir un essai entièrement non destructif est évident.

Dans les termes de la section 75 (inspection) de l'Acte de 1991, les compagnies doivent payer une somme annuelle basée sur la moyenne des travaux de l'année précédente et les autorités doivent inspecter au hasard 30% des travaux, mais ces 30% eux-mêmes sont subdivisés en 6% des inspections se faisant lors de cinq stades différents bien précis s'étalant depuis le début des travaux jusqu'à la fin de la période de garantie du travail de réfection.

Les dispositifs d'essai non destructif et le radar à impulsion en particulier, ont à jouer un rôle d'une importance majeure:les autorités en effet sont concernées par le fait que si seul le hasard décide des sites à inspecter, elles seront dans l'impossibilité de localiser avec précision les travaux qu'elles considèrent être d'un niveau en-dessous de la moyenne acceptable. La section 72 permet aux autorités d'inspecter là où elles pensent qu'il soit "nécessaire d'assurer que l'entreprise des travaux soit conforme aux normes". Il est donc tout à fait possible et parfaitement légitime et en accord avec les termes du nouvel Acte que les autorités utilisent le dispositif du radar à impulsion comme outil de contrôle de la qualité pour évaluer tout travail de restauration sur le chantier.

De leur côté, les compagnies sont tout à fait conscientes du fait qu'il est nécessaire de faire la restauration de façon convenable et leurs entrepreneurs sont en train de former de équipes spécialisées pour effectuer les excavations et la restauration. Les compagnies doivent non seulument encourir les frais de tout travail de mauvaise qualité qui est à refaire ainsi que les frais supplémentaires d'une nouvelle inspection des défauts, mais encore on peut leur imposer de lourdes amendes, et des listes de "performance" vont être publiées une décision qui a entraîné une importante compagnie britannique, British Gas, à déclarer que celle-ci "serait en tête de liste, et au cas contraire, voudrait savoir où elle a échoué".

Il est intéressant de noter le fait que British Gas a déjà investi beaucoup de temps et d'argent ces deux dernières années dans l'évaluation des bénéfices du radar à impulsion, à nouveau dans son rôle d'outil pratique de contrôle de qualité.

Les compagnies peuvent prendre l'initiative d'utiliser les dispositifs d'END, dans le même but que les autorités locales, pour s'assurer que leurs employés ou entrepreneurs ont réellement exécuté les travaux de restauration nécessaires dans les normes requises. Tout travail achevé pourrait en effet être remis entre les mains des autorités locales avec en quelque sorte l'équivalent d'un certificat de conformité basé sur l'examen indirect, non destructif mais bien concluant, du radar à impulsion.

Ceci pourrait permettre aux compagnies d'éviter plus tard des problèmes imprévus, et aiderait à bâtir une confiance et compréhension mutuelle entre les deux partis- critère généralement reconnu comme le plus important pour que le nouvel Acte soit appliqué avec succès.

Conclusion

Le radar à impulsion, en train de faire activement ses preuves en Grande-Bretagne, a un rôle important à jouer comme dispositif d'END à la fois pour localiser les canalisations ainsi que comme outil de contrôle de la qualité des travaux de restauration après le creusement de tranchées. La pression accrue sur les finances des autorités locales et des entrepreneurs, combinée à l'application rigoureuse des nouveaux critères pour les travaux de la route, entraînent que son usage va de plus en plus s'étendre et que cette technologie est en passe de devenir dans l'avenir proche le critère de facto pour tous les projets similaires.

Actuellement, son usage est sutout limité à l'opération et l'interprétation par des investigateurs expérimentés, étant très versés aussi bien en construction civile qu'en théorie de transmission radar, et il faut sélectionner avec attention la bonne main d'oeuvre afin d'éviter toute cause de mauvais renseignement: néanmoins, les avances qui se font actuellement dans le domaine de la technologie informatique vont très vraisemblablement changer cela au cours des prochaines années de telle manière qu'un outil relativement simple pourrait devenir disponible et rendre son usage plus général.

Le radar à lui seul restera cependant toujours un outil pour tracer des plans et ne sera jamais suffisemment sélectif pour identifier à coup sûr la différence entre une conduite en fonte transportant l'eau douce et une conduite en fonte pour les eaux d'orage: pour cela on aura toujours besoin de l'intelligence d'un cartographe.

No Trenches in Town, Henry & Mermet (eds) © 1992 Balkema, Rotterdam. ISBN 90 5410 085 0

The use of ground penetrating radar on a microtunneling site

M. Berosch
Compagnie Générale des Eaux, Paris, France

R. Foillard
Compagnie Générale de Géophysique, Massy, France

G. Le Ny
Entreprise Valentin, Alfortville, France

ABSTRACT : The scope of this paper is the presentation of an experience with a new underground investigation method on a microtunneling site for the installation of water mains in the Paris suburbs' water supply area. Considerations prior to the choice of an investigation technique are given taking into account the specific problems of the site. The radar and data treatment used are described and their first application on a microtunneling site in France discussed.

1 INTRODUCTION

The experience with ground penetrating radar on a microtunneling site in the water supply area of the suburbs of Paris is the subject of this paper. The Paris suburbs' supply area with its 4 million inhabitants is the most important network management contract of Compagnie Générale des Eaux that serves 24 million inhabitants in France. The 144 communities of the Paris suburbs water syndicate (Syndicat des Eaux d'Ile de France), France's largest water utility, are supplied by a pipe system with a total length of 8493 km (plus 3300 km of house connections). About 130 km of new pipes are commissioned each year, of which 105 km serve to renew the network (replacement ratio of 1.35 %), which grows about 34 km per year (average figures for the last 10 years). In the Paris suburbs' supply area there are 700 km of trunk mains (DN/ID ≥ 400) and 7800 km of distribution pipes (DN/ID ≤ 350). The usual depth of cover is 1.4 m for trunk mains and 0.8 to 1.3 m for distribution pipes. This zone is very congested by other utilities such as gas and cables. In order to reduce surface disruption and nuisance to residents trenchless techniques are used for pipelaying (Berosch and Revol 1991) and rehabilitation (Berosch, Ilotte and Randon 1992). Among the trenchless techniques for new installations guided drilling is the most applied and 6 km of small diameter distribution pipes were laid with it in 1991. However, it could be seen that a careful selection of the sites is necessary in order to complete jobs successfully (Ilotte and Schroeyers 1991). About one third of the guided drilling projects are abandoned because of unfavourable soil conditions. Many problems with geology and obstacles in the underground were also observed on microtunneling sites executed in the area of Paris, known for its extremely heterogenous underground conditions. Microtunneling started in France in 1989 and about 40 jobs were carried out until now. Since 1990, six microtunneling sites have been completed for the Paris suburbs Water Syndicate. For two recent jobs, carried out in the beginning of 1992 it was decided to try subsoil investigation using ground penetrating radar. The results are presented in this article.

2 SOIL INVESTIGATION FOR MICROTUNNELING SITES

Prior to microtunneling work it is recommended to examine the following points :

1. Geological conditions
2. Groundwater level(s)
3. Position of other utilities
4. Presence of underground structures (foundations, chambers, piles, galleries, etc.)

5. Risk to meet unexpected obstacles (concrete blocks, ancient or abandoned sewers or cellar walls, singular rocks, ancient landfills containing iron debris,etc.).

The list given in brackets of point 5 is based on examples from microtunnling jobs known to the authors that had come to standstill due to one of the mentioned obstacles.

The soil investigation should allow to get the maximum of information on the points mentioned above because it will determine :

1. The choice whether microtunneling or another technique is used (feasibility of the project).
2. The estimation of cost and duration of the project .
3. The choice of the microtunneling machine (auger or slurry type, machine with or without the possibility to be drawn back if blocked).
4. The lifespan of the laid main or sewer (especially in agressive soils, landfills, zones with cavities or risk of settlements).

A thorough investigation using traditional techniques only is often not possible. In an area with very heterogenous underground conditions such as the surroundings of Paris boreholes and standard penetration tests give very punctual information for a linear structure such as a pipeline if the frequency of investigation is not increased (up to every 10 m for example). This, however, would not be economical. In addition, singular obstacles cannot be detected systematically with these methods.

Traditional detection techniques are not always helpful to gain more information. Often, they are either too specialized (e.g. detection of metallic objects only) or too imprecise for the range of depth that is interesting for pipelaying (seismic, dynamic, geoelectric, and radiometric methods).

The information that can be gathered from other utilities or from local authorities is not always precise and up to date. In the underground however, more and more congested, there often is not enough space for "generous" security distances towards other utilities.

Very recently, the ground probing radar technique has become operational and allows to obtain pictures of the underground in order to identify nature and position of (metallic and non-metallic) objects, cavities and the depths of geological layers and the groundwater level. Today, this technique is used successfully for both the location of utilities (Nagashima et al 1992, Arioka et al 1992) and the detection of obstacles prior to the installation of utilities with trenchless techniques (Kathage 1991). Unfortunately, this technique is not applicable in very cohesive or saturated soils or underneath the groundwater level.

Summing up, the choice of the techniques for the underground investigation programme on a microtunneling site requires a thorough evaluation of cost and risk. The "cost" is the total amount to be spent for investigation whereas the "risk" comprises the cost of standstill of the site and resulting delays, the cost for additional pits to recover the microtunneling machine or the cost of retraction of the machine, etc. Both criteria have to be considered in relation with the total cost of the project. Thus, the use of an "expensive" technique can be justified even on a small site if the risk is high.

Finally, it is most important to define clearly, and in advance, the responsibility of both the client and the contractor for the investigation programme and for unforeseen difficulties due to the underground.

3 EXPERIENCE WITH MICROTUNNELING

The major characteristics of the six microtunneling jobs carried out for the Paris suburbs water supply syndicate are presented in table 1.

Table 1. Data of microtunneling jobs of the Paris suburbs water supply syndicate

No.	Length (m)	Depth (m)	ID (mm)	Type of pipe	Machine type
1	30	2.5	800	pressure	auger
2	70	3	500	sheath	slurry
3	40	4.5	500	sheath	slurry
4	43	5.5	500	sheath	slurry
5	96	4	800	pressure	slurry
6	98	11	800	pressure	slurry

The depths indicated are the maximum depths of cover.

Except of site no. 4 all jobs were above groundwater level. Unfortunately, many of the sites that had been completed previously (no.1

to 4), had met unexpected geological problems although systematic underground investigation had been carried out (borehole, penetration tests). On site no. 1, under a highway crossing, the soil (sand with clay) was too plastic for the auger type microtunneling machine and the working pace had to be reduced considerably. Site no. 2, a passage under a six lane motorway, suffered from unexpected obstacles. "Forgotten" reinforced concrete rubble was found on a length of 20 m between the starting pit and the beginning of the motorway's embankment. Microtunneling was then carried out in an excavated trench on this part. On site no. 3, a passage under a railway line, a big singular piece of rock on which the machine was gliding was met immediately after the starting pit. An additional pit was necessary to correct the machine's direction. There were no particular problems on site no. 4.

Unexpected problems on microtunneling sites in the area of Paris are very frequent. Therefore the contractor of the sites no. 5 and 6, Entreprise Valentin, decided to have an additional investigation by ground penetrating radar carried out.

4 THE RADAR USED

The principle of ground penetrating radar is well known today. Compagnie Générale de Géophysique (CGG) uses SIR 10 (TM GSSI) radar pulse reflectrometry acquisition systems, an 8-16 bit digital recording unit and the SIR 3 (TM GSSI) graphic recorder, which is a purely graphic stand-alone monitoring and display unit which can be used as a printing device for the SIR 10 system. With one to four recording channels, data originating from one to four transducers operating at different frequencies, investigation depth and time, gains, filterings, and offsets can be acquired simultaneously and independently. Multiple channels make multiple coverage acquisition possible (one reflection point is sounded by several different wave paths). Acquired data are stored on magnetic media with a capacity of 2.3 Gigabytes.

Data processing aims to highlight relevant information, by eliminating noise, and to improve resolution. Data processing procedures are specifically adapted to this target. Raw data, recorded on magnetic tape, are converted into time sections which show the discontinuities in the subsurface layers and the anomalies located (cavities, detachment faults, and various other objects). Data processing and interpretation can be carried out on site in cases where small volumes of measurements are being handled and when anomalies are being monitored along the ground. Large volumes of data are processed in CGG's data processing centers on Cray or Convex computers using a specially developed software. It reduces processing time by 10 to 20 in comparison with the on-site-treatment on a PC and offers rapid radar profile interpretation, especially for larger projects. A fast workstation which can operate on site was also developed.

A typical data processing sequence consists of different steps :

1. Filtering. Low-pass or high-pass filtering reduces high frequency (electronic and instrumentation background noise) and low frequency noise (geological and mechanical background noise).

2. Deconvolution. It reconstructs the pulse signal by taking into consideration factors related to the antenna and the nature of the subsurface. The process enhances resolution and removes multiple reflections created by high energy reflectors.

3. Hilbert transform. This transform calculates the energy, phase and instant frequency of a signal, in order to determine particular properties of the subsurface traversed by it.

4. Migration. It restores wave reflections to their true geological position.

5. Special transform. This operation transforms the amplitude spectrum of the signal without changing its phase. High frequencies are accentuated, improving resolution and amplifying certain signals.

6. Static corrections. Anomalies in the travel time, caused by variations in the topography and velocity of near-surface layers, are corrected.

7. Spatial filtering. The fast fourier transform is applied to devise a two-dimensional (phase and amplitude) filter which attenuates noise.

The principal technical data of the radars used are :

- Pulse frequency : 3.2 to 256 kHz
- Measurement density : 128 to 1024 samples per second
- Antenna frequencies : 100, 300, 500, and 1000 Mhz

- Daily working pace : 50 to 500 m when tracing anomalies
- Storage capacity (magnetic tape) : 2.3 Gigabytes
- Power : 60 to 80 W
- Power supply : 12 V DC, 220 V AC

5 THE SITE TEST AND ITS RESULTS

The sites chosen for radar investigation were in the northern part of the Paris suburbs. The considered section comprised open trench sites and a series of four microtunnel drives, two drives in each of two locations (site no. 5 and site no. 6 in table 1).

At site no. 5, two drives of 44 m and 54 m were required, passing under an important road intersection. As a result of borings the following geological composition was found : up to 2 m depth a clayey sand, from 2 m to 3.5 m depth a sandy and clayey soil and beneath clayey soil with gravels and rocks; no groundwater.

On site no 6, a passage underneath a bridge approach embankment of a four lane national road, there was one drive of 74 m (beneath the embankment) and one drive of 24 m (beneath an off ramp of the carriageway). The maximal depth indicated in table 1 refers to the depth beneath the road, which corresponds to a depth of 5 to 6 m beneath the ground surface. The following geological layers could be identified : up to a depth of 4 m sandy clay with layers of sand, from 4 to 4.4 m clay with chalky gravels and beneath marls with gypsum; no groundwater.

The site evaluation showed that site no. 6 was not suitable for an efficient radar investigation. Because of the marly soil the radar images were expected to be not precise enough. In addition, the risk of anomalies was not considered as high as on site no. 5.

On site no. 5, the risk was evaluated as high because of the presence of a multitude of utilities : gas, sewer, water, electricity, telephone and street lightning cabling. In addition, old cellar walls were discovered on a open trench section adjacent to the microtunneling site. The ground penetrating radar investigation required little preparation work. First, the axis of the microtunneling machine's route was drawn on the pavement and there was a length mark every metre. The radar was drawn by an operator. Two passages with two different antennas were carried out in order to obtain a good penetration. The length marks were registered electronically in order to localize objects and anomalies precisely.

The investigation profiles were carried out with 100 and 300 MHz transducers and the depth of investigation was about 5 m. The usable signal corresponded to an investigation depth of 4.5 m for the 100 MHz antenna but only to 1.75 m for the 300 MHz antenna. This meant that anomalies located deeper than 1.75 m could not be seen clearly with this tranducer because of the too strong attenuation of the signal due to the geological conditions (clayey soil). On the section examined, the route of the microtunneling machine being under a cover of 1 to 3.8 m, the interpretation was carried out using the images of the 100 MHz antenna. This antenna has a vertical and horizontal precision of 10 cm.

A first interpretation was carried out on site. It was found that a zone of approximately 5 m length showed many anomalies of reflection associated to two anomalies in a greater depth (2 to 2.90 m) and thus on the route of the microtunneling machine. These anomalies were located in an area where telephone and electricity cables were buried according to the maps. Two possible origins for the anomalies were considered :

1. Presence of a block or an utility deeper than indicated on the map (risk hypothesis).
2. Effect of multiple reflections generated by the high density of the utilities above (no risk hypothesis).

Therefore, complete data processing was necessary to make the definitive interpretation. The results, available the next day, showed that the use of ground penetrating radar was justified. On the one hand, it could be found that the deep anomaly was not an obstacle on the route of the microtunneling machine. On the other hand, it could be seen that the cables were buried deeper than indicated on the maps, but without any risk for the passage of the machine. The data treatment allowed to eliminate the different parasite reflections and the image finally obtained is presented on figure 1. A test digging carried out confirmed that the depth of the concrete bedding of the cables was in fact 50 cm greater then indicated on the map.

6 CONCLUSION

The installation of utilities with trenchless techniques requires a thorough underground investigation which should be applicable in usual depths of pipelaying and in all soil conditions. The use of ground penetrating radar for the detection of obstacles and geological anomalies can be an important step forward towards this objective. It could be seen, however, that the radar is not applicable everywhere. Furthermore, the antenna has to be chosen taking into account the target depth, the soil conditions and the required precision.. Summing up, the ground penetrating radar is not an universal investigation method, but where applicable and with the right equipment and a performant data treatment, the results obtained are reliable and satisfying.

REFERENCES

Arioka, R., Nagashima, Y., Kanno, A. and Masuda, J. 1992. New underground radar system. Proc. No-Dig 92 - 8th International Conference On Trenchless Technology : E3. Washington : ISTT

Berosch, M. and Revol, D. 1990. Trenchless technology for water supply. Proc. No-Dig 91 - 7th International Conference On Trenchless Technology : 261 - 274. Hamburg : CCH

Berosch, M., Ilotte, V. and Randon, G. 1991. The use of new techniques for water mains rehabilitation. Proc. IWEM East Anglian Branch Seminar on Water Mains Rehabilitation. Cambridge : IWEM

Ilotte, V. and Schroeyers, J. 1991. Guided drilling and potable water supply. Proc. No-Dig 92 - 8th International Conference On Trenchless Technology : P2. Washington : ISTT.

Kathage, A.F. 1991. Erkundung der unterirdischen Boden -und Leitungsstruktur mit dem Georadar für den grabenlosen Leitungsbau. Proc. No-Dig 91 - 7th International Conference On Trenchless Technology : 335 - 341. Hamburg : CCH

Nagashima, S., Suyama, K. and Kobori, T. 1992. Techniques for locating underground utilities. Proc. No-Dig 92 - 8th International Conference On Trenchless Technology : E2. Washington : ISTT.

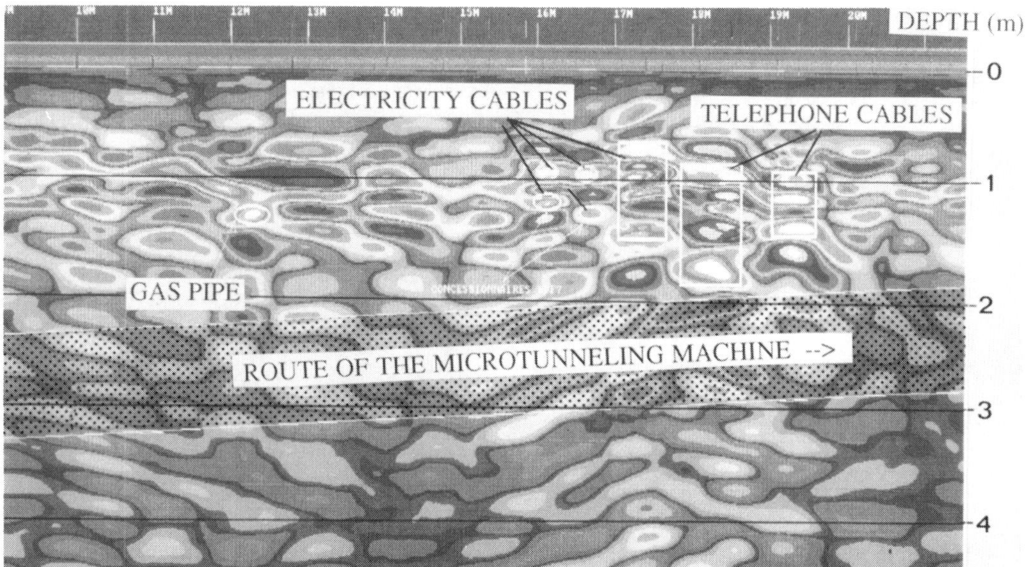

Fig. 1 Image of a zone with anomalies (after treatment)

No Trenches in Town, Henry & Mermet (eds) © 1992 Balkema, Rotterdam. ISBN 90 5410 085 0

The development of the 'RADARMAN' – A compact underground radar system

Jitsuo Koyabu
Distribution Engineering Center, Osaka Gas Co., Ltd, Japan

ABSTRACT: It is very important to develop a technique to locate various underground structures without excavation. Osaka Gas has developed the 'RADARMAN', a compact underground radar system, to locate accurately gas pipes and other underground structures such as water pipes and power cables.

The strong points of the 'RADARMAN' are as follows.

1. It is equipped with three-dimensional tower-shaped antennas, so it is possible to emit pulse radar waves into the ground at high efficiency. Accordingly, we can locate, at high accuracy, underground structures buried at less than 1.5 m depth. (The maximum locating depth is about 2.5 m.)

2. It is equipped with 'STC' (Sensitivity Timing Control), a high-frequency signal amplification unit, so it is possible to amplify signals at the best magnification for a given depth. Accordingly, we can get high locating sensitivity for nonmetal pipes such as polyethylene pipes as well as metal pipes.

3. With the electronic circuits, CRT display, memories, and operating unit mounted together on the antenna housing, the system is very compact, so it can be operated by one person.

As the field test results were very good, we introduced it for on-site use in 1990. We have already obtained good results in various underground surveys at more than 6,000 work sites.

1 PURPOSE OF DEVELOPMENT

The electromagnetic-induction type locator has been used so far for locating gas pipes in Osaka Gas. However, it was confined to electrically conductive gas pipes and was not applicable to gas pipes and other underground pipes made of nonmetallic materials such as polyethylene or connected with insulation joints. Under these circumstances, the development of an underground radar locator was undertaken.

2 OUTLINE OF THE DEVELOPED UNIT ('RADARMAN')

The 'RADARMAN' is an underground radar locator developed in 1988 by Osaka Gas Co., Ltd. An outline of the 'RADARMAN' is given below.

2.1 Appearance and construction of the unit

Photo 1 and Fig. 1 respectively indicate the appearance and construction of the unit.

2.2 Principle

The underground radar is used to locate underground objects (pipelines, cables, cavities and so on) by means of pulse waves. Fig. 2 shows its principle, and Fig. 3 indicates an example of images of located objects.

Pulse waves are generated by supplying pulse currents of very short time intervals to a transmission antenna. The generated pulse waves propagate underground and are reflected by a buried object. A reception antenna receives these reflected waves and the object's buried depth (H) is obtained from the time (T) elapsed between the transmission and reception of the pulse waves. For instance, if a buried pipe is to be located, transmission and reception are repeated continuously while moving the ocator across the ground surface as indicat in Fig. 2. When the reflected signals are converted into an image, an image like a parabola is obtained.
The peak (point D) of the parabola indicates the position just above the under-

ground pipe and the vertical distance at this position is the buried depth of the pipe.

$$H = VT/2 \quad [V = V_o/\sqrt{\epsilon r}]$$

Photo 1. RADARMAN.

where

V : Underground propagation speed of pulse waves
V_o: Speed of pulse waves in vacuum (3 × 10^8 m/s)
ϵr: Dielectric constant of earth (= 5 to 40)

2.3 Specifications

Table 1 indicates the major specifications of the 'RADARMAN'.

3 TECHNICAL ADVANTAGES OF THE UNIT

3.1 Three-dimensional tower-shaped antenna

The 'RADARMAN' transmits base-band pulse waves and receives the waves reflected by underground objects. The pulse signal has a wide band-width (from 0 Hz to 1 GHz). In order to transmit the pulse waves efficiently, the antenna requires a very high gain and sharp directivity over that band-width.

We conducted many empirical studies. First, we measured the wave-forms of transmitted signals in air and soil, using cross-dipole antennas to find the band-width useful for sensing. Next, resistance-loaded two-dimensional antennas, such as the wire antenna and plane antenna

Table 1. Specifications.

Item	Performance
Emission	Impulse type (base band)
Display mode	Grey level images (400 length × 342 width × 8 bits)
Transmitted pulse width and voltage	1.2 n sec ±10%, 97.5 Vp-p ±10%
Received wave band	10 to 500 MHz
Display	9-inch monochromatic CRT
Memory capacity	Frame memory: 512 k bytes (three frames) Character memory: 16 k bytes
Recording	Monochromatic video signal hard copy
Output terminal	GP-IB terminal (IEEE standard 488-1978)
Dimensions	650 mm length × 540 mm width × 680 mm height
Weight	60 kg
Power source	DC 12 V lead storage battery

shown in Photo 2 (a) and (b), were examined for this purpose.

Finally, referring to a horn antenna structure, we developed the three-dimensional tower-shaped antenna loaded with resistance shown in Photo 3, which transmits pulse waves with sharper directivity and has less unwanted ringing noise caused by the waves reflected by the antenna itself and the ground surface.

3.2 Sensitivity Timing Control (STC)

As electric waves propagate underground, their signal level attenuates considerably because earth, in which electric waves can travel, has conductivity and absorbs them as resistance loss and changes it into heat.

The attenuation volume (A) of electric waves is obtainable from the following formula:

$$A = e^{-\gamma \ell}$$

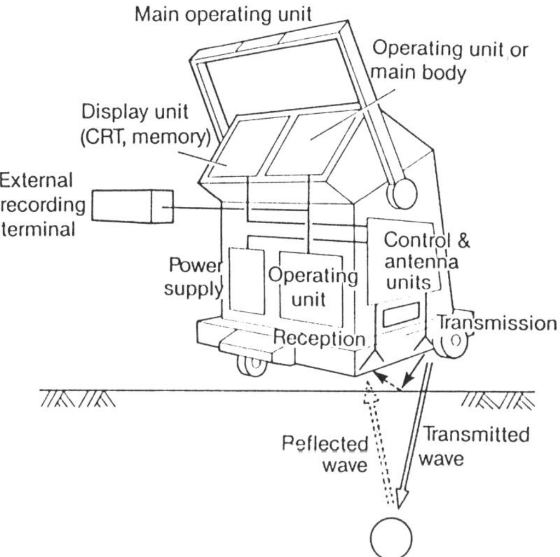

Fig. 1. Construction of locator.

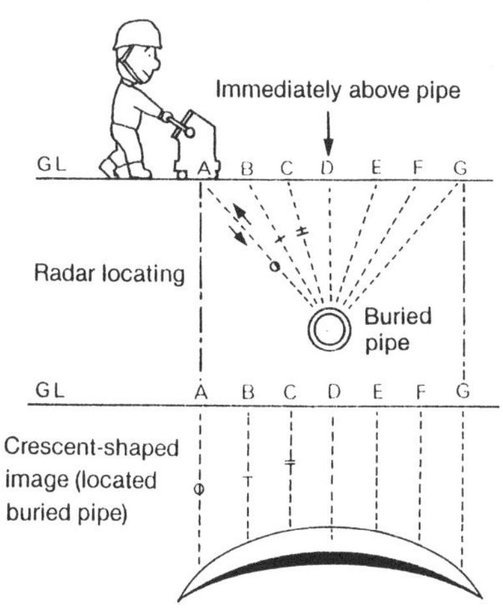

Fig. 2. Locating principle.

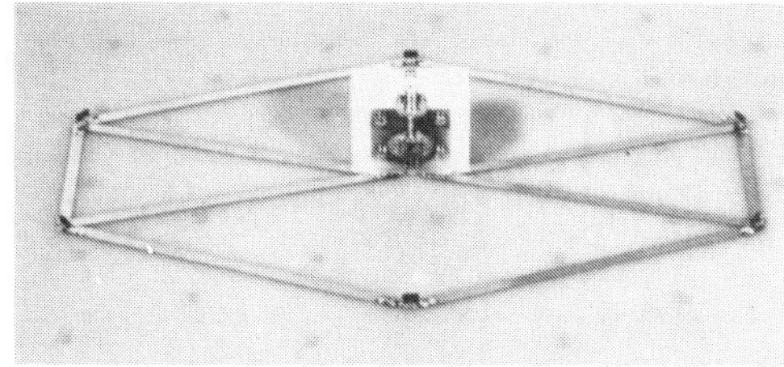

(a) Wire antenna

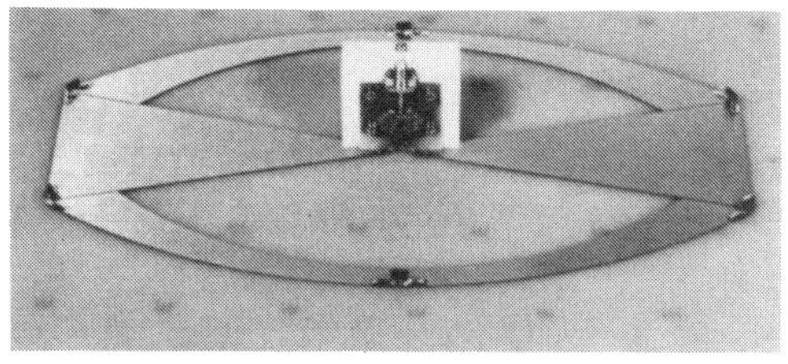

(b) Plane antenna

Photo 2. Two-dimensional antennas.

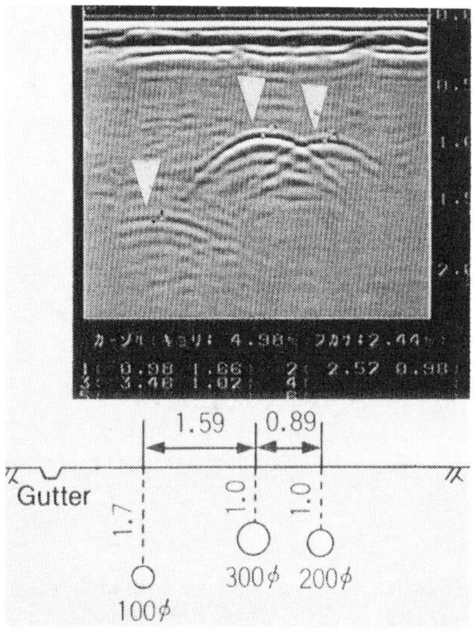

Fig. 3. Example of located pipe image.

$$\gamma = \alpha + j\beta$$

$$\alpha = 2\pi f \left[\frac{\epsilon\mu}{2} \sqrt{1 + \left(\frac{\sigma}{2\ \pi f \ell}\right)^2} - 1 \right]^{\frac{1}{2}}$$

$$\beta = 2\pi f \left[\frac{\epsilon\mu}{2} \sqrt{1 + \left(\frac{\sigma}{2\ \pi f \ell}\right)^2} + 1 \right]^{\frac{1}{2}}$$

where

ϵ : Permittivity of earth
σ : Conductivity of earth
μ : Permeability of earth
f : Transmitting frequency
ℓ : Traveling distance of electric waves

Now, as amplification is required for the detection of the faint signals reflected by the target, we tried to improve the S/N ratio by adopting a system (STC) for gradually increasing the amplification factor as time lapses from the start of transmission and optimizing the straight line of amplification factor (Refer to Fig. 4). Further, while a method to make the amplification factor change rapidly for every transmitted pulse within the limited range of observation time (approximately 100 ns) has been adopted generally so far, the method requires the change to be made at a very high speed and the realization of such changes is very difficult by the current electronic circuit technology. Therefore, we adopted the new STC system in order to reduce the processing time and synchronize the change in the amplification factor with a trigger signal for sampling. By the adoption of this new system, the impedance characteristics and linearity of the STC circuit were improved and noise was considerably reduced. Figure 5 indicates a comparative example of picture image (a) when the STC system was not used and (b) when it was use.

Photo 3. Three-dimensional tower-shaped antenna.

3-3 Integration of antenna and display unit

In order to integrate the antenna and display unit without deterioration in locating performance, experiments were conducted to reduce the size of the antenna shielding case and studies were made to clarify the effect of noise when the antenna and display unit were placed nearer to each other. Based on the results of these experiments and studies, an integrated unit was designed. Further, the unit was made compact with better operability by the use of a

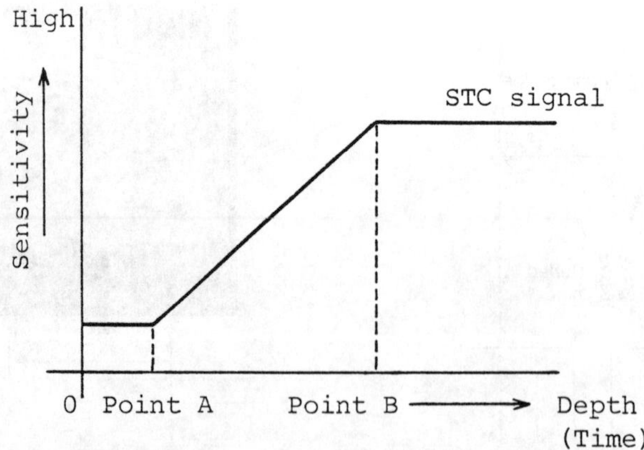

The point A indicates the point where sensitivity begins to increase, and the point B indicates the point where sensitivity reaches the maximum.

Fig. 4. Sensitivity timing control (STC).

lead accumulator of small size and long time durability and a reliable, easy-to-operate power switch.

4 APPLICATION TO WORK SITE USE

4.1 Results of field tests

We, at Osaka Gas Co., Ltd. performed field tests for 3 years from 1989, during which period evaluations were made in respect of the unit's applicability to work site use. The results of the tests are described below.

(1) Locating rate for underground objects

Defining the ratio between the number of underground objects confirmed by excavation and the number of underground objects located by the 'RADARMAN' as the "locating rate," locating was conducted at work sites of various soil types. As a result, a locating rate of approximately 80% was obtained for 593 excavated objects in total. In addition, the locating rate when the soil consisted of clay was approximately 50%, indicating a lower value. This was because clay has a higher electric conductivity and a larger electric wave attenuation factor. Also, the survey revealed that there was a considerable amount of dispersion in the locating rate, depending on depths of the buried objects.

Figure 6 shows the locating rate by depth. When the depth is less than 1.5 m, it generally indicates locating rates better than the mean locating rate (approximately 80%) and when in particular the depth is 0.6 to 0.9 m, it indicates a locating rate as good as 95%. However, it is shown that when the depth is more than 1.5 m, the locating rate decreases because

Table 2. Locating rates when soil is clay.

Depth of buried object	Locating rate
When depth is less than 1.0 m	80%
When depth is more than 1.0 m	36%

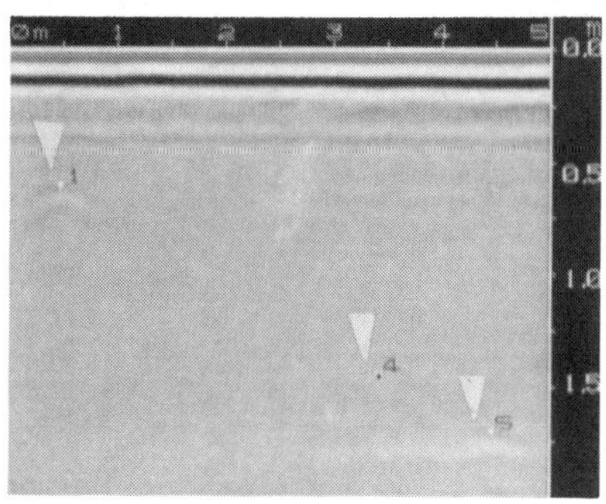

(a) Output image of non-STC

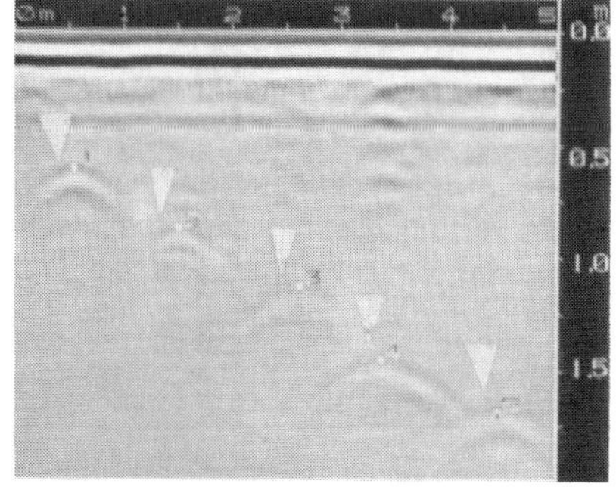

(b) Output image of adjusted STC

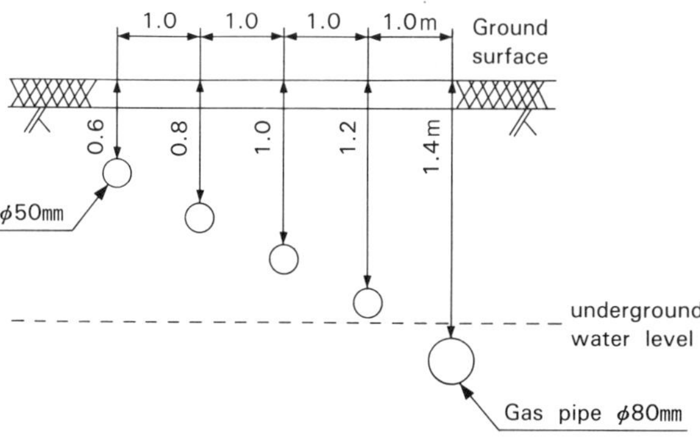

Fig. 5. Output underground images.

the pulse waves emitted into the ground rapidly attenuate with depth, considerably reducing the signal level. In clay in particular, as Table 2 indicates, the locating rate is significantly reduced if the depth is more than 1.0 m. Strangely, it is noted that the locating rate for underground objects less than 0.3 m deep is somewhat lower. It is because in some cases a strong multiple image due to the reflection of pulse waves by the surface of an asphalt pavement is superimposed on the image of the buried object.

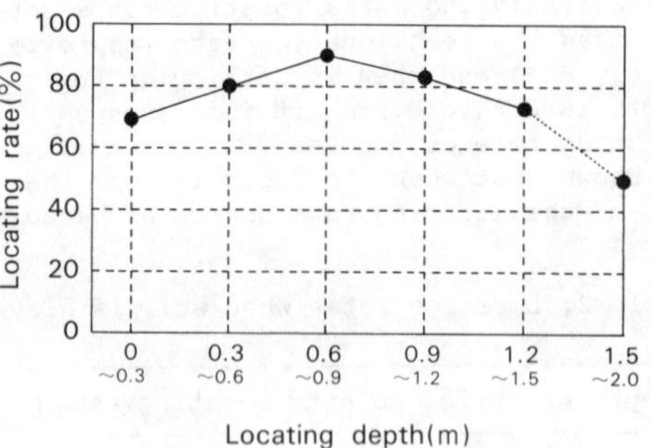

Fig. 6. Locating rate by depth.

(2) Accuracy of reading buried object image

Locating by means of the 'RADARMAN' is conducted by having the signals reflected by a buried object displayed as an image on a CRT screen and reading the image. Normally, the ease of reading the image depends on the unevenness in the properties of soil, the degree of underground congestion, and so on. In this investigation, we noted a relationship between the ease of reading the image and the accuracy of reading. As Table 3 indicates, compared with an image that is easy to read (Refer to Fig. 7), an image that requires skill in reading (Refer to Fig. 8) involves more reading error. For 82% of buried objects detected during this investigation, the error in the horizontal direction was within ±20 cm. Further, the rate of successful reading of an image with an error in the vertical direction of 20 cm was 74%.

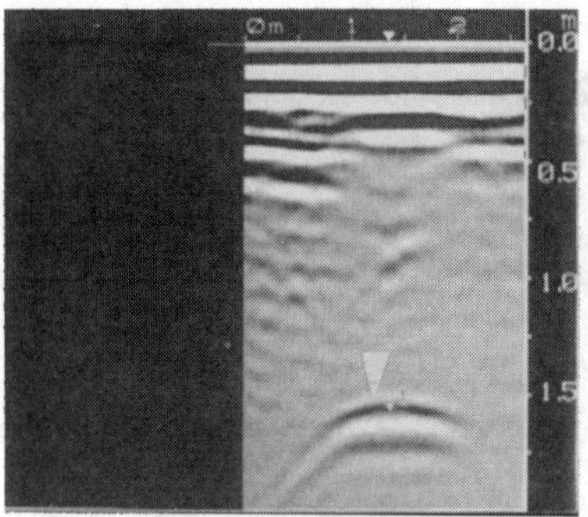

Fig. 7. Image easy to read.

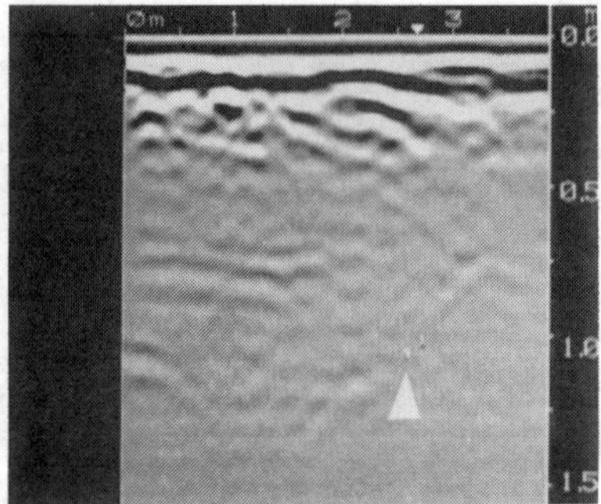

Fig. 8. Image requiring skill to read.

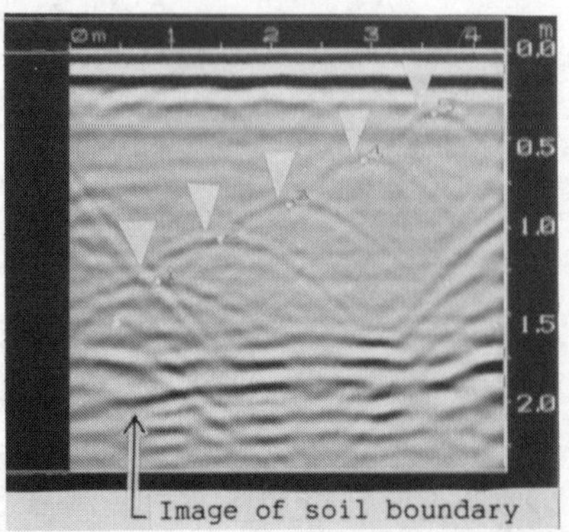

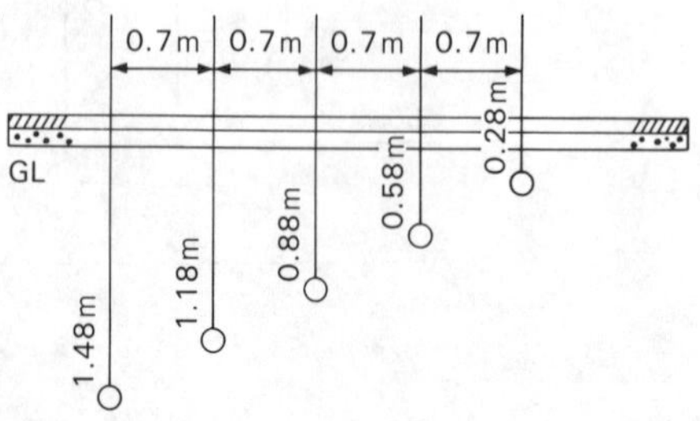

Fig. 9. Polyethylene gas pipes (50-mm diameter).

(3) Readability of images

As already mentioned, the 'RADARMAN' has a mean locating rate of 80%. The ratio between easy-to-read images and those requiring skill to read is approximately 1 : 1. Therefore, it is required that the operator should be well used to reading the images when using the 'RADARMAN'.

(4) Examples of images

Figures 9 to 11 show some typical examples of displayed images.

4.2 Status of introduction to work sites

With successful results of field tests, we have introduced the 'RADARMAN' to our work sites since 1990 to use it mainly for the following surveying work. We have already deployed 18 units of the 'RADARMAN' for use at more than 6,000 work sites and are achieving good results through their operation.

(1) Investigation of missing gas pipes (100 sites)

The 'RADARMAN' is particularly effective in locating missing gas pipes which are difficult to locate with the electromagnetic-induction type pipe locator or through trial excavation. The 'RADARMAN' can be used to determine trial excavation sites or to correct unreliable drawings and ledgers.

(2) Investigation of buried pipes prior to pipeline design and construction (1300 sites)

The 'RADARMAN' can be used to locate any buried object. Investigation prior to pipeline design and construction is primarily for checking if sufficient underground space exists, thereby providing reference data for trial excavation site selection. The 'RADARMAN' is particularly useful at sites where excavation is restricted owing to road conditions.

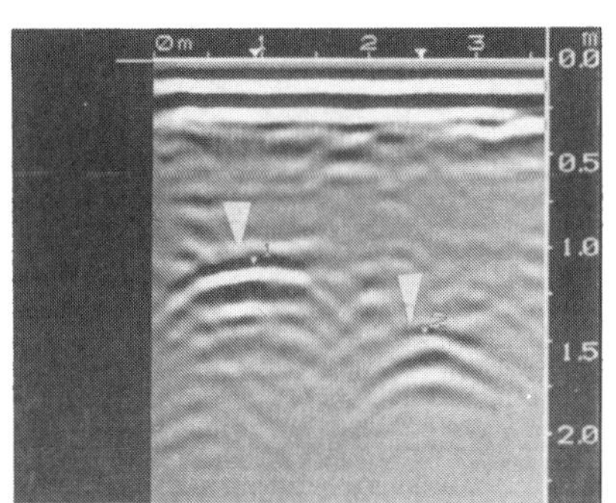

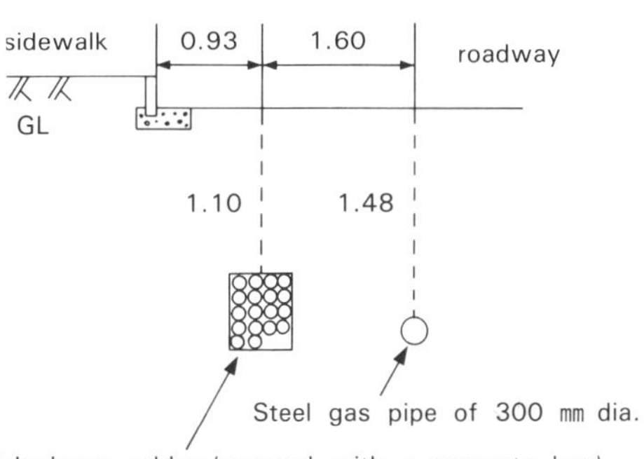

Fig. 10. A steel gas pipe and telephone cables.

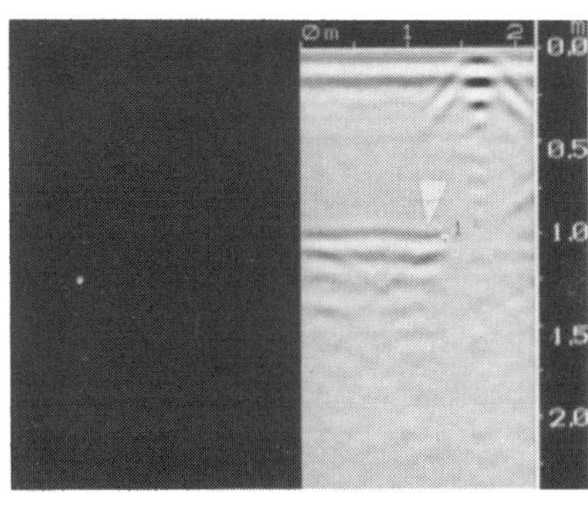

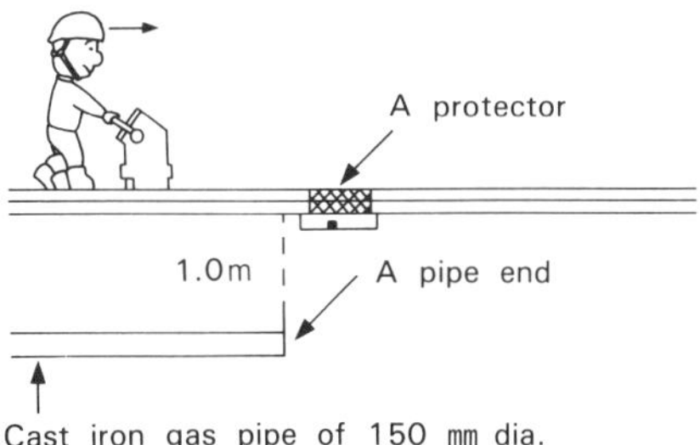

Fig. 11. A pipe end.

Table 3. Relationship between ease of reading and accuracy of reading.

Ease of reading an image	Rate of successful reading within error of ±20 cm
Image easy to read	91%
Image requiring skill to read	74%

Table 3. Development targets for an improved radar locator.

		Improved model	Current
Location depth		1.5 m (100% locating rate)	1.5 m (80% locating rate)
		2.0 m (80% locating rate)	2.0 m (Less than 50% locating rate)
Accuracy:	Horizontal	±10 cm	±20 cm
	Depth	±10%	±20%

(3) Indication of gas pipe location for work to be performed by others (5,000 sites)

If the 'RADARMAN' is used for indicating the location of gas pipes for which the electromagnetic-induction type pipe locator cannot be used, it will be very effective for the safety of work to be performed by others.

5. FUTURE SCHEDULE

Owing to the 'RADARMAN's' excellent basic performance, we are now planning, instead of full-scale introduction to work sites, the research and development of an improved radar locator, which requires no skill in reading the image and which can accurately locate gas pipes and other underground structures at greater depth. The research and development of such an improved unit already started last year. In this respect, we have set the development targets as indicated in Table 3.

Reference

I. SUGIMOTO, T. KIKUTA, Y. HAYASHI and M. TAKAGI, "COMPACT UNDERGROUND RADAR SYSTEM," The 6th Scandinavian Conference on Image Analysis Oulu. Finland June 19, 22, 1989

No Trenches in Town, Henry & Mermet (eds) © 1992 Balkema, Rotterdam. ISBN 90 5410 085 0

Le développement du 'RADARMAN' – Un géoradar compact

Jitsuo Koyabu
Distribution Engineering Center, Osaka Gas Co., Ltd, Japon

RESUME. Il est très important de développer une technique de localisation de structures variées enterrées sans escavations. Osakagaz a développé le radarman, un géoradar compact pour localiser précisément les conduites de gaz et autres structures souterraines telles que canalisations d'eau, câbles de puissance.

Les points forts du RADARMAN sont les suivants :

1. Il est équipé d'une antenne tourelle en trois dimensions. Ainsi il est possible d'émettre des impulsions d'ondes dans le sol avec une haute efficacité. Ainsi, nous pouvons localiser avec une grande précision des structures enterrées à une profondeur inférieure à 1,50 m (la profondeur maximale de localisation est d'environ 2,50 m).

2. Il est équipé avec "STC" (sensibility timing control), une unité d'amplification du signal haute fréquence ainsi il est possible d'amplifier le signal à la meilleure amplitude pour une profondeur donnée. Ainsi nous pouvons avoir des hautes sensibilités de localisation pour des tuyaux non métalliques tels que les tuyaux polyéthylènes aussi bien que pour les tuyaux métalliques.

3. Avec les circuits électroniques, affichage CRT, mémoire unité opérationnelle, fixés ensemble sur l'antenne, le système est très compact et peut être utilisé par une seule personne.

Comme les essais sur le terrain ont donné de très bons résultats, nous avons introduit le système en conditions opérationnelles en 1990. Nous avons déjà obtenu de bons résultats dans différentes investigations souterraines sur plus de 6000 sites.

1. OBJET DU DEVELOPPEMENT.

Le système de localisation par induction électro-magnétique a été utilisé pour la localisation des conduites de gaz par Osaka Gaz. Par ailleurs, ces applications étaient limitées aux tuyaux de gaz métalliques mais n'étaient pas applicables aux tuyaux non métalliques tels que polyéthylène ou tuyaux connectés avec des joints d'isolation. En conséquence, le développement d'un géoradar a été entrepris.

2. DESCRIPTION DU RADARMAN.

Le RADARMAN est un géoradar de détection développé en 1988 par Osaka Gaz Cy Ltd. Une description du RADARMAN est donnée ci-dessous.

2.1. Apparence et construction de l'unité.

Photo 1. et Fig. 1. indiquent respectivement l'apparence et la construction de l'unité.

Table 1. Specifications.

Item	Performance
Emission	Impulse type (base band)
Display mode	Grey level images (400 length × 342 width × 8 bits)
Transmitted pulse width and voltage	1.2 n sec ±10%, 97.5 Vp-p ±10%
Received wave band	10 to 500 MHz
Display	9-inch monochromatic CRT
Memory capacity	Frame memory: 512 k bytes (three frames) Character memory: 16 k bytes
Recording	Monochromatic video signal hard copy
Output terminal	GP-IB terminal (IEEE standard 488-1978)
Dimensions	650 mm length × 540 mm width × 680 mm height
Weight	60 kg
Power source	DC 12 V lead storage battery

Photo 1. RADARMAN.

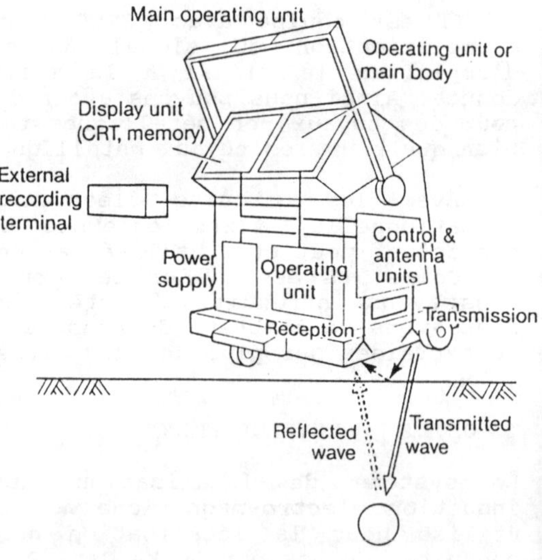

Fig. 1. Construction of locator.

2.2. Principe.

Le géoradar est utilisé pour localiser des objets souterrains (tuyaux, câbles, cavités, etc...) au moyen d'ondes impulsionnelles. La Fig. 2. montre son principe et la Fig. 3. donne un exemple d'images d'objets détectés.

Les ondes impulsionnelles sont générées en fournissant des courants impulsionnels à de très courts intervalles à une antenne de transmission. L'onde impulsionnelle

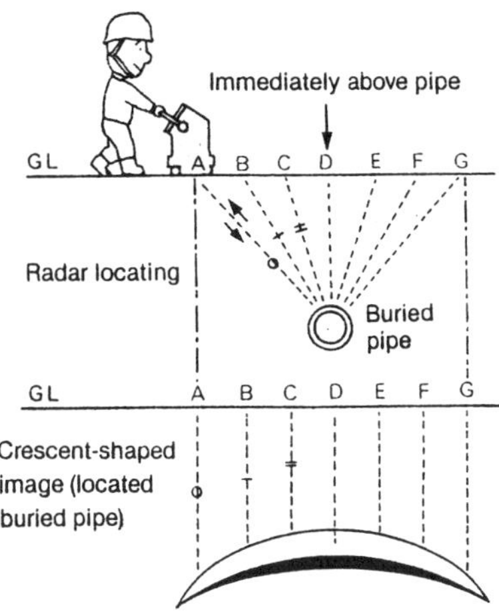

Fig. 2. Locating principle.

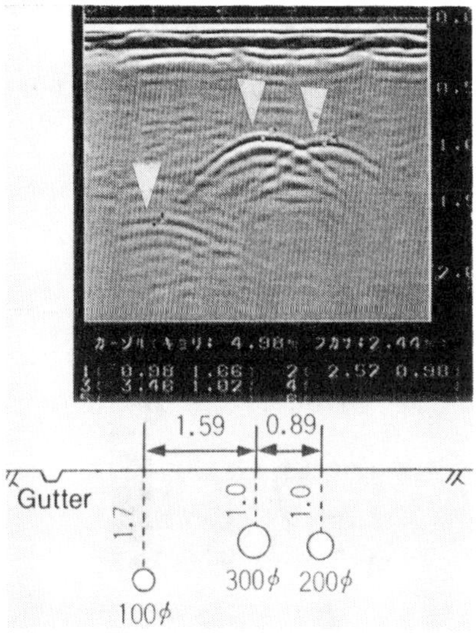

Fig. 3. Example of located pipe image.

générée se propage dans le sol et est réfléchie sur les objets enterrés. Une antenne de réception reçoit ces ondes réfléchies et la profondeur des objets enterrés (H) est obtenue à partir du temps (T) entre l'instant d'émission, de transmission et de réception de l'onde impulsionnelle. Par exemple, si un tuyau est localisé, la transmission et la réception se répètent en continu lorsque l'on promène le détecteur à la surface

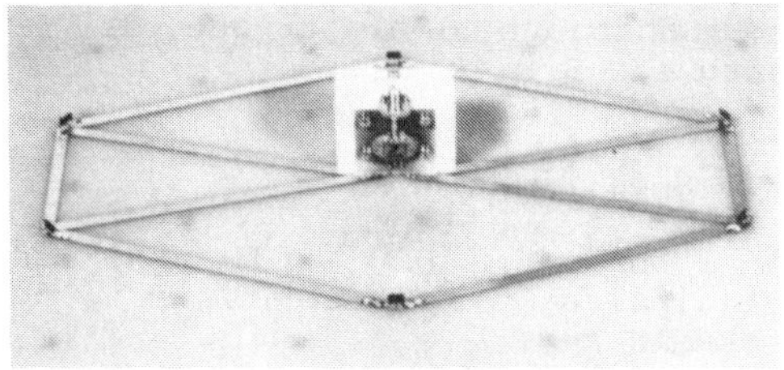

(a) Wire antenna

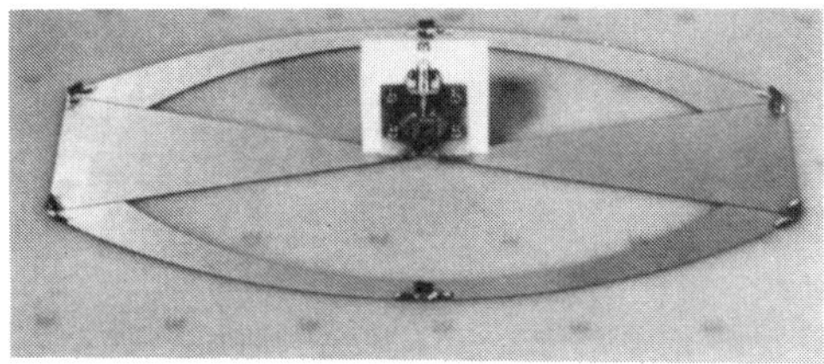

(b) Plane antenna

Photo 2. Two-dimensional antennas.

Photo 3. Three-dimensional tower-shaped antenna.

du sol comme indiqué à la Fig. 2. Lorsque les signaux réfléchis sont convertis en une image, une image ressemblant à une parabole est obtenue.

Le sommet (point D de la parabole) indique la position verticale du tuyau et sa profondeur.

$H = VT/2 \quad [V = V_0/\sqrt{\varepsilon r}]$

où :

V : vitesse de propagation des ondes impulsionnelles
Vo: vitesse des ondes impulsionnelles dans le vide

εr: Constante dielectrique de la terre (de 5 à 40).

2.3. Spécifications

Le Tableau 1. indique les principales spécifications du RADARMAN.

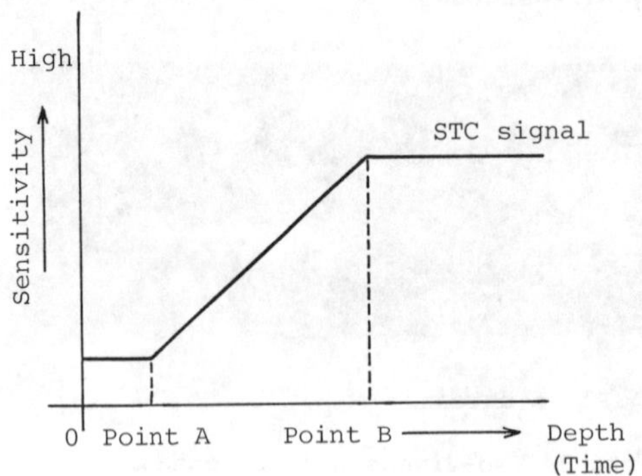

Fig. 4. Sensitivity timing control (STC).

3. AVANTAGES TECHNIQUES DE L'UNITE.

3.1. Antenne tourelle 3D.

Le RADARMAN transmet une onde impulsionnelle dans une bande donnée et reçoit les ondes réfléchies par les objets souterrains. Le signal impulsionnel a une très large bande (de 0Hz à 1GHz).

Pour transmettre efficacement les ondes impulsionnelles, il est nécessaire que l'antenne ait un gain élevé et une direction étroite sur toute la bande.

Nous avons conduit beaucoup d'études empiriques. Tout d'abord nous avons mesuré les formes des ondes, des signaux transmis en l'air et en sol pour trouver la largeur de bande utile pour l'exploration. Ensuite, des antennes bi-dimensionnelles telles que les antennes à fil et les antennes planes montrées dans les photos 2.a et 2.b ont été examinées pour cette raison.

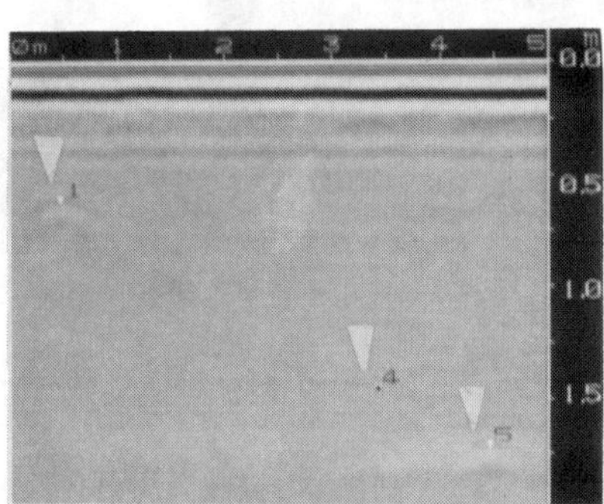

(a) Output image of non-STC

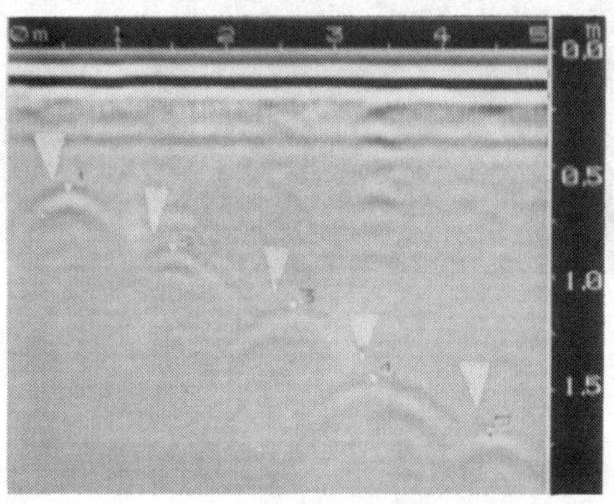

(b) Output image of adjusted STC

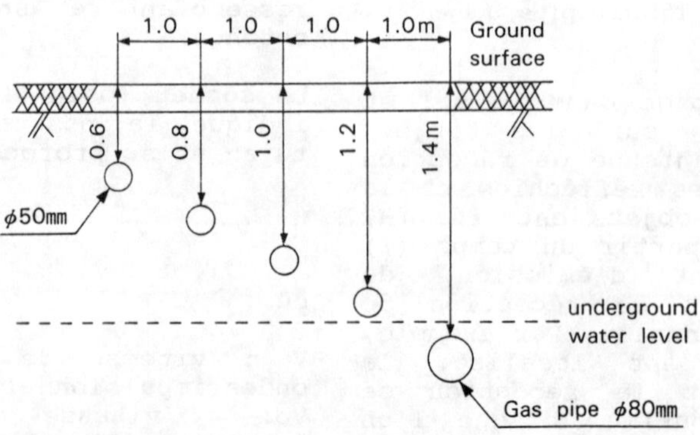

Fig. 5. Output underground images.

Finalement, en se référant à une structure d'antenne en forme de corde, nous avons développé une antenne tourelle 3D (photo 3.) qui transmet des ondes impulsionnelles avec direction étroite et qui a le minimum de bruits de fond causés par les réflexions de l'antenne elle-même et de la surface du sol.

3.2. Sentivity Timing Control (STC)

Lorsque les ondes électriques se propagent dans le sol, leur niveau de signal s'atténue considérablement parce que la terre que les ondes électriques peuvent parcourir a une conductivité et les absorbe lorsque la résistance baisse et les change en chaleur.

Le volume d'atténuation (A) d'une onde électrique est obtenu par les formules suivantes:

$$A = e^{-\gamma \ell}$$

$$\gamma = \alpha + j\beta$$

$$\alpha = 2\pi f \left[\frac{\epsilon\mu}{2} \sqrt{1 + \left(\frac{\sigma}{2\pi f \ell}\right)^2} - 1 \right]^{\frac{1}{2}}$$

$$\beta = 2\pi f \left[\frac{\epsilon\mu}{2} \sqrt{1 + \left(\frac{\sigma}{2\pi f \ell}\right)^2} + 1 \right]^{\frac{1}{2}}$$

où :

: permitivité de la terre
: conductivité de la terre
: perméabilité de la terre
: fréquence
: distance de parcours de l'onde.

Maintenant, comme une amplification est nécessaire pour la détection du signal évanescent réfléchi par la cible, nous avons cherché à améliorer le rapport S/N en adoptant un système (STC) pour augmenter progressivement le facteur d'amplification comme laps de temps à partir du début de la transmission et par optimisation de la ligne droite du facteur d'amplification (Fig. 4). En outre, une méthode pour changer rapidement le facteur d'amplification pour chaque impulsion transmise à l'intérieur d'une plage

Table 2. Locating rates when soil is clay.

Depth of buried object	Locating rate
When depth is less than 1.0 m	80%
When depth is more than 1.0 m	36%

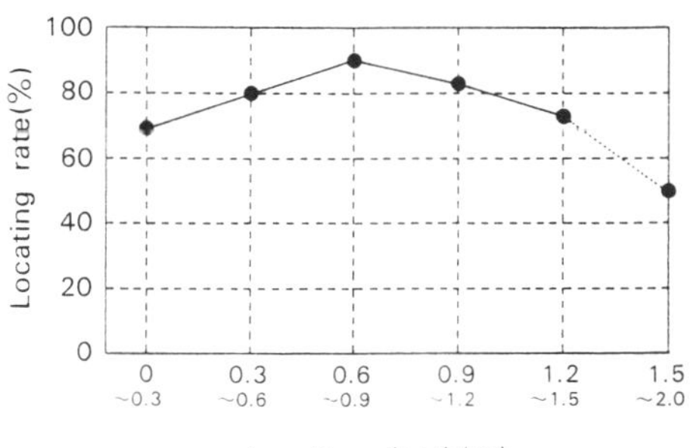

Fig. 6. Locating rate by depth.

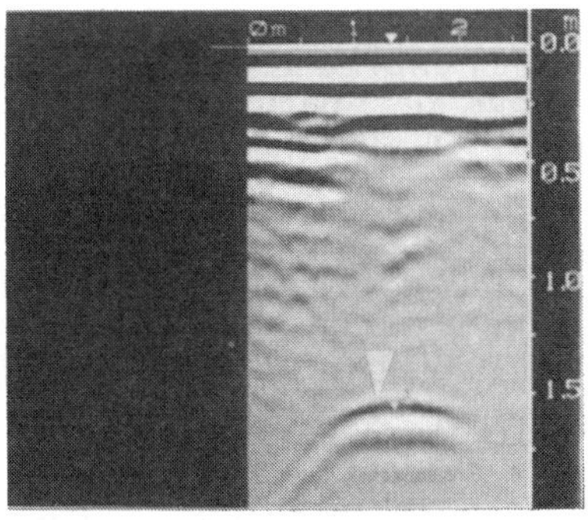

Fig. 7. Image easy to read.

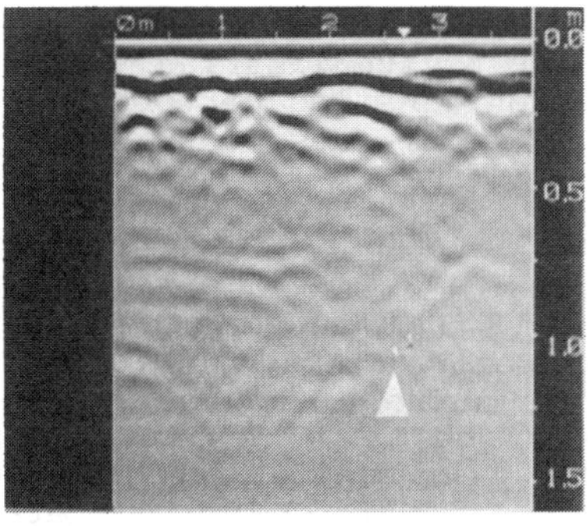

Fig. 8. Image requiring skill to read.

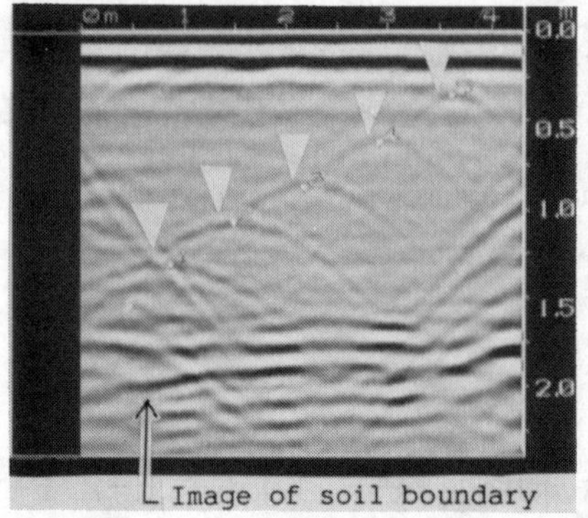

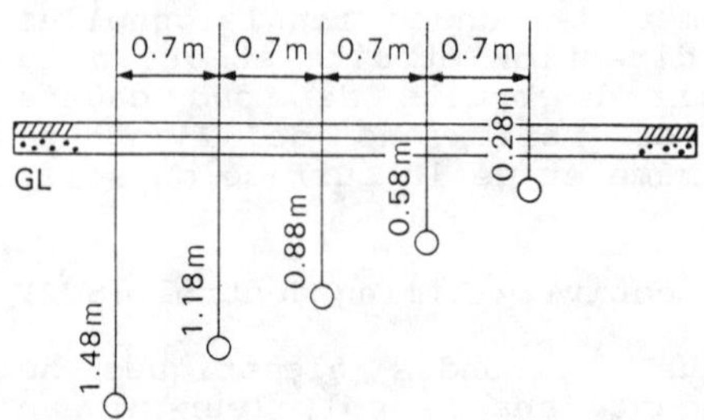

Fig. 9. Polyethylene gas pipes (50-mm diameter).

d'observation (approximativement 100 ns) a été généralement adoptée jusqu'à ce que la méthode nécessite un changement à une grande vitesse. La réalisation de cette modification est très difficile avec la technologie des circuits électroniques actuels. Cependant nous avons adopté le nouveau système STC en vue de réduire le temps d'exécution et de synchroniser le changement du facteur d'amplification avec un signal de déclenchement pour échantillonnage. Par l'adoption de ce nouveau système, des caractéristiques d'impédance et la linéarité du circuit STC ont été améliorées et le bruit a été considérablement réduit. La Fig. 5. montre un exemple comparatif d'images (a) lorsque le système STC n'est pas utilisé et (b) lorsqu'il est utilisé.

3.3. Intégration de l'antenne et unité de visualisation.

En vue d'intégrer l'antenne et l'unité de visualisation sans détérioration dans les performances de localisation, des expériences ont été conduites pour réduire la taille de la protection de l'antenne et des études ont été faites pour éclaircir les effets de bruit lorsque l'antenne et l'unité de visualisation sont rapprochées l'une de l'autre. Basée sur les résultats de ces expériences et de ces études, une unité intégrée a été conçue. En outre, l'unité a été rendue compacte avec une meilleure manoeuvrabilité par l'utilisation d'accumulateurs au plomb de petite taille et de longue durée et d'un commutateur fiable et facile à manoeuvrer.

4. APPLICATION A L'UTILISATION SUR SITE.

4.1. Résultats des essais sur site.

A Osaka Gaz Co. nous avons effectué des essais sur sites durant 3 ans à partir de 1989, période pendant laquelle des évaluations ont été faites au regard de l'utilisation de l'unité in-situ. Les résultats des essais sont décrits ci-dessous.

(1) Taux de localisation pour les objets enterrés.

En définissant le taux de localisation par le rapport entre le nombre d'objets enterrés confirmés par escavation et le nombre d'objets détectés par le RADARMAN, des essais de localisation ont été menés sur sites et pour des natures de sols différentes. La conclusion est que le taux de localisation est approximativement de 80% pour les 593 objets excavés au total. En outre, le taux de localisation est approximativement de 50% (valeur minimale) lorsque le sol est composé d'argile. Ceci est causé par le fait que l'argile a une conductivité électrique élevée et un facteur d'atténuation d'ondes important. Aussi la prospection

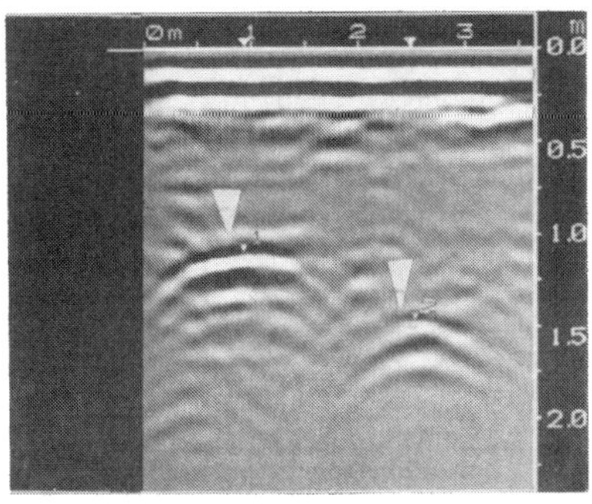

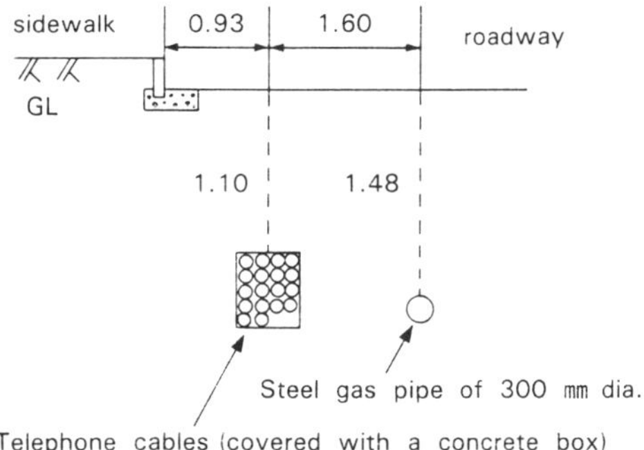

Fig. 10. A steel gas pipe and telephone cables.

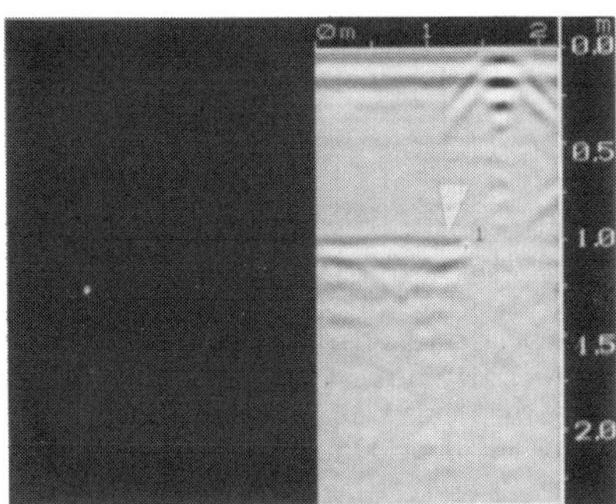

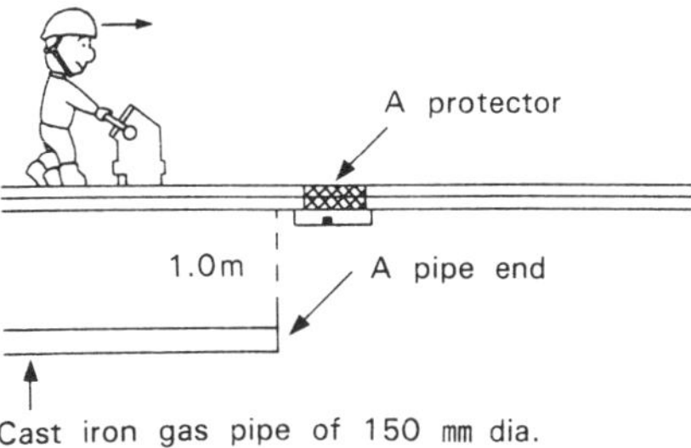

Fig. 11. A pipe end.

révèle qu'il y a une très forte dispersion dans le taux de localisation en fonction de la profondeur des objets enterrés.

La Fig. 6. montre la dépendance du taux de localisation avec la profondeur. Lorsque la profondeur est inférieure à 1,50m, on remarque que le taux de localisation est supérieur au taux de localisation moyen (approximativement 80%) et en particulier lorsque la profondeur est comprise entre 0,6 et 0,9 m, ce taux de localisation est de l'ordre de 95%. Cependant, il est montré que lorsque la profondeur est supérieure à 1,50m, le taux de localisation décroit car les ondes impulsionnelles émises dans le sol s'atténuent rapidement avec la profondeur réduisant considérablement le niveau du signal. Dans l'argile en particulier, comme le tableau 2. l'indique, le taux de localisation est fortement réduit si la profondeur est supérieure à 1m. De façon étrange, il faut remarquer que le taux de localisation pour des objets enterrés à des profondeurs à 0,3m est quelque peu inférieur. Ceci peut être dû dans quelques cas à des images multiples dûes à des réflexions des ondes impulsionnelles par la surface de la chaussée superposées aux images des objets enterrés.

La détection au moyen du RADARMAN est effectuée par obtention du signal réfléchi par l'objet enterré s'affichant comme une image sur un écran CRT et la lecture de cette image. Ordinairement, la facilité de lecture de l'image dépend de l'absence d'aléas dans les propriétés du sol, le degré d'encombrement du sous-sol, etc... Dans cette recherche, nous avons noté une relation entre la facilité de lecture de l'image et la précision de lecture. Comme l'indique le tableau 3., comparer à une lecture d'une image facile à lire, une image qui nécessite une

Table 3. Relationship between ease of reading and accuracy of reading.

Ease of reading an image	Rate of successful reading within error of ±20 cm
Image easy to read	91%
Image requiring skill to read	74%

Table 3. Development targets for an improved radar locator.

		Improved model	Current
Location depth		1.5 m (100% locating rate)	1.5 m (80% locating rate)
		2.0 m (80% locating rate)	2.0 m (Less than 50% locating rate)
Accuracy:	Horizontal	±10 cm	±20 cm
	Depth	±10%	±20%

lecture avec interprétation (voir Fig. 8.), conduit à plus d'erreurs de lecture. Pour 82% des objets enterrés détectés durant cette recherche, l'erreur dans la direction horizontale était inférieure à plus ou moins 20 cm. En outre, le taux de lecture fructueuse d'une image avec une erreur dans la position verticale inférieure à 20cm était de 74%.

(3) Facilité de lecture des images.

Comme il a déjà été mentionné, le RADARMAN a un taux de localisation moyen de 80%. Le taux entre les images faciles à lire et celles nécessitant une expérience de lecture, est approximativement 1/1. Cependant il est demandé que l'opérateur soit accoutumé à la lecture des images du RADARMAN.

(4) Exemples d'images.

Les Fig. 9 à 11 montrent quelques exemples typiques d'images sur écrans.

4.2. Méthodes d'introduction sur sites.

Avec les résultats fructueux des essais sur sites nous avons introduit le RADARMAN dans nos travaux depuis 1990 pour l'utiliser principalement dans les investigations suivantes. Nous avons déjà développé 18 unités du RADARMAN pour utilisation sur plus de 6000 sites, lesquels ont donné de bons résultats.

(1) Recherche de conduites de gaz perdues (100 sites).

Le RADARMAN est particulièrement efficace dans la localisation de conduites de gaz oubliées qui sont difficiles à localiser avec la méthode d'induction électro-magnétique ou par excavation. Le RADARMAN peut être utilisé pour positionner des tranchées de recherche ou pour corriger des plans non fiables.

(2) Recherche de tuyaux enterrés avant la conception et la construction de réseaux (1300 sites).

Le RADARMAN peut être utilisé pour localiser n'importe quel objet enterré. Des recherches avant la conception de réseaux et la construction sont primordiales pour vérifier s'il existe suffisamment d'espace souterrain et de ce fait procurer des informations de référence pour la sélection des sites d'excavations. Le RADARMAN est particulièrement utile sur les sites où l'excavation est restreinte en fonction des conditions de circulation.

(3) Indication de positionnement de tuyaux de gaz pour des travaux à réaliser par d'autres concessionnaires (5.000 sites).

Si le RADARMAN est utilisé pour indiquer la position des tuyaux de gaz pour lesquels la méthode d'induction électro-magnétique ne peut être utilisée, il sera très efficace pour la sécurité des travaux devant être réalisés par d'autres concessionnaires.

5. PROGRAMME FUTUR.

Puisque le RADARMAN a d'excellentes performances de base, nous prévoyons maintenant, plutôt qu'une introduction à grande échelle sur le site, la recherche et le développement d'un radar efficace servant à la localisation qui ne nécessite pas d'expérience dans la lecture d'images et qui pourrait localiser précisément des conduites de gaz et autres structures souterraines à de plus grandes profondeurs. La recherche et le développement d'une telle unité évoluée a déjà commencé l'année dernière. A ce tire, nous avons prévu les performances indiquées dans le tableau 3.

Reference

I. SUGIMOTO, T. KIKUTA, Y. HAYASHI and M. TAKAGI, "COMPACT UNDERGROUND RADAR SYSTEM," The 6th Scandinavian Conference on Image Analysis Oulu. Finland June 19, 22, 1989

No Trenches in Town, Henry & Mermet (eds) © 1992 Balkema, Rotterdam. ISBN 90 5410 085 0

Non-destructive testing of bridge, highway and airport pavements

Gary J.Weil
EnTech Engineering, Inc., USA

ABSTRACT : Billions of dollars have been spent over the last 50 years in order to build the world's concrete infrastructure. Bridges, highways and airport pavements comprise the major uses to which this money has been allocated. Unfortunately, these major pieces of our infrastructure are coming to the end of there design lives and very little funding exists for their maintenance or replacement. Therefore new technologies must be found and adapted for evaluating these structures so that their maximum value and life can be realized. Once a structure can be evaluated as to its areas in need of maintenance or replacement, proper economical decisions can be made for its rehabilitation.

Two nondestructive techniques, Infrared Thermography and Ground Penetrating Radar, have been developed, that when used either separately or together, can quickly, efficiently and economically evaluate large concrete pavement areas, structures and subsurface infrastructure comprised of soil, fluid pipelines and support soils. This technical paper will describe each of these technologies and illustrate their applications with case studies in which they were used, both individually and in unison.

In the inspection of metals and metal-based materials nondestructive testing is an accepted practice. For example, radiographic and ultrasonic techniques are routinely used to identify anomalies in steel pipelines and there are recognized national and international standards on their use. However, in the inspection of concrete the use of nondestructive testing is relatively new. The slow development of nondestructive testing techniques for concrete is that unlike steel, concrete is highly non-homogeneous composite material. Apart from precast concrete units which like steel products, are fabricated at a plant, most concrete is produced in ready-mixed plants and delivered to the construction site. The placing, consolidation and curing of concrete takes place in the field using labor which is relatively unskilled. The resulting product is, by its very nature and construction method, highly variable and does not lend itself to testing by nondestructive methods as easily as steel products.

In addition, concrete slabs are often supported by soil as in airport taxiways and runways. If voids develop beneath these concrete slabs such as, erosion caused by existing drainpipes, support can be lost and eventually the slabs may collapse.

Despite the above drawbacks there has been progress in the development of nondestructive methods for testing concrete and in recent years several methods have been standardized by the American Society for Testing and Materials (ASTM), the International Standards Organization (ISO), and the British Standards Institute (BSI).

Two nondestructive methods, Infrared Thermography (IRT) and Ground Penetrating Radar (GPR), have been developed to locate voids and delaminations in concrete structures such as bridge decks, highways and airport pavements. Their ability to locate voids and delaminations in and below the slabs, gives the structural maintenance engineer the ability to measure the actual cracking and weakening of concrete pavements, before catastrophic failures can occur. Their repeatability also allows engineers to set up databases which can be used to monitor ongoing deterioration which may take place over years or decades.

Although both infrared thermography and ground penetrating radar can be used to locate subsurface and internal voids and delaminations they have striking differences in their abilities. Infrared thermography has the ability to investigate large areas very efficiently. It can locate and image the horizontal dimensions of voids and delamination in pavements 12, 24 or 36 feet wide while moving forward at speeds approaching 10 miles per hour. It cannot though, tell the depth or thickness of these anomalies. Ground penetrating radar, on the other hand, can characterize anomalies as to their depth and thickness, although it is not very accurate or efficient at locating the anomalies or in discerning their horizontal dimensions.

Although many of the capabilities of infrared thermography and ground penetrating radar overlap, they should be considered more complimentary to each other rather than as interchangable replacements. The strong points of infrared thermography are: 1) the ability to investigate large areas very rapidly and very efficiently. Because of infrared thermography's testing techniques of using wide angle optical lenses to assist in the data collection, large pavement areas can be inspected very rapidly. Pavements as wide as 36 feet have been inspected at up to 10 miles per hour with 100% coverage. 2) its ability to give an image of the subsurface anomalies as if they were projected on the surface. This allows determination of the edges of each problem area along with its dimensions. 3) investigations can be performed almost 20 hours per day, either during the day or during the night. This allows testing during times of low traffic at airports, on highways, or over bridges, and decidedly reduces the cost and inconvience associated with normal traffic control. This technology normally gives clients 90% of the information that they require when evaluating pavements. The only answers not to be derived from the infrared thermographic testing are the depths and thickness of the located anomalies.

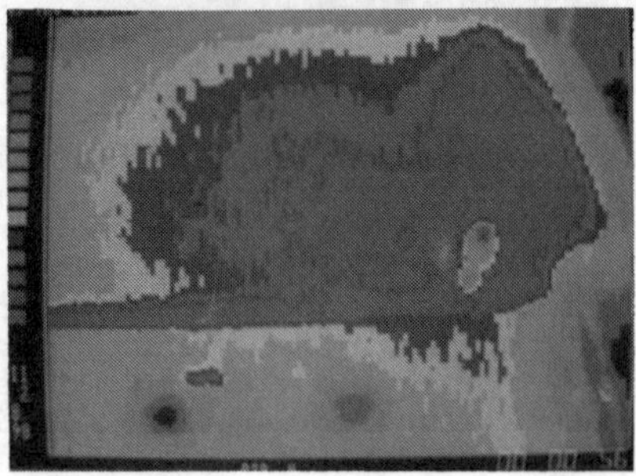

Figure 1. Visual image and thermogram of "bad" expansion strip. Red/white area depicts subsurface delaminations. Green area depicts water penetration.

On the other hand, ground penetrating radar can help determine the approximate depth and width of located anomalies. But it sacrifices convenience, speed of data collection, data analysis efficiency, 100% inspection areas and traffic control costs. Since ground penetrating radar is a point test procedure, where a single transmitting and receiving point is used to collect data, this collection is very slow and must utilize a semi matrix methodology, if used by itself.

To overcome these weaknesses EnTech Engineering, Inc. developed the complimentary methodology of using both infrared thermography and ground penetrating radar. A first pass over a pavement is made with the infrared thermographic equipment. Exact void sizes and locations are determined and plotted on a drawing to scale. A second pass is then made over only the areas previously determined to contain anomalies. This second pass is made with a ground penetrating radar unit attached to the same mobile van. During the second pass, these specific anomalies are then characterized as to approximate depth,

Figure 2. Airport pavement collapse caused by soil erosion near leaking subsurface drain pipeline.

Figure 3. Airport pavement collapse caused by soil erosion near leaking subsurface drain pipeline.

thickness and whether they are to be considered as to have been caused by overlay debonding, half depth problems or full depth problems. This last step gives the client 100% of the answers that he needs to set repair specifications while using both technologies in their most efficient and most economical manners.

2. CASE HISTORIES

2.1 Illinois Department of Transportation Bridge Deck Pavements

EnTech Engineering, Inc. has performed over 100 bridge deck inspections for the Illinois Department of Transportation over the last several years. These decks have ranged from some of the world's largest, such as the Poplar Street Bridge complex over the Mississippi River, to some of the country's smallest, such as those in rural areas of upstate Illinois. To our knowledge, they have had the nation's longest and most comprehensive NDT program for bridge deck pavement inspection ever officially established.

EnTech was brought in as the main testing consultant. It suggested that both infrared thermography and ground penetrating radar be combined and that physical corings be eliminated. EnTech's improved referencing system, tied into its custom designed mobile van transported infrared system, proved to be accurate to within 1% on any of the encountered bridge decks.

Multiple year repeat inspections on many of the deck pavements have confirmed the repeatability of the various technologies and the need for physical corings or major traffic control have been eliminated.
(See Figure 1.)

2.2 Manchester, New Hampshire International Airport Taxiway Pavement

On May 2, 1990, at the Manchester, New Hampshire International Airport, the landing gear of a DC-10 carrying a full load of passengers fell through the taxiway pavement while approaching its unloading gate. The damage caused to the airplane approached half a million dollars and included areas of the landing gear, fuselage and leaks in the fuel system.
Upon removal of the passengers and containment of the leaking fuel, airport authorities removed the airplane. (See Figure 2 and 3.) During the removal process, it was determined that a 6 ft. by 6 ft. by 8 ft. deep void had formed beneath the pavement because of leaks and infiltration of the soil into a buried storm water drainage system. The drainage system was approximately 40 years old. When it was determined that the drainage system was located throughout the entire airport pavement system, airport authorities and their consultants concluded that more drainage system leaks and erosion areas probably existed. Airport authorities then requested that the consultants locate a method of determining the leaks and

Figure 4. Visual image of erosion void beneath airport pavement.

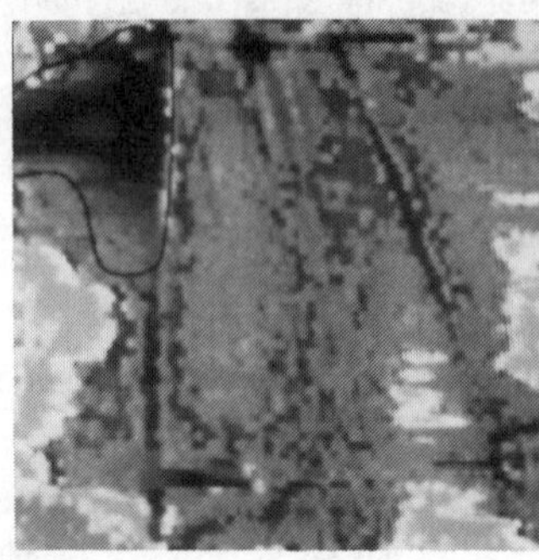

Figure 5. Thermogram image of erosion void beneath airport pavement.

possible voids, with 100% coverage, without interrupting airport traffic.

EnTech Engineering, Inc. was asked to utilize their patented, infrared thermographic leak and void pavement testing system.. The inspection of over 2,000,000 sq. ft. of pavement was conducted entirely at night, after 11:00 p.m., when air traffic and gate usage was at a minimum. The entire inspection took 3 nights and uncovered 12 subsurface voids of varying sizes. Several of these voids could have caused major damage to airplanes if the pavement above them had collapsed. (See Figure 4 and 5).

Due to the size and critical nature of the subsurface voids, ground penetrating radar was not needed, as all areas located with the IR system were rehabilitated.

3. CONCLUSIONS

1. Non-contact, nondestructive, infrared thermographic techniques can be used to detect voids and delaminations, very efficiently, in and below, concrete bridge, highway and airport pavements. It can also locate erosion voids and other types of anomalies below highway and airport pavements.

2. Nondestructive ground penetrating radar can be used to efficiently characterize the voids and delaminations located by infrared thermography.

REFERENCE

1. Malhotra, V.M., Weil G.J. et Al. CRC Handbook on Nondestructive Testing of Concrete.
CRC Press, Boca Raton, Library of Congress Card Nr 90-2698, 1991.

2. Weil G.J., Review of the ASTM Standard on IR Testing of Concrete Bridge Decks, Proceedings of SPIE - The International Society for Optical Engineering, Thermosense XI, March 29-31, 1989.

3. ASTM Standard D4788.

Much of the technical informations contained in this article was obtained from the U.S. and Worldwide Patents pending held by its author, Gary J. Weil and En Tech Engineering, Inc. It is suggested that anyone attempting to use these technologies consult with the patent holders in order to make the most accurate use of these technologies.

2 Rehabilitation

2.1 Techniques

No Trenches in Town, Henry & Mermet (eds) © 1992 Balkema, Rotterdam. ISBN 90 5410 085 0

Assessment of the different methods of rehabilitation tested on sawn, lined, egg-shaped sewers

Claude Resse & Idriss Benslimane
ABROTEC, Poissy, France

ABSTRACT : The aim of this series of experiments, carried out and financed by the Département of the Val de Marne, was to compare the efficiency of different sewer rehabilitation methods. To this end, prefabricated, egg-shaped sewers, sawn through on one side so as to simulate a state of disrepair and then repaired using different lining methods, were subjected to external loading until total collapse occured. The method and techniques tested are described. Results are given including some correlation values and comments.

1 EXPERIMENTAL PROCEDURE

The sewer repair techniques which were tried out are based on the installation of a thin lining inside a reinforced, egg-shaped sewer one of whose side walls has been sawn.

Two sewers were lined for each repair method tested. These lined sewers and the control sewers, sawn and unsawn, were subjected to external loading on the side walls until the sewer collapsed.

Figure 1 indicates the principle of the test.

Standardisation tests were carried out on core samples ϕ 50 drilled out of control panels and the lined egg-shaped sewers.

These tests consist of :

* simple compression tests and density measurements,
* tests on the bonding between the sewer wall and the lining,
* abrasion tests.

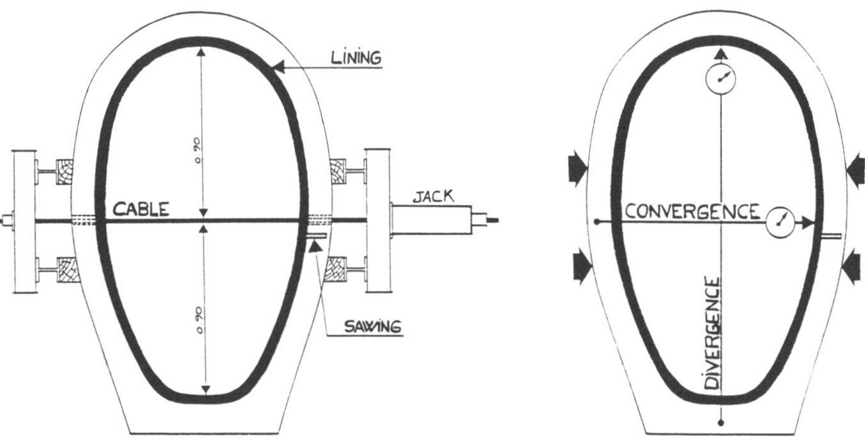

Compression equipment

Location of main measurements

Figure 1 : Diagram of the compression test.

Table 1 : Method characteristics

Method N°	Type of spraying	Type of concrete	Type of fibre	Lining thickness [cm]	Resistance to compression [MPa]
5	wet process	type A	Steel	5.7	26
2	wet process	type A	Stainless steel	5.6	30
4	wet process	type A	Polypropylene	5.5	24
7	dry process	type A	Cast iron	6.0	28
8	dry process	type A	Steel	5.7	55
10	wet process	type B	Polypropylene	5.7	55
9	dry process	type B	Polypropylene	6.0	40

Table 2 : Abrasion and bonding

Method N°	Type of spraying	Type of concrete	Type of fibre	Bonding constraint (mPa)	Abrasion value
5	wet process	type A	Steel	1.5	5.0
2	wet process	type A	Stainless steel	1.0	4.8
4	wet process	type A	Polypropylene	1.8	6.7
7	dry process	type A	Cast Iron	1.7	3.6
8	dry process	type A	Steel	2.4	2.2
10	wet process	type B	Polypropylene	2.6	2.9
9	dry process	type B	Polypropylene	1.6	5.5

Table 3 : Results of the tests involving side wall collapse.

Method N°	Type of spraying	Type of concrete	Type of fibres	Secant modulus kN/mm	Force required to cause cracks Fl [kN]	Force to cause total collapse Fm [kN]
5	wet process	type A	Steel	150	165	240
2	wet process	type A	Stainless steel	130(err=17%)	155(err=03%)	200(err=01%)
4	wet process	type A	Polypropylene	100(err=02%)	140(err=07%)	185(err=07%)
7	dry process	type A	Cast iron	120(err=01%)	130(err=02%)	200(err=05%)
8	dry process	type A	Steel	160(err=05%)	140(err=01%)	215(err=02%)
10	wet process	type B	Polypropylene	115(err=17%)	145(err=04%)	202(err=04%)
9	dry process	type B	Polypropylene	85(err=04%)	123(err=10%)	200(err=01%)
20	UNSAWN CONTROLS			120(err=24%)	115(err=03%)	160(err=06%)
22	SAWN CONTROLS			45(err=10%)	80(err=06%)	115(err=15%)

Repair techniques

A total of seventeen techniques were tested :

* techniques using the installation of a plastic lining bonded to the sewer wall with grout,
* spraying techniques using the dry or wet process with concrete alone or fibre reinforced concrete. Various fibres were tested (polypropylene, steel, stainless steel, cast iron).

This series of experiments has been described in an extensive report dealing with the results of every technique as well as the degree of correlation between different parametres.

In this article only concrete spraying techniques and their salient results are dealt with.

2 DEFINITION OF REHABILITATION METHODS

Table 1 gives the main characteristics of the methods investigated.

The linings installed are approximately 6cm thick.

The results of simple compression tests obtained from the core samples drilled out of the control panels showed the existence of two distinct groups of resistances.

* 25 to 30 MPa : techniques n° 5, 2, 4 and 7
* 40 to 55 MPa : techniques n° 8, 9 and 10.

3 RESULTS OF THE STANDARDISATION TESTS : BONDING AND ABRASION

Table 2 gives the following results :

* The bonding strength determined by tension testing on core samples drilled out of lined sewers.
* The abrasion value which is obtained by measuring the imprint left on the sample by a wet sandblasting jet.

An examination of tables 1 and 2 does not indicate that the repair technique (dry process or wet process) has any bearing on the test results. However, the following correlations were observed :

- a low abrasion value (corresponding to a high level of abrasion resistance) and good bonding between the lining and the sewer wall correspond to an increased resistance when compressed (except in the case of technique 9).

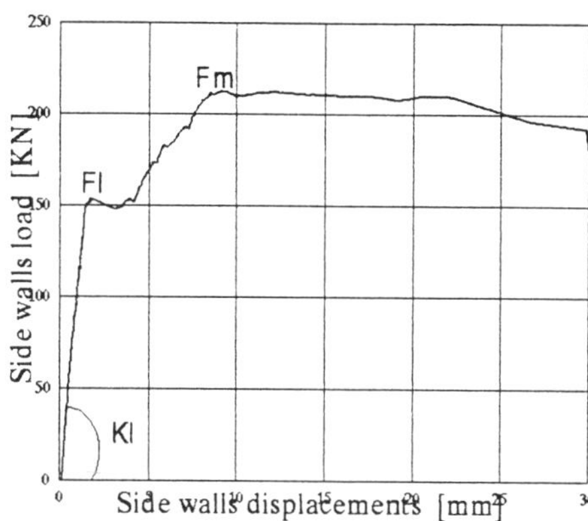

Figure 2 : Interpretation of test results

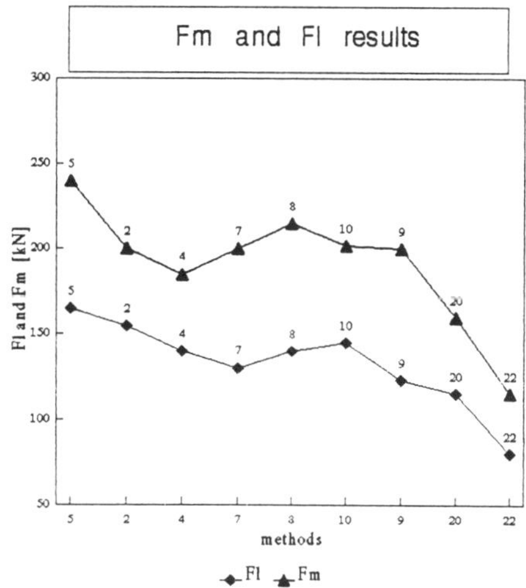

Figure 3 : Results of the compression tests. For technique 5 only one sewer was tested.

4 RESULTS OF THE COMPRESSION TESTS

Figure 2 shows the parametres used in this article to compare repair methods.

We have chosen three parametres :

* FL : The force applied corresponding to

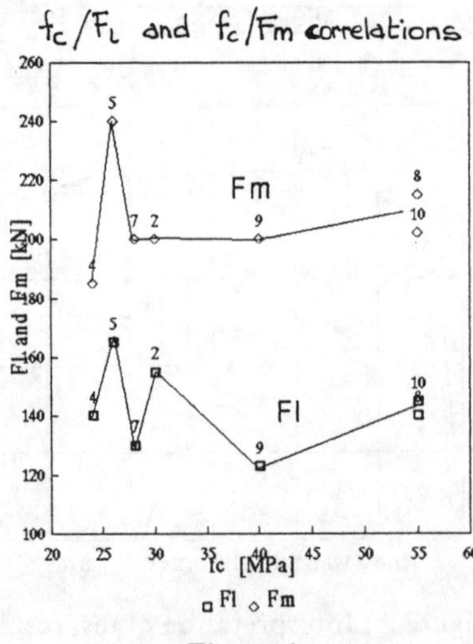

Figure 4

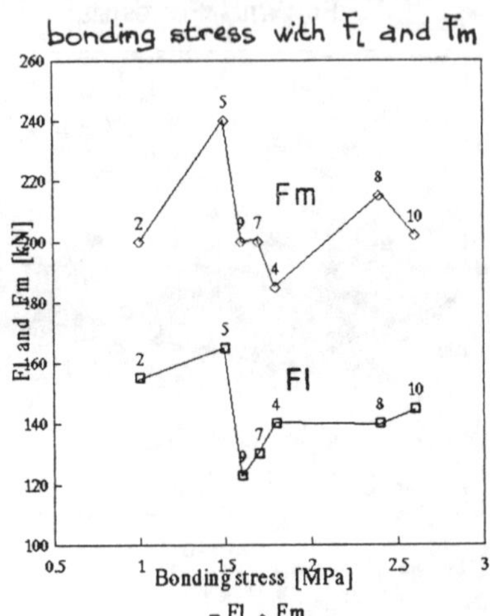

Figure 5

the end of the almost linear field of behaviour.

* KL1 : Secant rigidity of the linear field of behaviour.
* FM : The force required for the sewer under test to collapse completely.

Table 3 and Figure 3 give the values of three parametres obtained for all the techniques used.

Moreover, when the ruptures occured, we did not observe any differences between the methods of rupture due to the repair techniques used.

Comments for Table 3 and Figure 3:

Apart from the case of unsawn control sewers where the error on KL1 is high (25%),, a clear correlation is observed between the three parametres FL, FM and KL1.

Table 3 gives, for each technique, the average values of FM and FL obtained in the trials. N.B. the maximum error obtained for each technique appears in brackets.

* 100% gain
 - technique 5 - wet process - matrix A - steel fibres.
* 75% gain
 - techniques 2 and 8 - wet process and dry process - matrix A - stainless steel and steel fibres.
* 65% gain
 - all the other techniques.

It can be held that the matrices containing fibres with strong characteristics (steel type) give higher gains, however, it should be borne in mind that these additions entail a risk of corrosion. The effect of the spraying method was not detected.

5 STUDY OF SEVERAL CORRELATIONS

The following figures show the correlations which can exist between different parametres.

Contrary to what might be expected no correlation was observed between the characteristic values (bonding and resistance under compression) and the force required to cause total collapse (FM). However, it was observed that strong bonding and strong resistance do not mean that the force required to cause cracking was increased.

6 CONCLUSIONS

The results recorded in this article show that for each type of compression test carried out, spraying methods have little influence on the values of the forces required for cracking and complete collapse of the sewer.

Strong bonding or resistance constraints under high compression of matrices do not lead to greater force required for total collapse and cracking during the adopted test.

On the other hand, matrices containing fibres with strong characteristics (steel and stainless steel) require higher values.

No Trenches in Town, Henry & Mermet (eds) © 1992 Balkema, Rotterdam. ISBN 90 5410 085 0

Sewerage rehabilitation planning: The interface between theory and practice

D.I.Aikman
Babtie Shaw & Morton, Consulting Engineers, Preston, UK

ABSTRACT: Sewerage rehabilitation planning in the United Kingdom is a high technology orientated procedure which relies on large volumes of data collected under arduous field conditions. The ultimate success of the approach depends on verification of the data and the associated computer analysis at all key stages. The investigative procedures and the benefits which can result are summarised. Problems of obtaining reliable data and developing practical designs are discussed, along with relevant computer software. Finally, potential technological developments are reviewed.

1 INTRODUCTION

The majority of trunk sewerage systems serving the main urban areas of the United Kingdom (UK) are well over 100 years old and many are suffering structural deterioration and hydraulic overload. In addition, the provision of storm overflows to relieve hydraulic problems has resulted in pollution of watercourses.

The various techniques, developed in the UK over the last two or three decades, to assess the performance of sewerage systems and to design appropriate rehabilitation works, were brought together in the Sewerage Rehabilitation Manual (Water Research Centre and Water Authorities Association 1986) which was first published in 1984 and updated in 1986. The main objectives of the procedures set out in the manual are:-

1. to quantify the key structural, hydraulic and pollution problems;
2. to identify integrated solutions to all three problems;
3. to design rehabilitation works which minimise surface disruption and overall cost by making maximum use of the existing sewerage system.

The design procedures are summarised briefly in the following section.

2 GENERAL APPROACH

2.1 *Methodology*

Sewerage investigations may be initiated by poor performance of the existing system, proposed new development/redevelopment, or as part of a programme of asset management planning.

In any of these three cases, the first action is to check the quality of sewer records on which all future work will depend. If records are generally inadequate, a full manhole survey may be required.

Given adequate records, the core sewers (those for which pre-emptive maintenance is justified on economic grounds) are surveyed by closed circuit television (CCTV) or man entry methods. At the same time a computer model of the core sewers is built and its predictive accuracy verified by using actual rainfall events as input data and comparing the computer generated flow and level hydrographs with recorded conditions.

A series of design storms events are run to assess the performance of the sewerage system and identify hydraulic restrictions or areas which may be prone to flooding. In addition, the discharge from combined sewage overflows is determined for both individual storm events and for typical annual precipitation levels. The quality of the receiving waters and the impact of the potential pollution load is then assessed.

Sewers which have been subject to internal inspection are assessed and graded according to the defects noted, the surrounding ground conditions, and the operational regime.

The structural, hydraulic and water quality problems are considered jointly and a range of possible rehabilitation options are identified and modelled to ensure that they meet the required hydraulic and water quality standards.

Maximum use of the existing sewerage system and minimum surface disruption are guiding principles behind the selection of options. Ground conditions, the possible diversion of other services and of road traffic, are essential considerations at this stage. The most cost effective solution which meets the required standards is then adopted for the purpose of detailed design.

2.2 *Is it cost effective?*

The answer to this question depends on the accuracy and reliability of the data and modelling procedures, the costs, particularly those associated with the survey work, and the achievement of financial savings at the construction stage which exceed the cost incurred in the design process.

Babtie Shaw & Morton's experience of more than 20 projects over the last ten years has shown these procedures to be an effective method of accurately assessing the performance of sewerage systems and designing rehabilitation works. Furthermore, the construction costs derived from this approach can be expected to be 20% to 30% below those resulting from more traditional methods and these savings greatly exceed the cost of developing an integrated sewerage rehabilitation plan (Perfect and Aikman 1991).

3 SURVEY AND VERIFICATION

Errors can occur at any stage in a survey. The greater the volume of data the greater the statistical probability of errors. Conversely, those types of error which are least likely to occur and should be most obvious are those which are most likely to go unnoticed. In the following section the key methods available for survey and elimination of errors are discussed.

3.1 *Manhole survey*

The survey and the production of manhole records is prone to errors due to the difficult working conditions and the tedium of repetitive unskilled work. Between 20 and 50 items of data are typically required to define the key features of a manhole and the associated sewers (e.g. material, diameter, level etc.)

In a town of 100 000 population there may be around 10 000 manholes, which in turn means up to half a million items of data. A five percent error in the records could in this case result in up to 25 000 incorrect items of data; from past experience even this level of accuracy can be very difficult to achieve and many town records may contain a much higher percentage error.

Computer software packages, now available in the UK, can greatly reduce the problems by checking and validating field data, highlighting inconsistencies and producing sewer layout plans. Data may be recorded in the field directly onto a hand-held logger, downloaded to a desk top computer and then displayed on a geographic base which has been previously entered as a digitised record. However, as computers only work to fixed rules, a final visual assessment of the records is always valuable to pick out anomalies the computer may not recognise.

3.2 *Sewer survey*

In small diameter sewers, typically those less than 1000mm diameter, survey is by means of winched or self-propelled CCTV cameras which provide a continuous video record. Visible defects are manually logged on a micro computer as the survey takes place. In larger sewers, surveyed by many entry methods, defects may be logged electronically and recorded on still photographs.

Surveys are often hampered by high in-sewer water levels or poor access. To overcome these problems, survey vehicles have successfully been fitted with sonar equipment which operates under water and provides accurate profiles of the inner circumference of the sewer. In addition, tethered floating vehicles have been developed to provide a CCTV record above water level.

The survey data are normally validated and checked for inconsistencies as they are logged using appropriate computer programs. Recent developments allow the manhole and sewer survey computer files to be interfaced and cross checked for consistency; from practical experience this should greatly increase the reliability of the final records.

3.3 *Flow and rainfall survey*

Rain gauges with integral data loggers are sited at secure locations across the catchment. Combined velocity and depth monitors (pressure transducers) are located at key points in the sewerage network. The proposed location of monitors is carefully assessed in the light of possible unstable hydraulic conditions, extreme velocities, or silt problems. Where flows are required at or close to pump stations, a depth only monitor may be used in the pump sump as it may give more reliable results than a combined monitor when flows are constantly changing. All equipment is checked, calibrated and data downloaded at weekly intervals.

A typical survey may encounter any one or more of the following problems:-

1. No significant rainfall within the normal five week survey period or rainfall very unevenly distributed across the catchment.
2. Total failure of one or more flow monitors or rain gauges.
3. Sewer flow conditions outside the operating range of the equipment or silt or rags blinding the velocity sensor.

4. Hydraulically complex conditions making interpretation of the depth and velocity records impractical.

Depth and velocity data are subject to a "scattergraph" analysis (Water Research Centre 1987) to check the validity of the results and the velocity/depth relationship before the data are converted to depth and discharge/hydrographs. Before the individual results are used for model verification, a comparative analysis of data should be carried out to check for inconsistencies including results which have been incorrectly identified with regard to location or time frame.

3.4 *Model building and verification*

In the 1980s the most widely used sewerage modelling software in the UK was the Wallingford Storm Sewer Package, known as WASSP (Department of the Environment/ National Water Council 1981). This allowed the hydraulic operation of complete sewerage catchments to be modelled and the effect of proposed changes to be analysed under a range of rainfall conditions. WASSP has now been replaced by the WALLRUS model (Hydraulic Research 1991) which, amongst other improvements, can simulate free surface backwater effects, spatially varied rainfall and sediment deposits. For looped systems where reverse flow may be a feature a complementary module, known as SPIDA is available (Hydraulics Research 1992).

Detailed comment on these programs are available in Wallingford procedure user group notes and WALLRUS technical update sheets and will not be considered here.

Both the WALLRUS and SPIDA models have been developed for use under a wide range of hydrological conditions, not just those in the UK, and indeed parts of the Paris catchment have been successfully modelled for Ville de Paris in the course of a model trial.

The model comprises a simplified representation of the sewer network with, typically, approximately one pipe per hundred persons (catchment population) for a detailed model and rather fewer pipes for strategic models of major catchments. The model network and the program input data may be produced directly from computerised manhole survey records, thereby minimising data transfer errors.

The assessment of permeable and impermeable areas draining to each pipe is a time consuming exercise which may require field surveys to determine which surfaces or premises drain to the foul sewer and which to the surface water sewer; incorrect connections or illegal cross connections have been found to be a common occurrence.

Once the model is built, a number of simple checks are carried out to ensure that it is functioning correctly and giving valid results; these include a comparison of rainfall and outflow volumes and a check on flow depth/velocity relationship at key points.

To verify the representational accuracy of the model, measured rainfall data are used to generate flow and depth hydrographs which are then compared with the infield measured hydrographs. Failure to meet the specified tolerances requires investigation and corrective action. The most common problems are unrecorded changes at pump stations and unrecorded connections to, or overflows from, the sewerage system.

The final check involves a comparison of predicted areas of flooding with historical records or the results of questioning householders in suspect areas or, if the opportunity arises, carrying out surveys during actual storms. Experience with this type of survey has shown the above models to be extremely reliable predictive tools, often identifying problems not previously known to the municipal authorities.

4 ASSESSMENT AND OPTIMISATION

4.1 *Structural assessment*

Sewer structural condition grades are determined by analysis of the recorded defects, comparison with reference photographs of typical sewers in each grade, and consideration of ground conditions and surcharge frequency. A five grade system is used in which Grade 1 is for sewers in good condition, whilst Grade 5 is associated with partially or completely collapsed sewers. The initial analysis may be carried out by computer, thereby greatly reducing the scope for error, and the condition of sewers in Grades 3 to 5 checked manually.

4.2 *Hydraulic and water quality assessment*

A series of design storms with different return periods (1, 2, 5, 10 years etc.) are run through the verified computer model and the performance of the sewers, overflow etc. analysed in terms of frequency of surcharge and flooding and the volume of overflow spills. A standard series of 100 typical annual storm profiles may also be run to assess the likely annual volume and frequency of stormwater discharge. Comparison is made of the potential polluting load with the capacity of the river or stream to accept the discharge. Hydraulic inadequacies in the sewerage system and unacceptable storm overflows are then identified.

4.3 *Rehabilitation options*

Structural rehabilitation methods are selected in the light of identified hydraulic conditions, e.g. a thin walled though costly resin impregnated liner may be selected in preference to a thick walled lining system where hydraulic capacity is a limiting factor.

Hydraulic problems may potentially be relieved by the addition of on or off-line tanks, the addition of relief sewers or overflows. The number and design of overflows may be rationalised to meet higher water quality standards.

A number of possible solutions to the structural, hydraulic and water quality problems are identified and hydraulically optimised using the above model.

Each solution is costed on the basis of standard unit rates but taking due account of the cost of any service diversions required and the impact of surface works on traffic, commerce and domestic life. These latter considerations require a semi-detailed assessment of the proposed works and services at each potential construction site. Based on the above, the most cost effective solution is selected for detailed design.

4.4 *Case Studies*

An investigation of the Preston sewerage system in the north west of England identified eight previously unrecorded overflows. Computer modelling, followed by a survey of householders, showed that less than ten percent of those living in flood prone areas had reported known problems. Rehabilitation works, including extensive structural lining of sewers around the town centre and under the London to Edinburgh railway line have since been carried out with negligible impact on community life.

At Withernsea on the north east coast of England, a 5000m^3 detention tank was designed, using a WASSP model, to eliminate a pollution problem and avoid the need for a costly new outfall. The tank, located under a car park, is due to be commissioned this year.

5 FURTHER DEVELOPMENTS

Sewerage rehabilitation design has undergone a major evolutionary change over the last 20 years. Potential improvements are still being identified and developed, a number of which are briefly discussed below.

The physical survey of manholes is labour intensive, prone to error and would benefit greatly from the introduction of some form of electronic measurement and logging system.

Flow and rainfall surveys can be very expensive if there is prolonged dry weather, particularly in view of the relatively high equipment failure rate. More robust equipment could allow fewer monitors to be used for longer periods.

The assessment of sewer structural condition relies on assumptions with regard to ground support etc. Ground probing radar has been used to check for voids behind culvert walls but this technology has not yet reached the stage where it can form a routine part of survey work. Such development would improve the identification of high risk areas and should, as a result, reduce the overall cost of rehabilitation works.

The measurement of sub-catchment areas, and the separation of them into permeable and impermeable surfaces, is still largely a slow manual exercise. Technologies used for air and satellite photographic interpretation can both identify and measure different surface areas; although work has been carried out to adapt this technology to the needs of drainage design, it has not yet reached a satisfactory conclusion.

The one feature which consistently hampers the full use of trenchless technology in sewerage rehabilitation work is the inability to reliably remake existing side connections. If open excavation and manual reconnection is required then all the benefits of the technology are lost. There are a number of techniques available and a number of embryo technologies awaiting development but this problem has yet to be solved with a reasonably universal solution.

REFERENCES

Department of the Environment/National Water Council 1981. Design and analysis of urban storm drainage - the Wallingford Procedure. *National Water Council, Standing Technical Committee Report No. 28.*

Hydraulics Research Ltd. 1991. *WALLRUS users manual,* Fourth edition. Hydraulics Research, Wallingford.

Hydraulics Research Ltd. 1992. *SPIDA product brief.* Hydraulics Research, Wallingford.

Moys, Dr. G. 1992. *Application of SPIDA to Paris.* Paper presented at SPIDA launch. Hydraulics Research, Wallingford.

Perfect, H.G. and Aikman, D.I. 1991. Drainage area planning : from the Pennines to the sea. *J. Municipal Engineer:* 8 April 71-82.

Water Research Centre and Water Authorities Association 1987. *A guide to short term flow surveys of sewer systems.* Water Research Centre, Swindon.

Water Research Centre and Water Authorities Association 1986. *Sewerage Rehabilitation Manual.* Second edition. Water Research Centre, Swindon.

No Trenches in Town, Henry & Mermet (eds) © 1992 Balkema, Rotterdam. ISBN 90 5410 085 0

Water mains rehabilitation

C.D. Murray
Tate Pipe Lining Processes Limited, Denton, Manchester, UK

(a) Abstract

This paper is concerned with the more widely used methods for the rehabilitation or replacement of pipelines primarily used for transportation of domestic water supplies. These pipes, ranging in diameter from 80mm up to a massive 4,500mm, can be rehabilitated insitu or in Stockyard or Factory locations. The two methods described in the following transcription are Cement Mortar Lining and Pipe Bursting, for rehabilitation and replacement respectively. As previously stated, the reference is to domestic water supply pipelines and aqueducts where experience has shown, over the years, that any form of cleaning without subsequent internal protection creates problems of a serious nature. Cement mortar linings have proved to be economical, long durable lasting and highly satisfactory. These linings have been used for over 50 years both in the U.K. and overseas.

Where external corrosion is a serious problem, and consideration has to be given to replacement, the more recently developed Pipe Bursting method can be employed. This paper will describe one particular system in more detail.

(b) Introduction

The vast majority of insitu mains, worldwide, are constructed of cast iron and, in a large percentage of cases, have been in the ground for a long period of time. Corrosion of these mains occurs in varying degrees and is the main contributory factor leading to failure, leakage, poor water quality and reduced flows. Where these problems occur, the consumer will experience a poorer quality of service, and rehabilitation of the defective mains will help to overcome these problems. Relining is now considered the more important method of rehabilitation because of its cost-effectiveness compared to the more traditional methods of open cut replacement, thus avoiding all the unnecessary trenching and associated traffic delays caused by this method. Of course, as stated before, some pipes may be so badly corroded, or reduced in capacity, that replacement becomes the only solution.

This is where the pipe bursting method can be most usefully employed.

Of course there used to be a time when the only protection to pipes was a thin coat of bitumen paint which only afforded protection for a limited period of time. This prompted both end users and manufacturers to seek protection of a more permanent nature. Earlier solutions included a spun-concrete lining of 4.5mm thickness, still used in some factory sites today, which is described later in this paper. This solution, of course, only offered protection to short lengths of main and no solution was available for the thousands of kilometres of pipes already laid. In the U.K. for example, the cleaning of short lengths was carried out in badly affected areas until it became apparent that, far from curing problems, corrosion was actually accelerated by cleaning because of the water and rapid re-growth of encrustation. This, coupled with the problems of leakage, caused many consumer complaints and industry began to look for new methods of rehabilitation.

(c) Discussion

(1) Cleaning

There are several methods of cleaning corroded and encrusted pipes insitu, with the governing factors as to which should be employed being diameter and length.

In all methods, access to the pipeline is required by means of a small excavation at each end of the cleaning length. This length must be the maximum possible to alleviate the number of excavations together with their associated costs. Where possible excavations are located at bends.

Power Boring (75mm to 200mm)

A mechanically driven boring machine - known throughout the industry as a "Rack Feed Borer" - drives a series of steel rods with a flail head fixed to the leading end and is rotated within the pipeline. As more rods are connected, up to 160 metres of pipeline can be cleaned. Flow of water against direction of flailing is maintained to wash out debris.

(Fig. 1)

Drag Scraping (150mm dia. upwards)

Drag Scrapers are constructed of sprung steel blades mounted on a cylindrical chassis. These blades are mounted in multiple rows to give total coverage to the pipe wall. The scrapers, which resemble large metal pineapples, are attached to winch cables at both ends and dragged forwards and backwards until the encrustation is removed from the pipe wall.

Although large quantities of debris are removed by this method, the pipe must be finally cleaned with a plunging action using rubber disc tandem pistons to ensure that the main is free of fine residue corrosion and water.

(Fig. 2)

Pressure Scraping (Greater than 450mm)

Where possible, large diameter mains are "pressure scraped" rather than drag scraped. Again, as in drag scraping, the equipment has a resemblance to an oversized metal pineapple with many overlapping blades fixed to a cylindrical body.

This piston assembly is introduced into the pipeline and transported along by a pressurised flow of water, about 2 bar, which is controlled by special valves fitted to an "end box" assembly at the downstream end of the main. Each scrape can be as long as 5.0km.

A certain amount of water is allowed to escape past the scraper and this carries the encrustation and corrosion debris removed by the scraper to the discharge point. It is not usually acceptable to discharge this debris laden water into a public drainage system, so it must be passed into a large filter assembly placed between the end box connection and discharge point. The large filter contains baffle plates which retain all heavy deposits and allow supernatent water to run to waste into the foul sewer.

(Fig. 3)

Cement Mortar Lining - General

Many years ago it was stated by Hazen and Williams, whose formulae for flow co-efficients is internationally known and accepted, that "the gradual roughening of the interior of the cast iron pipe is one of the most familiar of water works phenomena. It is also one of the most difficult to compute. In a general way it may be said that in a series of years, which is not long compared with the total life of the pipe, the roughening of the surface and the reduction of the area through rusting and tuberculation reach such an extent that twice as much head is consumed in sending a given volume of water through it as was the case when the pipe was new".

This theory has been proved so many times that internal protection to guard against this is now universal.

Having dealt in the introduction with the history of cement mortar linings, the various methods of application are now detailed.

For the purposes of this paper there are two methods of cement mortar lining, these being for insitu, or existing mains and for stockyard or factory lining of new pipes prior to laying in the ground.

Insitu Lining (80mm to 200mm)

The cost effectiveness of small bore refurbishment is now widely accepted throughout the water industry. Actual cost comparisons with relaying the main is dependent upon the way that environmental and other on-costs are accounted for, however, it is generally agreed that refurbishment can be carried out for approximately one-third the cost of relaying.

Lining is carried out by pumping a mixture of 1:1 graded sand and Portland Cement from a surface mixer unit, known as a "Carousel", via a cement mortar hose to the lining machine. This method is important to ensure a constant rate of delivery.

(Fig. 4)

The hose and lining machine are withdrawn through the pipe by means of a winching hose reel mounted on the "Carousel". The air-driven lining machine at the end of the hose provides an even 360° "orange peel" finish coating. The surface finish can be further improved by pulling a conical trowel behind the lining machine.

Insitu Lining (225-660mm)

This method of lining work is carried out in a similar way to the smaller diameter work except that the lining machine and, where applicable, the drag trowel are carried back towards a larger Carousel unit by a slow motion carriage. This improves the quality of the finished lining, and enables a distance of up to 220 metres to be lined between excavations.

(Fig. 5)

Insitu Lining (Man Entry)

In larger diameter mains, where man-entry is possible, lining is carried out using a "train" of man operated machines, consisting of three vehicles: the lining machine, the intermediate storage hopper, and the Power Loader.

The cement mortar is mixed on the surface and introduced into the Power Loader located at a convenient excavation in the pipe. Once full, the Power Loader is driven along the pipe where the intermediate hopper and lining machine are located. The mixed mortar is then transferred into the intermediate hopper and pumped to the lining machine which applies the mortar to the pipe wall.

(Fig. 6)

The operator sits on the lining machine transporter unit, controls the rate of backward travel to achieve correct thickness and also operates the rotating trowels which follow the lining head and smooth the surface.

Stockyard Lining (600mm/larger)

Originally designed to meet the needs of the overseas market, the Tate designed "Bridge Plate" system has been developed into the best and most economical way of lining new, large diameter pipes. The key benefits of this system are minimum pipe handling and the compact design of equipment.

The system itself comprises of a central batching plant from which thoroughly mixed 1:1 sand and cement is transported by a high discharger or dumper to the lining site. From there it is transferred to a high capacity mortar pump mounted on the Prime Mover (i.e. a heavy tractor or rough terrain forklift). Also mounted on the Prime Mover is the "Bridge Plate" and lining machine which are located to the end of the pipe.

The lining machine travels rapidly to the far end of the pipe and returns to the unit at a controlled speed, applying the cement mortar as it does so. Rotating trowels attached to the front of the machine smooth the newly applied lining to ensure a high quality finish.

The "Bridge Plate" and lining machine which suit various diameters of pipe, can be quickly changed within the hour. Pipes with the largest of diameters can be lined using this method.

Factory Lining (80-600mm)

This involves the "orange peel" spray coating of individual pipes, whilst they are held at the "stinger station" by a lining machine on the end of the beam carrying a cement mortar feed hose and power line. The lining machine is placed, by fast travel, at the remote end of the pipe, where it is then fed with an even flow of 1:1 sand and cement mortar. The coating is performed during withdrawal and lining thickness is related to flow of mortar and speed of withdrawal.

Following application of the lining, each pipe in turn is spun at a rolling station in a spinning machine with safety lock-down driven rollers. During a rolling period, which is kept short to prevent mix separation, vibration is applied to give a high quality finish.

Each pipe is ejected to an output curing track. Curing is then achieved either naturally or, if rapid handling is required, in a specially designed oven. In either case end caps are fitted to retain moisture within the pipe.

General Details and Procedures

The curing of Cement Mortar involves a reaction with water which is temperature sensitive. To enable pipes to be lined, a minimum air temperature of 2°C and rising is necessary. If temperatures of 2°C (or 2°C and falling) prevail then pipes should not be lined.

The lining should be kept moist for the hardening process so that only a minimum of shrinkage cracks appear. Therefore the pipe should be filled with water not less than 0.02 metres deep and each section closed by using plastic end caps.

After lining and curing, cracks may appear in the lining. It is now well established that surface crazing and hair cracks repair themselves in service by two phenomena. Firstly, when lining is fully dried out, ensuring shrinkage is at a maximum. On flooding the line with water the coating will expand by approximately 0.1% to 0.15%. Secondly, any remaining cracks will normally heal, autogeneously, by the hydrating of unhydrated cement in the mortar and/or deposition of salts from the flowing water.

Due to shrinkage of the mortar lining, hairline cracks are unavoidable. Hairline cracks and sporadically occurring surface cracks are only allowed up to a width in accordance with A.W.W.A. C602.89.

Apart from aggregate grains which may stand out at the surface sporadically, all surfaces should be smooth and even. However, single waves or grooves are acceptable provided the minimum specified lining thickness is observed.

Any work to damaged areas should be carried out to A.W.W.A. specifications.

In some cases where damage occurs it may be necessary to remove to complete lining, clean out and re-spray. Again, all work should comply with specifications.

All mortar should be mixed in suitable equipment, with sand meeting required specifications and cement to British Standard Portland.

Any lining which is not produced in accordance with these specifications should not be accepted by the purchaser.

Replacement Moling or "Pipe Bursting"

This system of pipe replacement is to be encouraged by engineers who need to replace a defective main due to either continuing external corrosion, making a lining technique of little use, or the need to increase capacity of the original line.

The existing mains can be burst by one of three approved techniques. The first technique that may be used is a percussive mole, which consists of a horizontal reciprocating hammer acting within a cylinder fitted inside a tapered head. The head may be fitted with "stress raisers" to increase the breaking force whenever joints or service connections are encountered. The reciprocating action provides both forward motion and radial force to shatter the existing pipeline and compress the pipe fragments into the adjacent soil, which is itself locally consolidated. Forward motion of the mole is assisted by a winch.

The second technique is a further adaptation of the percussive action, but in the place of the "hammer" action a hydraulically powered "vane" is constantly lowered and raised, thus splitting the pipe on its travel along the main.

(Fig. 7)

Thirdly, and more recently developed, is the "Ram Rod" system, which consists of a hydraulic ram arrangement, located in the first excavation, which pushes through a series of rods with a splitter head attached.

(Fig. 8)

As the arrangement passes through the main, the pipe is again shattered with pieces compressed into the ground. On reaching the "away" excavation the "splitter" is taken off and a spreader piece, together with the new M.D.P.E. pipe and sleeve, are attached. The hydraulic unit is then set in reverse and the complete assembly is pulled back through; on the larger diameter mains bursting is achieved on this return journey.

The great advantage with this third method of moling is that there is no vibration set up within the pipe. This means that pipe bursting can now take place without disturbing other services that may be close by. Also, in the event of any breakdown, because the motive power is a "ram" system operated from outside the pipe, the need to recover the bursting unit by excavation is alleviated.

In all cases replacement pipe should be of butt fusion welded medium density polyethelyne (M.D.P.E.) or H.P.P.E. with welding carried out prior to the bursting operation, so that all welds can be checked. Site supervisors should be fully conversant with variables affecting the welding process, such as:-

* Temperature at the surface of the heating plate.

* Time during which the pipe ends are pushed against the heating plant with pressure applied.

* Pressure used to push the ends onto the plate.

* Maximum time allowed to elapse between the pipe ends and their contact.

After welding M.D.P.E. pipe should not be dragged over the ground and rollers or slings should be used for transportation on site with ends capped to prevent any ingress of foreign matter.

Service connections can be made with specially designed fusion fittings.

A "window" is cut in the M.D.P.E. or PVC sleeve pipe wall, following excavation at service connections, ensuring that all exposed surfaces are maintained for the fusion operation.

Finally, as with lined pipes, the new replacement M.D.P.E. must be thoroughly disinfected where recommendations are for at least 20mg/litre of free available chlorine with a contact time of 24 hours. Once satisfying the water quality criteria, the main can be returned to service.

However, this length of time out of service can be impracticable and can be reduced to two hours by using a higher does of chlorine. (25mg/litre).

Conclusions

While there exists other methods of rehabilitation within the U.K. water industry, such as Epoxy Lining, (yet to be approved outside the U.K.), Slip Lining and Swage Lining, the two previously described methods still remain on the worldwide "stage" as the main methods of rehabilitation and replacement.

Though replacement moling remains the "new baby" of replacement technology, there is still much more that can be written about this and Cement Mortar Lining.

Alas, space does not permit this; although discussions may well afford a better understanding of the subjects covered in this paper. It will be appreciated, however, that the present position now reached, where modern plant machinery is used extensively, has only been achieved by benefit of experience gained on the thousands of kilometres of pipelines so processed in many countries of the world. Difficulties encountered have been treated as a challenge and overcome with constant research and development.

In these times of high costs it is important for engineers to give full consideration to trenchless or low-dig methods of rehabilitation and replacement, which give financial benefit when compared with meeting the cost of laying new or secondary pipelines.

Where new lines have to be laid, then, these same engineers should consider internal protection prior to laying, to alleviate future problems.

(Fig 1)

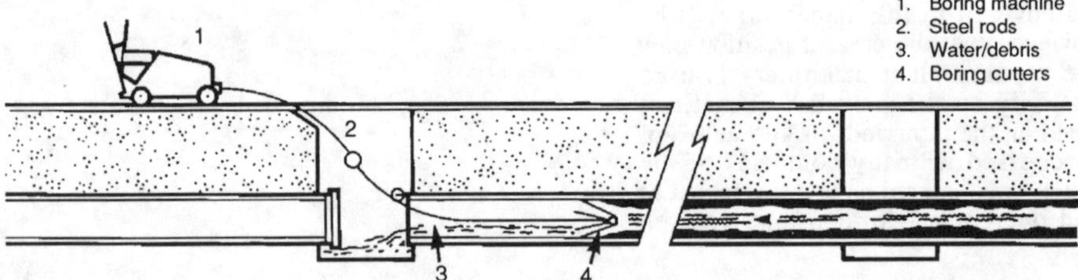

(Fig 2)

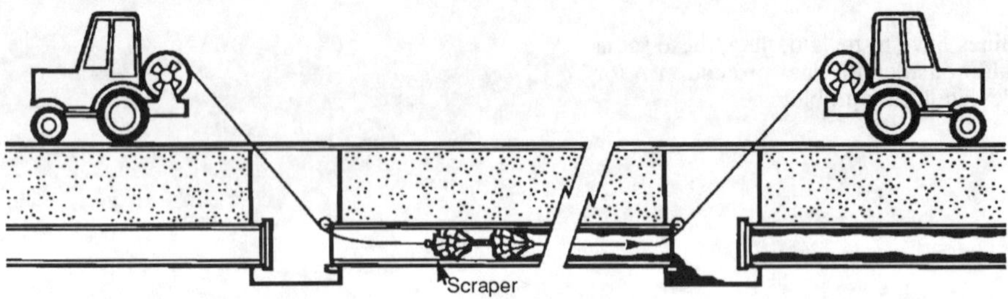

(Fig 3)

1. Make-up piece
2. Pressure scraper
3. Debris/water
4. Catch box
5. Control valve

1 2 3 4 5

(Fig 4)

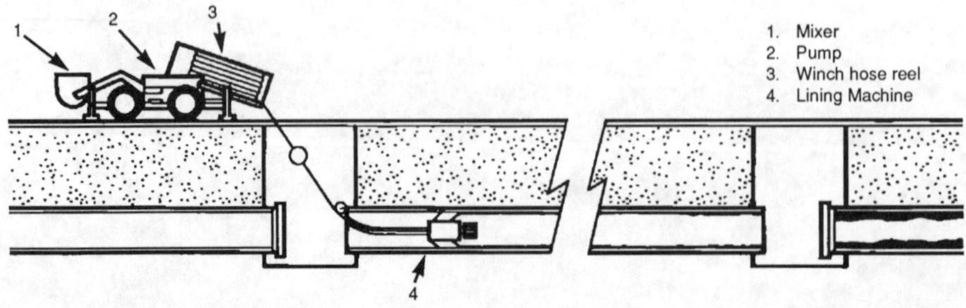

(Fig 5)

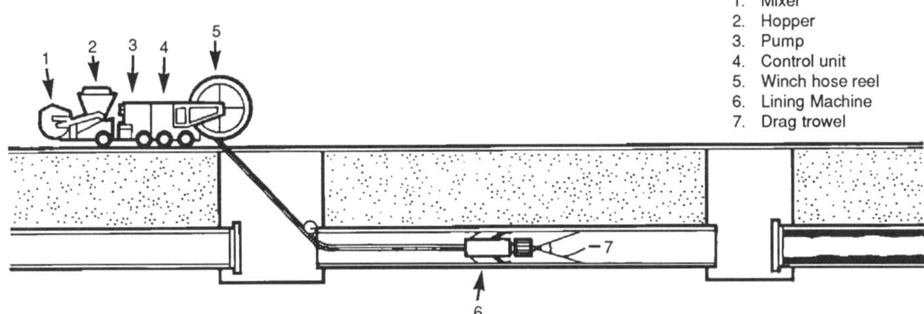

(Fig 6)

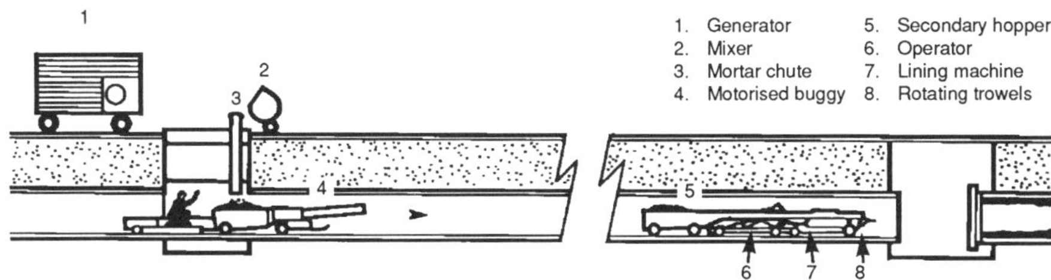

(Fig 7)

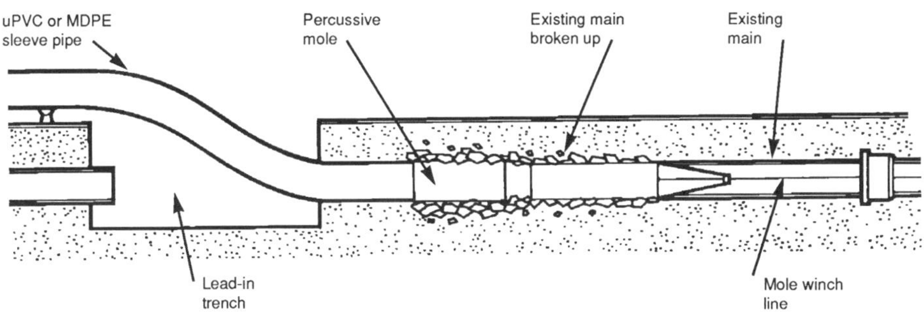

(Fig 8)

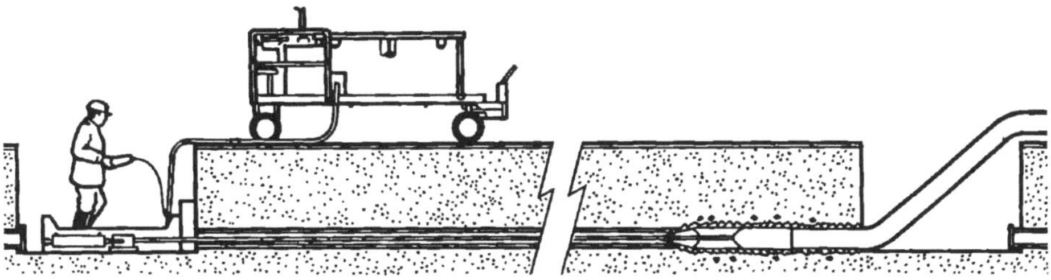

No Trenches in Town, Henry & Mermet (eds) © 1992 Balkema, Rotterdam. ISBN 90 5410 085 0

Rénovation des conduites de distribution d'eau

C. D. Murray
Tate Pipe Lining Processes Limited, Denton, Manchester, Royaume-Uni

(a) Resume

Cette présentation s'intéresse aux méthodes de rénovation ou de remplacement des pipelines les plus employées pour l'alimentation d'eau domestique. Ces conduites dont le diametre va de 80mm a un enorme 4500mm,peuvent être rénovées sur le terrain, dans un parc de stockage ou dans une usine. Les deux methodes decrites dans ce texte sont le "revatement par mortier au ciment" et l' " éclatement de tuyau" qui traitent respectivement de la renovation et du remplacement. Comme cela a été mentionné, il s'agit des pipelines et des aqueducs d'alimentation d'eau domestique où il a ete démontré, au fil d'années d'experience que toute forme de nettoyage sans protection interne ultérieure crée de graves problémes. Les revêtements par mortier au ciment se sont avérés éconoigues, durables et trés satisfaisants. Ces revêtements sont utilises tant en Grande-Bretagne qu'à l'étranger depuis plus de 50 ans.

Dans le cas où la corrosion est un problème sérieux et qu'il faut envisager le remplacement, l' " éclatement de tuyau" qui est la méthode la plus récente, peut être employé. Un système est décrit plus spécifiquement dans cette communication.

(b) Introduction

La vaste majorité des conduites existantes dans le monde entier sont en fonte et, pour la plupart, sont dans le sol depuis trés longtemps. Ces conduites sont corrodées à divers degrés et c'est là le facteur principal de leur défaillance, des fuites, de la mauvaise qualité de l'eau et des reductions de débit. Lorsque ces problémes se présentent, le service au consommateur n'est pas de bonne qualité, et la rénovation des conduites défectueuses contribuera à resoudre ces problémes. Actuellement, le revêtement est considéré comme la méthode de rénovation la plus importante à cause de son bon rapport efficacité/prix comparée avec les méthodes plus traditionnelles comme le remplacement à ciel ouvert, car on évite ainsi tout creusement de tranchée inutile ainsi que les ralentissements de circulation causés par cette méthode. Mais bien entendu, comme mentionné précédemment, certains tuyaux peuvent être très corrodés ou d'une capacité réduite au point que le remplacement est la seule solution.

C'est dans ces cas-là que la méthode de l'éclatement des tuyaux peut être la plus utile.

Il fut un temps où la seule protection des tuyaux était une mince couche de peinture bitumineuse qui ne procurait une protection que très limitée dans le temps. Ceci a poussé les utilisateurs et les fabricants à rechercher une protection de plus longue durée. L'une des premières solutions consistait en un revêtement de béton centrifugé de 4-5mm d'épaisseur ce qui est encore utilisé dans certaines usines et qui est décrit plus loin. Cette solution ne peut offrir une protection que sur de courtes sections de conduites et ne convient pas aux milliers de kilomètres de tuyaux déjà posés. En Grande-Bretagne par exemple, le nettoyage des sections courtes a été pratiqué dans des zones particulièrement affectée jusqu'à ce qu'on se rende compte que loin de résoudre le problème, la corrosion était en fait accélérée par le nettoyage à cause de l'eau et de la recrudescence des incrustations. Ceci, lié à des problèmes de fuites, a été la cause de nombreuses plaintes de la part des consommateurs et l'industrie s'est mis à chercher de nouvelles méthodes de rénovation.

(C) DISCUSSION

(l) Nettoyage

Il y a plusieurs méthodes pour le nettoyage sur le terrain des tuyaux corrodés et incrustés, le facteur déterminant celle qui doit être employée étant le diamètre et la longueur.

Dans toutes les méthodes, l'accès au pipeline est

obtenu par une petite excavation à chaque extrémité de la section à nettoyer. La section doit être la plus longue possible de facon à éviter le nombre d'excavations et par là-même les frais encourus. Dans la mesure du possible, les excavations doivent être situées sur des coudes.

Forage mécanique (de 75mm a 200mm)

Machine à forer mécanique - appeleé aussi "foreuse à crémaillère" qui entraine une série de tiges d'acier, avec tête hâcheuse fixée a l'extrémité menée et qui tourne à l'intérieur du pipeline. En raccordant des tiges supplémentaires, une section de pipeline pouvant aller jusqu'à 160 mètres de long, peut être nettoyée. L'écoulement de l'eau en sens inverse du hâchage permet de débarrasser les débris.

Fig. 1

Dragline (150 mm de diametre et au-dessus)

Les draglines sont composés de lames d'acier à ressort montées sur un bâti cylindrique. Ces lames sont montées en plusieurs rangées pour permettre de couvrir toute la paroi du tuyau. Les scrapers qui ressemblent à de gros ananas métalliques sont attachés à des cables à treuil à leurs extrémités et dragués d'avant en arrière jusqu' à ce que les incrustations soient éliminées de la paroi du tuyau.

Bien que de grosses quantités de débris soient extraites par cette méthode, le tuyau doit aussi être nettoyé par deux pistons-plongeurs à disque en caoutchouc pour éliminer complètement tout résidu, corrosion et eau de la conduite.

Fig. 2

Raclage sous pression (au-dessus de 450 mm)

Les conduites de grand diamètre sont, dans la mesure du possible, "raclées sous pression" et non pas raclées par dragage. Comme avec cette dernière méthode, l'équipement ressemble à un gros ananas métallique comportant des lames se chevauchant fixées sur un corps cylindrique.

L'ensemble piston est introduit dans le pipeline et transporté par un écoulement d'eau à une pression d'environ 2 bar qui est réglée par des clapets spéciaux fixés à un "boitier d'extrémité" en aval de la conduite. Chaque raclage peut être de 0,5 km de long.

Une certaine quantité d'eau passe le long du racleur ce qui permet d'emporter les débris d'incrustation et de corrosion extraits par le racleur vers le point de décharge. En général, il n'est pas permis de décharger cette eau chargée de débris dans le système du tout à l'égout et elle doit donc passer par un grand filtre situé entre le boitier d'extrémité et le point de décharge. Le grand filtre contient des déflecteurs qui retiennent tous les gros sédiments et permettent au liquide surnageant de se déverser dans le collecteur des eaux usées.

Fig. 3

Revêtement par mortier au ciment - Généralités

Il y de nombreuses années que Hazen et Willians dont la formule des coefficients d'écoulement est internationalement connue et acceptée, ont déclaré que "le fait que l'intérieur d'un tuyau en fonte devienne rugueux graduellement est l'un des phénomènes les plus courants affectant les conduites d'eau. C'est aussi l'un des plus difficiles à calculer. De manière générale, on peut dire que sur une série d'années, ce qui n'est pas long comparé à la durée de vie utile totale d'un tuyau, le fait que la surface devienne rugueuse et que la superficie diminue par rouille et tuberculisation puiffent atteindre le point où il faut une hauteur d'eau deux fois plus importante pour un volume donné que quand le tuyau était "neuf".

Cette théorie a été prouvée si souvent qu'il est reconnu universellement qu'une protection intérieure est la meilleure défense contre ce phénomène.

Après cette introduction sur l'historique du revêtement par mortier au ciment, voici une explication détaillee des diverses méthodes d'utilisation.

Dans le cadre de cette communication, les deux méthodes de revêtement par mortier au ciment decrites sont, d'une part celle qui est utilisée sur le terrain ou dans les conduites existantes, et d'autre part celle qui est utilisée dans les parcs de stockage ou dans les usines avant la pose dans le sol.

Revêtement sur le terrain de 80 mm à 200 mm)

La rentabilité de la rénovation des tuyaux de petit alésage est désormais reconnue par toute l'industrie de la distribution des eaux. La comparaison avec le prix de revient du remplacement d'une conduite dépend de la façon dont les frais dus à l'environnement et autres coûts supplémentaires sont calculés, cependant, il est généralement admis que la rénovation peut être effectuée pour environ le tiers du coût du remplacement.

Le revêtement est effectué en pompant un mélange 1/1 de sable calibré et de ciment Portland, d'un malaxeur en surface, appelé un "Carousel", a la machine à revêtir par l'intérmediaire d'un tuyau flexible. Cette méthode permet de garantir un débit constant.

Fig. 4

Le flexible et la machine à revêtir sont tirés dans

le tuyau au moyen d'un touret de flexible à treuil monté sur le "Carousel". La machine à revêtir pneumatique qui est au bout du flexible produit un revêtement régulier de finition "peau d'orange sur 360°. La surface peut être améliorée en tirant une truelle conique derrière la machine à revêtir.

Revêtement sur le terrain (de 225 mm à 660 mm)

Cette méthode de revêtement est semblable à celle pratiquée pour les petits diamètres excepté que la machine à revêtir et, le cas échéant, la truelle de raclage, sont ramenées vers un "Carousel" plus grand par un chariot à déplacement lent. Ceci améliore la qualité du revêtement fini et permet de revêtir une distance de 220 mètres entre les excavations.

Fig.5

Revêtement sur le terrain (avec accès pour homme)

Dans les conduites de très grand diamètre où l'accès d'un homme est possible, le revêtement est effectué en utilisant un "train" de machines opérées par un homme et qui est composé de trois véhicules: la machine de revêtement, la trémie de stockage intermédiaire et le chargeur mécanique.

Le mortier au ciment est mélangé à la surface et introduit dans le chargeur mécanique situé à une excavation pratique du tuyau. Lorsque le chargeur est plein, il est entrainé le long du tuyau à l'endroit où se trouvent la trémie intermédiaire et la machine à revêtir. Le mortier au ciment est alors transféré dans la trémie intermédiaire et pompé dans la machine à revêtir qui enduit le mortier sur la paroi du tuyau.

Fig. 6

L'opérateur est assis sur la partie transporteuse de la machine à revêtir, règle la vitesse de retour pour obtenir l'épaisseur correcte et fait aussi fonctionner les truelles rotatives qui suivent la tête de revêtement et égalisent la surface.

Revêtement dans parc de stockage (600 mm et au-dessus)

Conçu à l'origine pour répondre aux besoins du marché étranger, le "Bridge Plate" de Tate a été amélioré et est actuellement le système le meilleur et le plus économique pour le revêtement des tuyaux neufs de grand diamètre. Les principaux avantages de ce système résident d'une part dans la manipulation minimum du tuyau et d'autre part dans la taille réduite de l'équipement.

Ce système comprend un poste doseur central qui mélange le sable et le ciment 1/1 puis le ciment est transporté par un déchargeur ou tombereau vers le lieu du revêtement. De là, il est transféré dans une pompe à mortier de grande capacité montée sur le "tracteur principal" (c'est-à-dire un gros tracteur ou un chariot élévateur tout terrain). Sur le "tracteur principal" se trouvent également le "Bridge Plate" et la machine à revêtir qui sont situés du côté du tuyau.

La machine à revêtir circule rapidement vers l'extrémité du tuyau la plus éloignée et revient à une vitesse réglée en enduisant le mortier au ciment à ce moment-là. Les truelles rotatives fixées à l'avant de la machine égalisent la surface fraichement enduite pour donner une finition de haute qualité.

Le "Bridge Plate" et la machine à revêtir conviennent à divers diamètres et peuvent être changes rapidement en moins d'une heure. Les tuyaux de grand diamètre peuvent être revêtus en utilisant cette méthode.

Revêtement en usine (80-100 mm)

Une machine à revêtir qui se trouve au bout de la poutre portant le flexible d'alimentation en mortier et la ligne électrique, enduit par projection le revêtement "peau d'orange" sur le tuyau pendant que celui-ci est dans le "poussoir". La machine à revêtir est amenée, par déplacement rapide, à l'extrémité éloignée du tuyau, où elle est alimentée par un débit constant de sable et mortier au ciment 1/1. Le revêtement est effectué pendant le retrait et son epaisseur depend du debit du mortier et de la vitesse du retrait.

Aprés l'application du revêtement, chaque tuyau passe à une station d'enrobage dans une épandeuse centrifuge comportant des rouleaux récepteurs à fermeture de sécurité. La période d'enrobage est brève de façon à éviter la séparation du mélange et comporte des vibrations pour produire une finition de meilleure qualité.

Chaque tuyau est éjecté vers un rail de cure. La cure à lieu soit naturellement, soit dans un four spécial. Dans tous les cas, des obturateurs sont installés aux extrméités du tuyau pour maintenir l'humidité à l'intérieur.

Description et méthode

La cure du mortier au ciment passe par une réaction avec l'eau dont la température est critique. Les tuyaux doivent être revetus à une température ambiante de 2°C ou plus. si la température est de 2°C (ou a tendance à baisser), les tuyaux ne doivent pas être revêtus.

Le revêtement doit être maintenu humide pour l'opération de cure de façon à ce que les fissures de retrait soient minimales. Par conséquent, le tuyau doit être rempli d'eau jusqu'à un niveau de 0,2 mètre et chaque section doit être fermée à l'aide d'obturateurs en plastique.

Après le revêtement et la cure, des fissures peuvent apparaître. Il est désormais bien établi que les fendillements et les fissures capillaires se réparent d'elles-mêmes par le biais de deux phénomènes. D'abord, quand le revêtement est complètement sec, s'assurer que le retrait est à son maximum. Lorsque la canalisation est remplie d'eau, l'enduit va se dilater d'environ 0,1 à 0,15%. Deuxièmement, toute fissure restante se fermera de manière autogène par l'hydratation du ciment déshydraté du mortier et/ou le dépôt de sels provenant de l'eau qui s'écoule.

Le retrait du revêtement au mortier rend inévitable les fissures capillaires. Les fissures capillaires et les fissures apparaissant à la surface de manière sporadique ne sont admissibles que jusqu'à la largeur stipulée dans les normes de l'A.W.U.A. C602.89.

A part des granulats qui peuvent remonter à la surface de manière sporadique, toutes les surfaces doivent être lisses et unies. Cependant, quelques ondulations ou rainures sont tolérables à condition que l'épaisseur minimum spécifiée du revêtement soit respectée.

Tout travail effectué sur les endroits endommagés doit répondre aux normes de l'A.W.W.A.

Dans certains cas d'endommagement, il peut être nécessaire d'extraire tout le revêtement, de le nettoyer et de le repulvériser. Là aussi, toutes ces opérations doivent être effectuées suivant ces normes.

Le mortier doit être mélangé dans un équipement convenable, le sable répondant aux spécifications prescrites et le ciment étant de qualité Portland suivant les normes britanniques.

Tout revêtement qui n'est pas réalisé suivant ces spécifications ne doit pas être accepté par l'acheteur.

Remplacement souterrain ou "eclatement de tuyau"

Ce système de remplacement de tuyau doit être préconisé par les ingénieurs qui doivent remplacer une conduite défectueuse soit à cause d'une corrosion externe constante où la technique du revêtement est inefficace, soit parce qu'il faut augmenter la capacité du tuyau d'origine.

L'une des trois techniques suivantes agréées peut être utilisée pour faire éclater des conduites existantes. La première consiste à utiliser une taupe à percussion qui est composée d'un trépan horizontal à mouvement alternatif agissant à l'intérieur d'un cylindre monté dans une tête conique. Cette tête peut être équipée de "points de concentration des tensions" pour accroitre la résistance à la rupture aux raccords ou aux branchements. L'effet alternatif produit à la fois le mouvement en avant et les forces radiales qui brisent le pipeline existant et comprime les fragments du tuyau dans le sol qui est tassé à cet endroit. Le mouvement en avant de la taupe est aidé par un treuil.

La deuxième technique est une autre adaptation de l'effet percutant, mais à la place de l'effet de "marteau", c'est une "pale" à commande hydraulique qui est constamment abaissée et élevée et qui brise ainsi le tuyau à mesure qu'elle avance dans la conduite.

Fig. 7

Troisièmement, et plus récemment, le système "Ram Rod" consiste en un piston hydraulique situé dans la première excavation et qui est poussé par une série de tiges à tube-incise.

Fig. 8

A mesure que le piston passe dans la conduite, le tuyau est brisé et les morceaux sont comprimés dans le sol. Lorsque le tube-incise atteint l'excavation "éloignée", il est démonté et un épandeur avec le nouveau tuyau M.D.P.E. et fourreau sont attachés. Le dispositif hydraulique repart et tout le système est ramené en arrière; dans le cas de conduites de grand diamètre, l'éclatement à lieu pendant le mouvement de retour.

Le gros avantage de cette troisème méthode est qu'il n'y à pas de vibrations dans le tuyau. C'est-à-dire que l'éclatement peut désormais avoir lieu sans déranger les autres canalisations qui pourraient être à proximité. D'autre part, en cas de panne, comme la force motrice est un dispositif à piston qui fonctionne à l'extérieur du tuyau, le besoin de récupérer la machine d'éclatement par excavation est éliminé.

Dans tous les cas, le tuyau de rechange doit être en polyéthylène moyenne densité (M.D.P.E.) ou en polyéthylène haute densité (H.P.P.E.), soudé bout à bout par fusion, le soudage étant effectué avant l'opération d'éclatement de façon à ce que toutes les soudures puissent être vérifiées. Les responsables du chantier doivent connaitre parfaitement les variables suivantes dont dépend l'opération de soudage:

* Température à la surface de la plaque de chauffage.

* Durée pendant laquelle les bouts des tuyaux sont pressés contre l'appareil de chauffage.

* Pression utilisée pour pousser les bouts contre la plaque.

* Durée maximum admissible entre entre bouts leur contact.

Après soudage, le tuyau M.D.P.E. ne doit pas être trainé au sol; des galets ou des élingues doivent être utilisés pour le transporter sur le chantier, les extrémités étant protégées pour éviter la pénétration detout corps étranger.

Les raccords de branchement peuvent être effectués avec des pièces de fusion spéciales.

Une "fenêtre" est découpée dans la paroi du tuyau M.D.P.E. ou à fourreau P.V.C. à la suite de l'excavation aux raccords de branchement, garantissant que toutes les surfaces exposees soient maintenues pour l'opération de fusion.

Finalement, comme avec les tuyaus doublés, le nouveau tuyau M.D.P.E. doit être bien désinfecté selon les recommandations suivantes : au moins 20 mg/litre de chlore avec un temps de contact de 24 heures. Une fois que les critères de qualité de l'eau sont satisfaits, la conduite peut être rebranchée.

Cependant, cette durée hors service peut ne pas être pratique et peut être ramenée à deux heures en utilisant un dosage de chlore plus fort (25 mg/litre).

Conclusions

Bien qu'il existe d'autres méthodes de rénovation dans la distribution de l'eau au Royaume-Uni, par exemple le rêvatement époxy (qui n'est pas encore homologué en dehors du Royaume-Uni), le revêtement par coffrage glissant et le revêtement par matrice, ces deux dernières méthodes sont actuellement les plus couramment utilisées dans le monde.

Bien que le remplacement par taupe soit encore le "petit dernier" de la technologie de remplacement, on pourrait écrire bien plus à ce sujet ainsi que sur le revêtement par mortier au ciment. Malheureusement, les limitations imposées dans le cadre de cette communication ne le permettent pas, mais des discussions pourraient contribuer à mieux comprendre ces sujets. On appréciera que la position actuelle, où un outillage moderne est très largement utilisé, n'a pu être atteinte qu'au prix de l'expérience acquise par l'installation de milliers de kilomètres de pipelines dans de nombreux pays. Les difficultés qui se sont présentées n'ont pu être surmontées que grâce à un effort constant dans la recherche et le développement.

A une époque où les prix de revient sont élevés, il est important que les ingénieurs prennent sérieusement en considération des méthodes de rénovation et de remplacement sans creusement de tranchée ou avec un minimum d'excavation et qui sont avantageuses du point de vue financier comparées à la pose de pipelines neufs ou de récupération.

Dans le cas où des tuyaux neufs doivent être posés, les ingénieurs doivent également envisager une protection interne avant la pose pour éviter les problèmes ultérieurs.

(Fig 1)

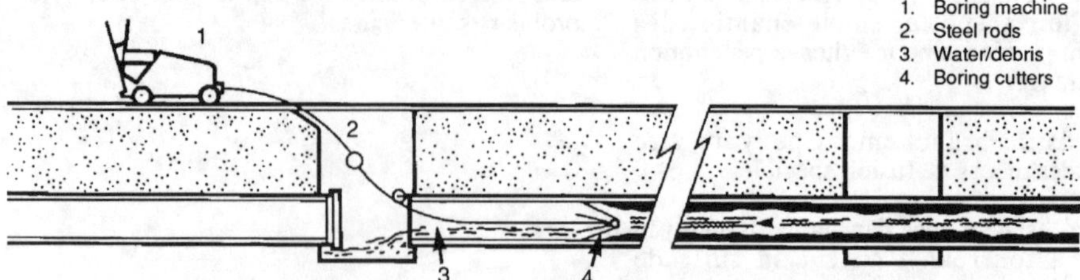

(Fig 2)

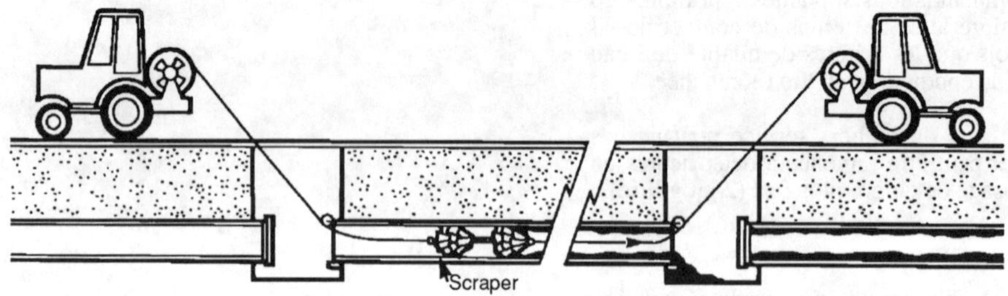

(Fig 3)

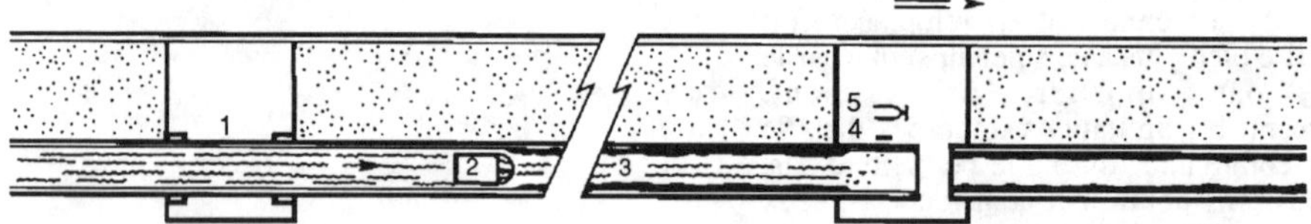

(Fig 4)

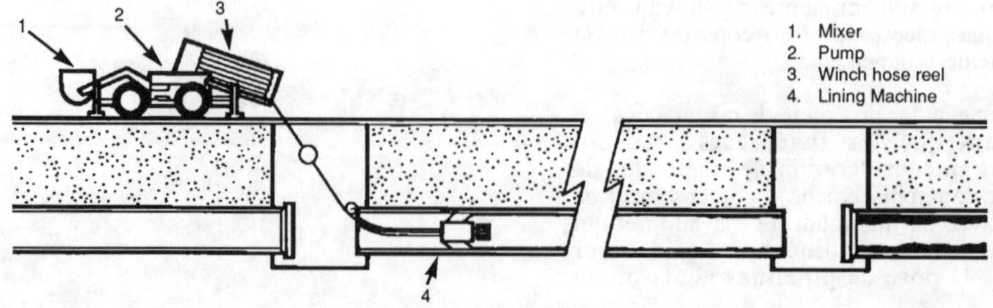

(Fig 5)

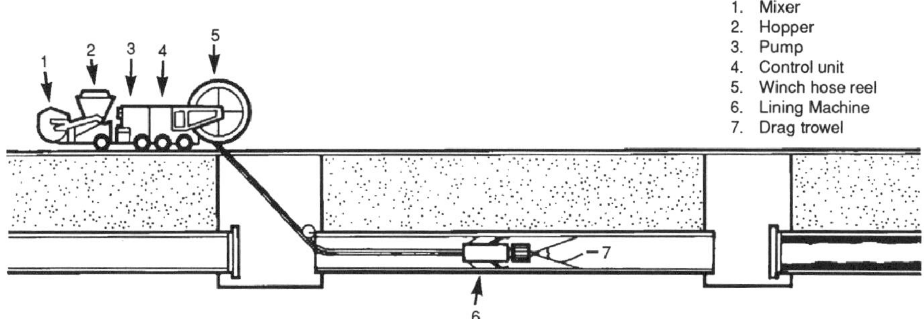

(Fig 6)

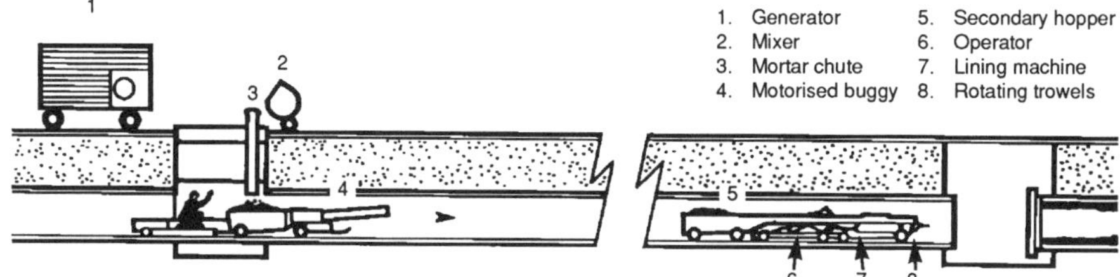

(Fig 7)

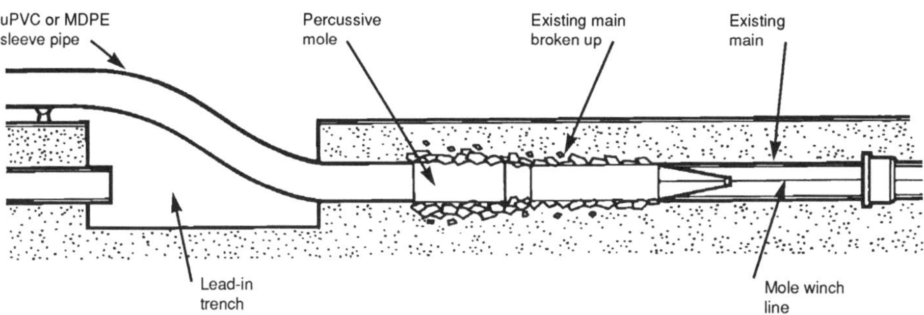

(Fig 8)

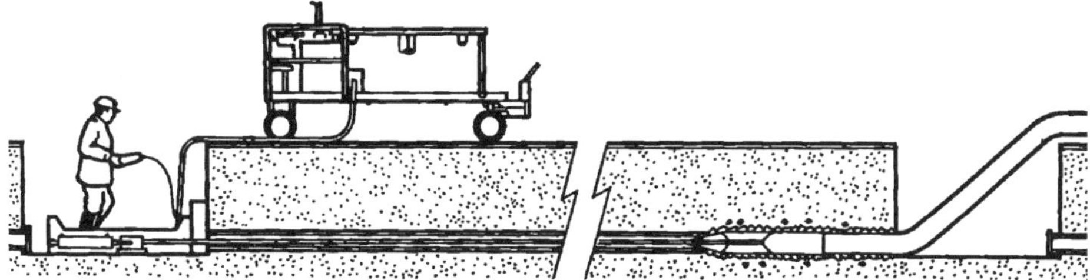

No Trenches in Town, Henry & Mermet (eds) © 1992 Balkema, Rotterdam. ISBN 90 5410 085 0

Benefits to society from experience gained in the use of an in situ repair system for man entry sewers and culverts

L. Decelle
Ferro Systems Europe, Belgium

P.J. Lancaster
Northumbrian Water, UK

G. Singh
University of Leeds, UK

A. B. Venn
Ferro Monk Systems and Sewer Renovation Federation, UK

ABSTRACT: This paper highlights the benefits of insitu sprayed ferrocement since its significant reintroduction in 1984. This is a natural follow-up to the two previous presentations at the 1988 and 1989 No DIGS. The paper shares with the delegates the results and benefits of repairing sewers and other underground structures in varying stages of dereliction, arrived at by numerous field experiences in cities and towns throughout the UK, Brussels, regions of Paris and as far south as New Zealand. In the first part various issues relating to the surveying, operational, technical and safety aspects are addressed. The second part describes through the case studies, the aspects that are of particular interest to the the clients, i.e. maximising existing sewer sizes, minimising excavations in sensitive trafficed and pedestrian areas,and readily dealing with obstructions. The various case studies are illustrated with photographs.

1. INTRODUCTION

At NO-DIG 88 two of the authors (1) pointed out the construction and design problems with the existing materials and methods which produce adverse effects on the end use, performance and cost. They followed this up with a presentation at the NO-DIG 89, discussing the criteria for the structural design of linings and coatings and highlighting the pitfalls of misconceptions and misuse of theoretical ideas (2).

It has long been accepted that effective sewer renovation requires systems that will add to and improve the structural design, flow and storage capability of the original system and therefore save and use again the existing 'hole in the ground'. Most main carrier sewers, normally large enough for man to enter, follow the line of old streams giving irregular changes in direction, level and size. It is now well estabilished that these sewers must maintain their size and/or rates of flow to suit modern requirements and to save the cost and disruption of renewal or laying additional pipework. The design of ferrocement has been explained in previous papers (1,2) and is now well detailed in WRc's Sewerage Rehabilitation Manual Addendum No.2(3). The Ferro Monk system of ferrocement renovation is an established process having its own W.I.S. Number - 4-12-06 (3) and has been used in many miles of repair giving confidence to all its users. An additional fact is that this material has little to no rebound and gives a minimum capacity saving of 25% on most other systems.

Following the publication of British Standard No. 8005 Part 5(4), Safety in the rehabilitation of sewers was well highlighted and acted upon by the UK's Health and Safety Executive and the newly formed Sewer Renovation Federation(5). Ferrocement conforms to all the material requirements and the application method further highlights its own safety regime in that existing manholes can be used and subsequently renovated, and by using specialist pre-formed invert units over-pumping is not a requirement, thus seeing the end of open cut excavations together with the hazards and disruptions caused by overpumping pipework being layed along footpaths and roads.

The objective of this paper is to

highlight the benefits of insitu sprayed ferrocement since its significant re-introduction in 1984 and it forms a natural follow-up on the two previous presentations. It shares with the delegates the results of the authors' experience with the benefits of repairing sewers and other underground structures, in varying stages of dereliction , which have been arrived at through numerous field experiences in cities and towns throughout the UK, Brussels, Paris and other parts of the world including New Zealand. In the following are presented first a few of the numerous case studies describing, briefly, the aspects that are of particular interest to the clients. These include minimisation of (a) size reduction , (b) excavations in the sensitive and busy areas, and (c) efforts required to overcome obstructions.

SOME CASE STUDIES

1. Newcastle upon Tyne - Agent for Northumbrian Water plc

This area has particular problems due to it being founded geologically on superficial sands and clays which vary from 30m to 0 in depth together with coal measures consisting of alternative bands of sandstones and shales with interbedded seams of coal totalling 1100 metres in depth at the greatest thickness of strata. The superficial deposits were eroded by several fast flowing watercourses which flowed southwards to the deep Tyne gorge. As the city expanded the watercourses were culverted and the valleys, up to 130m wide, infilled to form the general pattern of this modern city.

The culverts, which form the backbone of the current trunk sewerage system were constructed piecemeal. Their direction both horizontally and vertically deviated wildly between manholes and probably more importantly from a renovation point of view, had numerous instances of major dimension and shape changes within the sewer lengths. Statutory services also gave major problems by simply passing straight through sewers at whatever angle or level they were encountered. Ferrocement thus proved to be an ideal material for the problems of Newcastle. A typical sewer problem is highlighted in Plates 1 & 2.

The 1150mm x 750mm two brick thick egg shaped sewer at Redheugh Bridge Road with a stone invert was in an unstable state and was in need of urgent renovation. Vibrations from heavy traffic some 4m above were causing movement of bricks in the soffit and the invert which had failed throughout its length of 300 m. This situation, together with initial cleaning works to the brickwork revealed a 100 year old, 600mm dia. gas main some 1m above and parallel to the sewer at a point almost midway in its length where the sewer changed direction to go under a main line railway bridge. The soffit failure was localised and in consultation with British Gas the sewer, either side of the gas main, was stabilised using ribs and lagging. The whole sewer was then re-inverted using precast ferrocement units and backgrouted into the original stone invert. The area around the gas main and for a length of 20m. either side of it was repaired using a Type II insitu 22mm thick ferro-cement coating reinforced with two layers of 12mm x 12mm x 16 gauge galvanised mesh. The remainder of the sewer was similarly renovated but designed on a Type 1 basis with a total thickness of 19mm using one layer of reinforcement.

Additional problems on this length included three major shape changes, two service pipes passing through and across the sewer and it was tidally influenced for the last 20m. of its length. The repair method and system guaranteed maximum storage capacity for the sewer and easily dealt with all the obstacles and design change requirement.

2. Rue d'Angleterre, Brussels, Belgium

This road is in the centre of Brussels and is heavily trafficed. There are a number of small business workshops, offices and a school for the blind along its length making it extremely important that the trenchless technology used would not interfere with the traffic or pedestrians. All work had to be done from existing manholes with minimum working areas. The sewer was brick egg shaped 1200mm high x 800mm wide and the damaged section was 90 m. long approximately twothirds of the length of the manhole to manhole length. Construction of a new section of the Metro had been responsible for failure of the sewer which had been propped to stop any settlement of the road as shown in Plate 3. This propping gave great problems to the Authority in selecting a suitable repair system as removal had to be minimal until repairs had been carried out. The only solution was the use of precast and insitu ferrocement, and following discussions with the Authorities and the Engineers for the Metro's insurance company, the final scheme was that the props were removed in 8 m. lengths whilst the invert units were placed and back grouted. The insitu ferrocement, designed as a Type 1 system was 25mm thick and had two layers of 12.5mm x 12.5mm x 16 gauge and was completed in the 8m lengths. The props were reinstated until the repair system developed its working strengths.

3. Quai Blanqui in Ivry sur Seine, Paris, France

This sewer was approximately 150 m. in length, egg shaped, 2m high x 1m wide and constructed in a material called 'Meuliere' a small chalk stone cobble which is typical in Paris. The problems were a failed invert causing further failure in the walls of the sewer. This sewer was being used to check the suitability of other renovation systems and the major problems experienced on this contract came from the number of people passing through the work to carry out these different systems and schemes. The design was based on a Type 1 requirement and was for two layers of 12.5mm square mesh of 16 gauge and a total thickness of 25mm. The pre-cast invert units were manufactured in England to take the dry weather flow conditions and shipped out to the site. The work was controlled by the 'Direction des Services de l'Eau et de l'Assainissement' of the Val de Marne with the assistance of 'Abrotec', and ferrocement was proved to have the best results of all the tried systems when comparing price, speed of execution and when considering the elimination of overpumping, and disruption to the public.

DISCUSSION

Numerous other cases, which cannot be described here because of space limitations, have lead the authors to develop strong confidence in the technical and financial benefits of this insitu repair system. This confidence is backed by detailed analytical studies, extensive laboratory investigations and the successes of the numerous completed contracts.

Whilst experience from the field is accumulating further fundamental analytical and experimental work is being carried out at Leeds University with the aims of extending the applications of this versatile material as well as of improving its durability in more severe chemical and mechanical working environments.

There is no need to carry out expensive surveying operations and the requirement to carry out difficult three dimensional proving of the sewer size for system acceptability. Survey is limited simply to finding the size of invert required to take the normal dry weather flows and the limiting circumferencial dimensions from which to order and measure the reinforcement. Silt in the invert does not have to be removed prior to carrying out the repair, which again saves costs and disruption.

CONCLUSIONS

Successful application of Ferro Monk Systems, using ferrocement, in many miles of sewers has led to the acceptance of the system as an established one with its own Water Industry Specification No. 4-12-06.

The versatility of the material and the Ferro Monk System allows it to be designed for uses in a very wide variety of applications in and outside the limitations of the WRc Manual. This is highlighted by the fact that the system has been used for sizes starting from as small as 800 x 500mm and upwards, the largest todate being a 10m. wide culvert, 7m. high.

The system has been shown to provide social, economical and technical benefits of much greater magnitude than provided by other systems.

REFERENCES

1. Venn, A.B. and Singh G. The development of ferrocement for use in sewers and tunnels as a renovation and rehabilitation material, NO-DIG 88, Washington, USA, October 1988.

Figure 1. Detailing obstructions encountered due to services crossing sewer.

Figure 2. Detailing changes of shape and size.

Figure 3. Typical propping system used in Brussels.

2. Singh G., Venn A.B., lp, Lilian and Xiong G.J., Alternative Material and Designs of Renovating Man-Entry Sewers, NO-DIG 89, London, 1989.

3. WRc Sewerage Rehabilitation Manual Addendum No.2 Ferrocement linings and coatings (Ferro Monk Systems) 1990.

4. B.S.I. 8005 Sewerage Part 5 - Guide to Rehabilitation of Sewers 28th October 1990.

5. Sewer Renovation Federation - Manual of Working Practice, Warrington, UK July 1992.

No Trenches in Town, Henry & Mermet (eds) © 1992 Balkema, Rotterdam. ISBN 90 5410 085 0

Localised sewer repairs in Northumbrian Water

Brian Syms
Northumbrian Water Limited, UK

Paul Hayward
Amtec Bunting Limited, UK

ABSTRACT: A method of assessing the suitability of various trenchless sewer renovation techniques has been devised by Northumbrian Water, and trials have been conducted of several local repair systems. The procedure has shown the effectiveness of local repairs in appropriate situations.

THE CLIENT'S VIEWPOINT

NWL is responsible for approximately 13 120 km of sewers of varying size, depth and material and some 240 000 manholes, the operation and maintenance of which lies with twenty three Sewerage Agents. Approximately 70% of the sewers are 225mm diameter or less and some 25% lie in the next size range up to 450mm dia. Analysis also indicates that over 90% of sewers are made of clayware or concrete. Similar to any other U.K. sewerage utility our region covers rural to urban populations and has a population of 2.56m connected to the public sewerage system, which represents 98% of the region's total.

In cognizance of the Sewerage Rehabilitation Manual (SRM) (1), Northumbrian Water instigated a programme whereby all man-entry sewers (i.e. sewers greater than 900mm diameter) were surveyed in the period 1985 to 1987 after which a programme to survey all critical or strategic sewers was started. In this region, critical sewers (A and B) represent 33% of the whole public sewerage system. In 1989 the rate of survey was stepped to survey 600km per year with the aim to complete all category A and B sewer surveys by 1993. More recently the survey programme has been now complimented by the survey of 40 000 manholes per year in order to survey all known manholes by 1997. In order to manage such a comprehensive survey, it is essential that a unique manhole referencing system is used. Such a system has been adopted and is used together with pipe length reference (PLR), both of which are described in the STC Report No. 25 (2).

Surveys are analysed by computer to provide a structural grading in line with the SRM. The grades shown in Tables 1 and 2 are as used by Northumbrian Water Limited and have been produced by careful consideration of SRM descriptions of defects and situations together with our own more recent observations and experience.

The package also enables the production of concise studies for particular areas so that both large and small contracts can be organised. The current project is limited to pipe sewers in the size range 225mm to 800 mm diameter, which also have a worst case of structural grade 4 defects, these parameters being part of the computer analysis. The software programme is used to produce a list of prospective lengths for rehabilitation in PLR order so that a technical assessment of defects can be made and a repair option identified. The plotting of repair lengths onto plans in PLR order may also influence the final rehabilitation process e.g. if consecutive lengths are identified as being in need of renovation or repair, it may be more economical to consider continuous manhole to manhole renovation.

To assist the technical assessment a number of tables have been produced to enable a simple but reasonably accurate cost breakdown, which can be appraised for the various methods being considered. It should be pointed out that it is not always the cheapest option that is preferred since the location of the sewer may preclude an open-cut option, for instance if the sewer lies under a river, a motorway or a busy cross-roads. Table 3 shows a number of the assumptions and parameters provided to assist the analysis. Note that only sewer lengths with not more than 25% of their lengths affected by defects are included in the current programme of work and in order to assist towards the identification a structural analysis is also provided. Simple arithmetic quickly identifies those lengths which qualify. Table 4 shows one section of sewer (Sewer A) where defects (breaks, fractures, etc) represent up to 100% of the whole length, and another section (Sewer B) where defects represent 6%. The use of full CCTV reports, photographs and video recordings complete the technical assessment for each of the chosen lengths.

Generally grade 5 defects are not considered for renovation by the methods being considered. Consequently, as stated above, grade 4 defects have been identified as the main area of interest, however, grade 3 and 2 defects are also considered when comparing localised repairs against manhole to manhole renovation.

Using these guidelines etc. the analysis then proceeds to identify which method of repair is to be used. There are four basic methods being considered:

- localised open-cut repairs
- localised robotic repairs
- localised patch repairs
- manhole to manhole renovation

although other techniques are continually being assessed, such as re-rounding and joint testing and sealing.

N.B. Full length renovation may be considered even where defects total <25% of the sewer length if technical or economic reasoning justify such a decision.

It should be noted that a single defect in a sewer length will affect the structural grade of the whole sewer, but one localised repair will restore the designation of the entire length. Now that the privatised U.K. water and sewerage companies are mindful of their asset values which are reflected in their balance sheets, localised repairs are a quick and economic means of restoring infrastructure value whilst at the same time providing a sound engineering solution to deteriorating pipework.

Over the past 18 months site experience has shown that robotic methods are not appropriate in certain situations and for the following defects we would insist on patch repair or local excavation :-

i) Multiple cracks or fractures where the number of cracks or fractures exceeds two.

ii) Holes or breaks where the depth of the defect is greater than the pipe thickness or exceeds 100mm x 150mm in surface area.

Once the analysis has been completed, the results are appraised so that separate contracts can be awarded or the area of work forms a part of an annual term contract. In order to give each of our Sewerage Agents a "flavour" of the work each year, the current term contracts are broken down into two week periods.

Before the work on site commences, the CCTV video recording and reports are checked by the contractor to ensure that all identified defects can be repaired. In addition, a quick CCTV resurvey is completed to determine whether defects had deteriorated or that an additional number of defects had not materialised. Our experience has shown that critical sewers are deteriorating, structurally, at such a rate that only CCTV surveys 3 years old or less are considered for this project. If either occurs, the length is re-appraised by the client Engineer.

THE CONTRACTOR'S VIEWPOINT

The objective of the contractor should be to offer the customer a range of services to solve a variety of problems. There is no cure-all solution. Many companies offered, and even more now offer, manhole-to manhole relining techniques, but there were relatively few local repair methods available. Rather than looking at the renovation systems available, we started by assessing the customers' problems of which there were several different types:-

a) The removal of intrusions e.g. intruding laterals, steel reinforcement, roots, encrustation, etc.

b) Repairing circumferential cracks/fractures, isolated defective joints, short longitudinal cracks, and poorly made connections.

c) Repairing breaks, multiple cracks/fractures, and longer longitudinal cracks/fractures.

d) Re-rounding and repairing pipes with deformation in excess of ten percent.

e) Testing and sealing pipe joints where the pipe structure was otherwise generally sound.

Robotic Repair

Problems (a) and (b) indicated the need for robotic grinding or drilling equipment to remove intrusions and to mill out cracks and fractures, followed by a structural repair system to restore the pipe to at least its original strength. After evaluating techniques which claimed to meet these objectives, we selected the KA-TE system which appeared to offer the best combination of technical competence and productivity.

KA-TE uses one robot for drilling and grinding, and another for the application of an epoxy filler which is designed to bond to wet substrates. The equipment is powerful and has a high productivity rate. The robots can work in sewers of between 200 and 800mm diameter.

The first stage in the process is to cut off intrusions, and to mill out cracks or fractures to form a slot of approximately 25mm depth and 30mm width. This offers a good key for the repair material.

The second stage is to fill the milled-out cracks with a specially formulated epoxy compound which provides greater strength than the original pipe. Any gaps between defective connections and the main sewer can also be filled with this material.

If necessary, the repaired areas can be burnished by the grinding robot once the epoxy has cured.

Sleeve Repair

Problem (c) is more difficult to deal with by robotic methods which work by re-bonding the existing pipe fabric. Defects which are more serious or extensive can often be better dealt with by installing a structural sleeve within the existing pipe.

Various such systems exist, but the one used most extensively within Northumbrian Water is known as Econoliner. This uses a composite material comprising a layer of needle-felt sandwiched between two layers of glass fibre woven rovings. The material is impregnated with an epoxy resin, and is scrolled around an inflatable packer. When in position, the packer is inflated with water, and hot water is circulated to cure the resin. The system uses a patented flow-through packer so that sewage flows can usually be maintained during installation, and over-pumping is seldom necessary.

The thickness of the sleeve is less than 5mm, so there is minimal reduction of pipe bore. Sleeves of up to 3 metres long can be installed, and longer sections can be repaired by using two or more sleeves in series.

Pipes from 100mm diameter upwards can be repaired by the Econoliner method, though the size of the packer is a limiting factor with larger diameters.

Re-Rounding Systems

Pipes with more than ten percent deformation are generally considered unsuitable for renovation, mainly because the stresses in the liner would be much greater than in a circular pipe under similar load. The reduction in cross-sectional area is also a consideration.

A method of re-rounding pipes prior to relining or local repair has therefore been developed. As at June 1992, the working prototype has been used to re-round 300mm diameter pipes, and further production models are being built for other sizes.

The system, known as Sirkit, uses a hydraulic ram to expand a series of "petals" which act on the deformed pipe. The design of the expander device ensures that the petals move out equally, and the re-rounded pipe is therefore circular.

A stainless steel clip is scrolled around the device prior to expansion, and this clip acts as a permanent shield to retain broken pipe fragments until the sewer is relined or repaired. In practice, these clips have been left in position, without detriment, for several weeks prior to the completion of the repair.

Joint Testing and Sealing

Pipelines which are structurally sound may be defective due to leaking joints. This leads to overloading of the sewerage network and high treatment costs if there is infiltration, or to pollution of the subsoil and aquifers if there is exfiltration. In either case, the pipe bedding and surround may be eroded, causing unseen voids and the possibility of catastrophic failure.

The Amtec joint testing and sealing system is used to pressure test each joint individually to a pressure of 0.7 bar. Joints which fail are sealed by injecting a water-reactive gel, of which various types are available, and the joint is then re-tested. Test pressures are displayed on the monitor screen along with the CCTV image and chainage.

Depending on the extent of any voids around the pipe, and therefore the amount of sealant which has to be injected, the process is fairly quick and economical. When sealing against infiltration, the pay-back period is often quite short because of the reduced loading on pumping equipment and sewage treatment facilities.

The system can also be used to seal minor cracks, although it is not claimed to be a structural repair technique.

TABLE 1 - STRUCTURAL CONDITION GRADES

GRADE	IMPLICATION
5	Collapsed or collapse imminent
4	Collapse likely in foreseeable future
3	Collapse unlikely in near future but further deterioration likely
2	Minimal collapse risk in short term but potential for further deterioration
1	Acceptable structural condition

Economic Considerations

Unless the sewer is relatively shallow, outside the highway, and not overlain by other services, renovation is usually more cost-effective than open-cut replacement or repair. In the UK, the New Roads and Street Works Act, which is being progressively implemented, will increase both the cost and the aggravation of excavation in the highway.

Trenchless local repair is generally quicker and less disruptive than manhole-to-manhole lining, and the mobilisation costs are less. However, the economic assessment of the various techniques must take into account:-

- The nature of the defects, and the difficulty of repairing them by trenchless methods.
- The number of defects, and hence the number of robotic repairs and/or repair sleeves required.
- The overall length of the sewer, and the cost of a manhole-to-manhole liner.
- The number of connections which would need to be reintroduced if a full liner were installed, but could be avoided if using a local repair system.
- The location, and any access restrictions, which may render certain techniques unsuitable.
- The possibility of further defects appearing in unrepaired sections of the sewer.
- The real cost of open-cut repair or replacement, including highway reinstatement, compensation to landowners, and the time expended by the Engineer in design, supervision and administration.

Quality

The question is often asked, "How long will the repair last?" It is, of course, impossible to give an accurate answer, since renovation techniques in general, and trenchless local repair techniques in particular, have not existed for long enough to have the same track record as traditional methods.

However, some of the earliest Insituform linings, installed over twenty years ago, have recently been examined in detail and found to exhibit no sign of deterioration. Local repair methods such as the Econoliner system use materials which should be stronger and more stable than those used in soft lining systems, so their longevity should be at least as good.

It is also worth remembering that, if the durability of traditional pipe materials such as concrete and clayware was as good in practice as it is in theory, there would be fewer sewers in need of renovation. Many pipelines which are now being renovated are less than twenty years old.

Our opinion, which is necessarily based on qualitative rather than quantitative assessments, is that a properly renovated pipe should have at least the same life expectancy as a new pipe, and, depending on the type of renovation system chosen, may be far less susceptible to future damage.

TABLE 2 - DEFECTS : STRUCTURAL GRADES

Defect	Grade	Defect	Grade
Longitudinal Cracks	3	Light Encrustation	1
Circumferential Cracks	2	Silt Debris	1
Multiple Cracks	3	Grease Debris	1
Longitudinal Fractures	3	Other Debris	1
Circumferential Fractures	3	Obstructions	1
Multiple Fractures	4	Surface Damage: Spalling	1
Breaks	4	Surface Damage - Wear	1
Deformation	5	Faulty Connections	2
Small Displaced Joints	1	Faulty Junctions	2
Medium Displaced Joints	2	Connections	1
Large Displaced Joints	3	Junctions	1
Small Open Joints	1	Line Changes	1
Medium Open Joints	2	Level Changes	1
Large Open Joints	3	Collapses	5
Fine Roots	1	Branches	1
Mass Roots	1	Displaced Bricks	3
Tap Roots	1	Horizontal Deformation	5
Infiltration - Seeper	1	Dropped Invert	4
Infiltration - Dripper	1	Vertical Deformations	5
Infiltration - Runner	1	Missing Bricks	4
Infiltration - Gusher	1	Mortar Missing - Medium	3
Scaling Encrustation	1	Mortar Missing - Surface	2
Heavy Encrustation	1	Mortar Missing - Total	3
Medium Encrustation	1		

The Future

No-dig local repair systems are in their infancy, but are already proving valuable additions to the rehabilitation armoury. Within Northumbrian Water, they are proving an effective means of achieving rapid results.

New methods, particularly in the fields of lateral detection and cutting, are currently being developed, as well as enhancements to existing techniques. The increasing demand for these systems will inevitably lead to more innovation and lower costs, to the benefit of both the client and the contractor.

TABLE 3 - GUIDELINES FOR LOCALISED REHABILITATION OF SEWERS

CRITICAL SEWER REHABILITATION

GUIDE TO REMEDIAL WORK - SEWER SIZE 225mm TO 800mm

STRUCTURAL GRADE 4

MATERIAL - ANY EXCLUDING BRICK

TYPE OF DEFECT	TYPE OF REPAIR
Deformation <10%	Local by excavation or robotic repair or patch
FL,FC,CC,B	As above
Hole	Check video
FM,CM	Local by excavation or patch
>25% surveyed length	Relay or reline total sewer length

Videos should be checked in all cases. If there is a hole and if the depth of the hole is NOT obvious and/or the area of the hole exceeds 100mm x 150mm the type of repair is limited to either local excavation or to a patch type repair.

Intruding connections shall be indicated and costed either by robotic repair or by local excavation.

Patch Repairs:- Separate patches shall be used when adjacent defects are more than ONE metre apart.

All patch repairs shall be costed to the nearest metre.

Excavations:- Separate excavations shall be undertaken when adjacent defects are more than FIVE metres apart.

Each local excavation will extend ONE metre either side of the defect and will therefore be assumed to be THREE metres length per individual defect.

REFERENCES

(1) Water Research Centre / Water Authorities Association, 1986, Sewerage Rehabilitation Manual 2nd Edition.

(2) Department of the Environment / National Water Council, 1980, Standing Technical Committee Report No.25, Sewer and Water Main Records.

(3) Northumbrian Water Ltd, Sewerage Manual.

TABLE 4 - SEWER CONDITION COMPARISON

SEWER A

```
Identification : 0002/    /CM01     Survey Key No. is 615930
Authority :              Drainage Area Code : 20568596ST
Division : 1  District : 238  Pipe Length Reference : NZ19858915X
Op/Contr : NBA/SG01     Contract No :AV 2102  Job No : HR/20/NCSA
Survey Date : 04-Jun-91  Survey Time : 10:45  Search  :  Upstream
Location : BRIDGE STREET
Start MH No   : NZ19858904  Depth : 2.60   Cover Level  : 0.00
                                           Invert Level : 0.00
Finish MH No : NZ19858905  Depth : 0.00   Cover Level  : 0.00
                                           Invert Level : 0.00
Sewer Use : Combined         Size : 375mm      Shape     : Circular
Material  : Vitrified Clay          Lining :
Total Length :  51.5  Surveyed Length :  51.5  Pipe Length : 0.80
Year Laid : Not Known               Video Tape No :  2
Comments  :  CM01
Purpose   : 0    Weather : Dry    Loc'n :
Category  : A    Grades  :   43210     Override Grade  :
```

GRADE SUMMARY REPORT - Lengths are in Metres

```
Cracks                     21         Fractures                39
Breaks                      2         Deformations
Displ. Joints              51         Open Joints              51
Roots                                 Infiltration
Encrustation               14         Debris
Obstructions                          Surface Damage
Mortar Missing                        Bricks Missing
Displ. Bricks                         Dropped Invert
Connections :      1     Faulty :     1        Line changes  :    0
Junctions   :      5     Faulty :     3        Level changes :    0
```

SEWER B

```
Identification : 0386/    /CM01     Survey Key No. is 648584
Authority :              Drainage Area Code : 20863101PS
Division : 1  District : 238  Pipe Length Reference : NZ18868402X
Op/Contr : NBA/JN01     Contract No :AT 0899  Job No : HR/20/NCSB
Survey Date : 20-Sep-91  Survey Time : 11:01  Search : Downstream
Location : WANSDYKE                    LANCASTER PARK
Start MH No   : NZ18868402  Depth : 2.85   Cover Level  : 0.00
                                           Invert Level : 0.00
Finish MH No : NZ19858905  Depth : 3.55   Cover Level  : 0.00
                                           Invert Level : 0.00
Sewer Use : Foul             Size : 225mm      Shape     : Circular
Material  : Vitrified Clay          Lining :
Total Length :  69.7  Surveyed Length :  69.7  Pipe Length : 0.90
Year Laid : Not Known               Video Tape No : 17
Comments  :  CM01
Purpose   : 0    Weather : Dry    Loc'n :
Category  : B    Grades  :   43210     Override Grade  :
```

GRADE SUMMARY REPORT - Lengths are in Metres

```
Cracks                                Fractures                 3
Breaks                      1         Deformations
Displ. Joints              70         Open Joints
Roots                       3         Infiltration
Encrustation               14         Debris                   69
Obstructions                          Surface Damage
Mortar Missing                        Bricks Missing
Displ. Bricks                         Dropped Invert
Connections :      2     Faulty :     2        Line changes  :    0
Junctions   :      2     Faulty :     0        Level changes :    2
```

No Trenches in Town, Henry & Mermet (eds) © 1992 Balkema, Rotterdam. ISBN 90 5410 085 0

Réparations localisées d'égouts dans le district de la Northumbrian Water

Brian Syms
Northumbrian Water Limited, UK

Paul Hayward
Amtee Bunting Limited, UK

RÉSUMÉ: Une méthode d'évaluation de diverses techniques de rénovation d'égouts sans tranchées a été mise au point par la Northumbrian Water et plusieurs systèmes de réparation localisée ont été testés. Cette opération a montré l'efficacité des réparations localisées dans des situations appropriées.

LE POINT DE VUE DU CLIENT

NWL a sous sa responsabilité environ 13.120 km d'égouts de taille et de profondeur diverses, construits dans des matériaux variés et quelques 240 000 trous d'homme, vingt-trois entrepreneurs de maintenance de systèmes d'égout étant chargés d'assurer leur bon fonctionnement et leur entretien. Environ 70% des égouts ont un diamètre égal ou inférieur à 225 mm et 25% d'entre eux ont un diamètre situé entre 225 mm et 450 mm. L'analyse effectuée montre également que plus de 90% des égouts sont construits en terre cuite ou en béton. Comme tous les autres services des égouts dans le Royaume-Uni, notre région couvre des populations rurales et urbaines et le système d'égouts public dessert une population de 2,56 millions, ce qui représente 98% de la population totale de la région.

Ayant pris connaissance du Manuel de Réfection des Systèmes d'Égouts (SRM) (1), la Northumbrian Water a établi un programme grâce auquel tous les égouts à trous d'homme (c'est à dire les égouts d'un diamètre supérieur à 900 mm) ont été inspectés entre 1985 et 1987, après quoi un programme d'inspection de tous les égouts critiques ou stratégiques a été instauré. Dans cette région, les égouts critiques (A et B) représentent 33% de la totalité du système d'égouts public. En 1989, le programme d'inspection a été accéléré afin de couvrir 600 km par an avec pour objectif de terminer l'inspection de tous les égouts de catégorie A et B en 1993. Récemment, le programme d'inspection a été complété par l'inspection de 40.000 trous d'homme par an afin d'avoir terminé l'inspection de tous les trous d'homme connus en 1997. Pour parvenir à effectuer une inspection aussi complète, il est essentiel d'utiliser un système de référence unique pour les trous d'homme. On a donc adopté un tel système et on l'utilise avec un système de référence de longueur de tuyau (PLR), ces deux systèmes étant décrits dans le rapport STC N°25 (2).

Les inspections sont analysées par ordinateur pour obtenir une graduation structurelle conforme au SRM. Les degrés indiqués dans les Tableaux 1 et 2 sont ceux utilisés par la Northumbrian Water Limited et ils ont été obtenus en considérant attentivement les descriptions des défauts et des situations du SRM ainsi que nos propres observations plus récentes et notre propre expérience.

Le logiciel permet aussi la production d'études concises pour des endroits particuliers de façon à permettre d'organiser des gros et des petits contrats. Le projet actuel est limité à des tuyaux d'égout d'un diamètre allant de 225 mm à 800 mm, dont les défauts correspondent au pire au degré structurel de 4, ces paramètres faisant partie de l'analyse par ordinateur. On utilise ce logiciel pour produire une liste des longueurs susceptibles de devoir être rénovées en ordre PLR de façon à pouvoir faire une évaluation technique des défauts et de déterminer une option de réparation. Le tracé des longueurs à réparer sur des plans en ordre PLR peut également influencer la méthode finale de rénovation; par exemple, si on identifie des longueurs consécutives qui ont besoin d'être changées ou réparées, il peut s'avérer plus économique de considérer une rénovation continue entre trous d'homme.

Pour faciliter l'évaluation technique, un certain nombre de tableaux ont été produits pour permettre une décomposition simple mais précise des frais, qu'on peut évaluer pour les diverses méthodes considérées. Il convient de signaler que ce n'est pas toujours l'option la moins coûteuse qu'on préfère adopter étant donné que l'emplacement de l'égout peut empêcher une option de réparation avec excavation à ciel ouvert, par exemple si l'égout se trouve sous une rivière, sous une autoroute ou sous un carrefour très fréquenté. Le tableau 3 montre un certain nombre des suppositions et des paramètres fournis pour faciliter l'analyse. Il convient de noter que seuls des égouts dont la longueur affectée par des défauts n'est pas supérieure à 25% sont inclus dans le programme actuel de travail et que, pour faciliter l'identification, une analyse structurelle est également fournie. Une simple opération arithmétique permet d'identifier rapidement les longueurs répondant à ce critère. Le Tableau 4 montre une section d'égout (Égout A) où les défauts (cassures, fractures etc...) représentent jusqu'à 100% de la longueur totale et une autre section (Égout B) où les défauts ne représentent que 6%. L'utilisation de relevés complets par télévision en circuit fermé, de photographies et d'enregistrements magnétoscopiques permet d'achever l'évaluation technique de chacune des longueurs choisies.

Généralement, on n'envisage pas de réparer les défauts de degré 5 par les méthodes considérées. Par conséquent, comme indiqué plus haut, les défauts de degré 4 ont été identifiés comme étant ceux présentant le plus grand intérêt, bien que les défauts de degré 3 et 2 sont aussi considérés quand on compare des réparations localisées à une rénovation entre trous d'homme.

En utilisant ces directives etc..., l'analyse cherche alors à identifier la méthode de réparation à utiliser. Quatre méthodes fondamentales sont considérées:

- réparations localisées avec excavation à ciel ouvert
- réparations robotiques localisées
- réparations localisées par sections
- rénovation entre trous d'homme

bien que d'autres techniques soient évaluées d'une manière continue, comme la réparation de l'arrondi et la vérification et l'obturation des joints.

N.B. Il est possible qu'on envisage de rénover la longueur complète même quand les défauts représentent moins de 25% de la longueur de l'égout si des raisons techniques ou économiques justifient cette décision.

Il convient de noter qu'un seul défaut dans une longueur d'égout affectera le degré structurel de l'égout tout entier, mais une seule réparation localisée rétablira le degré structurel de la longueur complète. Maintenant que les compagnies privées des eaux et des égouts dans le Royaume-Uni font attention aux valeurs de leurs actifs qui sont reflétées dans leur bilan, les réparations localisées constituent un moyen rapide et économique de rétablir la valeur de l'infrastructure tout en fournissant une solution technique saine à la détérioration des tuyauteries.

Au cours des 18 derniers mois, l'expérience à pied d'oeuvre a montré que les méthodes robotiques ne convenaient pas dans certaines situations et, pour les défauts suivants, nous insisterions que des réparations par sections ou une excavation locale soient exécutées:

i) Fissures ou fractures multiples quand le nombre des fissures ou des fractures est supérieur à deux.

ii) Trous ou cassures quand la profondeur du défaut est supérieure à l'épaisseur du tuyau ou dépasse 100 mm x 150 mm de superficie.

Une fois l'analyse terminée, les résultats sont évalués afin de permettre l'adjudication de contrats séparés ou bien le travail à effectuer fait partie d'un contrat annuel. Afin de donner à chacun de nos entrepreneurs de maintenance de systèmes d'égouts une idée du travail à effectuer chaque année, les contrats actuels à long terme sont décomposés en périodes de deux semaines.

Avant que les travaux à pied d'oeuvre ne commencent, l'entrepreneur vérifie les relevés par télévision en circuit fermé, les enregistrements sur bande-vidéo et les compte-rendus pour s'assurer que tous les défauts identifiés peuvent être réparés. De plus, une nouvelle inspection rapide par télévision en circuit fermé est effectuée pour déterminer si les défauts n'ont pas empiré et si des défauts supplémentaires ne se sont produits. Notre expérience nous a montré que les égouts critiques se détérioraient structurellement à une telle vitesse qu'on ne peut considérer que les inspections par télévision en circuit fermée datant de 3 ans ou moins pour ce projet. Dans l'un ou l'autre de ces cas, la longueur à réparer est ré-évaluée par l'ingénieur du client.

LE POINT DE VUE DE L'ENTREPRENEUR

L'objectif de l'entrepreneur devrait être d'offrir à ses clients une gamme de services permettant de résoudre des problèmes divers. Il n'existe pas de solution universelle. Beaucoup d'entreprises offraient déjà des techniques de revêtement entre trous d'hommes, et encore plus le font maintenant, mais il y avait relativement peu de techniques de réparation localisée. Plutôt que d'examiner les systèmes de rénovation disponibles, nous avons commencé par évaluer les problèmes des clients qu'on peut classer sous plusieurs types différents:-

a) L'élimination d'intrusions, par exemple intrusions latérales de raccords ou de fossés d'écoulement, armatures en acier, racines, incrustation etc...

b) Réparation de fissures/factures circonférentielles, de joints défectueux isolés, de courtes fissures longitudinales et de raccords mal faits.

c) Réparation de cassures, de fissures/fractures multiples et de fissures/fractures longitudinales plus longues.

d) Réparation de l'arrondi et réparation des tuyaux dont la déformation est supérieure à 10%.

e) Vérification et obturation des joints des tuyaux quand la structure des tuyaux est saine.

Réparations robotiques

Les problèmes (a) et (b) indiquaient la nécessité d'utiliser un matériel robotique de meulage ou de perçage pour enlever les intrusions et pour éliminer par meulage les fissures et les fractures, puis un système de réparation structurelle pour redonner au tuyau au moins sa robustesse d'origine. Après avoir évalué des techniques qui prétendaient répondre à ces objectifs, nous avons sélectionné le système KA-TE qui semblait présenter la meilleure combinaison de compétence technique et de productivité.

Le système KA-TE utilise un robot pour percer et meuler et un autre robot pour l'application d'une matière de remplissage en époxy qui est conçue de façon à se lier aux substrats humides. Ce matériel est puissant et possède un taux de productivité élevé. Les robots peuvent travailler dans des égouts dont le diamètre se situe entre 200 mm et 800 mm.

La première étape du processus consiste à éliminer les intrusions en les taillant, en éliminant les fissures ou les fractures par meulage afin de former une encoche d'environ 25 mm de profondeur et de 30 mm de largeur qui présente une bonne prise pour la matière de réparation.

La deuxième étape consiste à remplir les fissures meulées avec un composé époxy formulé spécialement et qui donne une résistance supérieure à celle du tuyau d'origine. On peut également remplir avec cette matière tous les espaces entre les raccords défectueux et le tuyau d'égout principal.

Si besoin est, les portions réparées peuvent être polies par le robot de meulage une fois que l'époxy a pris.

Réparation avec manchon

Le problème (c) est plus difficile à réparer par des méthodes robotiques qui consistent à lier à nouveau le matériau existant du tuyau. Il est souvent préférable de réparer les défauts plus sérieux ou plus étendus en installant un manchon structurel à l'intérieur du tuyau existant.

Il existe divers systèmes pour ce type de réparation mais celui utilisé le plus souvent par la Northumbrian Water est appelé Econoliner. Ce système utilise un matériau composite comprenant une couche de feutre aiguilleté inséré entre deux couches de stratifil tissé en fibre de verre. Cette matière est imprégnée de résine époxy et elle est enroulée autour d'un dispositif gonflable de pose de revêtement. Quand il est en position, ce dispositif est gonflé d'eau et de l'eau chaude est circulée pour faire prendre la résine. Ce système utilise un dispositif de pose de revêtement breveté à travers lequel l'écoulement des eaux usées peut continuer lors de l'installation et un pompage supplémentaire est rarement nécessaire.

L'épaisseur du manchon est inférieure à 5 mm, ce qui signifie que la réduction du diamètre interne du tuyau est minime. On peut installer des manchons d'une longueur allant jusqu'à 3 mètres et il est possible de réparer des sections plus longues en utilisant deux ou plusieurs manchons en série.

On peut réparer des tuyaux de 100 mm de diamètre et plus avec la méthode Econoliner bien que la taille du dispositif de pose de revêtement soit un facteur limitatif pour les grands diamètres.

Systèmes de réfection de l'arrondi

Les tuyaux déformés de plus de 10% sont

TABLEAU 1 - DEGRÉS D'ÉTATS STRUCTURELS

Degré	IMPLICATION
5	Effondré ou effondrement imminent
4	Effondrement prévisible dans l'avenir
3	Effondrement peu probable dans un avenir proche mais plus ample détérioration probable
2	Risque minime d'effondrement à court terme mais possibilité de plus ample détérioration
1	État structurel acceptable

généralement considérés comme ne pouvant pas être rénovés, essentiellement parce que les contraintes dans le revêtement seraient plus grandes que dans un tuyau circulaire soumis à une charge semblable. La réduction de la coupe transversale du tuyau est aussi un facteur qui entre en considération.

Une méthode a donc été mise au point pour réparer l'arrondi des tuyaux avant de les revêtir ou de les réparer localement. En juin 1992, un prototype de travail a été utilisé pour réparer l'arrondi de tuyaux de 300 mm de diamètre et d'autres modèles de production sont en train d'être construits pour d'autres tailles de tuyaux.

Ce système, appelé Sirkit, utilise un vérin hydraulique pour ouvrir une série de "pétales" qui agissent sur le tuyau déformé. Le dispositif d'ouverture des pétales est conçu de façon à ce que l'ouverture des pétales soit uniforme et à ce que le tuyau arrondi soit donc circulaire.

Une bande circulaire en acier inoxydable est enroulée autour du dispositif avant l'ouverture des pétales et cette bande sert de bouclier permanent pour maintenir les fragments de tuyau jusqu'à ce que l'égout soit pourvu d'un revêtement ou soit réparé. Dans la pratique, ces bandes ont été laissées en position sans aucun effet néfaste pendant plusieurs semaines avant que la réparation soit terminée.

Vérification et obturation des joints

Des conduites qui sont en bon état structurellement peuvent être défectueuses à cause de joints qui fuient. Cela entraîne une surcharge du réseau d'égouts ainsi que de frais élevés de clarification des eaux d'égout s'il y a infiltration, ou bien une pollution du sous-sol et des aquifères s'il y a exfiltration. Dans les deux cas, il peut se produire une érosion de la couche d'assise du tuyau et du pourtour, causant des vides invisibles et la possibilité d'une défaillance catastrophique.

On utilise le système de vérification et d'obturation des joints Amtec pour vérifier chaque joint individuellement sous une pression de 0,7 bars. Les joints qui s'avèrent défectueux sont obturés par injection d'un gel réactif à l'eau dont il existe plusieurs types, et le joint est alors vérifié de nouveau. Les pressions d'essai sont affichées sur l'écran avec l'image de télévision en circuit fermé et la distance.

Selon l'ampleur des vides autour du tuyau et par conséquent la quantité de produit obturateur qui doit être injectée, ce processus est assez rapide et économique. Quand on obture des joints pour empêcher l'infiltration, la période d'amortissement est souvent assez courte à cause de la réduction du travail à fournir par le matériel de pompage et les installations de clarification des eaux d'égout.

On peut aussi utiliser ce système pour obturer des petites fissures bien qu'il ne prétende pas constituer une technique de réparation structurelle.

Considérations économiques

À moins que l'égout soit relativement peu profond, à l'écart de grandes routes et pas situé en dessous d'autres services, la rénovation est généralement plus rentable que le remplacement ou les réparations par excavation à ciel ouvert. Dans le Royaume-Uni, la nouvelle loi régissant les travaux sur les routes et dans les rues, qui est progressivement mise en vigueur, augmentera le coût et la difficulté de l'exécution d'excavations sur les routes.

Les réparations localisées sans tranchées sont généralement plus rapides

TABLEAU 2 - DÉFAUTS: DEGRÉS STRUCTURELS

Défaut	Degré	Défaut	Degré
Fissures longitudinales	3	Légère incrustation	1
Fissures cironférentielles	2	Débris de silt	1
Fissures multiples	3	Débris de graisse	1
Fractures longitudinales	3	Autres débris	1
Fractures circonférentielles	3	Obstructions	1
Fractures multiples	4	Dommages superficiels: effritement	1
Cassures	4	Dommages superficiels: usure	1
Déformation	5	Raccords défectueux	2
Petits joints déplacés	1	Jonctions défectueuses	2
Joints moyens déplacés	2	Raccords	1
Grands joints déplacés	3	Jonctions	1
Petits joints ouverts	1	Changements de direction	1
Joints moyens ouverts	2	Changements de niveau	1
Grands joints ouverts	3	Effondrements	5
Petites racines	1	Branchements	1
Racines massives	1	Briques déplacées	3
Racines principales	1	Déformation horizontale	5
Infiltration - suintement	1	Radier affaissé	4
Infiltration - égouttement	1	Déformations verticales	5
Infiltration - écoulement	1	Briques manquantes	4
Infiltration - jaillissement	1	Mortier manquant - Moyen	3
Incrustation à écailles	1	Mortier manquant - Surface	2
Incrustation épaisse	1	Mortier manquant - Totalité	3
Incrustation moyenne	1		

et moins incommodes que le revêtement des tuyaux d'égouts entre trous d'homme et les frais de mobilisation sont moins élevés. Cependant, l'évaluation économique des diverses techniques doit tenir compte des facteurs suivants:

- La nature des défauts et la difficulté de les réparer avec une méthode sans tranchée.
- Le nombre de défauts et par conséquent le nombre de réparations robotiques et/ou de manchons de réparation nécessaires.
- La longueur totale de l'égout et le coût d'un revêtement entre trous d'homme.
- Le nombre de raccords devant être refaits si un revêtement complet était installé mais qui pourraient être évités en utilisant un système de réparation localisée.
- L'emplacement et les restrictions d'accès éventuelles qui pourraient rendre certaines techniques inutilisables.
- La possibilité de l'apparition d'autres défauts dans des sections de l'égout qui ne sont pas réparées.
- Le coût réel de la réparation ou du remplacement par excavation à ciel ouvert, y compris la réfection des routes, les compensations aux propriétaires et le temps passé par l'ingénieur à déterminer et surveiller les travaux et à effectuer des tâches administratives.

TABLEAU 3 - DIRECTIVES POUR LA RÉFECTION LOCALISÉE D'ÉGOUTS

RÉFECTION D'ÉGOUTS CRITIQUES

DIRECTIVES POUR LES TRAVAUX DE RÉPARATION - TAILLES D'ÉGOUTS DE 225 mm À 800 mm

DEGRÉ STRUCTUREL 4

MATÉRIAU - N'IMPORTE LEQUEL SAUF BRIQUES

TYPE DE DÉFAUT	TYPE DE RÉPARATION
Déformation < 10%	Localisée par excavation ou réparation robotique ou réparation de section
FL, FC, CC, B	Idem
Trou	Vérifier la bande-vidéo
FM, CM	Localisée par excavation ou réparation de section
>25% de longueur inspectée	Remplacer ou appliquer un revêtement sur la longueur totale de l'égout

Il faut vérifier les bandes-vidéo dans tous les cas. S'il y a un trou et si la profondeur du trou n'est PAS évidente et/ou si la surface du trou dépasse 100 mm x 150 mm, le type de réparation est limité soit à une excavation locale soit à une réparation par section.

Réparations par sections:- Des sections séparées seront utilisées quand les défauts adjacents seront à plus d'UN mètre l'un de l'autre.

Le coût de toutes les réparations par sections sera évalué au mètre près.

Excavations:- Des excavations séparées seront entreprises quand les défauts adjacents seront à plus de CINQ mètres l'un de l'autre.

Chaque excavation locale s'étendra à UN mètre de part et d'autre du défaut et on suppose donc qu'elle aura TROIS mètres de longueur par défaut individuel.

Qualité

On pose souvent la question: "Combien de temps les réparations vont-elles durer?" Il est bien sûr impossible de donner une réponse précise étant donné que les techniques de rénovation en général, et les techniques de réparations localisées sans tranchées en particulier, n'existent pas depuis suffisamment longtemps pour qu'on puisse en évaluer les résultats comme ceux des méthodes traditionnelles.

Cependant, quelques-uns des premiers revêtements Insituform installés il y a plus de vingt ans ont récemment été examinés en détail et se sont avérés ne présenter aucun signe de détérioration. Les méthodes de réparation localisée comme le système Econoliner utilisent des matériaux qui devraient être plus résistants et plus stables que ceux utilisés dans les systèmes de revêtement tendres et leur longévité devrait donc être au moins aussi bonne.

Cela vaut aussi la peine de se rappeler que, si la durabilité pratique des matériaux traditionnels pour tuyaux comme le béton et la terre cuite était aussi bonne dans la pratique que dans la théorie, il y aurait moins d'égouts ayant besoin d'être rénovés. Un grande partie des tuyaux d'égout actuellement en train d'être rénovés ont moins de vingt ans.

TABLEAU 4 - COMPARAISON D'ÉTATS D'ÉGOUTS

ÉGOUT A

Identification: 00002/ /CM01 N° de Réf. d'Inspection: 615930
Service responsable: Code d'Aire de Drainage: 20568596ST
Section: 1 District: 238 Référence de Longueur de Tuyau: NZ1985891SX
Entrepreneur: NBA/SG01 Contrat N°: AV 2102 Ouvrage N°: HR/20/NCSA
Date d'inspection: 04-juin-91 Heure d'inspection: 10:45 Recherche: en amont
Emplacement: BRIDGE STREET
N° de trou d'homme de départ: NZ19858904 Profondeur: 2,60 Niveau de couvercle: 0,00 Niveau de radier : 0,00
N° de trou d'homme d'arrivée: NZ19858905 Profondeur: 0,00 Niveau de couvercle: 0,00 Niveau de radier : 0,00
Utilisation de l'égout: Combinée Taille: 375 mm Forme: circulaire
Matériau: Argile vitrifié Revêtement:
Longueur totale: 51,5 Longueur inspectée: 51,5 Longueur de tuyau: 0,80
Année de pose: Inconnue Bande-vidéo N°: 2
Commentaires: CM01
Temps: Sec Emplacement:
Catégorie: A Degrés: 43210

COMPTE-RENDU SOMMAIRE DE DEGRÉ - Longueurs indiquées en mètres

Fissures	21	Fractures	39
Cassures	2	Déformations	
Joints déplacés	51	Joints ouverts	51
Racines			
Incrustation	14	Débris	
Obstructions		Dommages superficiels	
Mortier manquant		Briques manquantes	
Briques déplacées		Radier affaissé	
Raccords: 1	Défectueux : 1	Changements de direction: 0	
Jonctions: 5	Défectueuses: 3	Changements de niveau : 0	

ÉGOUT B

Identification: 0386/ /CM01 N° de Réf. d'Inspection: 648584
Service responsable: Code d'Aire de Drainage: 20863101PS
Section: 1 District: 238 Référence de Longueur de Tuyau: NZ18868402X
Entrepreneur: NBA/JN01 Contrat N°: AT 0899 Ouvrage N°: HR/20/NCSB
Date d'inspection: 20-sept-91 Heure d'inspection: 11:01 Recherche: en aval
Emplacement: WANSDYKE LANCASTER PARK
N° de trou d'homme de départ: NZ18868402 Profondeur: 2,85 Niveau de couvercle: 0,00 Niveau de radier : 0,00
N° de trou d'homme d'arrivée: NZ19858905 Profondeur: 3,55 Niveau de couvercle: 0,00 Niveau de radier : 0,00
Utilisation de l'égout: Eaux usées Taille: 225 mm Forme: circulaire
Matériau: Argile vitrifié Revêtement:
Longueur totale: 69,7 Longueur inspectée: 69,7 Longueur de tuyau: 0,90
Année de pose: Inconnue Bande-vidéo N°: 17
Commentaires: CM01
Temps: Sec Emplacement:
Catégorie: B Degrés: 43210

COMPTE-RENDU SOMMAIRE DE DEGRÉ - Longueurs indiquées en mètres

Fissures		Fractures	3
Cassures	1	Déformations	
Joints déplacés	70	Joints ouverts	
Racines	3	Infiltration	
Incrustation	14	Débris	69
Obstructions		Dommages superficiels	
Mortier manquant		Briques manquantes	
Briques déplacées		Radier affaissé	
Raccords: 2	Défectueux : 2	Changements de direction: 0	
Jonctions: 2	Défectueuses: 0	Changements de niveau : 2	

À notre avis, qui est nécessairement basé sur des évaluations qualitatives plutôt que quantitatives, un tuyau rénové correctement devrait avoir au moins la même durabilité qu'un tuyau neuf, et, selon le type de système de rénovation choisi, il est possible qu'il soit moins susceptible d'être endommagé à l'avenir.

L'avenir

Les systèmes de réparation localisée sans excavation en sont à leur tout début mais elles s'avèrent déjà être des additions utiles à la panoplie des méthodes de réfection. Au sein du district de la Northumbrian Water, ils s'avèrent être un moyen efficace d'obtenir des résultats rapides.

De nouvelles méthodes, en particulier dans les domaines de la détection et de la taille latérales, sont en train d'être mises au point et les techniques existantes font aussi l'objet d'améliorations. La demande croissante de ces systèmes conduira inévitablement à d'autres innovations et à une réduction des frais dont bénéficieront à la fois le client et l'entrepreneur.

No Trenches in Town, Henry & Mermet (eds) © 1992 Balkema, Rotterdam. ISBN 90 5410 085 0

Replacing by extracting decayed mains of potable water supply

J.Schroeyers
Sade, Paris, France

ABSTRACT : Replacing mains ducting in urban areas is often very difficult when we have to draw the lay-out throughout a very busy underground and meet the requirements of district or municipal authority.
The idea of using the same place is a tempting one but two points have to be satisfied :
- maintaining the potable water supply whilst the job is being performed,
- material from the decayed pipe has to meet certain requirements of the "previous location replacement technique"...

During several years Compagnie Générale des Eaux and SADE have tried several techniques on the Syndicat des Eaux d'Ile de France network. Last year, some experiments have been carried out using Extracting Method which claims to be technically interesting and economically capable to compete with traditional open trench.
The pulling device has been developed, based on prestressed concrete strands with high yield point and Freyssinet hydraulic jack mounted on a frame capable of supporting the pulling forces whilst shored up to the pit end wall.
At the present time, this equipment has been successfully tested on grey cast iron pipe, internal diameter 60 to 150 mm (2" to 6"), replaced by HDPE pipes outside diameter 125 to 225 mm (4" to 8").

1 DISCUTION AND NEW OBSERVATION

The concept of using the location of the decayed pipe for the replacing one as been since a long time of great interest for network concessionary companies as well as for contractors. This interest is growing up as underground is getting more and more busy, specially in highly urbanized areas and as wishes are more and more put into words by the responsables for reducing the nuisances induced by road works.
The first works using "previous location replacement techniques" have used equipment descended from air impact mole to which the addition of bursting device and winch allowed bursting of pipe of certain materials (grey iron,not reenforced concrete, asbestos, cement, sandstone).
For the case of replacing the decayed pipe of potable water supply in grey iron, this technique failed when meeting repear couplings in duct iron; in such cases, opening a trench to remove the coupling were necessary.
In the face of the evidence, the Sade Company in the field of works it is doing for the "Syndicat des Eaux D'Ile-de-France", conceived a specific equipment for decayed grey iron extraction.

2 EQUIPMENT DESCRIPTION

The originality of the equipment is based on the "Freyssinet" hoisting jack and prestressed concrete strands to exert tractive effort. (pict. 1)
The prestressed concrete strands have the advantage of high strength limit which avoid the inconvenients due to the important elongation of standard cables.
The equipment is completed by :

pict.1 : extracting set including "Freyssinet" jacks

pict.2 : pushing piece with pulling head

pict.3 : spearhead piece for grey iron pipe bursting

- a pushing piece transmitting the tractive effort on the decayed pipe as well as pulling the new pipe (pict. 2),
- a piece with spearhead shape producing pipe bursting as pipe progress (pict.3),
- prestressed concrete strands barrels

3 PROCESS OPERATING PROCEDURE

The section of pipe to be replaced has one pit at each end, one will recieve the extracting equipment, the other will recieve the new pipe that will be introduced.
According to the decayed pipe diameter to remove, the section that can be up to 200 meters, will be divided into 15 to 25 meters sections.
This will be realised by removing 0,60 meter to 1,00 meter of pipe at the service branches locations.
Previously, in most majority of the cases, a temporary bypass will by installed to maintain water distribution.
The number of prestressed concrete strands that are fixed according to the pipe diameter, are introduced through the pipe to extract. This operation can be done either from the jack through the clams or from the other end.
The prestressed concrete strands end, opposit to the "Freyssinet" jack are fixed to the anchor plate by a wedge anchoring system and so will drive jack effort to the decayed pipe.
The pushing piece comes together with conical piece that will enlarge the hole left by the decayed pipe and also together with the piece that pull the new pipe.
At the other end, the prestressed concrete strands are going through the jack clams and are stored, as the work progress, in the barrels.
The "Freyssinet" jacks are mounted in a chassis able to drive the effort to the trench end wall and able to allow enough length for the alternative movement and for the last piece of pipe to be burst.
The jacks are working with an alternative movement of 300 mm.
As the decayed pipe is moving forward, it is burst on the spearhead that stand on the end of the following length of pipe that stays without pulling effort.
If some repair fittings are found during the pull, the are either removed (repair couplings), either cut (ductile iron, Steel fittings).

4 SCOPE

Considering the length, potable water network is mainly formed by diameter 60, 80, 100 and 150 mm pipes.
The machine developed by SADE as been desined for these diameters, and allows the installation of a greater diameter pipe.
The new pipe is HDPE pipe delivered on barrels.

5 FIRST EVALUATION OF THE SITUATION AFTER ONE YEAR OF WORKS

5.1. Reducing the nuisances

In addition the argument of using the same place in the ground and removing the decayed pipe, extractingtechnique is reducing the nuisances produced by pipe works, soils movements are reduced to these needed for the connecting pits and for the branche connections pits, noise and vibrations are only these of an insulated generator.

5.2.Cost and guarantee

The works realised since one year, essentially replacement of decayed pipes diameter 60 and 80 mm by HDPE diameter 125 mm, show that despite the installation of a temporary bypass, costs are roughly similar as a classical open trench technique for laying a pipe parallel to the decayed pipe.
By another way, extracting does not need any special items, pipes and fittings are these used for traditionnal open trench techniques.

5.3.Technical improvements

Furthermore improvement of some items to increase their reliability, some trials are going on to backfill the gap between the HDPE pipe and the bore left by the removed decayed pipe. For instance, in the case of pipes with same nominal diameters, for the diameter 100 mm cast iron , the tulip-shaped part as an outside diameter of 190 mm which is much greater than the outside diameter 125 mm of the HDPE.

No Trenches in Town, Henry & Mermet (eds) © 1992 Balkema, Rotterdam. ISBN 90 5410 085 0

Experience of boring from the inside

M. Hartenstein
Phoenix Treatment S.A., Penthaz, Lausanne, Switzerland

SUMMARY : Experiences and development works of the companies using the PHOENIX process regarding boring operations from the inside of consumers connections, after treatment of gas and water mains.

1. INTRODUCTION

The mains reconditioning from the inside is presently using, most of the time, the so called lining technologies.
These technologies cover the whole internal surface of the mains, either with material adherence or not on the pipe walls.

The distribution networks, gathering gas-, water- or sewerage pipings, include consumer connections which are consequently obturated.
The advantage of the reconditioning processes, a.o. is the possibility to avoid digging works, and as a matter of fact all costs and disturbances inherent thereto.

As we did not want to reduce these advantages, it was necessary to develop equipments able to carry out the required operations from the inside, focusing consequently on robotics which could enable reopening of the connections from the inside of the reconditioned mains.

2. ESTIMATION OF THE PROBLEMS TO SOLVE

Two main situations categories have to be considered :

2.1. Thick lining installation.

This type of mains reconditioning can be selected for sewerage networks as well as pressure mains (gas, water, and industrial pipings). Most of the time, the reconditioning material does not adhere to the mains. It is simply in contact or with a variable annular space. More rarely, the reconditioning linings are glued on the mains.

2.2. Thin lining installation.

In this category we can mainly find pressure mains. These processes have the particularity - and the advantage as well - to adhere to the treated mains.

When installed, the following problems will have to be solved :
- mains diameter variability and equipments miniaturization;
- connections location;
- information regarding the connections diameter;
- flowing of resin inside the connection section;
- drilling tool positioning;
- selection of the boring process :
  a) mechanical drilling;
  b) thermofusion
  c) hydraulic drilling.

3. SELECTION OF THE BORING PROCESS

Both thickness and nature of the lining will determine this selection.

For reconditioning processes with thin lining, the thermofusion boring system will be selected most of the time, as it gives the following advantages :
- auto-centering inside the consumer connection;
- "smooth" technology, preventing damages around the connection; this is particularly important on gas and water mains, as in Europe we can observe numerous material types, including lead; all boring operations of mechanical type on these mains materials have generated damages, even serious enough to bore through the connection itself;
- boring quality;
- more compact robotic equipments.

The treatment involving thick lining material require either a mechanical boring process, by milling, grinding and drilling, or hydraulic cutting system.

4. CONNECTIONS LOCATION

Necessary to perform robot positioning after treatment, the location is found by a means of a video camera visualization process, and an accurate circonferential and axial targetting.

5. MEASURE OF THE CONNECTION DIAMETER

While performing the location operations, a measure of the connection diameter can be done to be sure of the theoretical data, or maybe to determine the diameter, if this information is not available.

In this case again, robotics can give sophisticated solutions. Regarding the matter concerned, a visual record performed by means of millimetric compass appears as sufficient, avoiding this way important equipment overcosts and unnecessary sophistication.

6. FLOWING OF RESIN INSIDE THE CONNECTION

This problem has completely been neutralized regarding the treatment of sewerage networks.

Regarding treatments with thin lining in pressure mains, and the hot glueing process, this problem is still present, and in such a way that it is necessary to obturate the connection from the inside of the mains prior any treatment.

The main difficulties are most of all the high diversification of the connection types and cuttings in mains. We can meet indeed on the working mains :
- irregular pits (boring or groove cutting);
- connections rising above the mains;
- misaligned connections;
- screwed connections, or fitted with collars according to different processes, lead welded, etc.

Finally, a positional finding after treatment must :
- guarantee a perfect tightness of the connection by installing an obturation cap;
- give moreover a proper lining adherence around the opening after boring operation;
- enable an easy positional finding after treatment;
- enable an easy boring of the assembly lining + cap by means of a thermofusion process;
- to be approved regarding drinkability specifications;
- still be competitive economically.

An accurate development and adjustment has been essential consequently. A compound obturation product has been developed, answering all parameters mentioned hereabove.

7. INFLUENCE ON THE WORKING TIMES

Excepted the experimental works performed in laboratory and the work sites simulations, the whole of these processes, equipments and products have been developed on site :
- on gas networks with the support of GAZ DE FRANCE on the distribution centers of TOULOUSE (FRANCE);
- on water networks with the support of the 'Compagnie Intercommunale Liégeoise des Eaux' in Liège (BELGIUM).

All these works have given us the opportunity to limit the average treatment time to 12 minutes per connection, all operations involved included, and to reach a success level of 98 % of the treated connections.

8. FINANCIAL INCIDENCE

The development of such a technology has always taken into account not to alter the economical ratios which characterized the financial benefit to use such a reconditioning process.

On distribution networks, these ratios have still been improved as the operations on the mains from the outside have been suppressed.

Such performances are possible thanks to
- the selection of a standard equipment and unsophisticated robotics, free from any unnecessary features;
- the selection of materials among traditional products;
- a simple and fast working process.

9. CONCLUSIONS

We have developed a new system complementary to the NO-DIG processes, and which comes supporting our technics, increasing at the same time the possibilities range of our processes presently at the disposal of our customers.

No Trenches in Town, Henry & Mermet (eds) © 1992 Balkema, Rotterdam. ISBN 90 5410 085 0

Expériences en percement par l'intérieur

M. Hartenstein
Phoenix Treatment S.A., Penthaz, Lausanne, Suisse

RESUME : Développements et expériences réalisés par les exploitants du procédé PHOENIX en terme de percement par l'intérieur des branchements après traitement des canalisations de distribution de gaz et d'eau.

1. GENERALITES

La réhabilitation des canalisations par l'intérieur procède aujourd'hui le plus généralement au travers de techniques dites de chemisage ou de tubage. Ces technologies recouvrent donc toute la surface interne des canalisations, soit avec ou sans adhérence aux parois des tuyauteries.

L'ensemble des réseaux de distribution, qu'ils soient de gaz, d'eau ou d'assainissement disposent de raccordements d'abonnés qui se trouvent alors obturés.
L'intérêt des procédés de réhabilitation réside, entre autres, en le fait qu'ils évitent au maximum les travaux de terrassement, leurs nuisances et leurs coûts.

Si l'on ne veut pas aliéner ces avantages, il a donc fallu imaginer là encore des techniques et équipements permettant les travaux par l'intérieur et donc s'orienter vers une robotique permettant d'assurer les réouvertures des branchements ou dérivation depuis l'intérieur de la canalisation traitée.

2. EVALUATION DES PROBLEMES A SOLUTIONNER

Deux grandes catégories de situations sont à considérer :

2.1. Chemisages ou tubages épais.

Ce type de réhabilitation de canalisation se rencontre à la fois en canalisations gravitaires (réseaux d'assainissement) et en canalisations pression (réseau de distribution gaz et eau, voire canalisations industrielles).
Généralement, la technologie de réhabilitation n'adhère pas à la canalisation. Elle est simplement au contact ou présente un espace annulaire plus ou moins important.
Plus rarement, les techniques de réhabilitation adhèrent à la tuyauterie.

2.2. Chemisages minces.

L'on retrouve essentiellement dans cette catégorie les chemisages de canalisation pression. Ces techniques présentent la particularité - et l'avantage - d'être adhérentes à la conduite à traiter.

Ceci posé, il faudra faire face aux problèmes techniques suivants :
- variété des diamètres de canalisation et miniaturisation des équipements;
- localisation des branchements;
- connaissance du diamètre de ceux-ci;
- remontée des résines dans les branchements;
- positionnement de l'outil de percement;
- choix de la technique de percement :
  a) par forage mécanique,
  b) par thermo-fusion,
  c) par forage hydraulique.

3. CHOIX DES TECHNIQUES DE PERCEMENT

L'épaisseur et la nature du gainage ou du tubage seront directement déterminantes dans ce choix.

Aux traitements par gainage mince, nous aurons privilégié les forages par thermo-fusion qui présentent les avantages suivants :
- auto-centrage dans le branchement;
- technique dite "douce" évitant la détérioration de la prise de branchement; cet élément est particulièrement important en réseaux gaz et eau car l'on observe en Europe une multitude de types d'installations et équipements y compris au plomb; toute perforation à caractère mécanique a démontré, dans ce type de prises, le risque de détérioration voire de percement de celles-ci;
- qualité du percement;
- équipement robot plus compact.

Les traitements par chemisage épais ou par tubage amèneront quant à eux soit les techniques en percement mécanique par fraisage, meulage ou forage, soit les techniques par découpe hydraulique.

4. LOCALISATION DES BRANCHEMENTS

Indispensable à la performance du positionnement des robots après traitement, la localisation se fera par l'intérieur par visualisation caméra et repérage précis tant axialement que circonférentiellement.

5. MESURE DU DIAMETRE DU BRANCHEMENT

A l'occasion des travaux de localisation, une prise de mesure du diamètre du branchement peut être effectuée afin de s'assurer de l'exactitude des données théoriques, voire d'en déterminer le diamètre parce que cette donnée n'est pas disponible par ailleurs.

La encore, la robotique apporte des solutions plus ou moins sophistiquées. Pour ce qui nous concerne, une prise de mesure visuelle par approche d'un compas millimétrique a été jugée suffisante, évitant des surcoûts d'équipements importants et toute sophistication exagérée.

6. REMONTEE DES RESINES DANS LES BRANCHEMENTS

Ce phénomène a été entièrement neutralisé dans nos exploitations sur traitement de canalisations d'assainissement.

Par contre, en traitement par gainage mince en conduites pression et pour ce qui concerne les collages à chaud, le phénomène persiste au point de devoir obturer le branchement par l'intérieur de la canalisation avant son traitement.

Les difficultés à vaincre à cet égard sont essentiellement la très grande diversité de types de branchement et de découpe de la canalisation. Nous rencontrons en effet sur les réseaux d'exploitation :
- des orifices irréguliers (percements au bédane);
- des branchements dépassant;
- des prises de branchement décentrées;
- des prises vissées, colletées selon différentes techniques, soudées au plomb, etc.

Enfin, un repérage après traitement doit:
- réussir une <u>parfaite étanchéité</u> de la prise de branchement par placement d'un bouchon obturateur;
- permettre malgré tout une bonne adhérence du gainage en périphérie de l'orifice après percement de celui-ci;
- permettre un repérage facile après traitement;
- permettre une perforation aisée de l'ensemble gainage + bouchon par thermo-fusion;
- répondre aux critères de potabilité;
- conserver une compétitivité économique.

Une mise au point minutieuse a donc été nécessaire. Un produit composite d'obturation a été élaboré, apportant réponse aux paramètres décrits ci-dessus.

7. INCIDENCE SUR LES TEMPS D'EXPLOITATION

Outre les travaux expérimentaux réalisés en laboratoire et en simulation de chantier, l'ensemble de ces process, équipements et produits ont été développés sur site :
- sur réseaux gaz avec la collaboration de Gaz de France sur les centres de distribution de Toulouse (France);
- sur réseaux d'eau avec la collaboration de la Compagnie Intercommunale Liégeoise des Eaux à Liège (Belgique).

Ces travaux ont pu déterminer d'une cadence moyenne de 12 minutes par branchement toutes opérations incluses et d'un succès actuel de 98 % du nombre de branchements traités.

8. INCIDENCES FINANCIERES

La mise au point de cette technologie a été marquée par le souci de ne pas obérer les ratios économiques qui font des techniques de réhabilitation des produits réputés économiques.

En fait, sur réseau de distribution, ces rapports ont encore été améliorés puisque les interventions sur branchement par l'extérieur ont été supprimées.

Ces performances ont été atteintes grâce à :
- un choix d'équipements standard et une robotisation simple dépouillée de toute sophistication;
- un choix de matériaux recherchés sur le marché des produits traditionnels;
- un process d'exploitation simple et rapide.

9. CONCLUSIONS

Nous avons mis à disposition de notre clientèle et de nos techniques un outil supplémentaire et complémentaire aux techniques NO-DIG en augmentant ainsi la gamme d'utilisation de nos procédés.

No Trenches in Town, Henry & Mermet (eds) © 1992 Balkema, Rotterdam. ISBN 90 5410 085 0

Renovation of sewer lines with helicoidal bending pipes

J.M.Bordes
Sud Ouest Canalisation, Bordeaux, France

P.d'Heilly
Eternit Etex, Vernouillet, France

M.Gerber
SMCE Forage, Sierentz, France

J.P.Regnier
Barriquand, Compiegne, France

ABSTRACT: Every year in France, the number of sewer pipes renovated by an helicoidal bending system increase. It proves the interest that the designer and the district authorities shows to this system but also that they find in it all the answers and the safety that they need to achieve those projects.

1 DESCRIPTION OF THE ETR SYSTEM

ETR has been developed by the Eternit compagnies of five european countries : Austria, Belgium, France, Netherlands and Swizerland and since 1988. This system has been in use all over Europe.

The ETR system involves the in-situ formation of a structural liner from a PVC strip-form profile. the profile is fed into a specially designed machine, positioned at the bottom of a manhole or excavation, which transforms the strip into an helically wound cylindrical pipe. The ETR profile strip has interlocking edges in order to allow a mecanical jonction which tighness is given by two incorporates seals. This system is designed to reline pipes from DN 200 to DN 800 and even bigger.

1.1 Profiles

The profile are manufactured from impact resistant modified uPVC (usually coloured light grey) to DIN 8061.

As can be seen on fig 1, the extruded profile comprises a flat strip with a series of longitudinal T beams on one face which give the profile a high stiffness without using too much material. Furthermore the profil is wound with the T-beams external so as to give a good anchor to the surrounding grout, simultaneous with a smooth internal bore.

Opposite edges of the profile have male and female locks. The co-extruded primary sealing mechanism incorporated into the male lock is made of flexible PVC and the secondary seal which mates externally with the female lock is made of EPDM. These combine to give a very reliable water-tight joint. Test have been made and have shown that a pipe made with such a profile does not break until 12 bars. Furthermore the breaking

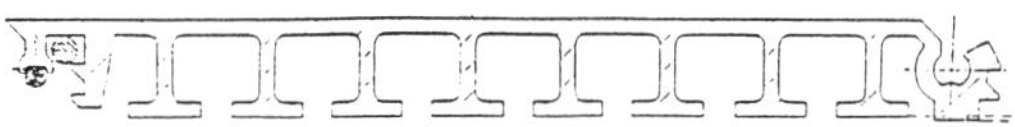

Figure 1. Cross section of the profile.

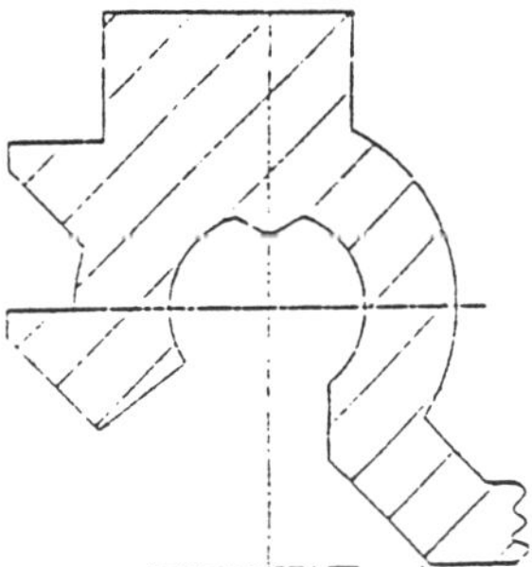

Figure 2. Cross section: female lock.

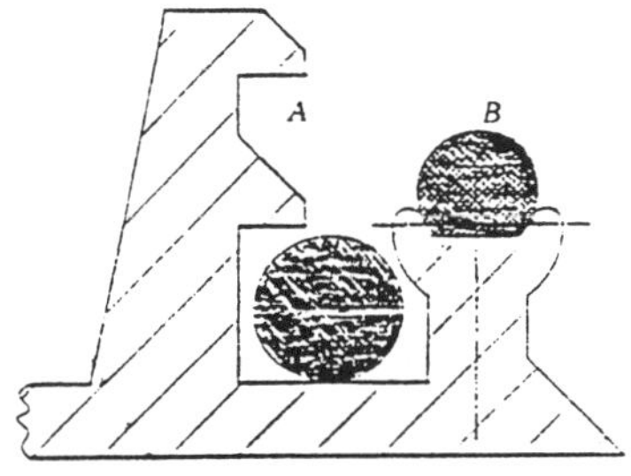

Figure 3. Cross section: male lock.
A: EPDM seal
B: PVC coextruded seal

occured on the wall of the profile and not at the jonction between profiles.

A range of four profiles of different widths and thicknesses is currently available to cover lining of sewers ranging from 200 mm to 800 mm and upper.

Table 1.

Sewer diameter range (mm)	Corresponding profile type Width/Thickness (mm)
200-250	50/5.5
200-400	60/7.5
350-600	90/9.5
600 and upper	140/14

1.2 Winding machines

The system comprises a feed mecanism and roller cages within which the liner is formed. The bars of the cages are conical and half of them are working parts in order to reduce the friction and to give to the pipe a rotating mouvement. Roller cages are changed to produce various diameters of lining. Changing a cage is easy and fast and the adjustements on site are limited to the winding angle - fonction of the width of the profile and the diameter of the lining - and to the winding speed - fonction of the condition of the pipe to be renovated. The machine is hydraulically driven from a power pack on the surface.

Cages are available for the following sewer/liner relationships.

Table 2.

Sewer diameter (mm)	Liner diameter (mm)
200	180
225	200
250	225
300	270
350	315
400	360
450	405
500	450
600	540
700 and upper	600
	700
	800
	900

2 ETR ON SITE

Thorough preparation, CCTV inspection and proving of the sewer are undertaken prior to lining.

Once the first adjustements made, the machine is lowered down the manhole and the profile introduce from above. Depending on the liner diameter and the

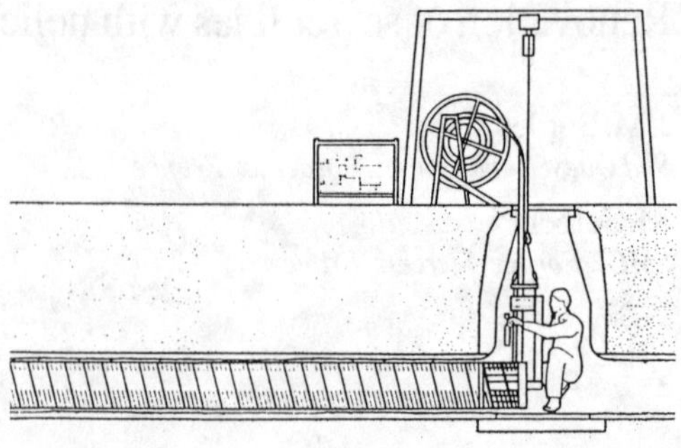

Figure 4.

condition of the sewer, the liner is produced at a rate of 1 to 2 meters per minute. The liner travels up the sewer aided by the corkscrew action of the winding process. When the liner reaches its destination, the liner is cut from the machine which is then withdrawn.

The end of the liner are trimmed off in preparation for annulus grouting. Various grout formulation are available to suit different site conditions. A very fluid grouting is favored in order to grout with low pressure and to assure a complete grouting.

3 CALCULATION

The calculation method used to verify that the lining will be structural is based on the Fascicule 70 edition 1992.

The calculation is made with the following hypotheses :

1. The remaining strength of the sewer is not taking into account.
2. The strength of the grouting is not taking into account exept as a laying bed
3. The new pipe must support all the loads including traffic loads and ground water forces.

4 EXEMPLES OF REALISATION IN FRANCE

All over France, the problems which the system had to resolve were differents : relining pipes layed under a high traffic road, in an-out-of-way place, under industrial buildings and so on... For all those projects, it was not possible to give back to the pipe its quality by any usual system.

4.1 Sete and Sceaux

In Sete, a 1949 pipe diameter 200 and 300 was a very bad shape. It was impossible to lay new pipes as they were installed under a heavy traffic road. With ETR the sewer was relined without any interruption of the traffic. The only indication, above ground, that a job was proceeding was a small van.

Because the formation of the liner does not rely on

Figure 5.

Figure 6.

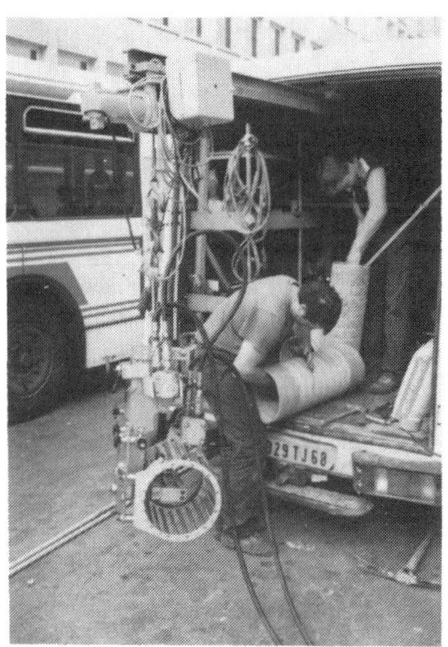

Figure 7.

in-situ chemical reaction, or adhesives that might be affected by water, the work can be done without any interruption of the effluent.

Renovation with ETR is a very cautious rehabilitation that pass unnotice from the neighborough. That was one of the reason why the ETR system was chosen for a DN 400 job in the Gardens of the Sceaux castle.

4.2 Annecy and Annemasse

On those two sites there was no possibility to reach the sewer's manhole with a van least a truck.

In Annecy the sewer was under a building and the manhole inside a warehouse . The ETR machine and the necessary profile was carried inside.

In Annemasse, the sewer was at the end of a non passable track. The profile was delivered with an all purpose vehicle and the machine carried up to the manhole. The job was made in winter with a temperature below zero degrees and allthough we had to reduce the winding speed because of the temperature, the weather had no influence on the quality and the resistance of the new pipe due to the fact that the system does not involve any chemical in-situ reaction.

5 CONCLUSIONS

The relining by helicoidal bending is a way to give back to the old pipe his strength and his tighness. The time on site is short, the size of the site is limited. Because of the easy use of the system, of the possibility to reline under effluents and of the small quantities of equipment needed, this system of renovation is unobstrusive with no influence on the life of the neighborough, to the pleasure of the districts that have adopted it.

No Trenches in Town, Henry & Mermet (eds) © 1992 Balkema, Rotterdam. ISBN 90 5410 085 0

No trenching necessary – U-Liner™ pipe: Trenchless technology the No Dig solution

Ivan C. Mandich
Delta US Engineering, USA

Stephen C. Catha
Pipeliners, Inc., USA

Craig S. Langworthy
Smith & Gillespie Engineers, Inc., USA

Michael J. Cawley
The US Department of the Air Force, USA

ABSTRACT. The proprietary U-LinerTM pipe system is a trenchless rehabilitation system, a real No Dig solution. As few as two people using the specified equipment can perform the installation from available manholes. The U-Liner pipe system can fit either within a confined area of an existing pipeline or at an installation site designed for this unique construction method. U-Liner Pipe requires no excavation debris at the site and normal activities are not interrupted.

The technological sophistication ofthis rehabilitation system of a pipe-within-a-pipe often elicits the question, "Is it as good as trench-installed pipe ? The answer is clear when the proven economic and technical advantages of this system are treated. The rehabilitation of dilapidated pipe structures must be done quickly, and because U-Liner Pipe fulfills this need so well, the amount installed within the past several years has increased exponentially. As a result, continuing prospects for innovation in U-Liner Pipe are realistic and forthcoming. Such an innovation is the fused U-shape pipe which is unique to the U-Liner Pipe system. Also, the remote-controlled fusing of a lateral or branch pipeline into the main line is now in the process of being developed for commercial use.

However, continuing technological advances require that engineers be able towork unhampered by the traditional politics of infrastructure debate. Trenchless technology will succeed in the future only if engineering science unites to free such technology to prospec on its own merit.

U-LINER™ PIPE - NEW GOODS AND SERVICES

The prosperity which exists today in trenchless technology can be attributed to the addition of the various systems for rehabilitation. The trenchless industry's success and growth is possible only through the creation of new goods and services developed through engineering innovation. Conversely, simple lateral expansion of a company that has dominated the rehabilitation market for 20 years without adding new goods or services will not be able to solve today's huge and pressing problem of a dilapidated infrastructure. Because of failing infrastructures, then, there is an increasingly higher demand for pipe rehabilitation. And demand for the advanced technologies of the trenchless pipe industry has ensured growth in this industry during the last decade even though competition for available products and installers became fierce at times.

TRENCH AND TRENCHLESS

The cost of pipe used is approximately 20% of the total in traditional trench installation. The predominant cost is in excavating and backfilling the trench. Such a process of construction with soils creates a load which is imposed directly on the buried pipe. Considering that there are varieties of soils used in the backfilling, a differential settlement of the freely buried structure occurs. The success of the installation can only be measured after consolidation takes place and after the pipeline is finally aligned.

If there are any imperfections in the new pipelines, they will occur during the first stage of installation, backfilling, and consolidation. The newly-buried pipelines are often found, after a period of time, misaligned and with open joints. Such structures could be improved with liners which can be installed immediately into a buried pipeline. This composite system would then compensate for the soils' unknown effect on the line and for the differential movements common to linear structures such as pipelines.

Besides being less costly to install, U-Liner Pipe is also more responsive than a single buried pipe during the pipeline's service life. The cost of the composite structure can be justified when the history of point repairs and maintenance costs is considered. The 12th Plastic Fuel Gas Pipe Symposium, September 24-26, 1991, contains a report on measurements of service loads on the buried pipelines. This first-of-a-kind finding indicates that remarkable features are evident from approximately 1 1/2 years of testing. The loads transferred onto the buried pipeline were benign because "the largest loads that have been observed are due to installation" and the installation loads "are the highest loads that the pipe is likely ever to see."

A question for engineers today is, in spite of conventional engineering, what are those cases wherein lining installation is superior to trench installation? The first advantage is that liners are immediately installed inside the new trench-laid pipelines. This type of design represents an unconventional design approach, where the pipe-within-a-pipe is considered a new installation. All of this is a step forward in quality and in establishing a modern infrastructure because a much more complex evaluation is required before an engineering decision can be made to install a liner. As a result, more is known about the peculiarities of that segment of rehabilitated pipe and the deeper consideration leads to longer usefulness and less upkeep of the pipeline in the future. In short, U-Liner Pipe makes for a better pipeline than does the common trench-installed pipeline. It would be beneficial to the pipeline and to the preservation of the homogeneous soils if a new pipeline could be installed without disturbing the soils during the excavation/backfilling procedure which is the most crucial point of the installation. The trenchless liner (tight-fit tunnel) represents a new pipe-within-a-pipe far surpassing in quality the trench-installed pipeline.

NEW AND OLD, HOW THEY COMPARE WHEN TESTED

Engineering reflections upon the performance of the trench-installed pipeline vs. a composite system, using a pipe-within-a-pipe in a tight-fit configuration, are based here on tests performed on the U-Liner Pipes further described in this article. The tight-fit configuration does not imply a perfect circular form in an existing and old circular pipeline but a form adaptable to the existing pipeline. This also applies when relining the egg-shaped pipelines with the U-Liner Pipe. As such, the U-Liner Pipe can be called a fit-into-shape pipe or a structure-within-a-structure. As far as the structure is concerned, the two pipes must fit together and have tight-fit clearances so that the composite structure can function under a load condition. However, a tight fit means that the pipes cannot be absolutely fitted until the pipeline is under stress. For example, the plastic door of an empty railway carriage was thought to be properly fitted by the engineers. But when the car was full or when, even empty, the car accelerated, the door would neither open nor close because the increased load had not been anticipated.

Tests were performed by the University of Cincinnati and by Tulane University on sections of 15-inch concrete pipe lined with U-Liner Pipe. The tested sections included the pipe processed in undersized pipe, where the surplus material formed a ridge, and the tightly fitted U-Liner Pipe, which was considered to be a standard installation. The objective was to find the following:

a) How does the structural capacity compare in undersized pipe vs. standard processed pipe?

b) Is the rehabilitated pipe capable of sustaining the original loads set by the design and/or criteria for the concrete pipe?

c) After a break of the concrete pipe, how

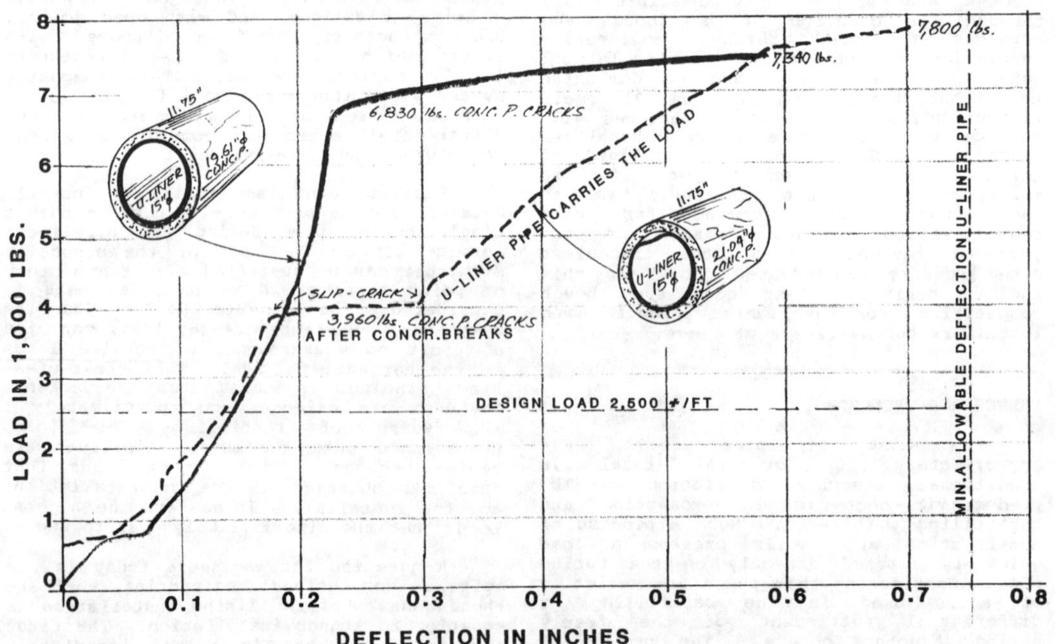

EXTERNAL LOAD CRUSHING STRENGTH TEST RESULTS

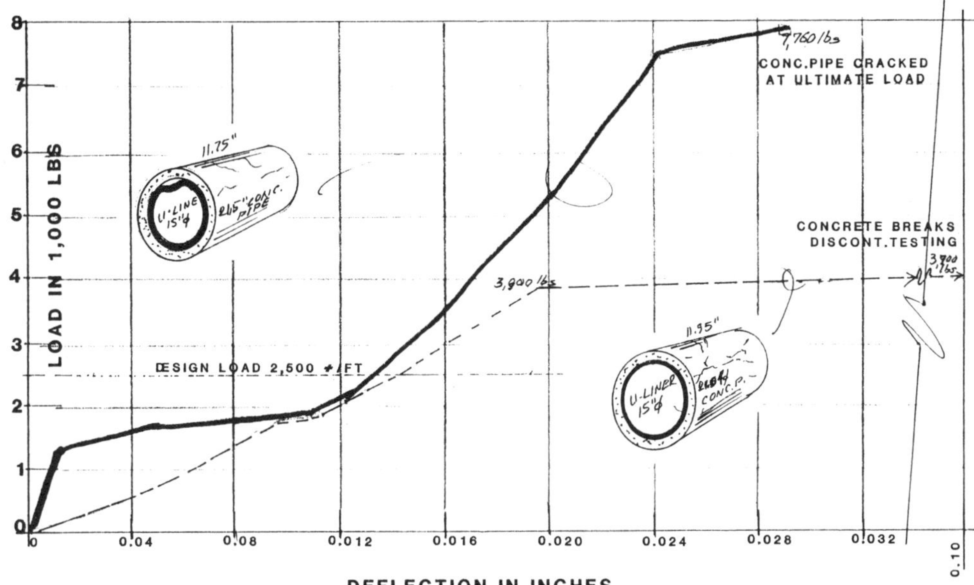

DEFLECTION IN INCHES

EXTERNAL LOAD CRUSHING STRENGTH

and to what degree is the structural performance carried on by the U-Liner Pipe? and

d) What is the general assessment of the adequacy of design for the composite pipe structures?

The results indicate that the U-Liner Pipe, within an old concrete pipe, performed excellently in answer to all of the above questions. The concrete pipe, as per Concrete Handbook 1959 Manual, applicable to the time of installation of this concrete pipe, was classified as a 15" I.D. pipe which is specified for three-edge testing at 1,750 lbs/ft. for standard strength and 2,750 lbs/ft. for extra strength. Both values were surpassed in testing. Additionally, the pipe with the ridge, which developed due to undersizing, shows no structural underperformance but rather a full structural performance.

When the composite structure was tested, the results obtained surpassed the original set of design criteria applicable to the originally installed concrete pipe; therefore, as far as its structural performance is concerned, the system proved to be equivalent to a new trench type of installation. As far as the rehabilitated pipe's hydraulic performance was concerned, the advantages were obvious: no infiltration, flexible adaptation to the changing soil and water conditions, and an increased flow.

A NEW ENGINEER EMERGES WITH TRENCHLESS TECHNOLOGY

From the most underdeveloped nations to the most developed nations, life depends generally on infrastructure. The global nations' economies, health, and welfare are generally based on the prediction of when the infrastructure will fail. Anywhere in the world, a failed infrastructure can hasten the spread of epidemics. As a matter of prevention, infrastructures must reach a state of health globally so they will not be a source of epidemics. The current status of antiquated infrastructures is well known and requires immediate rehabilitation. Government and public agencies will soon be urgently forced to implement rehabilitation programs. It is safe to predict that state and local budgets will be increased for the purpose of rehabilitation and that government policy will amount to a domestic "Marshall Plan" for this decade.

A new form of engineering practice will involve an international engineer who is an expert in trenchless technology and who can speak the language of the future rather than continuing to adhere to the political past. The present burden of regulations, inapplicable standards, computation methods, and specifications must be alleviated so that efficient infrastructure rehabilitation can take place.

WHO EXCAVATES AND WHO DOES NOT?

Various agencies and institutions, obsessed with categorization (such as sliplining, inversion linings, lining bags, etc.), have classified most rehabilitation systems incorrectly. Any new question posed to them results in a new categorization. Excavation categorization is an example. All methods that require an access pit equivalent to a manhole, a clean-out (or inlet) from the surface into the main pipeline, are considered trenchless rehabilitation. Those that require access trenches to guide the pipe and frequently excavated control pits, which may resemble trench installation, are departures from the actual intent of trenchless technology. Instead, the aim should be towards trenchless technology since it involves the least amount of interference in activities in the area and can be performed in a much shorter period of time.

Why is the U-Liner Pipe methodology of trenchless rehabilitation so valuable? Because it has unique qualities: a) as few as two technicians are needed, since compact equipment, containment of the pipe, and removal of the manhole cover are the only requirements for pipe insertion; b) inaccessible areas are approached with special portable equipment; c) the pipe is pulled in the cold state; d) the equipment operates without emission of toxic and hazardous fumes or gases; and e) the installation consists of one day of work, making it a real No Dig solution. For all these reasons, the U-Liner Pipe is the overwhelming choice.

Excavation is needed for sliplining systems and swedgelining systems (which use the round pipe insertion) in the form of entry pits and at the points of grouting controls. Excavation is also needed for drilled, installed, and segmented pipes larger than a manhole where insertion is required. The deformed U-liner Pipe - -the spiral liner, the folded pipe and the insitubag liner - - do not require entry pits; however, post-excavation is required where a liner's resin migration, clogging of the laterals and pipe removal is required. Excavation is necessary to remove the liner which is bonded to the existing pipe. Excavation will not be required when a liner is an independent pipe-within-a-pipe.

2,000 FEET OF THE U-LINER PIPE COILED IN AUSTRALIA

From the published information and data released at last year's trenchless technology conference in Kansas City, predictions for the trenchless methods and their use spread over the next five years are as follows:

SEWER REHABILITATION NEEDS PROJECTED IN THE USA (figures in $ millions) TYPE:	1992	1993	1994	1995	1996	RELATIVE % CHANGE
Grouting	84	93	103	114	126	6%
Inversion	105	94	84	75	70	-5%
Slip Lng.	152	169	188	203	219	10%
U-LINER™PIPE						
All other cumulative methods	86	118	150	195	254	24%
Drilling	9	14	18	23	28	3%
TOTAL:	436	488	543	610	697	37%

This table indicates that the relative change of a 10% increase can be expected in the category of sliplining and particularly in the subcategory of the tightfit lining, while the inversion lining shows a drastic decrease of 5%. Inversion linings will probably have exclusive jobs in very small sized pipelines. "All other methods" show an increase in the market and they will fulfill needs alongside other methods. Trenchless installation by drilling will capture a limited market due to this special application technology. Noticeable changes in the category of sliplining are not only because of economics, but because of the quality of the installation of these tightfit liners (excluding the round pipe sliplining). They are the foundation for the growth trend in this category.

As was represented in the 1991 Gas Pipeline Symposium in the U.S., use of polyethylene pipes in the gas rehabilitation infrastructure will predominate. The industry prioritization is in PE gas pipe. Most of the gas pipe networks are now polyethylene, covering over 640,000 km. or 400,000 miles in the United States alone. Annually 40,000 km. or 25,000 miles are added to the present system. Unquestionably engineers agree upon the value of implementing polyethylene pipe to replace the PVC pipe in gas lines and using polyethylene pipe for rehabilitation of an estimated 100,000 gas line failures, which amount to over $100 million in annual repair expenses. If rehabilitation by trenchless lining is performed instead of repairs, the amount of savings to the gas industry will be substantial. Clearly, the predominant factor in keeping this huge network functional at the least amount of cost and with the greatest benefit of uninterrupted services is trenchless installation. The result is a rehabilitated system with a pipe-within-a-pipe which should have been there in the first place. These systems, old and new, function best when trenchless technology is employed.

Considering that the length of the sewer network, both water and wastewater, surpasses the length of gas lines presently in operation, engineering will be challenged by a multitudinous demand for rehabilitation. For all the reasons mentioned above, not just for economics, the trenchless pipeline combination for new installations surpasses the benefits of a single pipe trench installation. The economic concept of "now vs. later" for rehabilitation of the infrastructure is a factor which the design and municipality engineers are compelled to work with. They should also realize that not only does the trenchless technology alleviate the pressing problem of infrastructure deterioration but it also adds to the quality of the new pipelines.

NEW SERVICES, NEW DESIGN DATA

Most infrastructure networks are so old that the original installation drawings, records, and specifications are for the most part unknown. The complexity of the system has grown so much over the years because of additions, repairs, and alterations that it can be assumed that little intelligent information is available from past historical records. Most of the trenchless engineering problems can be attributed to this factor alone. Insufficient information necessary for the preparatory work, such as estimating liner sizing, the correct pipe sizes, continuous lengthwise pipe profiles and grades, and other technical information makes for a nearly impossible task of determining the exact lining fit. Even though the size variability is considered substantial, sometimes this may pass as unobserved by simple CCTV procedures.

The question then arises as to whether this part of the engineering work should be performed by the contractor or by public agencies. Whoever takes responsibility for this part of the survey will be responsible for measuring the actual length and cross sections of the existing pipelines, and this information will be added to a roster of current information on pipe quality, pipe imperfection, infiltration, and damages. All of this is considered new engineering and should be set forth in legal and technical documents; video tapes should also be a part of the contract documentation. This kind of survey will be very costly if performed by a professional engineering organization as opposed to being handled by a trenchless pipe contractor. Because this work is such an integral part of the job, though, the trenchless pipe contractor should perform this work as an extra service. Finally, the equipment used by the contractor for measuring the pipe should become a permanent part of his inventory.

REVISING OLD RULES OF LIMITATION WHICH ARE NOT APPLICABLE TO THE TEMPORARY BENDING OF U-LINER PIPE

U-Liner Pipe installation requires bending the pipe during manhole insertion. Expectations of how much bending this pipe can withstand were, however, based on the performance of round pipe. Other limitations were set by manufacturers of various plastic materials. In addition, consideration was given to the long life of the pipeline and not to the pipeline's short-term need to bend, such as happens in temporary bending during manhole insertion. Trenchless technology imposes new requirements where temporary bending limitations are important and cannot be limited by bending codes appropriate to other types of pipe in other situations. The recent NASSCO specifications for Sewer Pipe Lining specifies that the bending radius for the round sliplined pipes, from which the size of the excavated trench for a guiding pit is determined, is 35 times the diameter of the liner. This is a substantial increase from the previously specified manufacturer's requirements.

Engineering for temporary bending can be evaluated as per ASTM D790 where flexural properties of plastics are specified. The maximum fiber stress of the plastic pipe material is used as a governing factor in determining the allowable stresses during bending. The formula is as follows:

$$S = 3PL/4bd^2$$

for a one-half support span. In the case of long-term, permanent installations of round polyethylene pipe, approximately 2% of the fiber stress could be safely applied to the criteria for bending. The criteria is based on avoidance of kinking in the round cross section of the pipeline. Computed by this method, a permanent safe bending of the round pipe is 20 to 25 D for SDR 11 to 13, 25 to 30 D for SDR 13 to 17, and 30 to 35 D for SDR 17 to 25. The semi-temporary coiling, such as for the irrigation pipe, is done at the level of 10 to 13 D for SDR 17 for polyethylene pipe. This criteria also applies to butt-fused pipe lines where the pipe is at least of the same strength at the fusion point as at the adjacent pipeline. The criteria for localized long-term bending in the areas of the fittings and branches is 100 D.

Gas Line Installation, Hamburg, Germany

The percent of fiber stress increases

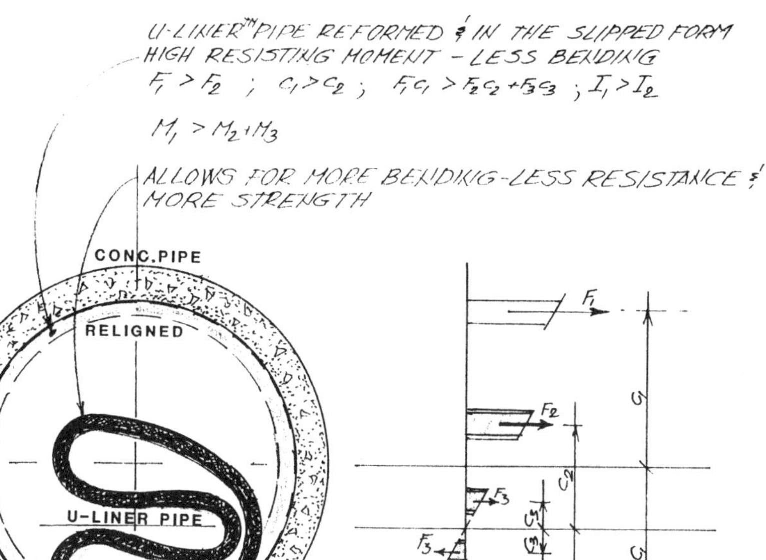

UNCONFINED PASSAGE FOR THE U-LINER PIPE
REDUCED CROSS SECTION
UP TO 1/2 OF THE AVAILABLE CONC.PIPE AREA.

U-LINER PIPE TEMPORARY BENDING CONDITION

from 2 to 10% during temporary bending of the polyethylene round cross section pipe. The following are the results for the long-term round pipe bending at 2% fiber stress and the short-term bending at 5 and 10%:

Fiber stress	at 2%	at 5%	at 10%
SDR 11 to 13	20 to 25*D	8 to 10*D	4 to 5*D
SDR 13 to 17	25 to 30*D	10 to 12*D	5 to 6*D
SDR 17 to 25	30 to 35*D	12 to 14*D	6 to 7*D

U-Liner Pipe is a very special case, different from the previously described cases, in terms of bending. It is a deformed pipe compressed in a U-shape and reduced to 45% from its round configuration. This cross section consists of the four U-Liner Pipe walls with increased moment of inertia and fiber stresses acting closer to the center of gravity for bending. As such, the pipe will surpass the previously-set criteria for bending.

Clear advantages in reducing fiber stress limitations during temporary bending conditions are attributed to the reduced cross section and double strength of the U-shaped U-Liner Pipe. To investigate the limitations of the U-Liner Pipe, Pipe Liners, Inc. tested their product at the laboratory and in the field. The following are test results performed at Southwest Research Institute in San Antonio on the U-Liner pipe for 8" diameter SDR 32 and for 24" diameter for SDR 26: In 1988 the tests were performed on an 8" SDR 32 U-Liner Pipe

SEWER REHABILITATION-U-LINER PIPE

in order to measure and observe the stress and strain induced on the U-shaped pipe section. The moment forces and curvatures were determined for testing on the logarithmic spiral bending shoe, which provided a variety of bending radiuses. U-Liner Pipe samples of low, medium, and high density polyethylene were tested. The objective of the testing was to determine whether a 30" diameter (15" radius) coiling drum would be acceptable for coiling of the U-Liner Pipe. Compared to the bending radius for this size round pipe (96" for 12 D and 15" set for testing), the ratio of the radius reduction was determined to be approximately 6 to 1. The bending radius and moment force measurements on the U-Liner Pipe are as follows:

In conclusion, the bending on the 30" diameter would be more than acceptable for reeling the U-Liner Pipe product without detriment to the polyethylene material. The laboratory experts also did comparable testing of round polyethylene pipes and concluded that the U-Liner Pipe behavior during bending is fundamentally different from the behavior of the round pipe. These results were predicted by theoretical engineering in evaluating the reduced cross section of the U-Liner Pipe within the double wall composition.

The second test performed on the 24" U-Liner Pipe, which has a wall thickness of 1.1", impressed the laboratory experts very much. The visual impression of the strength of the pipe sample during bending was far greater than the old criteria for bending, yet the result was surprising. The U-Liner Pipe, though technicians bent it extensively, conformed to very stringent criteria for bending. The same logarithmic spiral shoe used for testing the 8" U-Liner Pipe was also used on the 24" size. The effect of the creep on the bent pipe was not evident and the effect of the "I" moment of inertia decreased with the decrease of the radius of bending. The "I" decreases as the radius of the curvature decreases.

The beneficial effect of this flexible strength can be observed when the U-Liner Pipe is pulled through a manhole. Because the pipe is substantially smaller in size than the pipe through which the liner is being pulled, the forces of pulling are far reduced from the tension forces allowable for the material. The force concentration may drastically increase if the obstruction in pulling prevents the movement of the pipe. This is controlled not only by the pulling winch limitations but also by careful observation of the pulling and guiding mechanism. The protective sleeves installed at the inlet also minimize the friction of the pipe during the pulling process. The flexibility of U-Liner Pipe has turned out to be very useful in the adaptation of the pipe to given construction restraints. The material tends to adapt itself to the best position within a pipe as the stresses achieve a dynamic balance, therefore preserving the flexibility of the composite structure. Such a high degree of flexibility and

Bending radius in inches:									
11.3	12.7	14.3	16.0	17.9	20.1	22.2	22.6	23.6	25.3
Bending moment in ft-lbs: Low Density PE									
113	108	97	99	93	88	--	83	82*	78
Bending moment in ft-lbs: Medium Density PE									
292	288	281	260	264	249	--	227	--	219
Bending moment in ft-lbs: High Density PE									
--	--	--	--	--	--	--	--	--	559*
*at the point of kinking, the stresses were normal									

Bending radius in inches:					
17.9	20.1	22.6	25.3	28.4	31.8
Bending moment in ft-kips:					
22.7	24.9	25.6	27.8	26.8	29.0
24.2	25.1	27.2	26.9	23.8	18.4
24.7	24.7	23.8	22.4	21.2	19.2
Average ft-kips bending moment:					
23.9	24.9	25.5	25.7	23.9	22.2

strength is important in trenchless rehabilitation.

The above evaluation (theoretical, laboratory, and practical) confirms that the bending of U-Liner Pipe is not detrimental to the product, that a substantially small radius of bending is possible, that the U-Liner Pipe can be coiled and pulled through a manhole without damages, and that a temporary bending in excess of a very small radius is compensated for by the wonderful properties of the material - - its high elongation and flexibility.

From these two examples, the engineering evaluation for U-Liner Pipe sizes in the range from 4" to 24" are predictable. The sizes above 24" are subject to the arch method of evaluation, but it is feasible that with innovation in U-Liner Pipe technology, these sizes will also be included. At this time there is a comfortable level of experience and engineering to support the U-Liner Pipe commercial production and installation in a range of sizes.

ACCESS OTHER THAN MANHOLE

Where there is no manhole or cleanout for access for the U-Liner Pipe, an excavated pit equivalent to the manhole size can be used. This type of minimum excavation is usually found at industrial plant sites, under buildings, above buried lines under or below waterways, or at steep and inaccessible places. The U-Liner Pipe can easily handle these conditions with minimum construction alterations. Today gas lines require direct electrofusion connections, and this process requires an external installation into the pipe liner. Gas pipe service laterals are usually deeply buried pipelines, yet reinstallation requires no more than the work performed during excavation of the pit in place of the manhole.

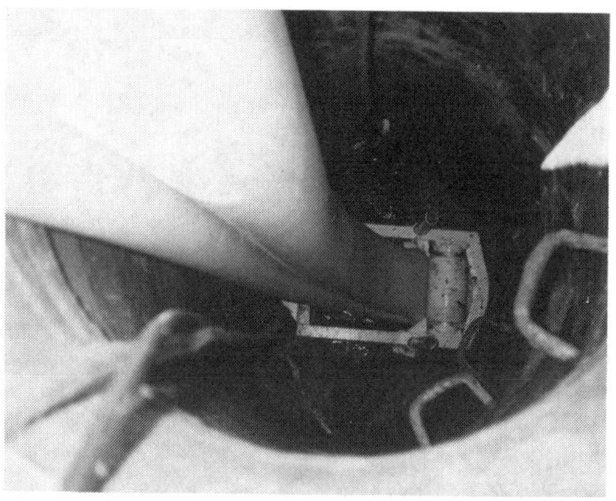

U-LINER PIPE NATURAL CURVE DURING THE PULLING PROCESS

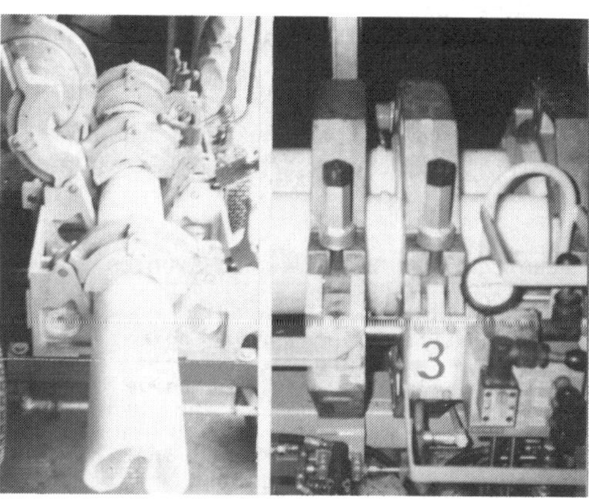

FUSION IN THE U-SHAPE

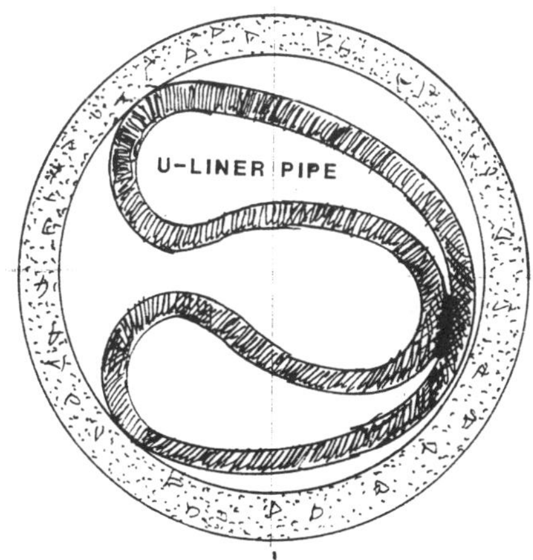

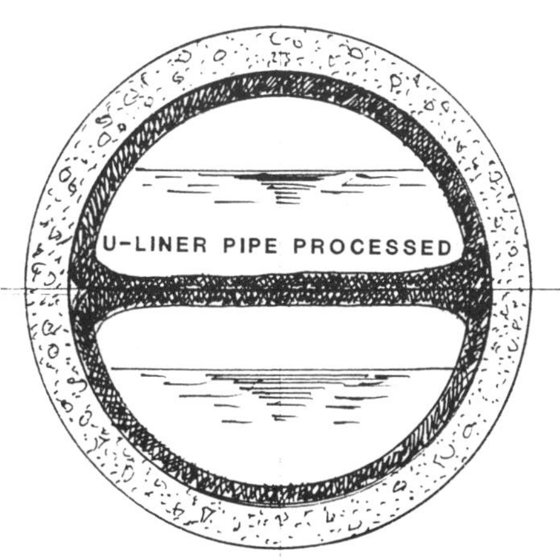

TRENCHLESS MAKING OF A DUAL SYSTEM OUT OF A SINGLE SYSTEM

TECHNICAL ADVANTAGES - CASE STUDIES

Not only does the installed U-Liner Pipe change the hydrological regime previously existing around the pipeline, but the hydraulic regime inside the pipe can be changed, too. The hydrological effect is known as the "French Drain" effect. U-Liner Pipe can be installed in a situation where the combined regime of the hydraulic flow needs to be separated into two totally self-operating systems. This type of case usually occurs in sanitary and industrial sewers. In the case of a smaller pipe in continuity with a larger pipe, a routine installation places a control above the point of overexpansion in the U-Liner Pipe.

The case involving the separation of the two regimes can also be solved by a unique U-Liner Pipe process wherein a mid-section, U-shaped fusion is performed, causing the U-Liner Pipe to become a single pipe with double cross sections. The pipe is processed at the same time with both compartments open and becomes a double pipe within an existing pipe. The advantage of this process is that the maximum area of the cross section of the existing pipe is utilized, therefore maximizing the flow capacities.

INSTALLATION IN HAWAII

U-Liner Pipe was installed in a totally submerged condition in the Pearl Harbor, Hawaii, project. Installation of any other lining process which would require chemical, thermal-melting, and curing procedures was not feasible. The amount of high-pressure water gushing at numerous breaks into the existing pipeline was greater than in any other underground system suitable for rehabilitation. The U-Liner Pipe installation was a risk in this situation, but because of proper engineering and job preparation, the risk was avoided. The installation was performed by totally flooding the pipeline to the top of the manhole but the U-Liner Pipe installation was successfully performed in this submerged condition nonetheless.

SYDNEY CITY HALL INSTALLATION

In this project, the U-Liner Pipe was successfully installed inside the vertical main. The polyethylene material and the pipe configuration remained intact during the installation process thanks to the crystallization memory point employed. This would not have been possible if the material had to be melted or cured in place. The success of the Sydney City Hall installation shows that where lining applications require rehabilitation of pipelines in areas having high slopes, the U-Liner Pipe's performance will be excellent.

UNDER BUILDING INSTALLATIONS

This installation was for a central Florida city requiring restoration of their 71 year-old VCP (vitrified clay pipe) sanitary sewer collection system, identified by the engineers' Sanitary Sewer Evaluation Survey (S.S.E.S.) analysis and report. The current status of deterioration was creating detrimental effects on collection, pumping, treatment, operations and maintenance cost. The complexity of the project involved environmental agency approvals, federal and state grant funding, environmentally sensitive areas, and restoration of inaccessible pipelines.

A bi-level approach was selected to solve the city's problems. Excavation and replacement was specified for pipelines requiring a change in slope, network re-routing, system expansion, and diameter enlargement. A trenchless technology was specified for the remaining pipelines with leaky or separated joints, structural damage, missing and partially dropped pipe segments, cross connections to storm water, road and railroad crossings, and non-excavational areas that include limited easements, legal or jurisdictional considerations, and under-building installations.

One building involved the county courthouse where judicial and governmental affairs are conducted daily. Built in 1921, it rests over approximately 150 lineal feet of the city's old 8-inch sewer. The buil ing's sewer provided several

challenges, not herein before identified. These were: top entry service connections under the inner 1/3 of the building; building settlement from subgrade loss due to sand immigration through joints and missing pipe segments; and vertical misalignment, causing pipe flooding and surcharging. The U-Liner technology, a pipe-within-a-pipe, provided a full range of SDR (standard dimensional ratio) ratings of a time-tested material suitable for wastewater. U-Liner was selected for its capability in providing the required structural integrity necessary, while retaining a property of ductility to allow flexibility in working with the dynamics of its installed environment. The installed U-Liner Pipe created an effective cross-sectional area for the existing 8-inch gravity sewer necessary for the required carrying capacity. The post-installation video showed no evidence of further pipeline flooding or surcharging.

This installation has successfully met both the performance criteria specified and the needs of the client.

Contributing author: Mr. Craig Langworthy, Technical Consultant, , Smith & Gillespie Engineers, Inc., Jacksonville, Florida. Mr. Langworthy is a recognized engineering authority on trenchless rehabilitation in the State of Florida.

GAS INSTALLATION

The installation of U-Liner Pipe in Hamburg received an outstanding evaluation because of its suitability for gas rehabilitation.

HONG KONG

This was very difficult terrain to work in especially because of the steep slopes. Here was the first challenge of the 23.5"-diameter U-Liner Pipe. Pipeliners, Inc.'s preparation and engineering evaluation predicted the possibility of this product development during the early stages.

LONG AND SUCCESSFUL EXPERIENCE WITH POLYETHYLENE PIPES AT THE U.S. AIR FORCE BASES

U-Liner Pipe was used on virtually every utility system supporting MacDill Air Force Base, Florida.

MacDill Air Force Base is a typical tactical Air Command installation housing a fighter wing, unified commands, and several other tenant organizations. The base is a microcosm of any city or municipality in the United States requiring the basic infrastructure facilities to perpetuate its existence.

MacDill Air Force Base has been in existence a little over 50 years and the utilities have experienced the normal degradation associated with use. As in similarly aged towns and cities, repairs to these systems in time and material became substantive. So that local engineering firms are not helpless when their city systems begin to degrade, it behooves the engineering community to stay current with the available technologies. The engineers at MacDill did just that and are pleased with the results of installing U-Liner Pipe.

Over the past 25 years, one of the most utilitarian materials developed has been polyethylene (PE) pipe. At MacDill, PE pipe is used on virtually every utility system supporting the base, including waterline, sanitary sewers, stormwater drainage, natural gas, and HVAC systems.

The water supply to MacDill is

U-LINER PIPE IN HONG KONG

supplied by the City of Tampa at three points. Because the source of the water is primarily dependent on the Hillsborough River, augmented by wells, the water contains a high level of calcium, causing significant flow and pressure problems. When low pressure becomes prevalent throughout the system, the natural tendency is to boost the pressure using additional higher head pumps. Yet boosting the pressure causes breakdown of the lead-filled joints which results in substantial repair time. To eliminate this problem, MacDill has elected to use PVC and PE pipe. Over the past 17 years, use of these materials has led to the significant upgrading of the system.

The sewage collection system was particularly adaptable for using PE pipe. Two of the largest pumping stations on the base had pumps burn out within a period of three months. The problem was to repair the station and convert the wet well/dry well station into one large wet well, all in two days without interruption of service. Using the thermoplastic property of PE, the pipe was first forced into a bypass configuration and connected to the force main. There using a check valve and knife gate valve in conjunction with the PE bypass, the new configuration was used to redirect the flow of sewage around the station. This procedure took approximately two hours. The reconfiguration of the rest of the station took less than two days.

The submersible pump bases were bolted in place and the PE pipe was fused and fitted to the bases to form the 16-foot-high risers, and the 40-foot-long headers (including the 90-degree elbows).

The remaining pipe configurations were made by fusing the pipe to conform to the variable angles necessary to connect to the force main. It might be noted that the whole PE pipe assembly can be held overhead by one person, replacing over 5,400 pounds of cast iron pipe and fittings. The ease of fabrication and handling accounted for the speed and accuracy of the installation, and thus the bypass configuration is now used on all of the base lift stations. The bypass allows a single gasoline pump to be used to bypass all lift stations for repairs, replacement, or maintenance without providing elaborate voltages for generators and supporting electrical equipment for providing backup support.

Two-thirds of MacDill Air Force Base is surrounded by Tampa Bay. Originally a low-lying land mass, the area was augmented with great amounts of fill, a portion of which unfortunately had pockets of soil containing various concentrations of sulfide. The gas mains, which were installed over 40 years ago, were not cathodically protected, and when exposed to the combined forces of the brackish water and sulfide soil, they corroded. Steel valves acting as anodes experienced varying degrees of pitting and became inoperable or very difficult to operate.

Using PE pipe and valves, MacDill virtually ended its maintenance problems on the repaired/replaced sections. Using insertion methods, the PE pipe was run through the existing cast iron and steel lines. The 32,000 linear feet of PE pipe and valves effected a 32% reduction in base gas use over a construction period of six months. An emergency repair of the gas mains servicing 50 single-family homes in the Military Family Housing area (which are totally dependent on gas for heating, cooking, hot water, drying clothes) was completed in approximately 30 days. Individual homes were interrupted only for the period of time that it took to transfer the respective service. The speed of replacement was due to the ease of handling PE pipe, installation, and fusion of the pipes and valves.

With the construction of the new base commissary, the old commissary building was slated for conversion to a "mall." The existing HVAC system serving the facility was over 20 years old and was kept in place to serve the new businesses that moved into the vacant areas of the building. Because the tenants required a more efficient HVAC unit, the existing 100-ton unit was scheduled to be removed; it was cost-prohibitive and maintenance-intensive. To support the proposed new 40-ton unit, new chilled water piping was required. Using PE pipe, the new chilled water system was lighter, could be installed faster, had better flow characteristics, did not necessitate use of the better insulation corrosion control chemicals, had a longer life with less maintenance, and could be fabricated using preconfigured branch takeoffs. This would supply future tenants with additional multi-ton chillers as they were required.

Contributing author: Michael J. Cawley, M.E., U.S. Department of the Air Force, MacDill Air Force Base, Florida. Mr. Cawley's extensive experience with PE products and his professional evaluations represent not only a credit to his successful career, but the data shared here is of significant technical value to the engineering community. Mr. Cawley was extensively involved in the evalution of the U-Liner Pipe system based on his experience with polyethylene.

CONCLUSION

Trench and trenchless pipe construction systems, by virtue of composition of the pipe and technologically advanced methods of installation, can be considered in many cases at least equal. But as the knowledge of trenchless technology propagates the lining industry, it will become a major part of all new pipeline construction as it is now in rehabilitation. As the trenchless rehabilitation of pipelines predominates and becomes a widely familiar technology, it will not be difficult to find a rationale for planning a brand-new trenchless pipeline.

Presently an enormous need exists for rehabilitation of the infrastructure, but powerful negative forces govern the market. As engineering philosophy and designs change, however, this market will mature and recognize fully the pipe liner industry. New products, inventions, and innovations will probably derive from the resulting competition, and they deserve the freedom from antiquated or inapplicable regulations in order to prosper and bring great benefits to the world's infrastructures.

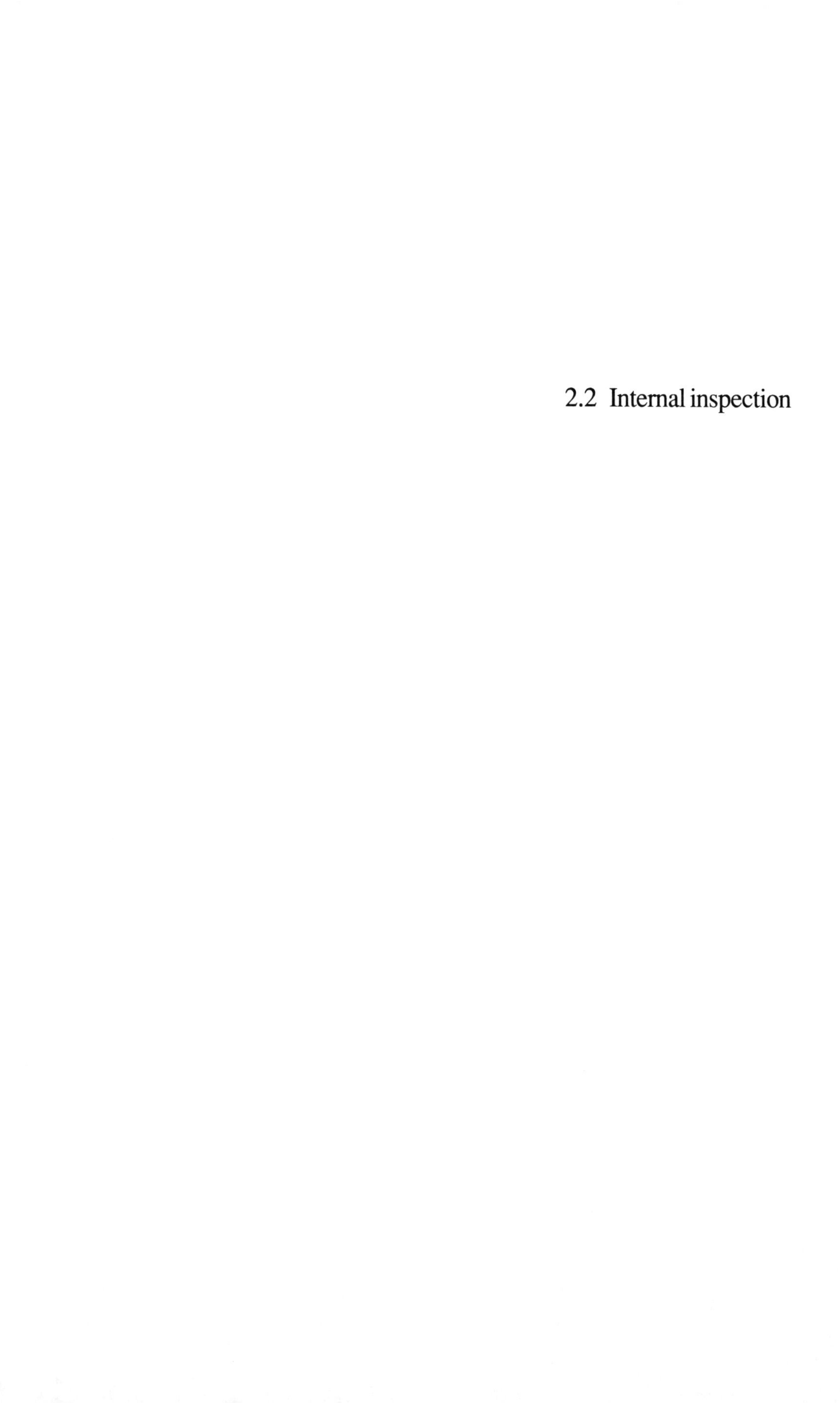

2.2 Internal inspection

No Trenches in Town, Henry & Mermet (eds) © 1992 Balkema, Rotterdam. ISBN 90 5410 085 0

A systematic development in designing a special inspection robot for use in private sewer lines

J. Schlattmann & J. Niewels
Laboratorium für Konstruktionslehre, Universität GH Paderborn, Germany

ABSTRACT: Leakage of private sewer lines is contributing considerably to contamining ground water with potential hazardous liquids. Therefore, environmental protection activities are more and more focussing on this part of sewage system. The essential step in eliminating the existing potential hazard is to provide reliable systems for a thorough inspection. Emerging from spotlighting the state of the art in this field, the paper addresses aspects of how a future robot insepction system specially designed for this task could look like.

1 INTRODUCTION

The necessity to check in particular the private sewer line systems is of crucial importance. The potential hazards are caused not only by improper installation, but also by external mechanical and geological effects.

A further problem is the superannuation of existing sewer pipelines with all its problems like obstructions caused by roots etc.

Nowadays action is only taken if a line is suspected to be faulty. There is almost no chance to detect exfiltration of toxides from the private sewer lines into the ground water so far. However, if a line is suspected to be defective it has either to be laid bare for the whole length or inspected from the inside.

The former is by far to expensive, the latter can be done by men only from a certain diameter on. Therefore a mechanical device has to be used for this task.Thanks to the advanced microelectronics video cameras have been miniaturized to such a degree that they can be used for this purpose. The standard application is either to push the camera into the pipe or to mount it on a carrier vehicle which then travels through the pipe.

The below stated classification also describes the differences between the two most common types of pipe inspection systems. Both types have gone through sophisticated development yet there are still limitations regarding their range of possible applications. In particular, the current systems are not able to cope sufficiently with the problem of inspecting sewer side line from the main pipe neither by quality nor quantity.

What is needed is the creation of new visionary approaches to this problem. Therefore the intention of this contribution is to systematically elaborate, state, and evaluate new principles of action in order to overcome above mentioned restrictions.

This paper has deliberately been restricted to inspection of private sewer side lines, yet the results may be applied to other pipe systems as well.

2 VISIONARY CONCEPTS FOR TECHNICAL HIGH-LEVEL INSPECTION DEVICES

Here the inspection of private sewer side lines from the premises to the sewer main is addressed. The goal is to detect either exfiltration or infiltration from the pipe to the ground water.

An abstract formulation of the design and development task to perform could be: Realization of an inspection unit for monitoring and documenting the system status automatically. Figure 1 shows the functional action structure of such a system.

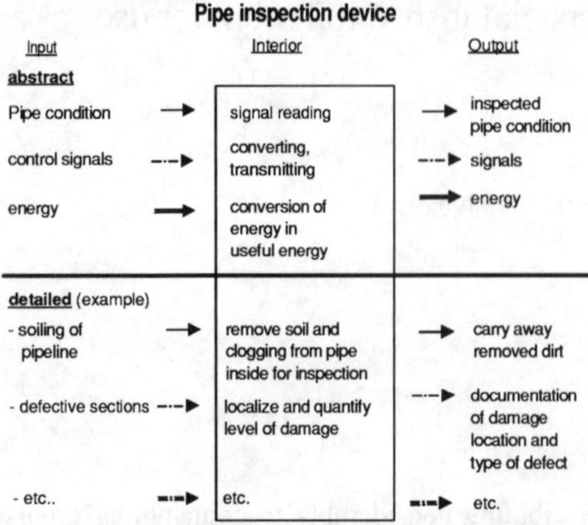

Figure 1. Functional action orientated description of a pipe inspection system.

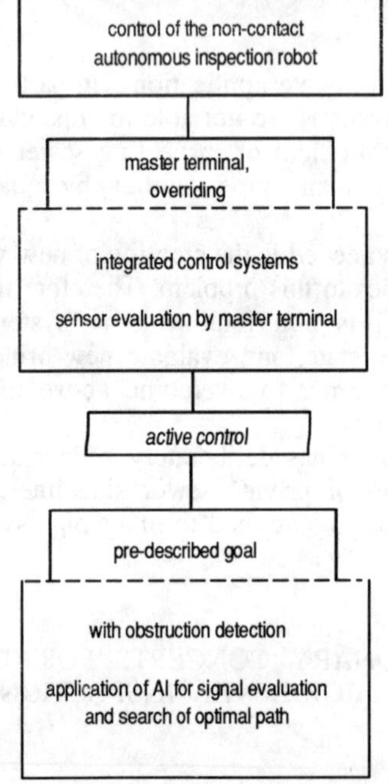

Figure 2. Control structure for a high level inspection system.

3 SYSTEMATICAL DESIGN EVALUATION

A possible way of detection is given by applying non contact sensor systems. These are e.g. infrared, radar or supersonic systems. Basically cameras belong to the group of non-contact sensors as well, yet they have to be inserted into the pipe and require a clean pipe surface to transmit reliable pictures.

All systems have one thing in common: They all have certain system related limitations. As an result the different types of action may be more or less applicable to different inspection task profiles. In general, camera inspection seems to be ahead of all other principles. The critical point of this system is to bring the camera in the sewer system and to control and operate the camera in a flexible, smooth, and reliable fashion.

Consequently, it is most apropriate to concentrate systematical improvement activities on the development of a flexible and reliable support and control system. Figure 2 describes the according control structure needed to operate a flexible autonomous inspection system effectively. Figure 3 represents a schematical overview of different action principles regarding transportation and displacement of the inspection device in the given environment.

4 CONCLUSION

The enormous increase in environmental awareness as well as alarming signals regarding the contamination of ground water with nitrate and other toxides are calling for thorough investigation and monitoring of the private sewer lines. Efficient inspection of these pipe systems can only be performed from the sewer mains. Due to the extremly difficult operating environment a systematical approach in designing a flexible system adapted to described conditions is required.

This task needs new unconventional approaches. The emphasis of this paper is focused on a fundamental design analysis in order to find well founded technical solutions helping to control the environmental threat.

REFERENCES

Kurokawa, Mitsuoka, Nakada. Development of pipeline inspection pigs. Proceed. of the Int. Mech.& Arctic Engineering Symposium. 1990.

Okada, Sanemori. Vehicles in pipe for monitoring inside of pipe. IEEE Journal of robotics and automation. 1987

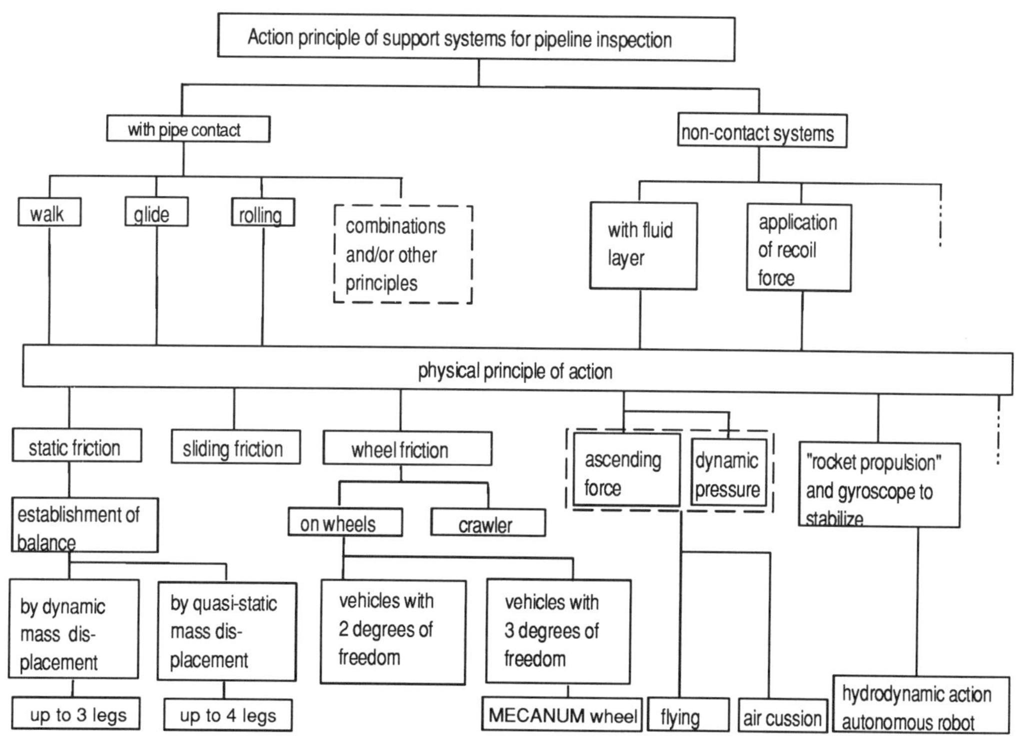

Figure 3. Action principles for a transportation transportation device of an inspection camera.

Douglas. A multi-task will inspect pipelines internally. Pipeline Industry.1986

Pahl, Beitz. Konstruktionslehre. Springer Verlag. 1986

No Trenches in Town, Henry & Mermet (eds) © 1992 Balkema, Rotterdam. ISBN 90 5410 085 0

Application of nuclear gamma-gamma and neutron-neutron log tools for sounding the immediate environment of non-inspectable sewer pipes

C. Schwarze
Centre de Recherches et d'Essais Appliqués aux Techniques de l'Eau, France

ABSTRACT: The causes of sewer pipe damages often results in geotechnical mecanisms leading to the creation of interface defaults between the pipe and the supporting soil: voids, non–compacted zones, wet points, hard points, ...At present, the traditional means of sewer diagnosis, essentially TV inspection, do not enable us to detect these types of anomalies in the immediat environment of sewer pipes. Consequently, a study was undertaken at C.R.E.A.T.E. (Colombes, France) to evaluate the possibilities of the application of different non–destructive investigation techniques used in geophysical prospecting, for sounding the environment near to pipes. Gammadensity and neutron logging probes, which are used in bore–holes to measure respectively the density and the water content of the formation croosed, are one of the most promising techniques. The experimentations carried out on several test sites and working networks with LCPC and BRGM probes showed that these tools were, in their present form, able to detect, locate and identify density defaults and excesses (voids, under–compacted zones, hard points) and excessive moisture contents (exfiltration at seals) adjacent to the pipe. The tests enabled us to find the limits of these techniques and their usage constraints in site conditions.

1 INTRODUCTION

Sewers are currently the most neglected of urban infrastructures. In France, the proportion of networks in poor condition requiring reconditioning is estimated 8% of the total length, which represents 8000 km of non–inspectable sewer pipes. Investigations carried out in the past on damaged sewer pipes showed that the causes of the defaults, mainly geometry problems (pipe uncoupling and offset, etc) were essentially external, and resulted in mecanisms leading to the creation of interface defaults between the pipe and the supporting ground: disturbance and varying settling at the bottom of the trench, damage to the laying bed and the in–fill by water circulation, causing uncompacted or even empty areas, wet points, etc.

The traditional means of diagnosis of sewer pipes do not currently allow external causes of problems to be identified, because of the non–availability of sounding techniques for the immediate environment (support ground + in–fill) of the pipe.

Many non–destructive continuously working techniques have been used for many years in geophysical prospection on the surface and in bore–holes, but, until now, they have not been applied in non–inspectable sewer pipes (diameters from 200 to 800 mm).

It is for this reason that the C.R.E.A.T.E. (Centre de Recherches et d'Essais Appliqués aux Techniques de l'Eau, Colombes, France) started a study in 1989 on non–destructive investigation techniques for sounding the immediate environment of non–inspectable sewer pipes. The purposes of this study are to find and experiment various existing non–destructive investigation techniques with a view to their application on sewer pipes, to evaluate their performances and limits under experimental conditions and, if necessary, to develop a sounding methodology suitable for this field of application. Among the various techniques tested, such as groung probing radar, geophysical resistivity measures, micro–seismology, ..., nuclear gammadensity and neutron logs, used in bore–holes and measuring respectively the wet volumetric weight (density) and the volumetric water content (humidity) of the ground at depths up to 20 cm, have given good results.

The article presents the tests carried out with two gammadensity logging probes, from the BRGM (Bureau de Recherches Géologiques et Minières, Orléans, France) and the LCPC (Laboratoire Central des Ponts et Chaussées, Trappes, France) respectively, and a neutron logging probe from the LCPC on various experimental sites and an operating sewer pipe.

2 NUCLEAR GAMMADENSITY AND NEUTRON LOGGING PROBES

2.1 *Principle of gammadensity logging*

The ground is subject to gamma radiation from a radioactive source. The gamma photons collide with the material. According to their energy, three types of interaction may arise. Of these interactions, there is one, called the Compton effect, which takes place when gamma photons collide with electrons which are weakly linked to atoms. The photon is deviated and communicates a part of its energy to the electron in the form of kinetic energy.

The gammadensity logging probe measures, by back-scattering and at a given distance from the source, the intensity of the gamma photons given off by the Compton effect. This increases with the electronic density of the material, which is itself a function of the wet volumetric weight of the ground. An increase in gamma counts, registered by the tool, indicates, as an initial approximation, a reduction in the density of the ground, and vice versa.

2.2 *Principle of neutron logging*

The material is subject to radiation by fast neutrons. The neutrons emitted penetrate into the material, collide with the atom cores nuclei making up the material, are slowed down and lose part of their energy. As the hydrogen nucleus has mass comparable with that of a neutron, the hydrogen atoms present in the formation, mainly in the form of water molecules, reduce the most the neutrons transmitted. The neutrons, slowed down by successive collisions, change into the "thermal" state when their energy becomes comparable with the agitation energy of the atoms present, before finally being absorbed by the nuclei of certain elements present in the environment: chlorine, iron, etc.

The thermal neutron logging probe records the back- scattered flows of the thermal neutrons at a fixed distance from the source. In practice, this is a means of determining the volumetric water content in the ground. An increase in neutron measurement registered by the probe indicates an increase in the water quantity, and vice versa.

2.3 *Description of the probes tested*

2.3.1 *The BRGM gammadensity probe*

The BRGM gammadensity probe (made by Mount-Sopris USA) is a device with 2 detectors, one of 10 cm (Small Distance detector, SD), and the other of 35 cm (Long Distance detector, LD) from the source. These, consisting of from ^{137}Cs, emit with an intensity of 100 mCi. The source and the detectors are collimated on the same generating line. The length of the probe is 2.20 m, its diameter 57 mm and its weight 17 kg.

2.3.2 *The LCPC gammadensity probe* (Figure 1)

The LCPC gammadensity probe (made by LCPC) consists of 2 detectors, 20 cm (SD) and 40 cm (LD) from the ^{137}Cs source, which emits with an intensity of 60 mCi. The length of the probe is 1.50 m, its diameter 46 mm and its weight 8 kg.

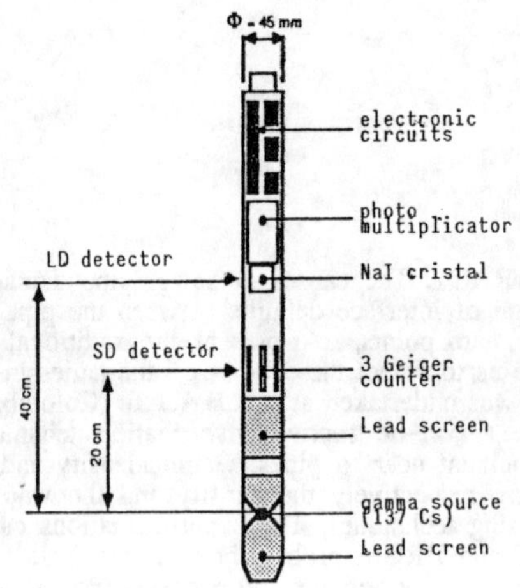

Figure 1. Diagram of the gamma probe of LCPC

2.3.3 *The LCPC neutron probe* (Figure 2)

The LCPC neutron probe (made by LCPC) has a set of 3 ^{241}AmBe sources (3x200 mCi) and 2 ^{3}He detectors. The SD detector is centered around the source, and the LD detector is situated 40 cm from the source. The length of the probe is 1.70 m, its diameter 46 mm and its weight 5 kg.

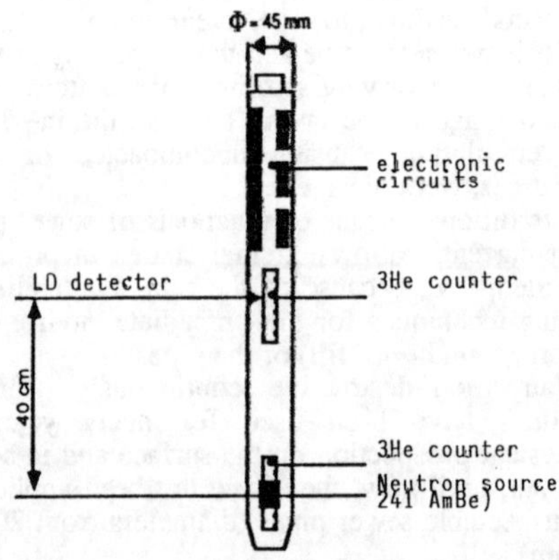

Figure 2. Diagram of the neutron probe of LCPC

The remaining volume of the different probes is occupied by electronic circuits supplying the probes and processing signals, transmitted via the traction cable to the data acquisition and processing unit.

The operation equipment consists schematically of a variable speed motor–driven winch, a coding wheel for the acquisition, through the cable, of the distance (depth) covered by the probe and an acquisition and processing unit controlled by a microcomputer. All this equipment is carried on a light motor vehicle.

3 TESTS OF GAMMADENSITY AND NEUTRON LOGGING PROBES ON CREATE EXPERIMENTAL TRENCHES

The LCPC and BRGM nuclear probes were tested on 2 experimental trenches at CREATE. The purpose of this first phase of experimentation was to establish the feasibility of their use in their existing form in non– inspectable sewer pipes, as a mean of sounding the immediate environment of the pipes. This was done by determining if the gamma and neutron radiation passed through the pipe material and provided information about the characteristics of the in–fill, taking into account the pipe material, the thickness of the walls, the type of in–fill and the presence of water in the pipes.

N.B. For these (and the following) tests, we took into consideration only the gamma and neutron count rates, which allow a qualitative interpretation of the changing soil conditions. Quantitative measures of volumetric weight and water content of the soil surrounding the sewer pipes require calibration of the tools on pipe samples for each material and diameter.

3.1 *Description of the test site*

2 trenches with the dimensions 15x1x1.5 m (Lxlxh) were made in the C.R.E.A.T.E. test ground. One of the trenches was filled in with mason sand, and the other with gravel. In each trench, the in–fill was compacted in layers over a length of 5 m starting from one end using a vibrating plate, in conformity with the LCPC/SETRA recommendations on the compacting of in–fills in trenches. Over the remaining length, the in–fill was not compacted. A plywood board separated the 2 compartments.

Artificial defaults were created at different points in the non–compacted part of the in–fill of the 2 trenches, to create "targets" and simulate the presence of:

* voids: blocks of expanded polystyrene
* uncompressed points: plastic bags containing a 50/50 mixture of in–fill material and expanded polystyrene balls
* water points: cans of water
* wet points: plastic bags containing the in–fill material saturated in water
* buried pipes: 200 mm dia. asbestos–cement pipe and 95 dia. cast–iron pipe.
* hard points: blocks of granite

A side–view showing the position of all the defaults in the sand trench is shown on figure 3.

In each trench, a pipe of about 15 m long, 400 mm diameter and made of the same material (reinforced concrete, asbestos cement, sandstone) was laid in the in–fill and filled in up to half of its height.

This arrangement does not correspond with the reality of buried pipes, but has the advantage of allowing the nature of the pipes (concrete, asbestos–cement or sandstone) to be changed and in this way several configurations of in–fill/pipe to be tested.

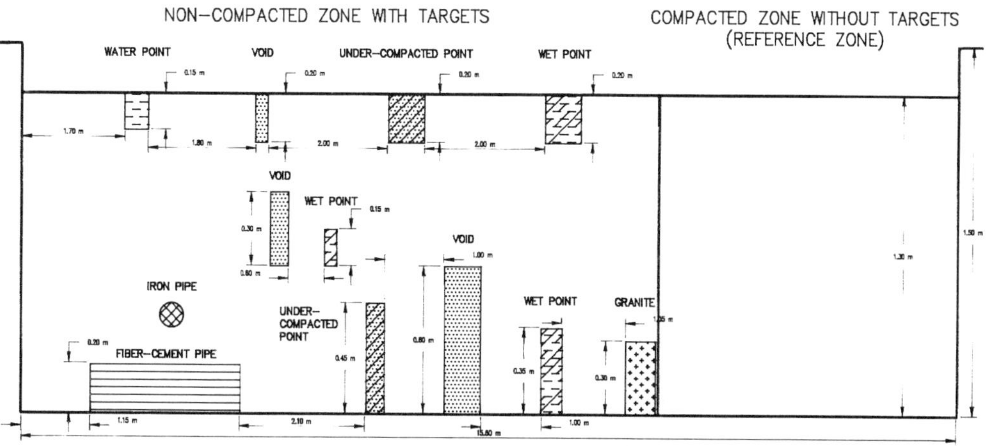

Figure 3. Side-view of sand-filled trench at CREATE showing target positions

3.2 *Test procedure*

The probes were pulled along the bottom from one end to the other of the different pipes, dry or filled with clean water (water height approximately 10 to 20 cm).

3.3 *Analysis of results*

The results are presented in the form of graphs representing the variations in counts (unit:counts per second) depending on the distances (unit:meters) run by the probes (measurement pitch: 10 cm) and showing, for the gamma probes, qualitative variations in the humid volumetric weight in place, and for the neutron probe, the qualitative changes in the volumetric water content.

The results obtained show that :

* The LCPC gamma and neutron probes are in their current form capable of detecting, localizing and of identifying the defects (and excess) of volumetric weight and water content of the ground adjacent to a 400mm diameter asbestos-cement pipe (Figure 4). The same results are obtained, although less clearly, for a sandstone pipe of the same diameter, whilst for a reinforced-concrete pipe the count variations do not allow the clear detection of the anomalies.

* The most powerful BRGM collimated gammadensity probe can detect defaults through 400mm diameter reinforced concrete.(Figure 5)

* The detection, localization and differentiation of defaults still remains possible when the pipe is filled with water and the probes immersed (Figure 5).

* The gamma probes allow the compacted and non-compacted zones in the trenches to be discerned (Figure 4).

* The neutron probe appears to allow the detection of water exfiltration through the sealings of a pipe filled with water.

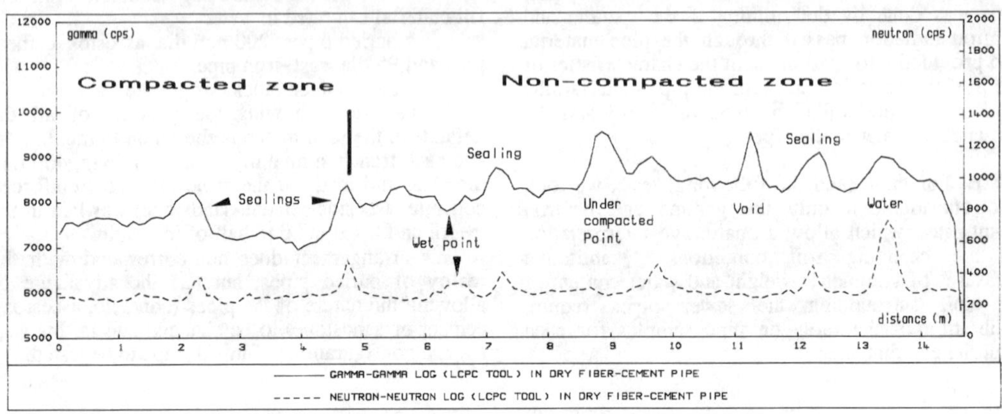

Figure 4. Gamma and neutron count profiles obtained in dry fiber-cement pipe

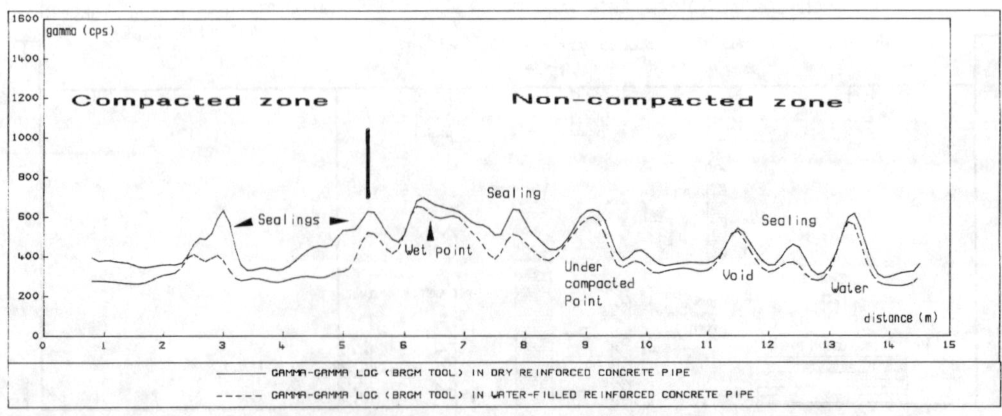

Figure 5. Gamma count profiles obtained in reinforced concrete pipe

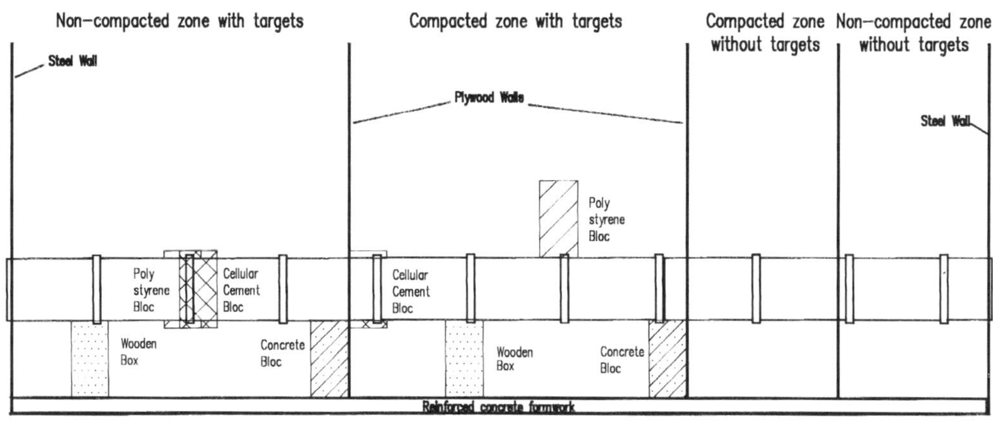

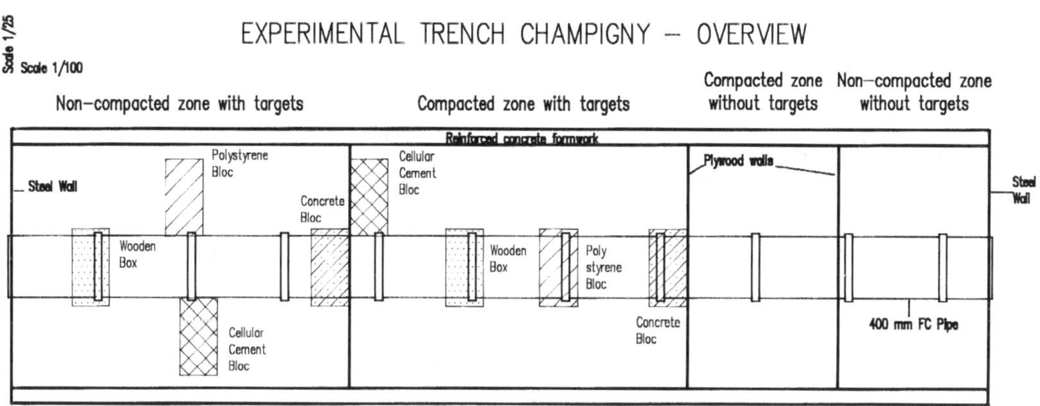

Figure 6. Experimental trench at Champigny

4 GAMMADENSITY AND NEUTRON LOGGING PROBE TESTS AT THE EXPERIMENTAL TRENCH AT CHAMPIGNY s/MARNE

The tests were continued with the BRGM and LCPC gammadensity and neutron logging probes on an experimental pipe, built by the DSEA 94 (Direction des Services de l'Eau et de L'Assainissement du Val–de–Marne) at Champigny sur Marne.

The aim of this second phase of experimentation was to strengthen the results obtained in CREATE's trenches, to evaluate the performance of these tools in conditions approximating reality and to determinate the parameters of optimum measurement and investigation procedure.

4.1 *Description of the test site*

The DSEA experimental trench at Champigny s/Marne is constructed in elevation from reinforced concrete casing, measuring about 26 meters in length, 1.5 meters in width, and 2.4 meters in height. The ends of the trench are shut by metal plates. The unit's watertightness is ensured by pointing in epoxy resin, bitumen sheets, and silicon mastic.

A 400mm diameter asbestos–cement pipe is placed 50cm from the ground in a sand embankment. The trench is divided into 4 zones: zones with compacted or non–compacted in–fill and with or without defaults.(Figure 6)

The defaults, simulating defects or excess of volumetric weight, are adjacent to the pipe, all of identical size (1x0.6x0.5 m), and positioned under (position 6.00am), on (position 12.00am) and at the sides (positions 3.00 and 9.00am) of the pipe.

An aquifer with variable water table can be simulated.

4.2 *Test procedure*

The probes were pulled along the pipe at generating

lines spaced every two hours : 12.00, 10.00, 8.00, 6.00, 4.00, 2.00 am, and at speeds varying from 1.5 to 20 m/min according to the test series. In order to achieve the passage of the probes along generating lines other than at 6.00am (simple traction along the bottom of the pipe), they were mounted upon a probe–carrying trolley, which kept them at the desired position during the passage.

4.3 *Analysis of results*

The results are presented in profile form showing the variations in the counts obtained from the passage along of the 6 generating lines, as well as in the form of iso–count rate cartographies.

An example of each result representations is given by the figures 7 and 8, showing the counts recorded by the BRGM gamma probe in the dry pipe.

We observes here that each default adjacent to the pipe and simulating a defect or punctual excess of density, is detected and localized in linear and hourly positions.

The presence of compacted zones (with a higher density) results in a reduction of the average level of gamma counts of around 300 cps, visible on all the profiles carried out, except that at 6.00am which agrees with the reality, it being given that the pipe's laying bed was not compacted. On the isovalue map, the compacted zones show themselves by the zone, drawn more darkly, situated between 10 and 22 meters.

Parallel to this experimentations, tests were carried out on half shells of isolated asbestos–cement and concrete pipes of different diameters (300 to 800 mm), layed in a mini–trench filled with sand and containing Wooden boxes, simulating dry voids or voids filled with water, with volumes corresponding to the probe's theoretical investigation volume, or a half or a quarter of this.

The aim of this tests was to

* determine in controlled conditions the performances (default detectability) and the limits of these techniques regarding material and thickness of pipe, in the presence of water as well as when the probes are detached from the pipe's wall.

* test the efficiency of a screen focussing the radiation emitted and received by the LPCP neutron probe with the view to increase its performance regarding this.

* establish clearly whether the neutron probe detects and localizes water leakage through the seals

5 GAMMADENSITY AND NEUTRON LOGGING PROBES ON A SEWER PIPE IN USE

LCPC gamma and neutron probes were used in a 500 mm diameter asbestos–cement sewer pipe at Igny, Essonne, France. A geotechnical survey had been carried out for this pipe, 3 years old, previous to the construction works, so as to understand the geotechnical environment. It had also been the

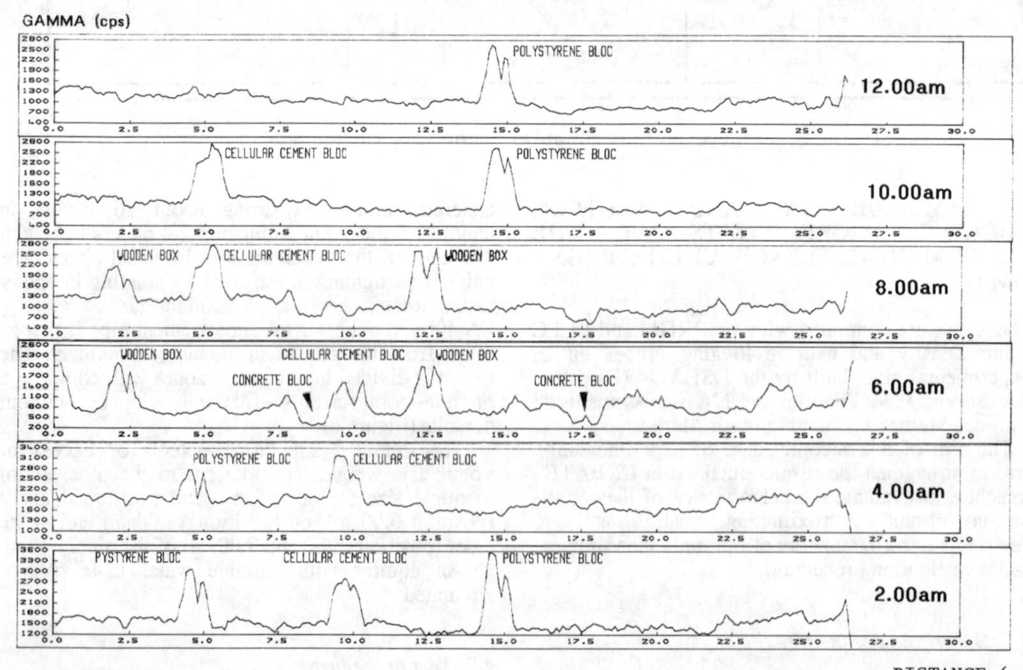

Figure 7. Gamma count profiles obtained in test pipe at Champigny site

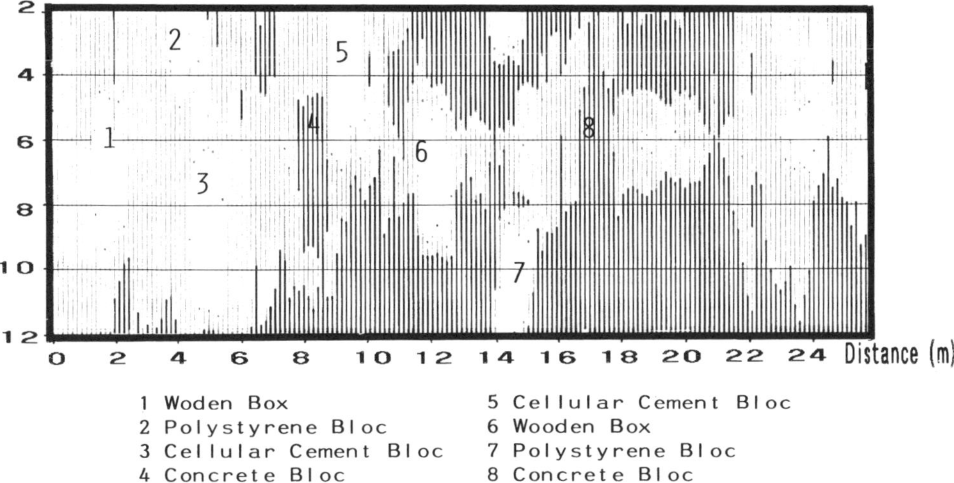

Figure 8. Iso-gamma counts cartography obtained in test pipe at Champigny site

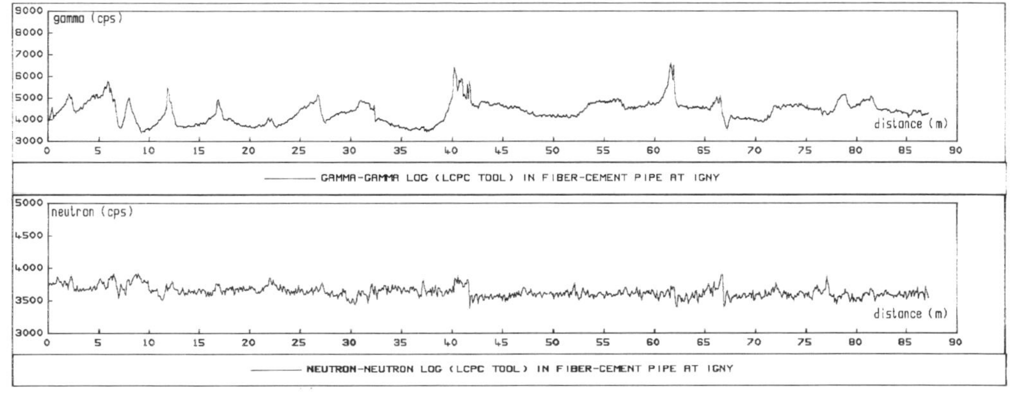

Figure 9. Gamma and neutron count profiles obtained at Igny site

subject of investigations to measure the geometry of the network, in order to know the fluctuations along its length profile, to map its degree of water-tightness and the degree of compacting of the infill.

All the information collected showed that zones existed where the pipe had sunk due to the effect of settling at the bottom of the trench, which was decompressed when the trench was opened.

5.1 *Test procedure*

The gamma and neutron probes were introduced into the pipes via inspection doors and pulled along the bottom at a speed of 2.5 m/min.

In total, 6 sections were passed, totalizing a length of 90 m. The tests lasted around 6 hours.

The pipe remained in service during these tests. The height of the water level was around 10 cm.

5.2 *Analysis of results*

The gamma and neutron count profiles are shown on figure 9. We observes :

* several gamma count peaks, corresponding to a reduction in volumetric weight all along the profile and almost certainly due to uncontrolled displacements of the gamma probe.

* an increase in the average level of gamma counts, resulting in an decrease in volumetric weight, passing from the section 0 to 40 meters to the section between 40 and 85 meters, with a light diminution of the average level of neutron counts at

the same points. A first interpretation consists in saying that the zone between 0 and 40 meters is a zone where the subsequent settling of the bedding soil of the pipe has been particularly significant.

6 CONCLUSIONS

The tests carried out with the LCPC and BRGM gammadensity and neutron nuclear logging probes on experimental sites and on a pipe in use, have shown the capacity and interest of these tools for detecting, localizing, and identifying punctual and extended defaults or excesses of volumetric weight (voids, hard points, decompacted zones), and of the volumetric water content (humid zones, exfiltration points at sealings) of the ground surrounding small diameter sewage pipes.

The tests showed that these sounding techniques can be carried out when the pipe is in use.

REFERENCES

CSTB (1989) : Résultats de l'enquête relative aux travaux de réhabilitation des réseaux d'assainissement ; AGHTM

LCPC (1985) : Influence des conditions géotechniques sur le comportement des canalisations d'assainissement ; Agence de Bassin Loire–Bretagne

BARON J.P., CARIOU J., THORIN R. (1989) : Les diagraphies nucléaires développées par les Ponts et Chaussées ; LCPC, Bull.Liaison.Labo. P.et Ch., n°164, pp.17–24, nov–déc.1989

No Trenches in Town, Henry & Mermet (eds) © 1992 Balkema, Rotterdam. ISBN 90 5410 085 0

Application des sondes de diagraphies nucléaires gamma-gamma et neutron-neutron a l'auscultation de l'environnement immédiat de canalisations d'assainissement non-visitables

C. Schwarze
Centre de Recherches et d'Essais Appliqués aux Techniques de l'Eau, France

RESUME: Les causes de la dégradation des canalisations d'assainissement résultent souvent de l'action de mécanismes géotechniques conduisant à la création de défauts d'interface entre la canalisation et le sol support: Vides, zones non–compactées, points humides, points durs,...A l'heure actuelle, les moyens traditionnels, essentiellement l'inspection TV, ne permettent pas de détecter ce type d'anomalies. C'est pourquoi, une étude a été entreprise au C.R.E.A.T.E. (Colombes, France) afin d'évaluer les possibilités d'une application de diverses techniques d'investigation non–destructive, employées en prospection géophysique, à l'auscultation de l'environnement proche des canalisations d'assainissement. Les sondes de diagraphies nucléaires gamma–gamma et neutron–neutron, qui, employées en forage, mesurent respectivement la densité et la teneur en eau des formations traversées, constituent une des techniques les plus prometteuses. Des essais effectués sur des sites expérimentaux et réseaux en service avec des sondes du LCPC et du BRGM ont montré, que ces sondes étaient, dans leur forme actuelle, capables de détecter, localiser et identifier des défauts et excès de densité (vides, zones sou–compactées, points durs) et excès d'humidité (exfiltrations aux joints) du terrain adjacent à une conduite. Les essais ont permis de déterminer les limites de ces techniques et leur contraintes d'utilisation en conditions de terrain

1 INTRODUCTION

Les réseaux d'assainissement sont, actuellement l'infrastructure urbaine la plus négligée. En France, la part des réseaux dégradés et nécessitant une réhabilitation est estimée à 8 % du linéaire total, ce qui représente 8000 km de canalisations non–visitables. Il ressort des enquêtes menées dans le passé sur des canalisations d'assainissement dégradées que les causes des désordres, essentiellement des défauts de géométrie (déboîtements et décalages des tuyaux, etc) sont presque toujours d'ordre externe et résultent de mécanismes géotechniques créant des défauts d'interface entre la conduite et le sol d'assise : remaniement et tassements différentiels du fond de fouille, désorganisation du lit de pose et du remblai d'enrobage des tuyaux par des écoulements de nappe, créant des zones décompactées voire des vides, des points humides etc.

Les moyens classiques de diagnostic des réseaux d'assainissement ne permettent pas, actuellement, de remonter aux causes externes des désordres, faute de techniques d'auscultation de l'environnement proche (sol d'assise + remblais d'enrobage) des canalisations.

De nombreuses techniques d'investigation non–destructives et travaillant en continu sont employées depuis longtemps en prospection géophysique de surface et en forage, mais n'ont pas, jusqu'à présent, fait l'objet d'applications dans des canalisations d'assainissement non–visitables (diamètres 200 à 800 mm).

C'est pour cette raison, que le C.R.E.A.T.E. (Centre de Recherches et d'Essais Appliqués aux Techniques de l'Eau, Colombes, France) a débuté en 1989 une étude sur les techniques d'auscultation non–destructives de l'environnement proche des canalisations d'assainissement non–visitables. Les objectifs de cette étude sont de rechercher et d'expérimenter différentes techniques d'investigation non–destructives existantes en vue de leur application en canalisations, d'en établir les performances et limites en conditions expérimentales et de développer le cas échéant une méthodologie d'auscultation adaptée au domaine d'application. Parmi les différentes techniques expérimentées telles que le radar géologique, la résistivité géophysique, la micro–sismique, ..., les sondes de diagraphies nucléaires gamma–gamma et neutron–neutron, employées en forage et mesurant respectivement le poids volumique (densité) humide et la teneur en eau volumique (humidité) du terrain sur des profondeur d'au maximum 20 cm, ont donné de bon résultats.

L'article présente les essais, effectués avec 2 sondes de diagraphie gamma–gamma, respectivement du BRGM (Bureau de Recherches Géologiques et Minières, Orléans, France)) et du LCPC (Laboratoire Central des Ponts et Chaussées, Trappes, France), et une sonde de diagraphie neutron–neutron du LCPC sur différents sites expérimentaux et une canalisation en service.

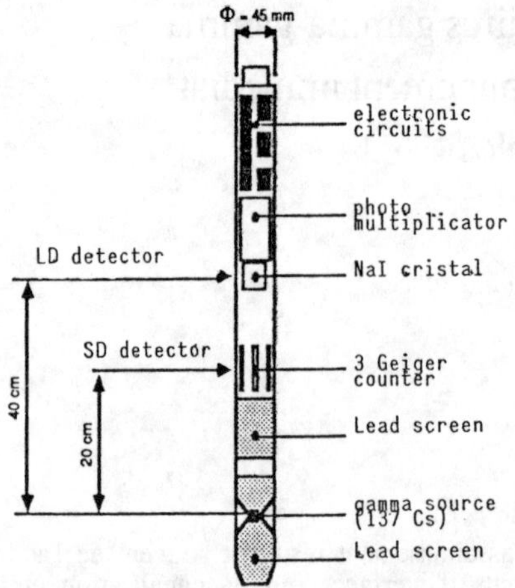

Figure 1. Schema de la sonde gamma du LCPC

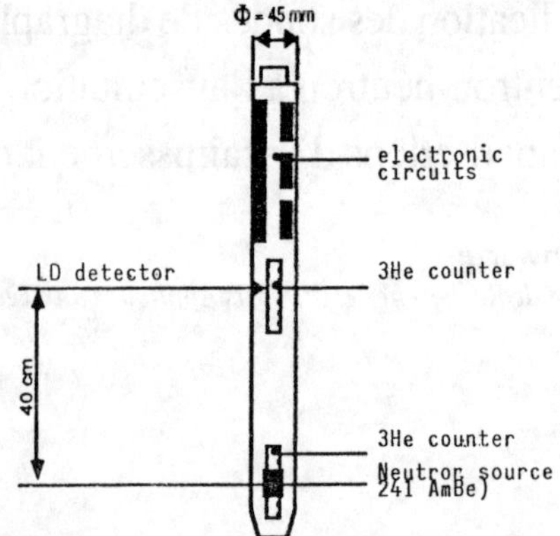

Figure 2. Schema de la sonde neutron du LCPC

2 LES SONDES DE DIAGRAPHIES NUCLEAIRES GAMMA-GAMMA ET NEUTRON-NEUTRON

2.1 *Principe de la diagraphie gamma–gamma*

Le terrain est soumis à un rayonnement gamma provenant d'une source radioactive. Les photons gamma entrent en collision avec la matière. Selon leur énergie, 3 types d'interaction peuvent se produire. Parmi ces interactions, il y a celle, appelée effet Compton, qui se produit lors de la collision des photons gamma avec des électrons peu liés des atomes. Le photon est dévié et communique une partie de son énergie, sous forme d'énergie cinétique, à l'électron.

La sonde de diagraphie gamma–gamma mesure, par rétro–diffusion et à une distance donnée de la source, l'intensité des photons gamma diffusés par effet Compton. Celle–ci croît avec la densité électronique de la matière, elle–même fonction du poids volumique humide du terrain. Une augmentation des comptages gamma, enregistrés par l'outil, signale, en première approximation, la diminution de la densité du terrain et vice versa.

2.2 *Principe de la diagraphie neutron–neutron*

Le matériau est soumis à un rayonnement de neutrons rapides. Les neutrons émis vont pénétrer dans la matière, entrer en collision avec les noyaux des atomes constituant celle–ci, être ralentis et perdre une partie de leur énergie. Le noyau d'hydrogène ayant une masse comparable à celle d'un neutron, ce sont les atomes d'hydrogène, présents dans la formation essentiellement sous forme de molécules d'eau, qui ralentissent le plus les neutrons émis. Les neutrons, ralentis par les collisions successives, passent à l'état dit thermique quand leur énergie devient comparable à l'énergie d'agitation des atomes, avant d'être finalement absorbés par les noyaux de certains éléments présents dans le milieu : Chlore, Fer,...

La sonde de diagraphie neutron–neutron thermique enregistre le flux, rétro–diffusé, de neutrons thermiques à une distance fixe de la source, ce qui revient, dans la pratique, à déterminer la teneur en eau volumique du terrain en place. Une augmentation des comptages neutron, enregistrés par la sonde, équivaut à une augmentation de la teneur en eau et vice versa.

2.3 *Description des sondes expérimentées*

2.3.1 *Sonde gamma–gamma du BRGM*

La sonde gamma–gamma du BRGM (de fabrication Mount–Sopris, USA) est un outil à 2 détecteurs, l'un à 10 cm (détecteur Petite Distance, PD), l'autre à 35 cm (détecteur Grande Distance, GD) de la source. Celle–ci, constituée de ^{137}Cs, émet avec une intensité de 100 mCi. Source et détecteurs sont collimatés sur une même génératrice. La longueur de la sonde est de 2.20 m pour un diamètre de 57 mm et un poids de 17 kg.

2.3.2 *Sonde gamma–gamma du LCPC* (Cf.Fig.1)

La sonde gamma–gamma du LCPC (fabrication LCPC) comporte 2 détecteurs à 20 cm (PD) et 40 cm (GD) de la source ^{137}Cs, qui émet avec une intensité de 60 mCi. La longueur de la sonde est de 1.5 m, son diamètre de 46 mm et son poids de 8 kg.

2.3.3 *Sonde neutron–neutron du LCPC* (Cf.Fig.2)

La sonde neutron–neutron du LCPC (fabrication LCPC) comporte un ensemble de 3 sources

SAND–FILLED TRENCH – TARGET POSITIONS
(SIDE VIEW)

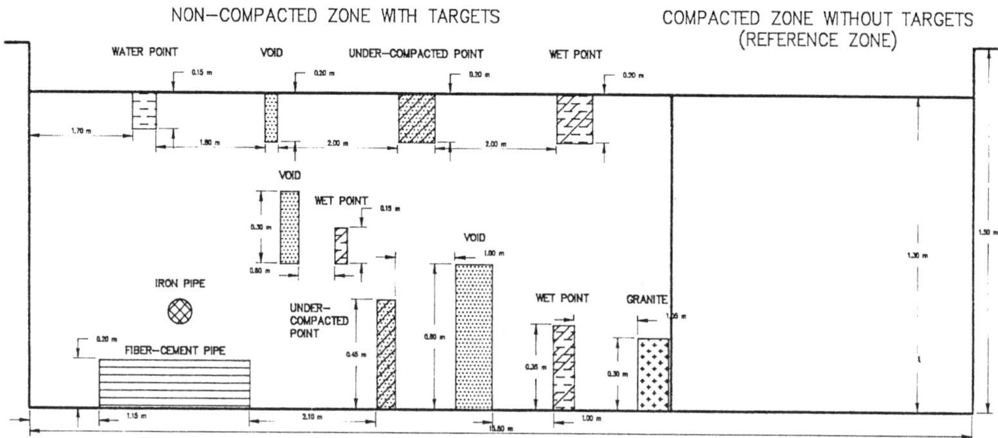

Figure 3. Vue de profil de la tranchee sable du CREATE

^{241}AmBe (3x200 mCi) et 2 détecteurs ^{3}He. Le détecteur PD est centré autour de la source, le détecteur GD est situé à 40 cm de la source. La longueur de la sonde est de 1.76 m pour un diamètre de 46 mm et un poids de 5 kg.

Le volume restant des sondes est occupé par les circuits électroniques d'alimentation des capteurs et de conditionnement des signaux, envoyés via le câble de traction vers l'unité d'acquisition et de traitement.

Le matériel de mise en oeuvre comprend schématiquement un treuil motorisé à vitesse réglable, une roue codeuse pour l'acquisition, via le câble, de la distance (profondeur) parcourue par la sonde et une unité d'acquisition et de traitement des signaux piloté par micro–ordinateur, le tout embarqué à bord d'un véhicule utilitaire léger.

3 ESSAIS DES SONDES DE DIAGRAPHIES GAMMA-GAMMA ET NEUTRON-NEUTRON SUR LES TRANCHEES EXPERIMENTALES DU CREATE

Les sondes nucléaires du LCPC et du BRGM ont été testées sur 2 tranchées expérimentales au CREATE. L'objectif de cette première phase d'expérimentation était d'établir la faisabilité de leur utilisation telles quelles dans des canalisations d'assainissement non–visitables, comme moyen d'auscultation de l'environnement proche de celles–ci, en déterminant en particulier si le rayonnement gamma et neutron franchissait le matériau de la conduite et renseignait sur les caractéristiques du remblai d'enrobage, compte–tenu du matériau des tuyaux, de l'épaisseur des parois, de la nature du remblai et en présence d'eau dans la canalisation.

N.B. Pour ces essais (et les essais suivants) seuls ont été pris en considération les taux de comptages gamma et neutron, qui permettent une interprétation qualitative des variations des caractéristiques du sol. Des mesures quantitatives du poids volumique et de la teneur en eau du sol encaissant la canalisation requierent un étalonnage des sondes sur des échantillons de tuyau pour chaque matériau et diamètre.

3.1 *Description du site d'essai*

2 tranchées de dimensions 15x1x1.5 m (Lxlxh) ont été mises en place sur le terrain d'essai du C.R.E.A.T.E. L'une des tranchées a été remblayée avec du sable de maçon, l'autre avec une grave. Dans chaque tranchée, le remblai a été compacté par couches sur une longueur de 5 m en partant d'une extrémité, à l'aide d'une plaque vibrante, conformément aux recommandations LCPC/SETRA sur le compactage des remblais en tranchées. Sur la longueur restante, le remblai n'a pas été compacté. Une planche de contre–plaqué sépare les 2 compartiments.

Des anomalies artificielles ont été aménagées ponctuellement dans le remblai des parties non–compactées des 2 tranchées, pour constituer des "cibles" et simuler la présence de :

* vides : blocs de polystyrène expansé
* points décomprimés : sacs plastiques renfermant un mélange 50/50 de matériau de remblai et de perles de polystyrène expansé
* points d'eau : bidons remplis d'eau
* points humides : sacs plastiques renfermant le matériau de remblai saturé en eau
* conduites enterrées : tuyau fibro–ciment 200 mm et tuyau en fonte 95 mm

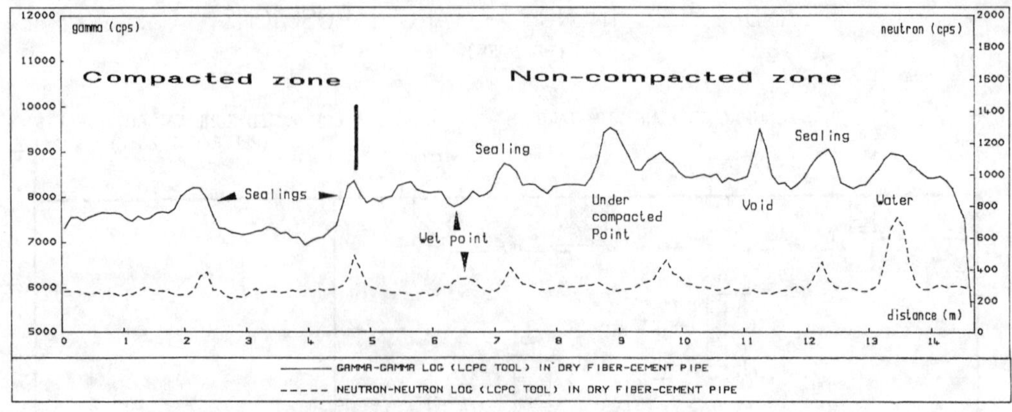

Figure 4. Profil des comptages gamma obtenu dans la conduite fibro-ciment a sec

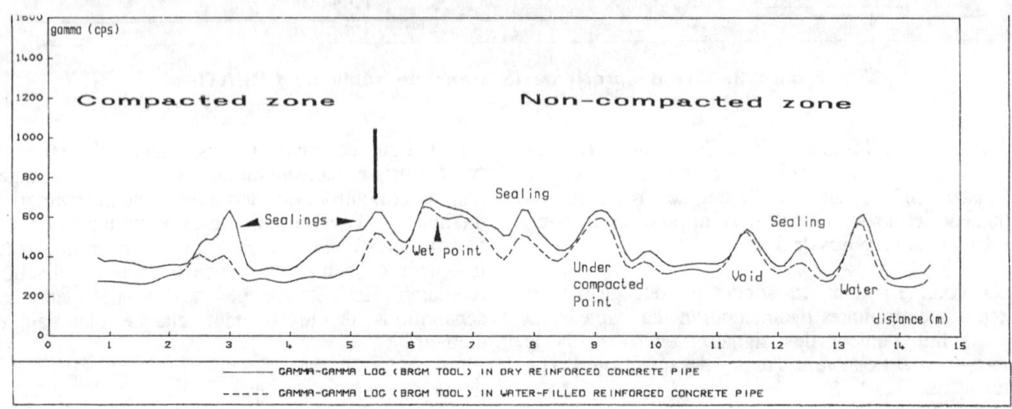

Figure 5. Profil des comptages gamma obtenu dans la conduite en beton arme sec/en eau

* points durs : blocs de granite

Une vue de profil de la disposition de l'ensemble des anomalies de la tranchée sable est représentée sur la figure 3.

Dans chaque tranchée, une conduite, d'environ 15 m de long, de diamètre 400 mm et constituée d'un même matériau (béton armé, amiante–ciment, grès) a été posée et remblayée jusqu'à mi–hauteur.

Cette disposition de la canalisation ne correspond pas à la réalité des canalisations enterrées mais a l'avantage de permettre de changer la nature des tuyaux (béton, amiante–ciment ou grès) et de tester ainsi plusieurs configurations remblai/tuyau.

3.2 *Déroulement des essais*

Les sondes ont été tractées le long du radier d'une extrémité à l'autre des différentes conduites à sec ou remplies d'eau propre (hauteur d'eau environ 10 à 20 cm).

3.3 *Analyse des résultats*

Les résultats se présentent sous la forme de graphiques représentant les variations des comptages en fonction de la distance parcourue par les sondes (pas de mesure : 10 cm) et traduisant, pour les sondes gamma, les variations qualitatives du poids volumique humide du sol en place, et, pour la sonde neutron, les variations qualitatives de la teneur en eau volumique.

Les résultats obtenus montrent que :

* les sondes gamma et neutron du LCPC sont, dans leur configuration actuelle, capables de détecter, de localiser et d'identifier des défauts (et excès) de poids volumique et de teneur en eau du terrain adjacent à une canalisation en fibro–ciment 400 mm (Cf.Fig.4). Les mêmes performances sont obtenus, quoique moins nettement, en présence d'une conduite en grès de même diamètre, tandis que dans une conduite en béton armé, les variations des comptages ne permettent pas de déceler clairement la présence des anomalies.

EXPERIMENTAL TRENCH CHAMPIGNY – SIDE VIEW

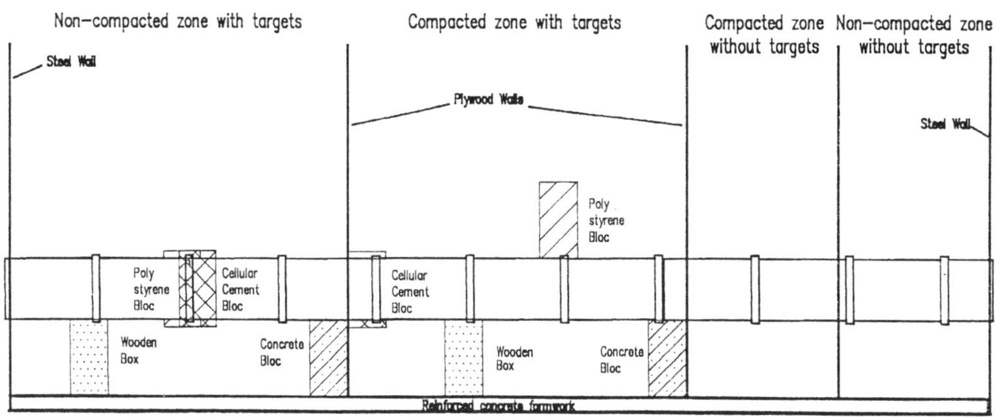

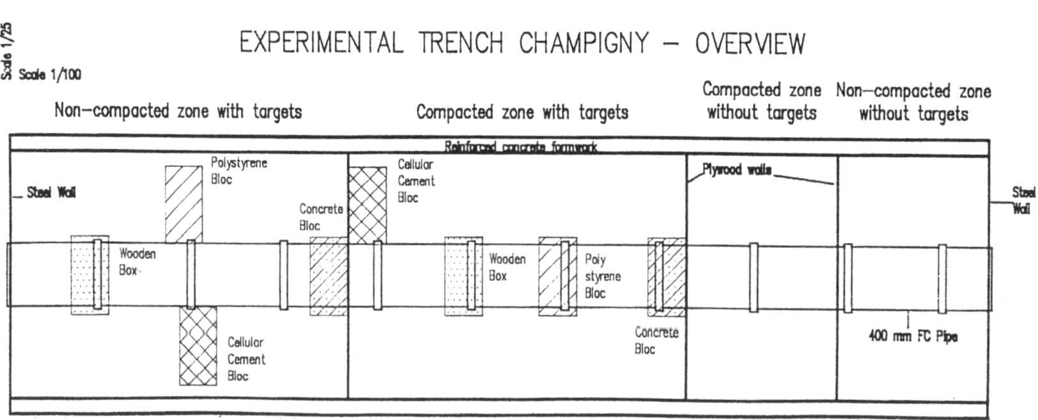

Figure 6. Vue de profil et plan de la tranchee experimentale a Champigny

* la sonde gamma–gamma collimaté et plus puissante du BRGM permet la détection des anomalies à travers la canalisation en béton armé 400 mm.(Cf.Fig.5)

* la détection, localisation et différenciation des anomalies demeurent possible quand la canalisation est remplie d'eau et les sondes immergées.(Cf.Fig.5)

* les sondes gamma permettent de distinguer les zones compactée et non–compactée des 2 tranchées (Cf.Fig.4).

* la sonde neutron semble permettre la détection d'exfiltrations d'eau aux joints d'une canalisation en eau.

4 ESSAIS DES SONDES DE DIAGRAPHIES GAMMA-GAMMA ET NEUTRON-NEUTRON SUR LA TRANCHEE EXPERIMENTALE DE CHAMPIGNY S/MARNE

Les essais des sondes de diagraphies gamma–gamma et neutron–neutron du BRGM et du LCPC ont été poursuivis sur une canalisation expérimentale, construite par la Direction des Services de l'Eau et de l'Assainissement du Val–de–Marne (DSEA 94) à Champigny s/Marne.

L'objectif de cette deuxième phase d'expérimentation était de conforter les résultats obtenus sur les tranchées du CREATE, d'évaluer les performances de ces outils en conditions s'approchant de la réalité et de déterminer les paramètres de mesure et les conditions de mise en oeuvre optimales.

4.1 *Description du site d'essai*

La tranchée expérimentale de la DSEA à Champigny s/Marne est construite en surélévation à partir d'un coffrage en béton armé d'environ 26 m de long, 1.5 m de large et 2.4 m de haut. Les extrémités de la tranchée sont fermées par des plaques d'acier. L'étanchéité de l'ensemble est assurée par le

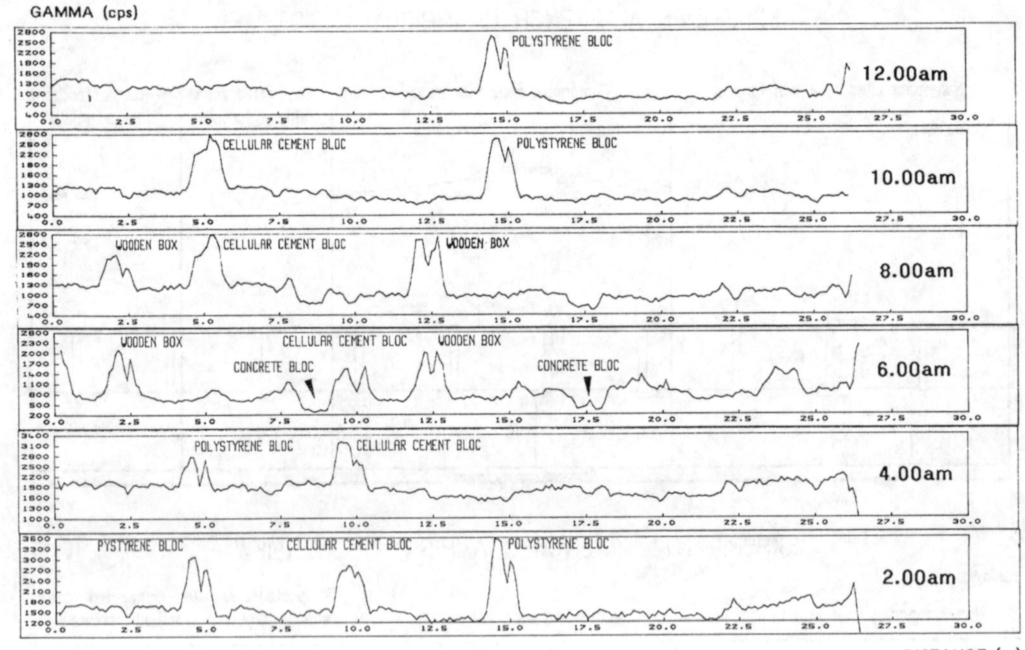

Figure 7. Profils des comptages gamma obtenus dans la canalisation d'essais a Champigny

jointoiement des éléments par résine époxy, feuilles de bitume et mastic silicone.

Une canalisation en fibro–ciment 400 mm est posée à 50 cm du sol dans un remblai de sablon. 4 zones sont aménagées dans la tranchée : zones avec le remblai compacté et non–compacté et comportant des anomalies ou non.(Cf.Fig.6)

Les anomalies, simulant des défauts et excès de poids volumique, sont adjacentes à la canalisation, toutes de taille identique (1x0.6x0.5 m) et disposées sous (position 6 h), sur (position 12 h) et aux cotés (positions 3 et 9 h) de la canalisation.

La présence d'une nappe à niveau variable peut être simulée.

4.2 *Déroulement des essais*

Les sondes ont été tractées dans la canalisation le long de génératrices espacées toutes les 2 heures : 12,10,8,6,4,2 heures et à des vitesses allant de 1.5 à 20 m/min suivant les essais. Afin de réaliser le passage des sondes aux génératrices autres que celle de 6 heures (simple traction le long du radier), celles–ci ont été montées sur un chariot porte–sonde, tracté dans la conduite et maintenant les différentes sondes à la position voulue durant le passage.

4.3 *Analyse des résultats*

Les résultats sont présentés sous la forme de profils montrant les variations des comptages obtenus en fonction du linéaire parcouru et pour les 6 génératrices auscultées, ainsi que sous la forme de cartographies iso–valeurs des comptages.

Un exemple de chaque type de représentation des résultats est donné par les figures 7 et 8, montrant les comptages enregistrés par la sonde gamma du BRGM dans la canalisation à sec.

On y observe que chaque anomalie, adjacente à la canalisation et simulant un défaut ou excès ponctuel de densité, est bien détectée et localisée en positions linéaire et horaire.

La présence des zones compactées (de densité plus forte) se traduit par une diminution du niveau moyen des comptages gamma d'environ 300 cps, visible sur tous les profils réalisés, excepté celui à 6 heures ce qui concorde avec la réalité, étant donné que le lit de pose de la conduite n'a pas été compacté. Sur la carte isovaleurs, les zones compactées se signalent par la zone, dessinée en plus foncée, située entre 10 et 22 m environ.

Parallèlement à ces expérimentations, des essais ont été effectués sur des demi–coquilles de tuyaux isolés en fibro–ciment et béton de différents diamètres (300 à 800 mm), posées dans une mini–tranchée remblayée avec du sable et dans lequel des caissons en bois ont été aménagés, simulant des vides secs ou remplis d'eau, avec des volumes correspondants au volume théorique d'investigation des sondes, de la moitié et du quart de celui–ci.

Le but de ces essais était de :

* déterminer en conditions maîtrisées les performances (détectabilité des anomalies) et limites

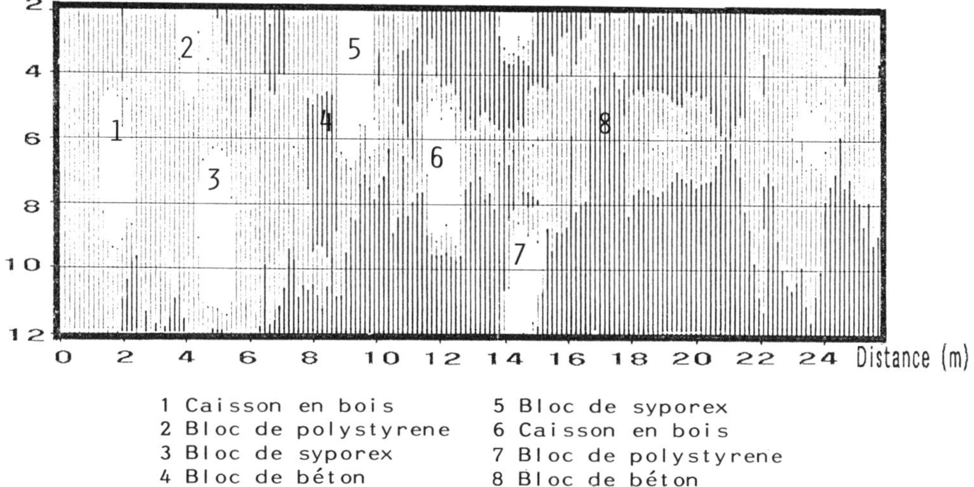

Figure 8. Cartographie iso-comptages gamma obtenue dans la canalisation experimentale a Champigny

de ces techniques en fonction du matériau et de l'épaisseur des tuyaux, en présence d'eau ainsi qu'en cas de décollement des sondes de la paroi de la conduite.

* tester l'efficacité d'un écran focalisant le rayonnement émis et reçu de la sonde neutron du LCPC en vue d'en augmenter les performances.

* établir clairement si la sonde neutron détecte et localise des fuites d'eau aux joints.

5 ESSAIS DES SONDES DE DIAGRAPHIES GAMMA-GAMMA ET NEUTRON-NEUTRON SUR UNE CANALISATION EN SERVICE

Les sondes gamma et neutron du LCPC ont été employées dans une canalisation d'eaux usées en fibro-ciment de diamètre 500 mm à Igny, Essonne, France. Pour cette canalisation, agée de 3 ans, une étude géotechnique préalable au projet de construction avait été entreprise afin de connaître l'environnement géotechnique de celle-ci. De plus, elle a fait l'objet d'investigations permettant d'établir la géométrie de l'ouvrage, à savoir ses fluctuations du profil en long et du tracé en plan, son degré d'étanchéité et le degré de compactage du remblai.

Tous ces informations réunies ont révélé qu'il existait des zones où la canalisation s'était affaissée sous l'effet d'un tassement du fond de fouille, décomprimé au moment de l'ouverture de la tranchée.

5.1 *Déroulement des essais*

Les sondes gamma et neutron ont été introduites dans la canalisation par les regards de visite et tractées en radier à une vitesse de 2.5 m. Au total, 6 tronçons ont été parcourus, totalisant d'une longueur de 90 m. Les essais ont duré environ 6 heures.

Pendant ces essais, la canalisation est restée en service. La hauteur d'eau était d'environ 10 cm.

5.2 *Analyse des résultats*

Les profils des comptages gamma et neutron sont présentés sur la figure 9. On y observe :

* plusieurs pics de comptages gamma, traduisant une diminution du poids volumique, tout le long du profil, et vraisemblablement dus à des déplacements incontrôlés de la sonde gamma.

* une augmentation du niveau moyen des comptages gamma, traduisant une diminution du poids volumique, en passant de la zone de 0 à 40 m à celle comprise entre 40 et 85 m, accompagnée d'une légère diminution du niveau moyen des comptages neutron aux mêmes points. Une première interprétation consiste à dire que la zone comprise entre 0 et 40 m est une zone où les tassements ultérieurs (à la pose de la conduite) du sol d'assise ont été particulièrement importants.

6 CONCLUSIONS

Les essais réalisés avec des sondes de diagraphies nucléaires gamma-gamma et neutron-neutron du LCPC et du BRGM sur plusieurs sites expérimentaux et sur une canalisation en service, ont démontré la capacité et l'intérêt de ces outils pour détecter, localiser et identifier des défauts et excès

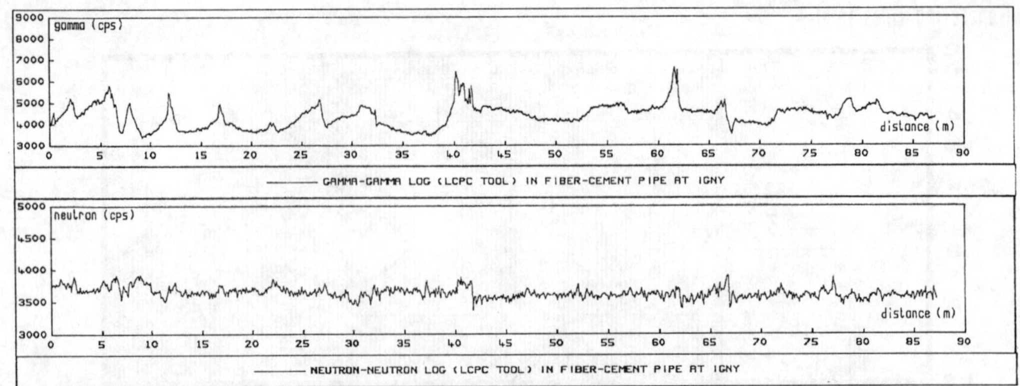

Figure 9. Profils des comptages gamma et neutron obtenues dans la conduite a IGNY

ponctuels ou étendus de poids volumique (vides, points durs, zones décompactées) et de teneur en eau (zones humides, points d'exfiltration aux joints) du terrain adjacent à une canalisation d'assainissement de petit diamètre.

Les essais montrent que l'auscultation peut être réalisé quand la canalisation est en service.

BIBLIOGRAPHIE

CSTB (1989) : Résultats de l'enquête relative aux travaux de réhabilitation des réseaux d'assainissement ; AGHTM

LCPC (1985) : Influence des conditions géotechniques sur le comportement des canalisations d'assainissement ; Agence de Bassin Loire–Bretagne

BARON J.P., CARIOU J., THORIN R. (1989) : Les diagraphies nucléaires développées par les Ponts et Chaussées ; LCPC, Bull.Liaison.Labo. P.et Ch., n°164, pp.17–24, nov–déc.1989

No Trenches in Town, Henry & Mermet (eds) © 1992 Balkema, Rotterdam. ISBN 90 5410 085 0

New ground penetrating radar acquisition methodology for non accessible pipe networks used to define and evaluate the degree of compaction of trench soils

G.Clement & R.Foillard
Compagnie Générale de Géophysique, Massy, France

M.Darras & F.Dubreucq
Services de l'Eau et de l'Assainissement, Val de Marne, France

ABSTRACT : The use of ground penetrating radar (G.P.R.) from the inside of non accessible pipes is now a reality. Il permits the detection of reflection anomalies located in the soil environment of pipes. However it does not make it possible to estimate the degree of compaction of trench fill. The simultaneous use of multiple antennas for scanning underground solves this problem.
This new radar data acquisition technique is currently used in reflection seismic and known as multiple fold profiling or the Comon Depth Point technique. This methodology provides continuous analysis of radar wave velocity from the interior of a pipe to the interface between the trench fill and natural soil.
Variations in the propagation velocity of electromagnetic waves in trench fill are linked to variations in soil compaction and soil moisture content. Knowledge of these physical parameters is very important for determining the origin of disorders in sewer pipes. The use of ground radar in multiple fold, and the processing applied to the data are described in this paper.
The methodology applied and results obtained on the experimental sewer pipe built by Val de Marne Water and Sewerage Authority will be presented and discussed here.

1 INTRODUCTION

Many experiments have been carried out with Ground Penetrating Radar over the last two years from the inside of non accessible pipes. They normally use a single transducer to visualize quickly reflection anomalies in the fill of pipe trenches.
These anomalies represent exfiltration zones, voids or cavities which are often the cause of disorders affecting pipes.
Radar imaging interpretation is used to make a qualitative diagnosis of the state of the immediate surroundings of the pipe and to propose appropriate curative and preventive renovation techniques. This kind of use for G.P.R. cannot be used to evaluate the degree of trench soil compaction, which is nevertheless of great significance. Indeed, compaction defects are reputed to be the cause of the development of exfiltration zones and voids in the immediate surroundings of trench soils which can unfortunately affect sewer pipes with collapses, dislodgements, ruptures...
Because radar and seismic techniques are very similar in principle and in practice, we thought of using G.P.R. with non traditional acquisition geometry, known as Commom Depth Point geometry (C.D.P.).
This multiple fold method has been developed and used for geological exploration by seismic reflection for years.
It has already been successfully used many times with G.P.R. for geological, hydrogeological and mining exploration to determine the moisture content of soils as a function of depth and to evaluate the distribution of porosity in soils.
The application of such an acquisition methodology will therefore produce significant results concerning the degree of compaction of trench soils around non accessible pipes, by mere analysis of wave propagation velocities in the medium crossed.

2 BASIC PRINCIPLES AND RADAR METHODOLOGIES

The radar method uses electrical energy to generate electromagnetic (radio) waves which are radiated into the ground.
It is currently operated in single fold data acquisition using a single transducer for underground exploration from the surface or from the wall of subterranean works or pipes.
A very short time impulse (ns) is generated at a very high frequency (80 MHz to 1 GHz) and radiated by an antenna, (each impulse is considered as a "shot").
When the signal encounters an anomaly, it is reflected and picked up by a receiver which transmits it to a magnetic and graphic recorder ; this constitutes a "scan" or radar echo.
The waves reflected by anomalies in the subsurface are observed successively with the regular displacement of antennas along the profile studied. The shot frequency can vary from 20 shots per meter to 150 shots per meter.
The data is presented as a "cross section" or radar profile, see figure 1.
The depths of exploration vary between 1 and 20 meters for both natural soils and construction materials.
The time cross-section can be transformed into a depth cross-section if the velocity function of wave propagation in the crossed medium is known. This is not usually the case when using standard radar data acquisition.
It is the multiple fold radar data acquisition alone which allows continuous and correct velocity analysis of wave propagation along a radar profile.
This particular technique requires the use of a transmitting antenna to radiate the radar signal and several receiving antennas at variable distances from the transmitter.
Variable distances are called offsets and are chosen as a function of the depth of the interface target.
The geometry of the array and the number of both transmitter-receiver units define the multiple stacking fold of the data acquisition.
So, operating with one transmitter linked up with two receivers along a profile with regular data acquisition spacing constitutes a 200 % fold acquisition.
We therefore obtain two radar echos from two different wave propagation travel times coming from the same deep reflection point (C.D.P.).

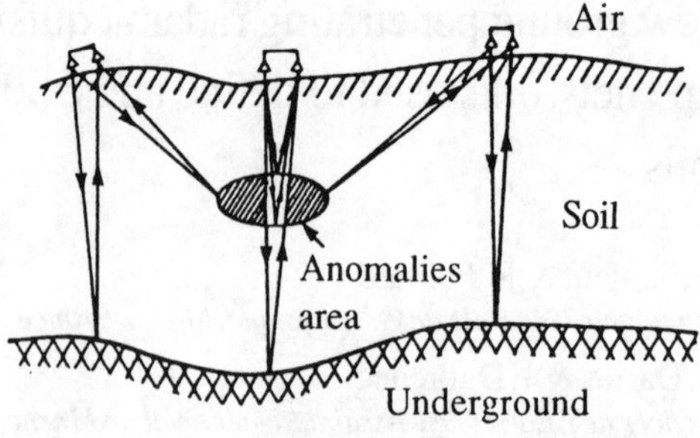

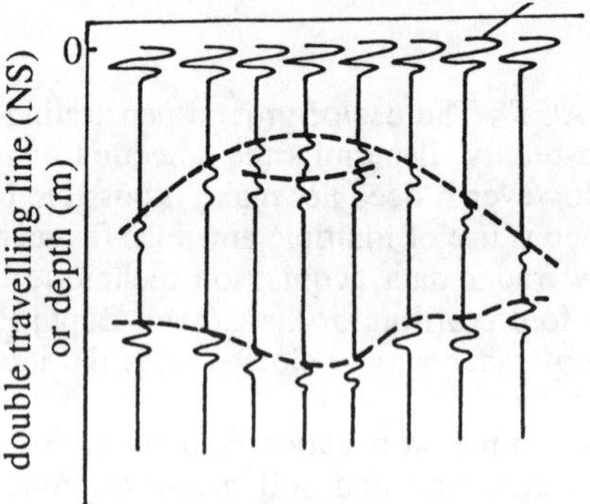

Theoric radar diagram (cross section)

Figure 1: operating principle

Knowing the two travel times of wave propagation from the same depth point allows us to calculate the radar wave propagation velocity in the medium (Vm) and is given by the formula:

$$Vm = \frac{X}{\sqrt{t_2^2 - t_1^2}}$$

3 EXPERIMENTAL 200 % FOLD RADAR DATA ACQUISITION CASE HISTORY IN AN ACCESSIBLE SEWER PIPE

A 200 % fold radar data acquisition experiment should provide :

- continuous analysis of radar wave propagation velocity in the trench fill around a pipe;
- the proof that velocity variations in wave

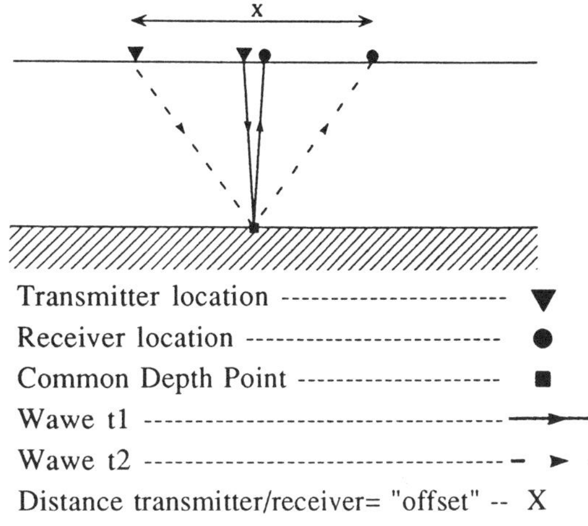

Figure 2: theorical data acquisition procedure in twofold profiling

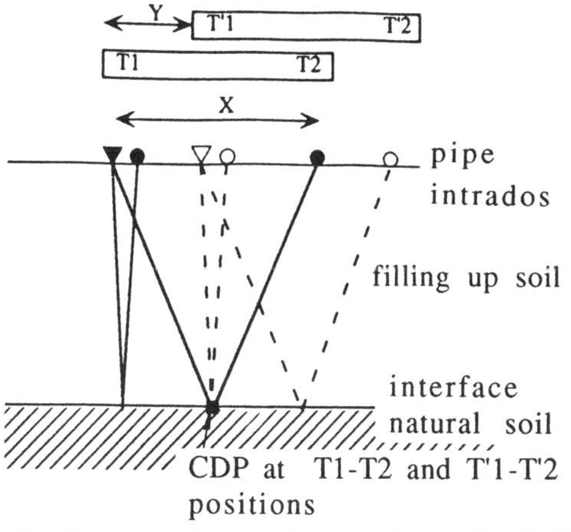

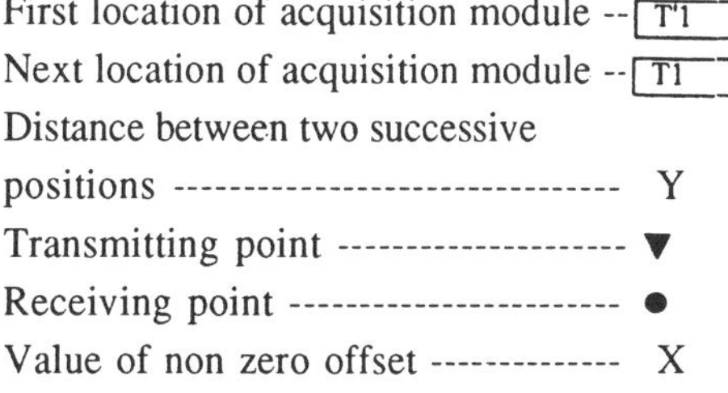

Figure 3: radar data acquisition geometry used.

propagation depend on the porosity of the medium and as a consequence, its degree of compaction.

This experiment was undertaken in collaboration with the Direction des Services de l'Eau et de l'Assainissement du Val de Marne (D.S.E.A.) on the site of its experimental sewer pipe at Champigny sur Marne.
Two transducers were used, one of them as a transmitter-receiver unit with zero offset and the other as a receiver with a 58 cm offset distance from the transmitting antenna, see figure 3.

The group of transmitting and receiving radar antennas formed a solid acquisition unit T1-T2, which was dragged along the bottom of the pipe. The shot number was one hundred per linear meter, equivalent to the C.D.P. sampling.
The reflection target was the interface between the trench fill/ natural soil of the pipe.
This interface is plane and supposed to be at a constant distance of 60 cm from the pipe, see figure 4.

4. DATA REDUCTION AND INTERPRETATION

Raw data collected in the experimental sewer pipe at Champigny were processed on Geovecteur seismic software (T.M. of the C.G.G.), suitable for radar signal processing.

Raw GPR data cannot be interpreted easily, due to the bad signal/noise ratio generated by multiple reflections from direct wave and electromagnetic noises.
The basic processing sequence applied to enhance raw data is as follows :

- static corrections
-resampling-averaging
- gain equalisation
- filtering-deconvolution
- plotting

This processing sequence enhances cross-sections on which it is then possible to see the reflection target which is the interface between trench fill/ natural soil situated 60 cm below the pipe, see figure 4.
This interface is plane and considered equidistant from the pipe.
Because, the trench fill soil is supposed to be an homogeneous material, the propagation velocity in this soil must be constant. So the reflection corresponding to the trench fill/

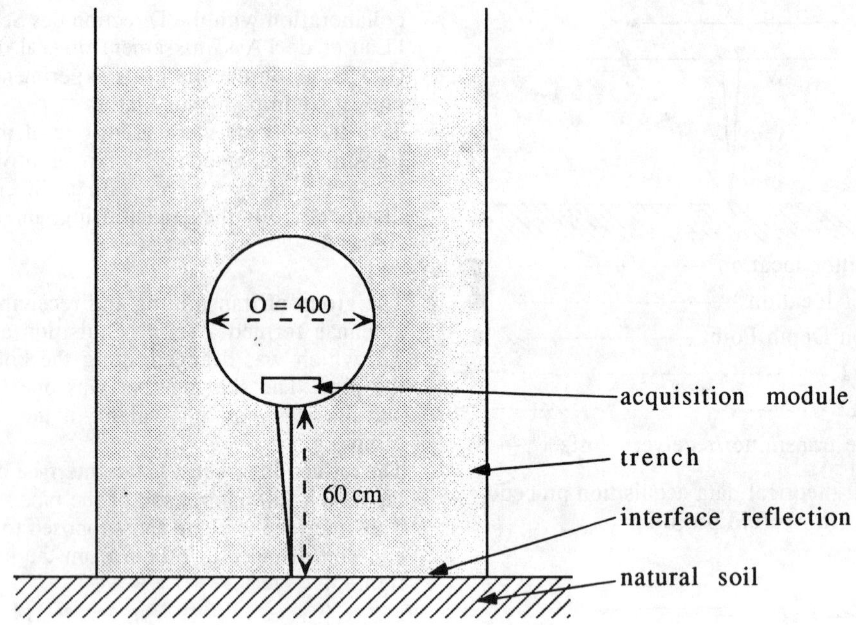

Figure 4: location of radar profile in sewer pipe.

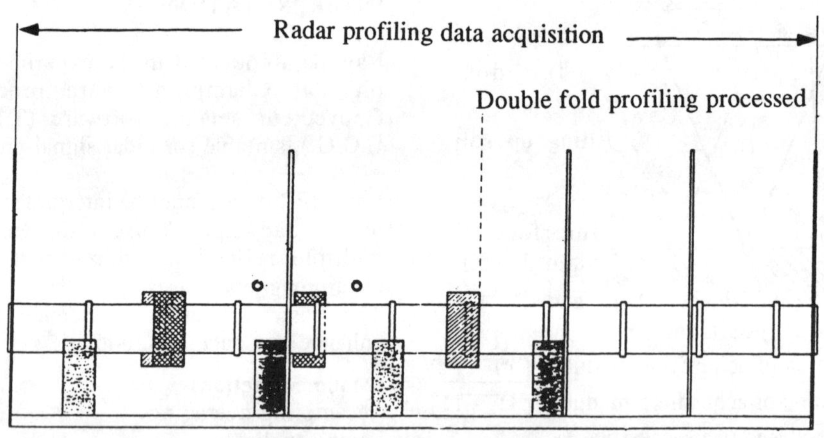

Figure 4b: Experimental Sewer Pipe at Champigny

natural soil must also be plane on radar cross-sections.

This is not the case on the radar cross section shown in figure 5, where we see significant temporal ondulations from the deep reflection. This means that velocity variations occur in radar wave propagation inside the trench fill because the distance of the target interface is constant in the experimental pipe along the profile according to the velocity formula :

$$V = \frac{\text{distance}}{\text{propagation time}}$$

We can conclude therefore that the soil below the pipe has variable porosity because of a constant moisture.

Considering that the fill material is homogeneous, such variations in porosity surely come from a the materials variable degree of compaction.

DSEA NEAR FX MCDEC
OFFSET NUL

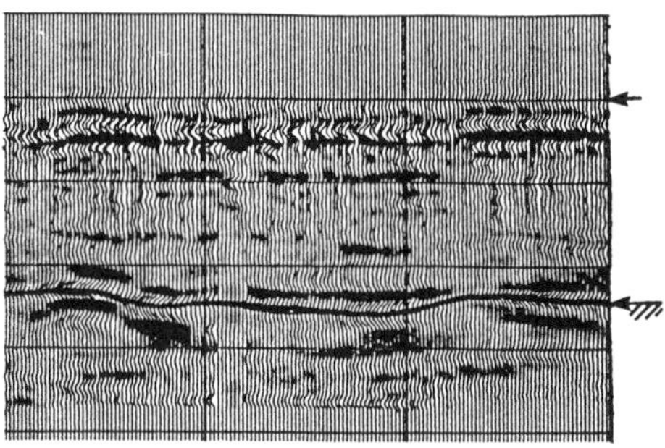

DSEA FAR FX REFOR INV
OFFSET = 58cm

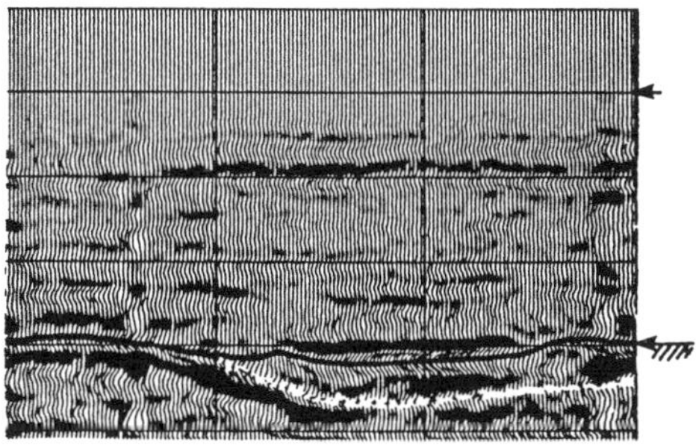

← Wall surface

← Interface fill/natural soil

Fig.5 RADAR CROSS-SECTION

The more the propagation velocity increases, the higher the porosity is and the less the material is compacted (high position of the deep reflection).
By contrast, the low position of the deep reflection indicates the high compaction of the material. The existence of variable compaction in the fill soil is demonstrated directly by visual analysis of deep reflections on cross-sections.
These conclusions are available because of a constant distance between the pipe and interface represented by the fill soil/natural soil inside the experimental pipe we worked on. This is not generally the case and usually the interface of the fill soil/natural soil is situated at a variable distance from pipes.

This is why continuous velocity analysis of radar wave propagation must be calculated from multiple fold radar data acquisition.

This calculation needs complementary data processing:
-wavefield identification;
-antenna calibration;
-spatial filtering;
-continuous velocity analysis;
-stacking.

By using this sequence of typical seismic data processing it is still possible to obtain a continuous velocity analysis of radar waves from the pipe to the target interface whatever its variations in depth.
The results of this process are shown on the diagram in figure 6.
The mean velocity calculated is around 100000 km per second, close to the one measured with radar tomography equal to 104000 km per second. The velocity diagram indicates a low velocity zone numbered II in the center of the diagram reaching 92000 km/s. This zone is situated between two higher velocity sectors I and III with 104000 km/s and 98000 km/s medium velocity.
There is a variation in velocity of more than 10 % which can be considered to be indicative of a change in compaction in zone I, II and III.

As we can see, the velocity variations calculated in the diagram from two fold acquisition are perfectly comparable to those variations in the reflection time position of the target.
It means that the velocity calculation was successful.

5 CONCLUSION

We can reasonably estimate that the study of the fill soil of the experimental pipe with the multifold radar exploration method achieved its objective.
Velocity variations calculated from this kind of radar data acquisition are significant because they are similar to those observed in the time position of the deep reflection target which is situated at a constant distance from the pipe and reproduced on the pseudo-section of the radar single fold data acquisition.

We are inclined to believe that these observed

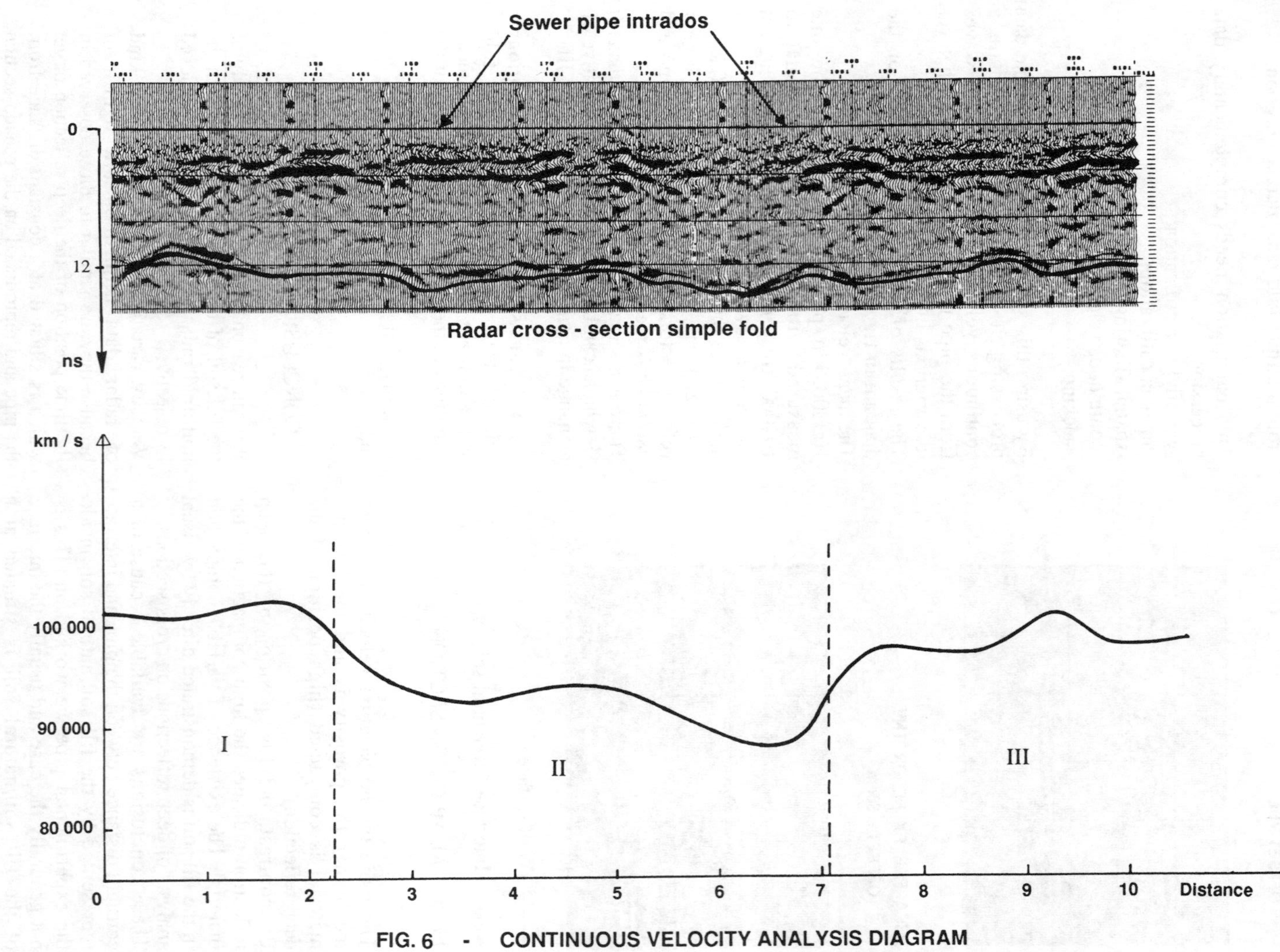

FIG. 6 - CONTINUOUS VELOCITY ANALYSIS DIAGRAM

and calculated propagation velocity variations are related to porosity variations inside the dry fill soil of the trench being studied:

- an increase in the propagation velocity soil is linked to an increase in its porosity and as a consequence to a decrease in its compaction;
- a decrease in propagation velocity in soil is linked to a decrease in its porosity and to an increase in its soil compaction.

Other applications could be found for such a multi fold radar data acquisition methodology and could be successful for measuring moisture content with accuracy.
Indeed, trench fill soil zones with high moisture content are often associated with defects in water tightness that it is useful to highlight.

Note that radar acquisition using multiple fold geometry is not more expensive than single fold acquisition, however it is more costly to process the data.
Sophisticated seismic data processing resources are required which have a high cost, four times more expensive than basic processing of single fold data acquisition.

REFERENCES

Bushforth, M. and Koppenjan, S. 1992. Ground Penetrating Radar applications for non destructive inspection of construction works. E.A.E.G. International Meeting; G 005 Paris 1992.

Chignell, RJ. 1992. A multi-channel surveying Radar System. E.A.E.G. International Meeting; G 023 Paris 1992.

Davis, J.L., Annan, A.P. and Vaughan, C. Placer exploration using Radar and seismic methods 1984. 54th Annual International Society of Exploration Geophysicists Meeting in Atlanta, Georgia, December 1984.

Mermet, M., Dubreucq, F., Foillard, R. and George, B. 1992. Method for diagnosis of state of sewer networks. Proc. NO-DIG 92- 8th International Conference on Trenchless Technology. Washington: I.S.T.T.

Meunier, J., Foillard, R. and Charachon, B. 1992. Radar imaging enhancement using seismic software. E.A.E.G. International Meeting; G 031 Paris 1992.

Topp, G.C., Davis, J.L. 1992 (a) Measurement of soil water content using time domain reflectometry. Canadian Hydrology Symposium: 82, Assoc. Comm. on Hydrol., National Research Council of Canada, Ottawa, out. pp. 269.287.

Topp, G.C., Davis, J.L. and Annan, A.P. 1982 (b) Electromagnetic determination of soil water content using T.D.R.I. applications to wetting fronts and steep gradients. Soil Sci. Soc. Am. O. 46; 672-678.

No Trenches in Town, Henry & Mermet (eds) © 1992 Balkema, Rotterdam. ISBN 90 5410 085 0

Nouvelle méthodologie d'acquisition radar en canalisation non visitable pour préciser et évaluer l'état de compaction des remblais encaissants

G.Clement & R.Foillard
Compagnie Générale de Géophysique, Massy, France

M.Darras & F.Dubreucq
Services de l'Eau et de l'Assainissement, Val de Marne, France

RESUME: L'utilisation du radar géologique (G.P.R.) comme méthode d'auscultation non destructive de l'environnement pédologique des canalisations non visitables est désormais courant. Elle permet la détection d'anomalies situées dans les sols environnants de la canalisation, et réputées préjudiciables à l'ouvrage. La mise en oeuvre courante à partir d'une antenne d'auscultation unique ne renseigne pas sur les paramètres physiques des sols qui ont été apportés en remblai lors de la pose de la canalisation. L'utilisation simultanée de plusieurs antennes d'auscultation permet de résoudre ce problème. Cette technique d'acquisition de données est couramment utilisée en sismique réflexion et connue sous le nom de "couverture multiple". En canalisation, elle permet l'analyse continue des vitesses de propagation d'ondes émises depuis la paroi de la canalisation, qui se propagent jusqu'à l'interface généralement continu du remblai de tranchée/sol naturel. Les variations de la vitesse de propagation des ondes électromagnétiques dans les remblais de tranchée traduisent les variations de compaction des sols et de leur teneur en eau. La connaissance de ces deux paramètres physiques est importante pour le diagnostic de l'origine des désordres affectant certaines portions de canalisation. L'utilisation du radar géologique en couverture multiple, et le traitement appliqué aux données sont décrits dans cette communication. L'application de la méthodologie et les premiers résultats obtenus sur la canalisation expérimentale des Services de l'Eau et de l'Assainissement du Val de Marne sont ici présentés et discutés.

1 INTRODUCTION

De nombreuses expérimentations du radar géologique (G.P.R.) ont été faites depuis un à deux ans pour l'auscultation de l'environnement des canalisations non visitables. Les expérimentations sont faites depuis l'intérieur des canalisations à partir d'une antenne émettrice/réceptrice ou transducteur unique.Elles permet-tent de visualiser rapidement les anomalies de réflexion d'ondes radar des remblais de canalisation. Ces anomalies représentent généralement des zones d'exfiltration, des vides ou cavités souvent à l'origine de désordres qui affectent ou affecteront la canalisation à plus ou moins brève échéance. L'imagerie radar permet alors de faire un diagnostic qualitatif de l'état de l'environnement proche de l'ouvrage et de préconiser des techniques de réhabilitation curatives ou préventives.

L'utilisation de la technique d'auscultation "Géoradar" traditionnelle à partir d'une antenne émettrice/réceptrice unique ne permet cependant pas d'apprécier quantitativement l'état de compaction des sols environnants et en particulier les sols utilisés pour remblayer la tranchée où repose l'ouvrage.

La connaissance de la compaction différentielle de ces remblais de tranchée apparaît pourtant être importante. En effet, les défauts de compaction sont réputés être à l'origine probable du développement de zones d'exfiltration et de vides situés dans l'environnement de l'ouvrage et donc à l'origine de futurs désordres pouvant affecter les canalisations posées en tranchée, tel qu'affaissement, déboitement, rupture....

Parce que les techniques radar et sismique sont très proches en ce qui concerne leur principe et

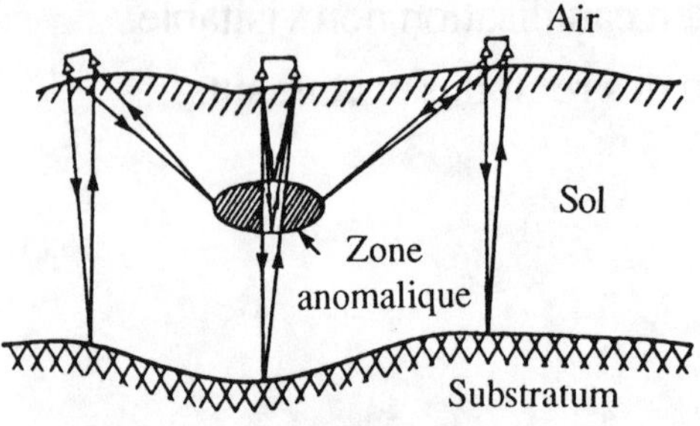

Illustration de l'utilisation du radar en mode de profilage continu sur le sol

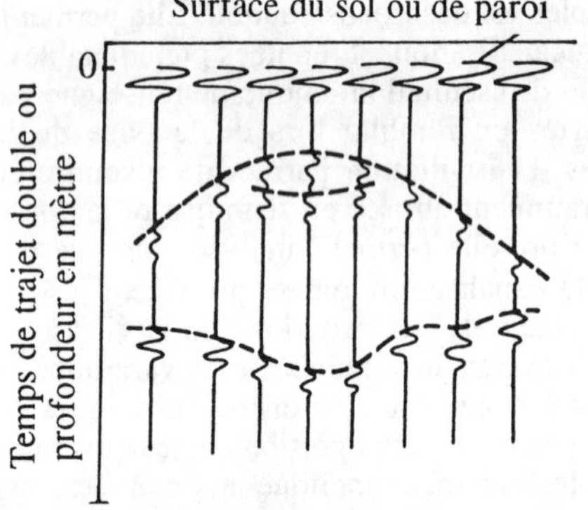

Coupe-temps théorique correspondant à l'illustration

Figure 1: Principe de fonctionnement

leur mise en oeuvre, nous avons donc imaginé d'appliquer au radar géologique, une méthode d'acquisition originale connue sous le nom de "couverture multiple" ou CDP (Common Depth Point). Cette méthode a été développée et est utilisée pour l'exploration géologique par méthode sismique réflexion depuis de nombreuses années. La technique de couverture multiple a déjà été appliquée à maintes reprises à l'acquisition "Géoradar" en prospection géologique, hydrogéologique et minière où elle a été utilisée avec succès pour déterminer la teneur en eau des sols en fonction de la profondeur et les variations de distribution de granulométrie généralement associées à des changements de porosité.
L'application de cette méthodologie d'acquisition devrait donc permettre d'évaluer le degré de compaction des sols remblayés de l'environnement des canalisations non visitables par la simple analyse des vitesses de propagation d'ondes dans le milieu.

2 PRINCIPE DE BASE ET METHODOLOGIE GEORADAR

La technique radar utilise l'énergie électrique pour générer des ondes électromagnétiques qui sont radiées dans le sol.
Elle est essentiellement utilisée en couverture simple (utilisation d'un seul transducteur) pour l'auscultation du sous-sol à partir de la surface du sol ou de tout ouvrage souterrain accessible à l'outil.
Une impulsion de très courte durée (quelques nanosecondes), est émise à cadence très élevée (100 MHz à 1 GHz) et radiée par une antenne (chaque émission correspond à un "tir").
Lorsque le signal rencontre une discontinuité, il est réfléchi et capté par un récepteur qui l'envoie vers un enregistreur graphique et magnétique, cela constituant un scan ou trace radar.
Les échos, correspondant aux différentes discontinuités rencontrées, sont observés successivement lors du déplacement régulier des antennes le long d'un profil à ausculter, la cadence des tirs pouvant varier de 20 à 150 tirs par mètre (figure 1)
Les traces reconstituent une coupe-temps ou profilage radar qui permet de visualiser des événements, discontinuités ou anomalies intéressant la tranche de sol auscultée qu'il conviendra d'interpréter. Les profondeurs d'investigation s'échelonnent de 1 à 20 m pour des matériaux pouvant être soit des terrains naturels, soit des matériaux de construction.
La coupe-temps peut être transformée en coupe-profondeur si la loi de vitesse de propagation d'onde est connue dans le milieu traversé, ce qui n'est généralement pas le cas pour l'utilisation du radar en couverture simple. Seule, la technique de couverture multiple permet d'obtenir une analyse de vitesse de propagation correcte et continue le long d'un profil.
La technique de couverture multiple requiert l'utilisation d'une antenne radiant le signal radar et de plusieurs antennes disposées à des distances variables de l'émetteur et destinées à la réception des échos en provenance du sous-sol.

Ces distances appelées "offset" sont fonction de la profondeur de l'objectif (interface) à atteindre.
La géométrie du dispositif et le nombre de couples émetteur-récepteur enregistrés simultanément, défini l'ordre de couverture de l'acquisition des données.
L'utilisation d'une antenne émettrice associée à deux antennes réceptrices situées à distances fixes et variables de l'émetteur défini une couverture d'ordre deux. Le déplacement de ce dispositif d'antennes émettrice et réceptrices le long d'un profil d'auscultation permet l'acquisition de données radar à intervalles réguliers.
On obtient deux échos radar à partir de deux trajets de propagation différents en provenance du même point de réflexion situé en profondeur (point miroir), pour un dispositif à deux récepteurs (figure 2). La connaissance des temps de trajet différents rapportés au même point miroir permet le calcul de la vitesse (Vm) de propagation effective des ondes dans le milieu traversé et est donné par :

$$Vm = \frac{X}{\sqrt{t_2^2 - t_1^2}}$$

3 MISE EN OEUVRE EXPERIMENTALE D'UN DISPOSITIF RADAR EN COUVERTURE DOUBLE DANS UNE CANALISATION

L'utilisation d'un dispositif d'acquisition radar en couverture double devrait permettre de :
- maîtriser la vitesse de propagation des ondes électromagnétiques dans les remblais de tranchée ;
- de démontrer que ces variations de vitesse sont intimement liées à la porosité du milieu et par conséquent à son degré de compaction.

Dans le cadre de cette étude méthodologique menée conjointement avec la Direction des Services de l'Eau et de l'Assainissement du Val de Marne (D.S.E.A.), nous avons donc testé un dispositif radar à géométrie de couverture double sur le site de la canalisation expérimentale de Champigny sur Marne construit par la DSEA.
Le dispositif utilisé se composait de deux transducteurs : l'un utilisé en émission-réception à "offset" nul, l'autre en réception

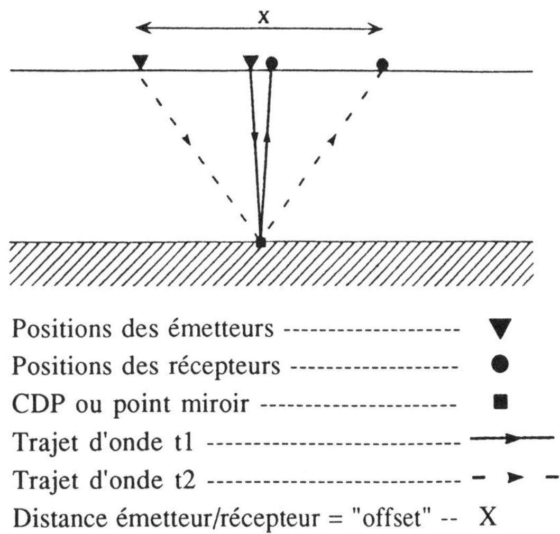

Figure 2: procédure d'acquisition théorique en couverture double

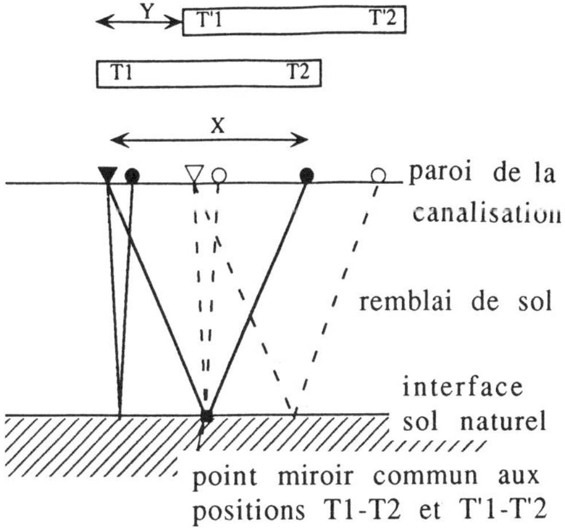

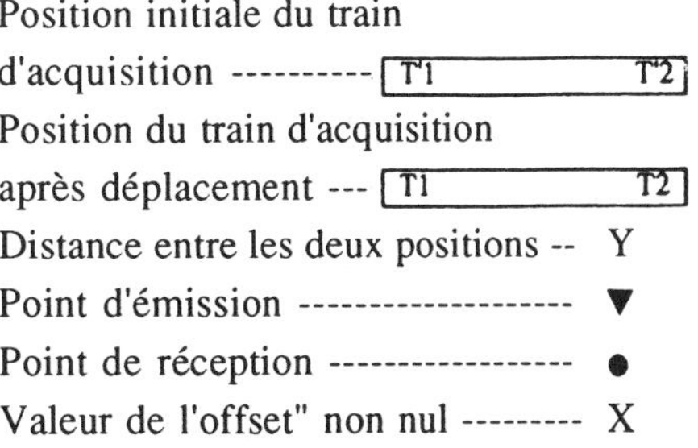

Figure 3: schéma du dispositif d'acquisition radar utilisé.

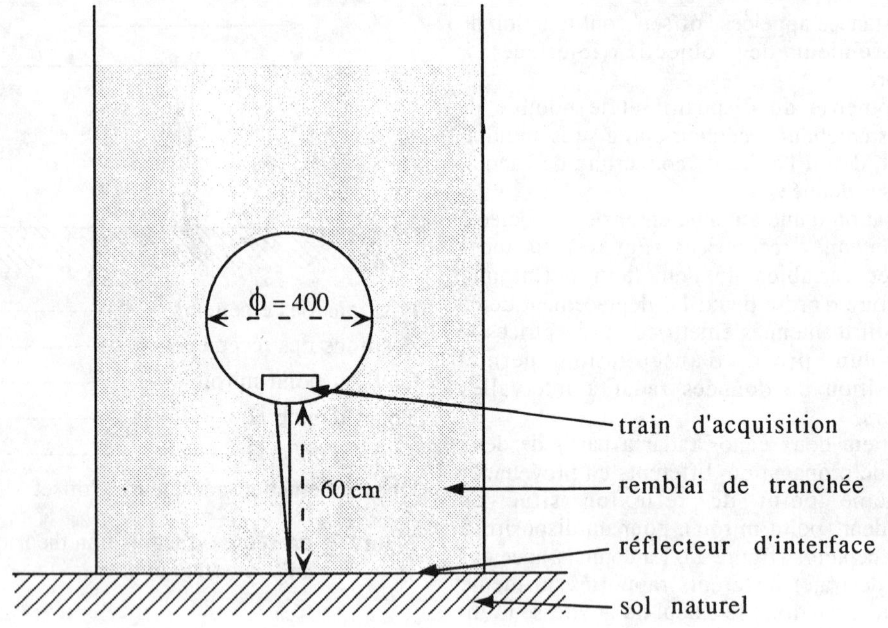

Figure 4 : schéma de positionnement du profil d'auscultation dans la canalisation.

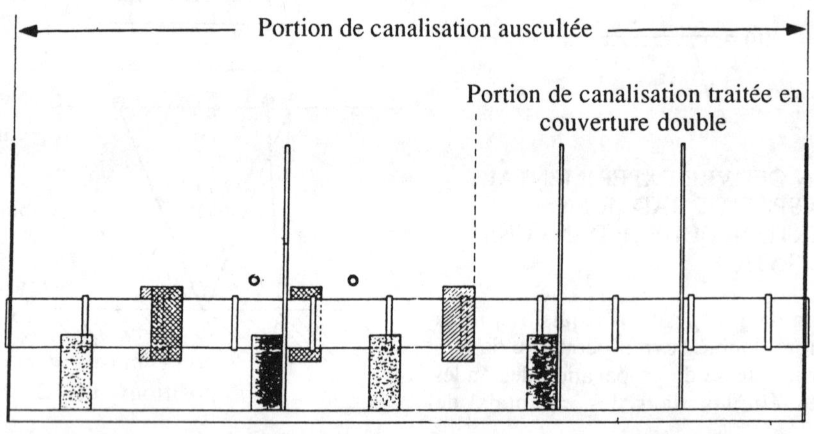

Figure 4b: Canalisation expérimentale de champigny

uniquement avec un "offset" de 58 centimètres (figure 3).
Cet ensemble solidaire constituait l'unité d'acquisition T1-T2 déplacé en radier d'ouvrage. La cadence d'auscultation était de 100 tirs radar par mètre linéaire soit un échantillonnage de points miroirs équivalent.
Le réflecteur cible de l'opération correspondait à l'interface remblai de tranchée/sol naturel de la canalisation, il était continu et plan dans les conditions de cette aquisition (figure 4 et 4b).
L'acquisition des données a nécessité l'emploi d'une unité d'enregistrement multicanaux SIR 10 de GSSI et deux transducteurs 1 gigahertz constituant le train d'acquisition. Les données radar ont été acquises et enregistrées sur environ 20 mètres linéaires d'ouvrage sur deux canaux d'enregistrement des données, l'un pour le dispositif émetteur à offset nul, et l'autre

pour le dispositif à offset long (58 cm).
On a ainsi obtenu deux pseudo-sections temps à offsets différents, de 15 nanosecondes de temps d'investigation soit environ 75 centimètres de profondeur ce qui était suffisant pour visualiser notre réflecteur cible situé à une profondeur constante de 60 centimètres sur la totalité du linéaire.

4 REDUCTION DES DONNEES ET INTERPRETATION

Les données radar brutes acquises dans la canalisation expérimentale du Val de Marne ont dû subir une série de traitements du signal réalisée sur le logiciel sismique Géovecteur™ adapté au traitement du signal radar.
En effet, les données brutes obtenues ne sont généralement pas intéprétables directement à cause de la présence de bruits électromagnétiques et réflexions multiples de l'onde radar directe.
La séquence de traitement de base appliquée aux données brutes des profils acquis avec offset nul et grand offset est du type :
- corrections statiques;
- rééchantillonnage;
- égalisation des gains;
- filtrage;
- déconvolution;
- sortie graphique.

Cette séquence de traitement permet de mettre parfaitement en évidence le réflecteur cible que constitue l'interface remblai/sol naturel situé à 60 centimètres de l'intrados de la canalisation .
Au niveau de la canalisation expérimentale, cet interface est rigoureusement plan et peut être considéré comme parfaitement parallèle et équidistant de la canalisation.
Dans l'hypothèse où le remblai présente des caractéristiques physiques homogènes, la vitesse de propagation dans le milieu traversé est constante. Le réflecteur correspondant à l'interface remblai/sol naturel doit donc être également plan et ne pas montrer de variations de position sur les pseudo-sections radar.
On remarquera que le réflecteur radar correspondant à l'interface remblai/sol naturel repéré sur les deux pseudo-sections temps présentées figure 5, montrent des ondulations temporelles non négligeables.
Elles indiquent des variations de vitesse de la

DSEA NEAR FX MCDEC
OFFSET NUL

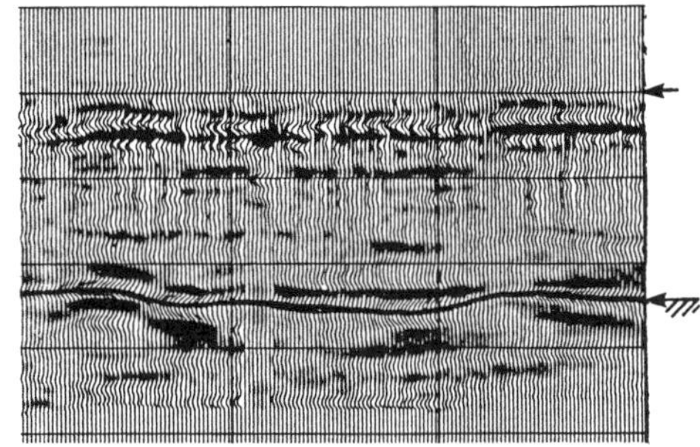

DSEA FAR FX REFOR INV
OFFSET = 58cm

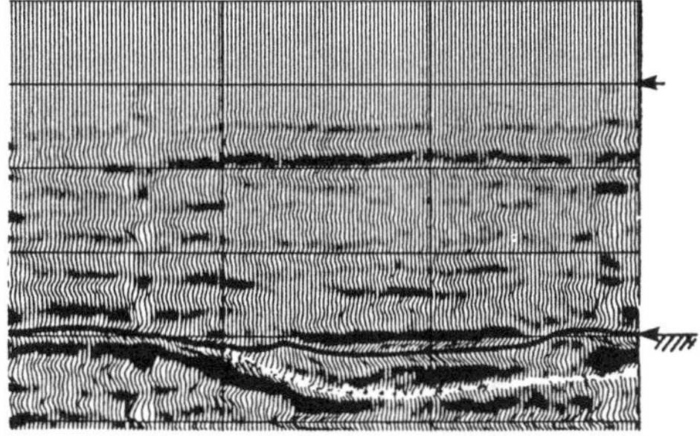

← Surface théorique de la paroi

← Interface remblai / sol

Figure 5: pseudo-sections radar

propagation des ondes radar dans le remblai traversé; des variations de profondeur de l'interface cible ou la conjonction des deux phénomènes en regard de la loi de vitesse :

$$\text{Vitesse} = \frac{\text{distance}}{\text{temps}}$$

Compte tenu qu'au niveau de la canalisation expérimentale, l'interface correspondant à notre réflecteur cible est rigoureusement équidistant de l'intrados de la canalisation sur la totalité du linéaire et que la portion de canalisation sélectionnée ne comporte pas d'anomalies d'environnement, <u>les variations de position temporelle du réflecteur enregistrées sur les sections ne peuvent que traduire des</u>

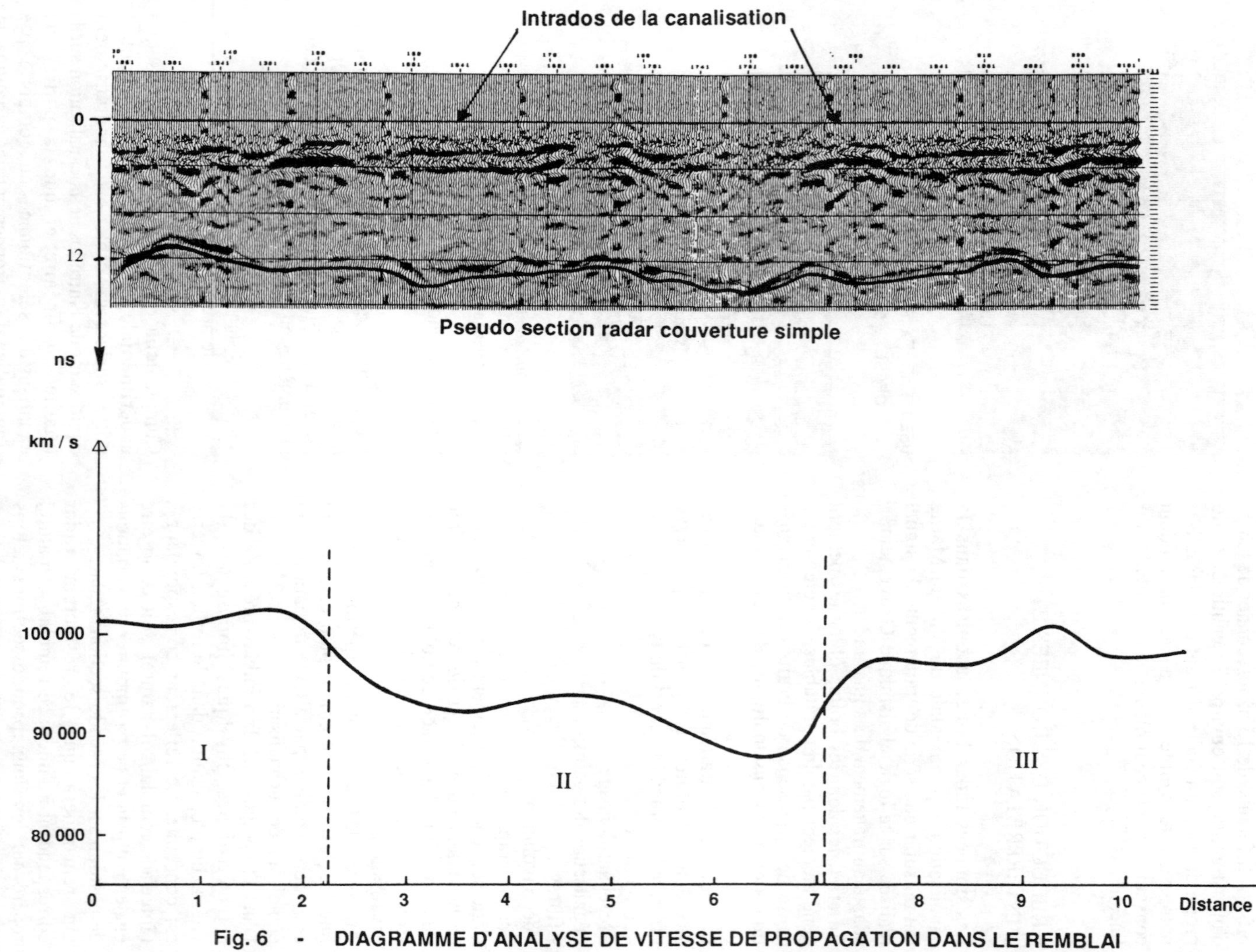

Fig. 6 - DIAGRAMME D'ANALYSE DE VITESSE DE PROPAGATION DANS LE REMBLAI

variations de vitesse au sein du milieu traversé. On peut donc affirmer en première approximation que le remblai situé sous la canalisation présente des degrés de porosité variables.
Pour un matériau de remblai de nature homogène, ces variations de porosité sont liées au degré de compaction du matériau. Plus les vitesses de propagation au sein de ce matériau sont rapides (position haute du réflecteur), plus l'indice de vide résiduel intergranulaire est important et moins le matériau est compacté. A l'inverse, les inflexions du réflecteur vers le bas indiquent un matériau plus lent avec un indice de vide intergranulaire plus faible, et donc un degré de compaction plus important. L'existence de variations de compaction du remblai sous la canalisation est donc démontrée directement par l'analyse visuelle des variations de position temporelle de notre réflecteur cible.
Ces conclusions ne sont valables que parce que le réflecteur cible de la canalisation de la D.S.E.A. du Val de Marne est plan et ne présente pas de variation de position en profondeur.
On doit considérer qu'il en va tout autrement pour une canalisation posée en tranchée, où l'on a alors toutes chances d'être en présence d'un interface tranchée/sol naturel à distance variable de intrados de la paroi.
C'est pourquoi l'analyse des variations de vitesse du remblai doit être faite à partir du calcul des vitesses de propagation d'ondes résultant d'une acquisition en couverture multiple.
Cette opération de calcul nécessite un traitement complémentaire des données consistant en :
- une identification du train d'onde primaire;
- une calibration d'antenne;
- un filtrage spatial;
- une analyse continue des vitesses;
- une mise en collection.

L'application de cette séquence de traitement sismique typique pour l'analyse continue des variations de vitesse de propagation d'ondes radar entre l'intrados de la canalisation et l'interface cible tranchée/sol naturel est présentée sur le diagramme d'analyse de vitesse du remblai (figure 6).
La vitesse moyenne calculée à partir de cette analyse est proche de celle que nous avons pu mesurer par transluminescence dans le remblai soit environ 100 000 km/s. Le profil de vitesse indique une zone II de vitesse relativement basse au centre du diagramme, environ 92000 km/s qui est encadrée de deux secteurs I et III où la vitesse moyenne atteint 104000 et 98000 km/s. La différence de vitesse enregistrée est de l'ordre de 10 %, qui peut être considérée comme significative d'un changement notoire de compaction du remblai. La comparaison des variations de vitesses calculées à partir de l'auscultation radar en couverture double et des variations temporelles de position du réflecteur cible de l'auscultation en couverture simple sont parfaitement comparables.

5 CONCLUSION

On peut donc raisonnablement estimer que l'étude menée sur le remblai de la canalisation expérimentale par acquisition radar en couverture double a atteint l'objectif assigné.
Les variations de vitesse calculées à partir de ce type d'acquisition sont significatives parce qu'elles sont en effet parfaitement corrélables avec les variations de la position temps du réflecteur cible tranchée/sol naturel équidistant de la canalisation et obtenue à partir de l'auscultation radar réalisée en couverture simple sans offset. On doit pouvoir démontrer que de telles variations de vitesse sont en intime relation avec des variations de porosité du remblai présenté sec :
- une augmentation de la vitesse de propagation dans le milieu correspondant à une augmentation de sa porosité et par conséquent à une diminution du degré de compactage du remblai;
- une diminution de la vitesse de propagation dans le milieu correspondant à une diminution de sa porosité et donc à une augmentation du degré de compactage du remblai.

Une telle méthodologie d'auscultation radar en couverture double doit trouver d'autres applications que la mesure du paramètre porosité et en particulier, cette technique devrait pouvoir être très utile pour mesurer la teneur en eau des remblais avec précision. Des zones restreintes de remblai à forte teneur en eau sont en effet souvent associées à des défauts d'étancheïté qu'il est utile de mettre en évidence.
Notons que l'acquisition radar en couverture double ne génère pas de surcoût particulier à l'acquisition des données, à l'inverse, le traitement des données est lourd. Il nécessite des moyens informatiques sophistiqués à base

de traitements de type sismique qui entraînent un surcoût évalué à un facteur 4 par rapport à un traitement du type couverture simple.

REFERENCES

Bushforth, M. and Koppenjan, S. 1992. Ground Penetrating Radar applications for non destructive inspection of construction works. E.A.E.G. International Meeting; G 005 Paris 1992.

Chignell, RJ. 1992. A multi-channel surveying Radar System. E.A.E.G. International Meeting; G 023 Paris 1992.

Davis, J.L., Annan, A.P. and Vaughan, C. Placer exploration using Radar and seismic methods 1984. 54th Annual International Society of Exploration Geophysicists Meeting in Atlanta, Georgia, December 1984.

Mermet, M., Dubreucq, F., Foillard, R. and George, B. 1992. Method for diagnosis of state of sewer networks. Proc. NO-DIG 92- 8th International Conference on Trenchless Technology. Washington: I.S.T.T.

Meunier, J., Foillard, R. and Charachon, B. 1992. Radar imaging enhancement using seismic software. E.A.E.G. International Meeting; G 031 Paris 1992.

Topp, G.C., Davis, J.L. 1992 (a) Measurement of soil water content using time domain reflectometry. Canadian Hydrology Symposium: 82, Assoc. Comm. on Hydrol., National Research Council of Canada, Ottawa, out. pp. 269.287.

Topp, G.C., Davis, J.L. and Annan, A.P. 1982 (b) Electromagnetic determination of soil water content using T.D.R.I. applications to wetting fronts and steep gradients. Soil Sci. Soc. Am. O. 46; 672-678.

No Trenches in Town, Henry & Mermet (eds) © 1992 Balkema, Rotterdam. ISBN 90 5410 085 0

Geometrical diagnosis of non-inspectable sewer pipes

C. Schwarze & J. Augarde
CREATE, Colombes, France

ABSTRACT: Comparative tests carried out at C.R.E.A.T.E. on experimental pipes and field conditions with several techniques, enabling the measure of vertical and horizontal deviations of non–inspectable sewer pipes:inclinometry, IR trajectometry and laser metrology, enabled us to evaluate the performances, advantages and disadvantages, limits and constraints of each method. Results given by inclinometry may be considered as satisfactory, provided that a computer program is available to readjust the longitudinal profile on the invert levels of the starting and arrival access points. The infrared trajectometry system provides accurate results in experimental conditions but its present configuration proves inadequate on the field. The laser metrology system is useful to check the quality of pipe by detecting vertical and horizontal misalignments of the pipes.

I INTRODUCTION

Our present knowledge of the geometry of non visitable sewerage systems is often still restricted to mere inspection ports, and measures can be made through topographical levelling means. We know nothing about fluctuations in longitudinal profiles and plane layouts of pipes in–between acccss holes whereas measuring such strains would be worthwhile on several accouts:

* to check compliance with project gradients when laying a pipe;
* to check how an actual system operates, and detect the gradient points of failure and the pipe misalignment points where deposits and loading may occur;
* to indirectly establish the geotechnical stresses (differential settlement) applied on underground systems.

The operator of a non visitable sewerage system only has one qualitative resource, ie. TV inspection cameras, to detect reverse gradients, as well as pipe disconnection and offset conditions; however, this does not allow for quantitative metering.

II SCOPE OF THE STUDY

The purpose of the "Geometrical Diagnotic" study conducted at the CREATE (Centre de Recherches et d'Essais Appliqués aux Techniques de l'Eau – Center of Research and Tests as Applied to Water Engineering, Colombes, France) from October 1988 to October 1989 (1) was to carry out, on experimental pipes and actual systems, research on and experimentation of several vertical and horizontal deviation metering systems for non visitable sewerage sytems.

The tests carried out on experimental pipes were meant to assess the accuracy and reliability of the selected techniques while the tests on actual systems were designed to establish the stresses and their limits under field conditions, as well as to demonstrate the relevance of geometrical diagnostic.

III SEWERAGE SYSTEM DETERIORATION

Numerous studies for the diagnostic of non visitable sewerage systems, notably undertaken with TV inspection means, demonstrate the existence of defects that often occur, including on extremely recent systems. Among these defects, geometrical faults such as pipe disconnection and offset conditions, surface depression and reverse gradient on ducts are by far the most frequent faults. A survey conducted in 1984 by the LCPC France on 77 instances of faulty networks (2) shows that the causes for geometrical defects result either from negligence in aligning or connecting the pipes when laying the system, or from geotechnical mechanisms such as differential settlement of the trench bottom disturbed during laying works, or fines from the pipe coating material carried away by downsliding nappes. The resulting distortions in the longitudinal profile and plane layout may jeopardize the structure joint tightness. By washing and carrying away coating materials in the duct or towards the subsoil, infiltration and exfiltration waters may then aggravate distorsions.

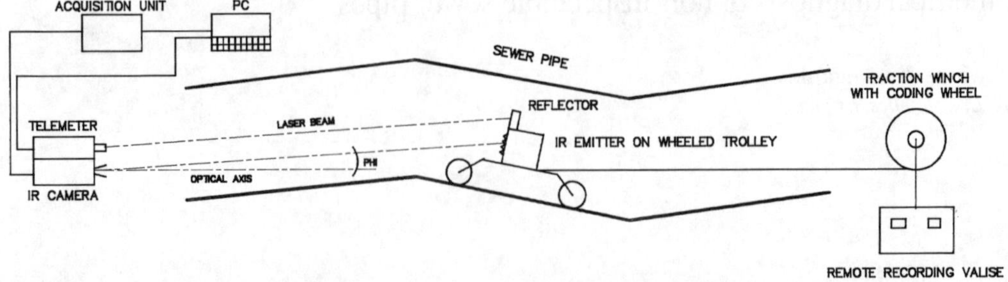

Figure 1. Schematic diagram of IR trajectometry system (Side view)

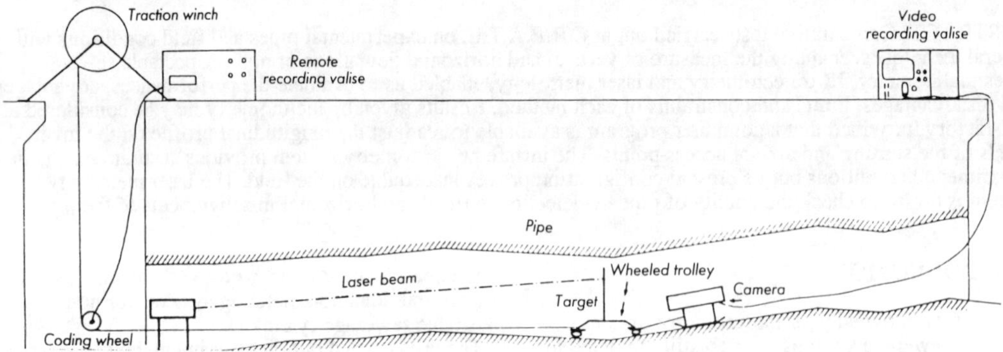

Figure 2. Schematic diagram of laser system (Side view)

IV COMPILING AND SELECTION METERING TECHNIQUES

A number of pick ups and metering techniques available on the market have been compiled. Three techniques have been selected:

* Banking Indicators. These pick ups measure gradient angles. They travel along the duct on board a moving support and they are associated to a meter that measures the distance covered, thus allowing to work out the longitudinal profile of a duct. It is presently the sole system fitted on TV inspection cameras for sewerage pipes.

* Infrared Trajectometry (Micromaine, France). This system includes an infrared transmitter travelling through the pipe on board a moving support, and an IR–sensitive stationary camera. The camera receives the IR radiation, follows the trajectory of the moving support, as a function of the duct deviations, and measures the relative displacements in angular coordinates thus providing a profile and plane layout of a pipe (see Fig. 1).

* YZ Laser Metrology. CREATE home–made equipment. The facility includes a pipe alignment laser system located at the inlet of the duct. The beam can be adjusted for gradient and alignment on the horizontal plane. It runs into a transparent target secured onto a mobile support that travels through the pipe. The XY position of the laser beam impact with the target, whose position depends on the pipe Z–vertical and Y–horizontal deviations, is displayed on an external screen by a video camera installed behind the target (see Fig. 2).

V TESTS ON EXPERIMENTAL PIPES

The three above–mentioned techniques have been tested on two surface–layed experimental pipes (length: 22.5 m and 25 m) made of 200 and 400 mm asbestos cement.

The plane layouts and profiles of both pipes were measured by topographical levelling to make up a set of references for the measures.

The accuracy of the techniques was measured taking into account the variation between the reference curves and the curves obtained based on the measures made on the equipment under test.

Measure reliability was assessed through the repetitiveness of the values found under identical testing conditions.

V.1 Analyzing the Results

V.1.1 Banking Indicator

Figure 3 shows the longitudinal profiles obtained

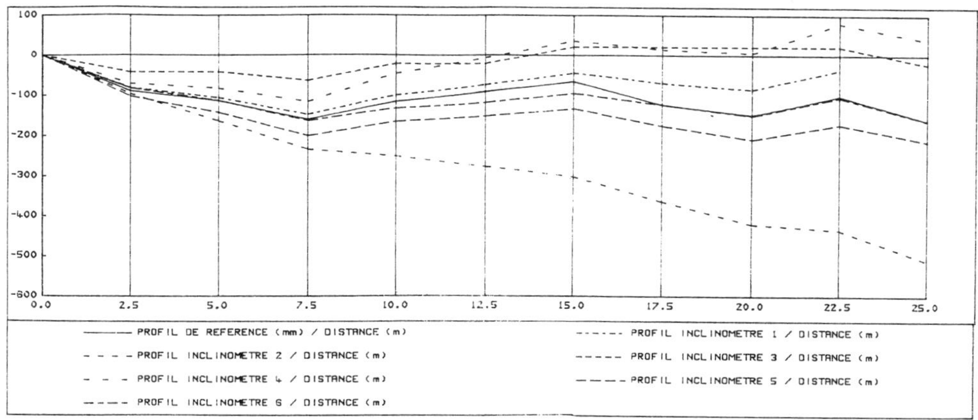

Figure 3. Long. profiles obtained in the test pipe using 6 diff. banking indicators

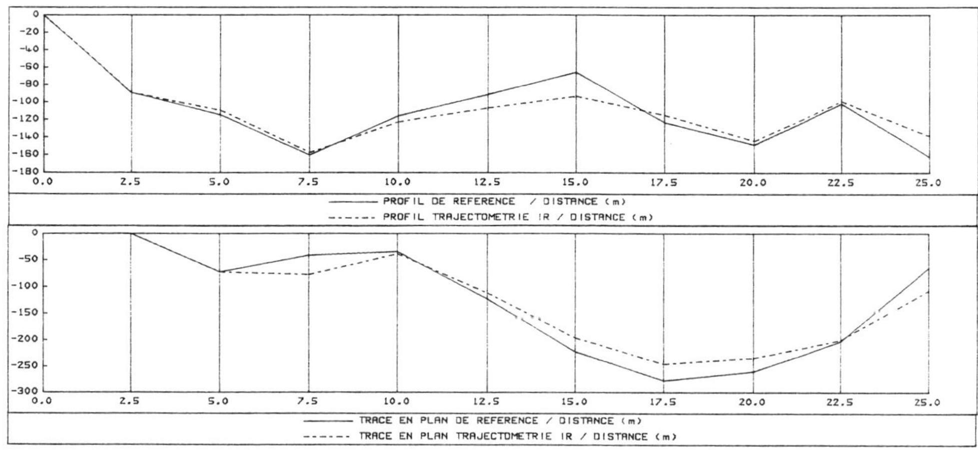

Figure 4. Long.profile and plane layout obtained in the test pipe using IR trajectometry

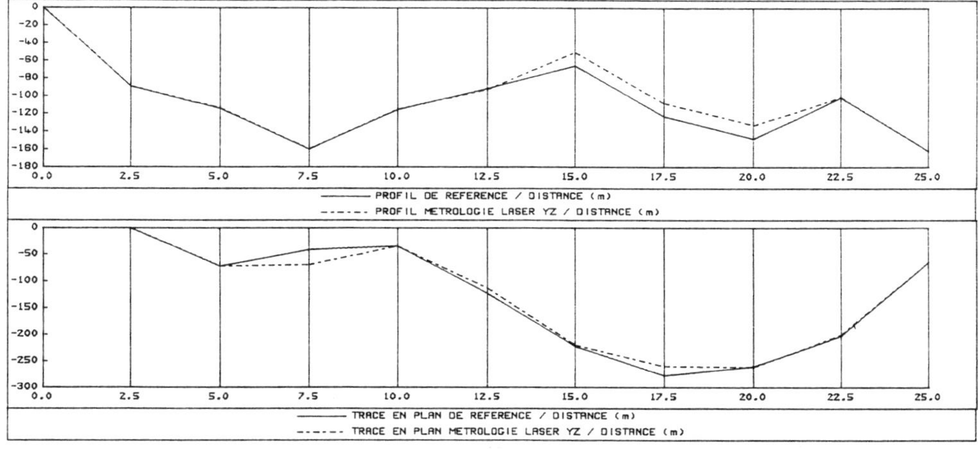

Figure 5. Long. profile and plane layout obtained in the test pipe by laser system

under identical conditions in the 400–mm test pipe using 6 different cameras.

Three of the devices provide satisfactory results with variations ranging from 30 to 50 mm at travel end between the reference profile and the profile read by the camera, while the run of the curve complies with that of the reference profile.
Pick up calibration and signal processing of the other three devices are faulty, and they produce profiles that are quite different from the reference profile (variations ranging from 100 to 400 mm) with a reversed average gradient, or not in compliance with the actual longitudinal profile run.

There is only one device (IBAK) that provides an opportunity to readjust the profile read on the invert levels for the camera starting and arrival points.

The major drawback involved in this metering technique is that errors on the gradient measure (and on the distance) caused by an offcentered carriage or by obstructions in the pipe result in variations between actual and obtained profiles, that increase with the distance covered.

V.1.2 IR Trajectometry System

Figure 4 shows an example of a longitudinal profile and plane layout obtained using an IR trajectometry system.

The curves do follow the run of the reference layouts. The variations between the curves obtained through IR trajectometry and the reference curves range from 0 to 50 mm. These deviations primarily originate in the fact that the moving support carrying the transmitter cannot be accurately centered throughout the pipe length. Measure repetitiveness is correct: variations are less than 0.1°.

The drawbacks identified for this system are:

* saturation of the camera IR pick up which compels a minimum distance between the transmitter and the camera and therefore makes it a priori difficult or even impossible to produce a complete profile and plane layout for a pipe system.

* the scope of the system depends on the camera visual field transmitter output.

V.1.3 YZ Laser Metrology System

Figure 5 shows an example of the longitudinal profile and plane layout obtained with this system. The variations between the curve obtained and the reference curve range from 0 to 35 mm.

VI FIELD TESTS

VI.1 Banking Indicators

The IBAK device has been selected because of the opportunity it provides to readjust the longitudinal profile obtained in relation to the invert levels for the access holes of the starting and arrival points, and it has been used in several downgraded pipe systems.

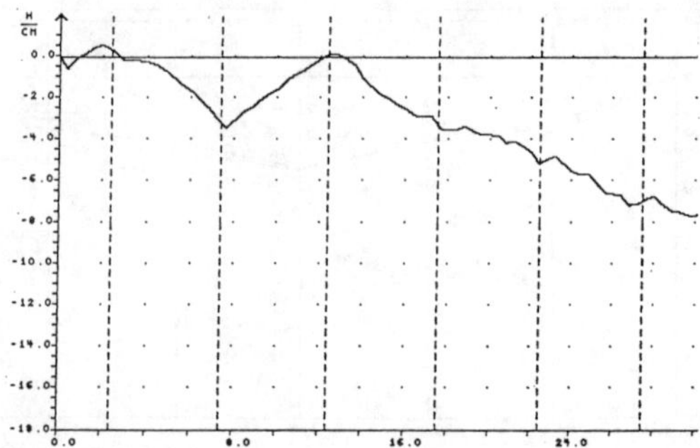

Figure 6. Example 1 of profile obtained using IBAK system

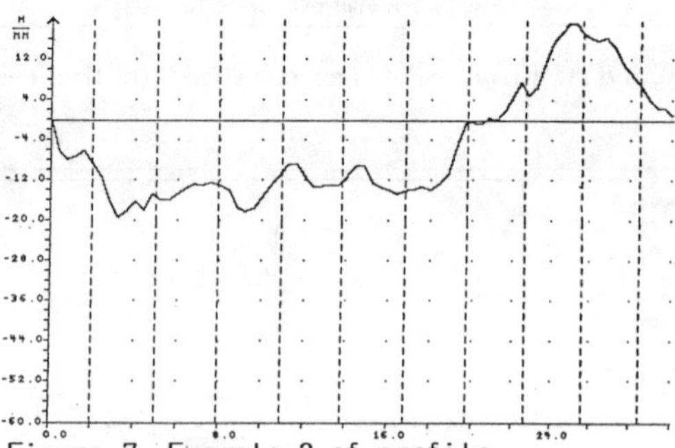

Figure 7. Example 2 of profile obtained using IBAK system

Example 1: Figure 6 shows the longitudinal profile obtained on a 200 asbestos cement pipe layed underneath the roadway, that lies in the Machemont nappe, Oise. The IBAK camera found a sharp reverse gradient caused by a 4–cm slump at the level of a joint between two consecutive pipes, therefore loading the pipe at this point.

Example 2: Figure 7 shows a profile drawn on a 200–mm PVC pipe system layed in the Bury nappe, Oise. There is almost no gradient at all for 200 m then a −16cm to +18cm difference over 6 m of duct, ie. a reverse gradient of over 5% which caused loading and clogging of the pipe.

VI.2 IR Trajectometry

The IR trajectometry system was tested on a 70–m long section of a pipe made of 500–mm asbestos cement in Buc, Essonne, France.

During the test, the stresses and limits of IR trajectometry for duct longitudinal profile and plane layout production clearly appeared:

* owing to the minimum distance required between transmitter and IR camera because of the saturation of the camera photoelectric pick up, the profile and plane layout could not be obtained on 17 m of the duct (out of a total of 70 m).

* The limit value for the IR transmitter detection distance by the camera was 30–40 m; as a result,

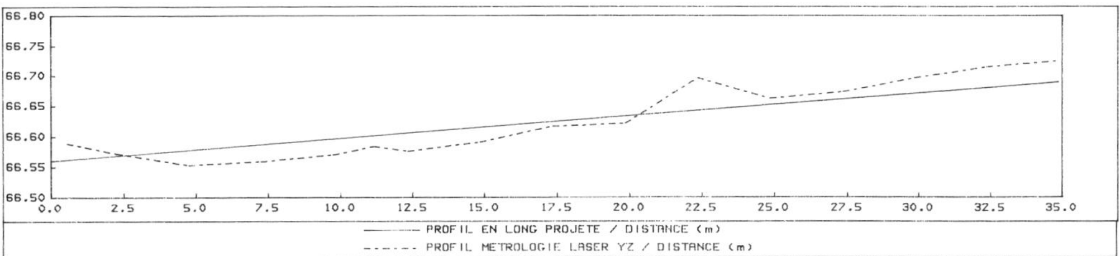

Figure 8. Differences observed between metered longitudinal profile and longitudinal profile scheduled by main contractor of the Igny pipe

reverse testing had to be resumed from the access hole on the other end of the section under test.

* It is impossible to establish the position in space of the camera optical axis whereas it is the reference mark for metering of the angular position of the transmitter that travels through the pipe. It was therefore not possible to obtain a longitudinal profile and a plane layout of the pipe vs. the invert levels of the upstream and downstream access holes in the section under study.

VI.3 YZ Laser Metrology

The YZ laser metrology system was used for measurements on a 500– and 600–mm asbestos cement pipe in Igny, Essonne, France, two months after construction was completed and before commissioning so as to have full control over compliance with the profile and plane layout specified by the main contractor. Approximately 300 running meters were inspected on the whole, with daily rates ranging from 100 to 150 m.

Concurrently, surface strains of the filling material for the ditch were observed using the LCPC.

YZ readings at target level were converted into altimeter dimensions (meters) as computed in relation to the corresponding invert levels so as to obtain the actual longitudinal profile and plane layout of the pipe and compare them to the project layouts.

The following applies:

* the tested sections usually do not include significant gradient gaps.

* The variations between metered longitudinal profile and the longitudinal profile scheduled by the main contractor are usually low, ranging from 0 to 40 mm at maximum (see Fig. 8).

* Conversely, the variations between the measured plane layouts vs. the scheduled layouts are greater, ranging from 0 to over 300 mm. This may result from inappropriate fitting of the pipes at the laying stage, using the power shovel.

* On various points of the pipe, surface strains (settlement) of the ditch fill match similar strains (slumps) on the longitudinal profile. This may result from the fact that the ground was broken without prior sheeting during the laying works; this caused decompression of the surrounding land, and more especially of the trench bottom which, once filled, settled again under the weight of the fill, driving the pipe along and causing strains on the longitudinal profile (see Fig. 8)

VII CONCLUSIONS

The tests carried out on experimental pipes and systems in operation with three metering techniques to measure the longitudinal profile and/or plane layout of a non visitable sewerage system lead to the following conclusions:

* measuring the longitudinal profile of a pipe using a banking indicator travelling on board TV inspection cameras is presently the sole operational technique on the field which is available on the market. The results may be considered as satisfactory, eg. to detect reverse gradients or check that the project gradients are complied with, provided that a computer program is available to readjust the longitudinal profile measured on the invert levels of the starting and arrival access points of the camera.

* The IR trajectometry system provides accurate results under experimental conditions but its present configuration proves inadequate on the field to measure the longitudinal profile and plane layout of a pipe which can be accessed through inspection holes. However, this system was selected as a cartography module for a sewerage system diagnostic and auscultation robot presently under being developed by Spie Trindel, France.

* The YZ laser metrology system is useful to check the quality of a pipe by detecting misalignments in the vertical and horizontal directions of the pipes, as well as local errors in the drain gradient. Development studies designed to turn it into an operational implement within the framework of construction work diagnostic and acceptance have not so far been undertaken.

REFERENCES

(1) Schwarze, C. 1990. Diagnostique géometrique des canalisations d'assainissement. CREATE.

(2) LCPC 1985. Influence des conditions geotechniques sur le comportement des canalisations d'assainissement ; Première partie : enquête sur les dégradations ; LCPC/Agence de bassin Loire–Bretagne.

No Trenches in Town, Henry & Mermet (eds) © 1992 Balkema, Rotterdam. ISBN 90 5410 085 0

Auscultation trails in an experimental sewer

M. Darras & F. Dubreucq
Val de Marne Administration for Water and Sewer Networks, France
C. Schwarze
Centre for Research Applied to Water Management Techniques, The Normandy-Seine Water Agency (CREATE), France

ABSTRACT

In-depth knowledge of the origins of the disorders responsible for the deterioration in sewer networks inaccessible to men is nowadays required to optimise their technical management.
The Val de Marne Administration County and CREATE have miniaturised, robotised and tested a certain number of techniques which are frequently used in accessible sewers.
These trials were carried out in an "experimental sewer", whose immediate surroundings have known abnormalities required in the validation of the tools used.
This article describes, on the one hand, the site of the experiments, and on the other, the protocol used, along with the results obtained and the methodology used to auscult the sewer networks.

INTRODUCTION

The Val de Marne Administration for Water and Sewer Networks and CREATE have combined their efforts to develop sewer network auscultation techniques for inaccessible sewers.
In fact, the first function of the network managers is to ensure the transport of effluent and to measure its quality and the quantity. Urban main sewers, required for the above mentioned rôles must therefore be operational and knowledge of their physical state is essential for efficient management.
Moreover, the procedure used to obtain in-depth knowledge of the origins of disorders responsible for the deteriorations observed in some sewers, is an original methodology based on the detection, the analysis and the understanding of the sewer's problems using observation, measurements, data processing and the interpretation of all this information. This methodology requires the development of tools and their validation in inaccessible main sewers, which is made possible thanks to the construction of an experimental sewer at Champigny/Marne in the Val de Marne.
The text hereafter briefly describes : the experimental protocol, the experimental phases, the results.

1 EXPERIMENTAL PROTOCOL

1.1 *DESCRIPTION OF THE EXPERIMENTAL SITE*

The experimental sewer is composed of ten pieces (o 400 mm) in asbestos cement. It measures 25 metres long and for technical reasons, could not be laid underground. It therefore placed in a U-shaped support composed of 13 prefabricated pieces and is closed off at each end metal sheets.
The internal joints of this U-shaped structure, covered in resin and aluminium sheets, ensure that it is tight. A drain in the bottom of this structure evacuates the water which has collected in the immediate surroundings of the sewer during the trials.
The sewer is placed on a bed of scouring sand at an incline of 5/1000.
This "trench" is divided up into four sections separated by wooden walls :
- the first is a control section, free of abnormalities, which contains uncompacted banking-up,
- the second is a control section, free of abnormalities, which contains compacted banking-up,
- the third has abnormalities and contains compacted banking-up,
- the fourth has abnormalities identical to the third, and contains uncompacted banking-up.
The abnormalities consist of concrete blocks, polystyrene blocks and blocks of cellular concrete as well as boxes which simulate cavities, sections of anothe sewer structure and different compacting densities (Fig 1).

1.2. *EXPERIMENTAL PHASES*

The series of experiments carried out in this sewer requires variations in the sewer's environment and in auscultation techniques. The experiments are therefore carried out in several phases.

1.2.1. *First phase*
The immediate surroundings of the sewer, composed of scouring sand, which may or may not be compacted and

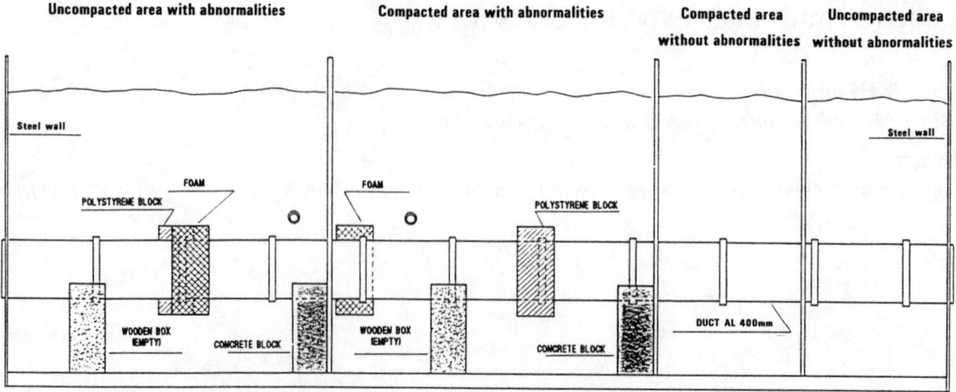

Fig.1 **EXPERIMENTALE SEWER AT CHAMPIGNY/MARNE SIDE VIEW**

which has abnormalities, is dry.

The auscultations carried out consist of visualisations on a screen, radar and nuclear diagraphies using neu and gamma-gamma probes.

1.2.2. *Second phase*

The immediate surroundings remain dry but the sewer is filled to a height of 10 cms. The conducitivity of this "effluent" is that of water and is therefore considered to be low. The auscultations are then carried out according to the established protocol.

1.2.3. *Third phase*

The same operations as in phases one and two are carried out but an aquifer is artificially created in the sewer's immediate surroundings.

At the same time, humidity measurements are regularly taken in the soil so as to follow the spread of the humidity.

Thus, in each of these phases, the dielectric conditions of the media used were varied.

1.3. ***TRIALS***

So as to be able to investigate state of the main sewer and its environment, several techniques were adapted and tested both from within the sewer and, in certain cases, from the surface.

In fact, although many tools are presently being used in the accessible networks, the constraints imposed by their use do not allow them to be used in inaccessible sewers. Therefore some of them were modified so as to be usable everywhere in the network.

The tools used in this series of experiments :

- identify the origin of certain disorders using knowledge of the terrain adjacent to the sewer,
- obtain a better command of certain auscultation techniques and the data processing methods which accompany them.

1.3.1. *Implementation*

For the majority of the techniques used, the implementation is more or less the same. When the inspections are longitudinal or punctual, the sewer section is compared to a clock so we can take our measurements and carry out the desired manipulation at predefined positions (hours). The hours used are 0.00pm, 2.00pm, 4.00pm, 6.00am, 8.00am and 10.00am.

When the investigations have to be carried out transversally, their position is defined along the length of the sewer.

Whatever technique is used, the first time the tool passes through the sewer is a "blind" passage. This means that no information shall be communicated to the tool operators, not only while the measurements are being taken but also during the data processing. A first examination is carried out and followed by a discussion of the results. During this first examination, longitudinal measurements are also taken every "two hours" as well as transversal measurements at the joints and at any other spot chosen by the operator, as he thinks fits after examination of the data.

1.3.2.*Auscultation using subsurface radar*

1.3.2.1. *Principle*

This type of auscultation which is frequently carried out in accessible networks, is used, on the one hand, to obtain more detailed knowledge of the structure and the environment of a sewer and, on the other hand, to detect and, in favourable conditions, measure and identify abnormalities.

The principle consists of sending an impulse of a short duration (several nanoseconds), emitted at a very high frequency (50KHz) by an antenna (100 MHz at 1 GHz). When the signal meets a discontinuity it is reflected and captured by a receiver which sends it to a recorder. The echos, which correspond to the different discontinuities, are recorded one after the other, as the antennae move at a regular pace (1 to 3 km/h) along the profiles to be ausculted, whose number depends on the precision required. A time cut is thus created, which can be

transformed into a depth cut. The depth of investigations can range from 0.1 to 20 metres for construction materials. These depths depend on :
- the capacity of the equipment,
- the frequency of the waves emitted,
- the dielectric properties of the materials.

The readings can be interpreted immediately, but they often require processing which is adapted to the objective.

1.3.2.2. *Trial procedure*

A miniaturised carriage, drawn or pushed by a mobile camera, is used to inspect the sewer from within.

The inspections are carried out in sewers with a diametre of 400 mm, nonetheless the equipment was designed to pass through sewers of 200 mm in diametre.

Measurements can be taken at any point along the length of the sewer, that is on the longitudinal or transversal traces.

Because of the position of the radar's antenna, in front of the camera lens, its postition in the sewer can be seen on a control screen on board the "camera van". This means that the position of the antenna can be corrected at any moment using a dial which indicates its angle (photo 1).

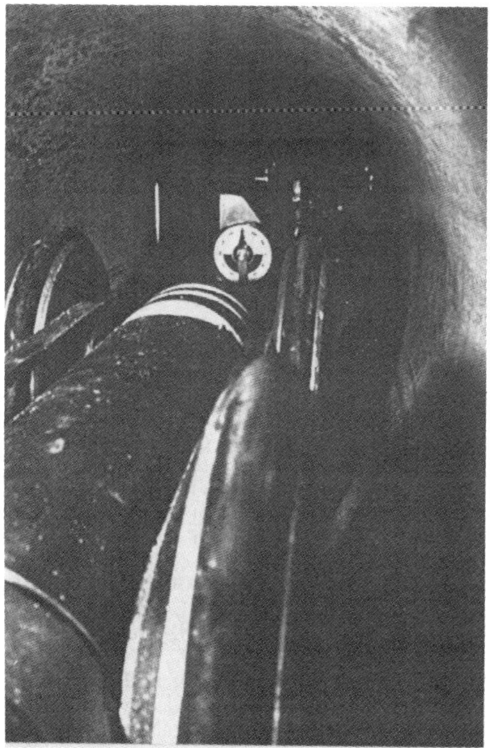

Photo 1 Radar and camera in a sewar with visualisation of the antennas position

1.3.3. *Auscultation using gamma-gamma and neutron-neutron nuclear diagraphies*

Trials were carried out in the experimental sewer at Champigny-sur-Marne using gamma-gamma and neutron-neutron diagraphy probes developed by the LCPC ("Laboratoire des Ponts et Chaussées")and BRGM (Bureau de Recherche Geologique et Minière).

The aim of these trials was to assess the interest of these tools, normally used in geological prospection, for non-destructive, continuous auscultation of the immediate surroundings of a sewer duct with a small diametre over a depth of 0 to 20 cm, so as to detect and situate abnormalities such as cavities, localised hard rock and decompacted areas.

1.3.3.1. *Principles and equipment*

* Using retrodiffusion at a given distance from a source of ^{137}Cs, gamma-gamma diagraphy measures the intensity of the gamma photons diffused by the COMPTON effect. It has been shown that the intensity of the photons decreases when the electric density of the matter, which is itself a function of the volumetric weight of the terrain increases. Gamma-gamma diagraphy measures therefore the density of the banking up when humid.

The BRGM gamma-gamma probe (manufactured by Mount-Sopris) contains a source of ^{137}Cs of 100 mCi of intensity and possesses 2 detectors, one situated 10 cm from the source (short distance detector), the other 35 cm (long distance detector). The source and the detectors are adjusted to the same generating line. When prosopecting, an arm is used to place the probe against the wall of the hole. The length of the tool is 2.20 metres, the diametre, 57 mm and the weight, 17kg.

The LCPC gamma-gamma probe (manufactured by LCPC in France) has a radioactive source of ^{137}Cs of an intensity of 60 mCi. The tool has 2 detectors, one 20 cm from the source (short distance detector) and the other 40 cm (long distance detector). The length of the probe is 1.5 m, the diametre, 46 mm and the weight, 8 kg (Fig 2).

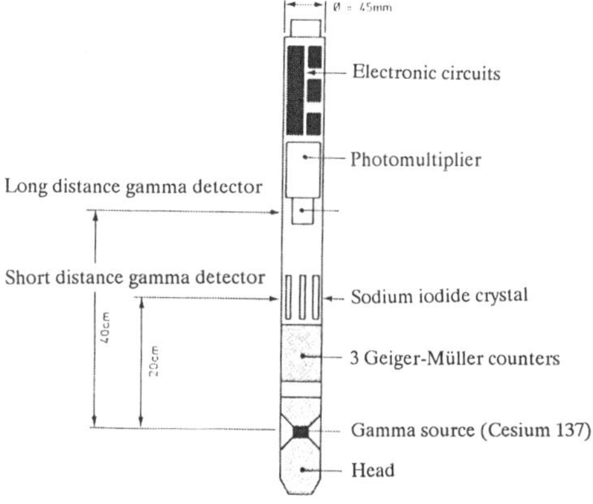

Fig 2 Diagram of the principle of the gamma-gamma probe developed by the LCPC

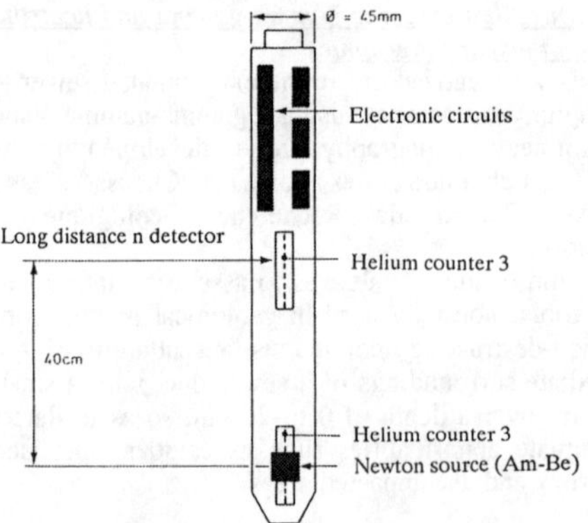

Fig 3 Diagram of the principle of the neutron diagraphy probe

* The nuclear, neutron-neutron diagraphy records, at a given distance from a source of 241 AmBe, the retrodiffused flow of thermal neutrons, that is fast neutrons which have lost a part of their initial energy by successive collisions with matter, essentially with the hydrogen atoms present in the medium. In practical terms, this means being able to measure the volumetric water content of the terrain ausculted.

The LCPC neutron-neutron probe (manufactured by LCPC) has 3 sources of fast AmBe neutrons of an intensity of 3 x 200 mCi. The SD detector is placed next to the source whilst the LD detector is 40 cm away (Fig. 3). The probe is 1.76 cm long, with a diametre of 46 mm and weighs 5 kg.

The remaining volume of the probes is taken up by electronic circuits which transform the readings into electric current, amplify it and send it to the data processing unit on the surface via the trolley cable.

1.3.3.2. *Trial procedure*

The different probes were drawn through the experimental sewer along generating lines spaced out at 2 hour intervals, at speeds ranging from 1.5 to 20 m/min according to the type of tools and the probe used.

So as to be able to auscult other generating lines than those situated at 6.00pm (tunnel floor), the probes were mounted on a specially adapted probe bearing trolley, which allowed them to be maintained in the desired position at least 2 cm from the wall of the duct as the trolley passed through it (photo 2).

Other examinations were carried out on the tunnel floor (generator line 6.00pm) of the sewer when filled with water, the height of the water varying between 7 cm at the highest point of the sewer and 30 cm at the lowest.

II RESULTS

Radar auscultations and nuclear diagraphy results are discussed here.

2.1. RADAR AUSCULTATION

Data acquisition is carried out continuously for each of the longitudinal or transversal profiles. The antenna used (1 GHz) was chosen because of its small size. The recordings were carried out simultaneously using two auscultation durations, 15 and 30 nanoseconds, on each profile. These times corresponding to average investigation depths of 75 and 150 cm respectively behind the sewer wall.

After the data from the signal has been processed, radar auscultation shows up many discontinuities present in the sewer environment.

Figure 4 corresponds to the scanning of a longitudinal profile carried out on the tunnel floor, that is at 0.00pm, from joint n°5 to joint n°7, that is approximately 5m. From the top to the bottom we can identify :

- a continuous interface corresponding to the surface of the wall,
- small abnormalities every 2.5m corresponding to the joints between each piece of the sewer,
- a discontinuity created by a series of reflectors of strong amplitude around joint n° 5, which is approximately one metre across. It corresponds to the presence of an artificial cavity of 0.6 x 1 m underneath the sewer,
- from joint n°6 to joint n° 7 a hyperbolic reflexion correspond to the pipe situated over the experi canalisation.

The position of an interface varies in propagation time from the waves of joint n° 7 (between 4 and 6 nanoseconds) which corresponds to the variations of the propagation speed of the electromagnetic wave in the medium crossed, in this case, the soil underneath and the air.

The radar auscultation technique was therefore used to locate and identify some abnormalities (concrete blocks, cavities, sewers, etc...)

2.2. *Nuclear diagraphies*

The results are presented as profiles with, on the y axis, the cut readings and on the x axis, the distance in metres covered by the probe in the duct.

In conformity with subsequent experiments only the following readings are taken into consideration. Those from:
- the LD detector of the LCPC GG probe,
- the SD detector of the BRMG GG probe,
- the SD detector of the LCPC NN probe.

Figure 5 presents the profiles of readings from the BRGM gamma probe, obtained along the different generator lines ausculted. A significant increase in the readings indicates a decrease in the volumetric weight of the medium, whilst their decrease indicates

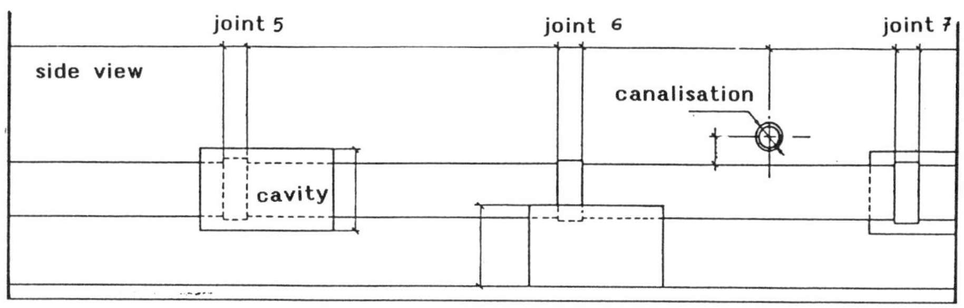

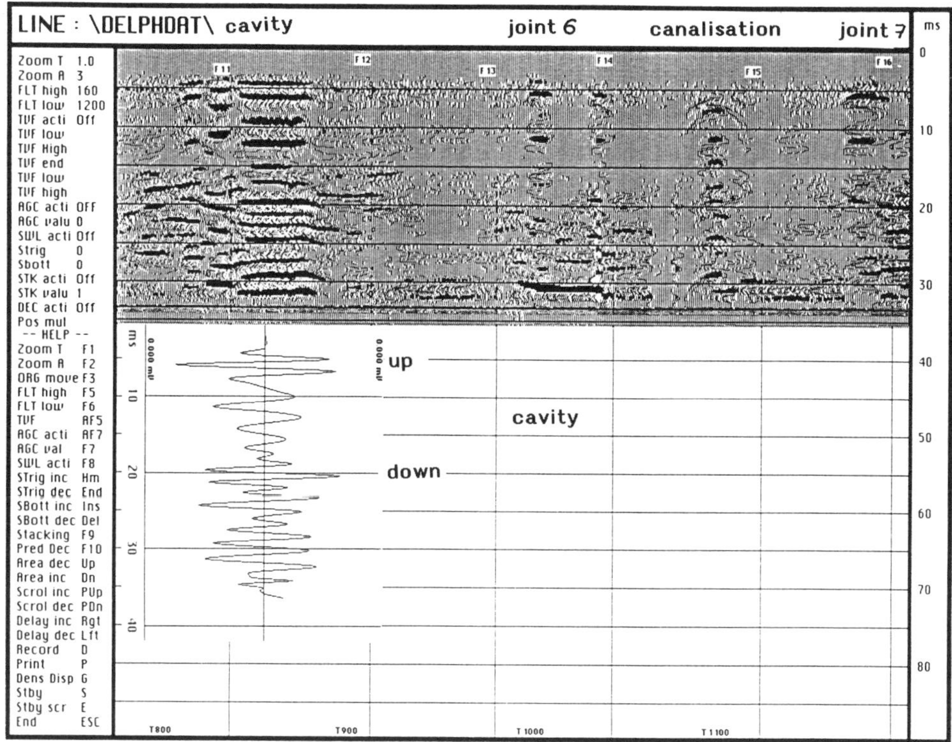

Fig 4 Experimental sewer- profile view-and radar pseudo-section of the tunnel at 0.00 am.

the presence of spots of a greater density. The following were observed.

A significant increase in the gamma readings indicating the probable presence of a cavity or spot of lesser density at approximately :
- 3.5 metres on the 8.00am and 6.00am profiles, indicating the cavity (wooden box) placed in this spot under the lowest generator line of the sewer,
- 7.5 metres on the 10.00am and 8.00am profiles, indicating the presence of a block of cellular concrete at 9.00am,
- 6.5 m on the 4.00am and 2.00am profiles indicating the presence of the polystyrene block buried at 3.00am,
- 11.5m on the 4.00am and 2.00am profiles (diminishing at 6.00am) indicating the block of cellular concrete placed at 3.00am,
- 16 m on 0.00am, 10.00pm and 2.00am which corresponds precisely to the presence of a polystyrene block placed on the highest generator line of the sewer.

* A significant decrease in the readings indicating the probable presence of a spot of greater density (localised hard rock) at 10.5 and 19.5 metres on the 6.00pm and 8.00pm profiles which corresponds to the position of the 2 concrete blocks placed under the lowest generator line of the sewer.

The compacted areas of the experimental sewer situated between 9 and 22 metres show up on the

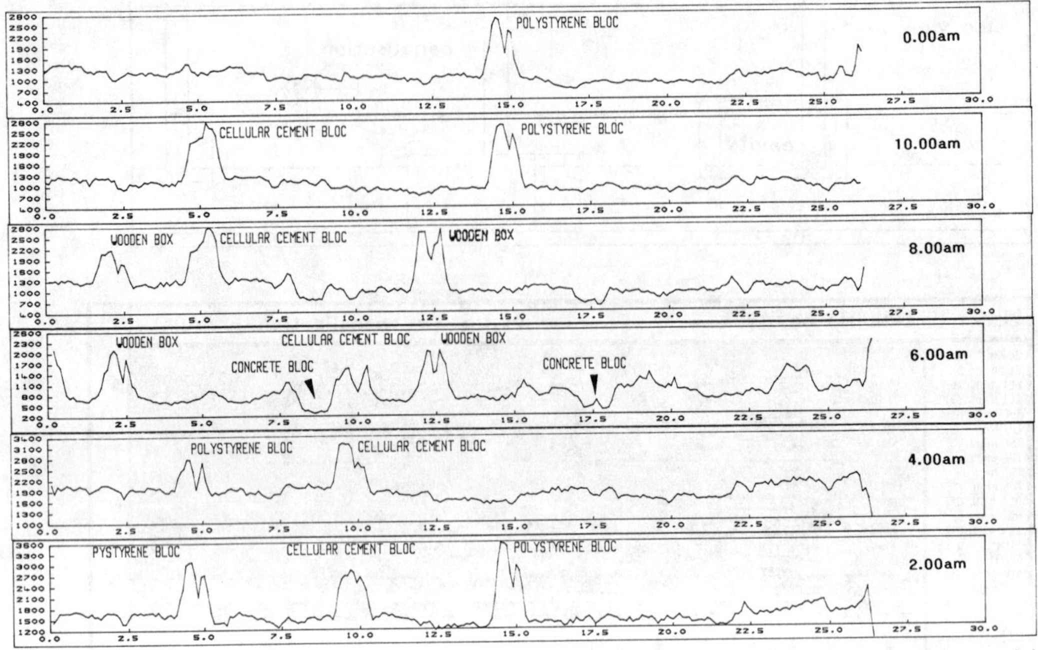

Fig 5 Profiles of the reading from the LCPC gamma probe

Tab.1 Methodology of the inspections in the Val de Marne.

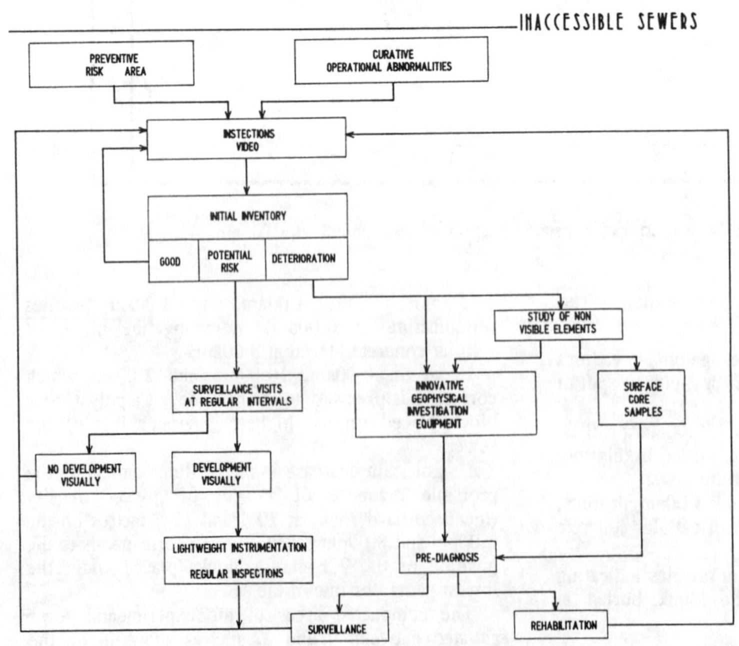

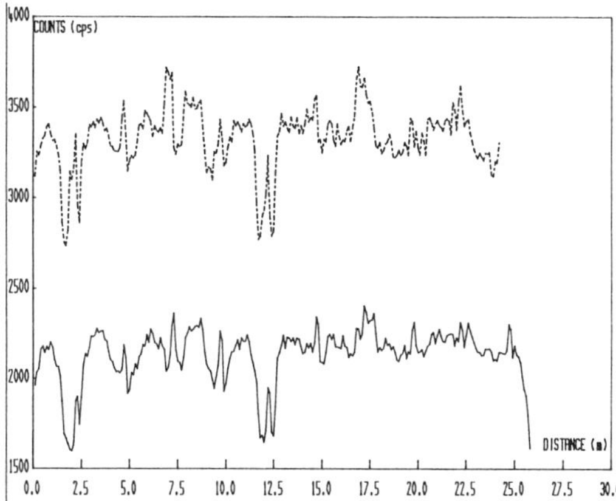

Fig 6 Profiles of the readings from the LCPC neutron probe

profiles of 0.00pm, 10.00am and 2.00pm by a reduction of the average level of the readings when we pass from uncompacted to compacted areas (arrows).

Figure 6 shows two neutron profiles of compacting from the LCPC neutron probe, carried out on the 6.00am generator line in the sewer when empty and when filled with water. The presence of cavities at 1.7 and 11.7 metres entails a significant reduction in the neutron readings, whilst the denser spots (8 and 17 metres) appear as a decrease of the neutron readings.

CONCLUSION

The selection criteria of the adapted and tested techniques for the auscultation of inaccessible sewers were their performances and complementary qualities.

These tests carried out in an experimental sewer fitted with control areas and areas presenting known abnormalities shall be used to obtain a better command of the tools used as well as the processing of their data.

The first results obtained have shown the possibility of adapting certain tools and envisaging their use, in the short term, and after several modifications, in the networks in service.

All of these results, along with visual inspections, will allow the application of the methodology used in the Val de Marne (table 1).

No Trenches in Town, Henry & Mermet (eds) © 1992 Balkema, Rotterdam. ISBN 90 5410 085 0

Essais d'auscultation dans une canalisation expérimentale

M. Darras & F. Dubreucq
Services de l'Eau et de l'Assainissement du Val de Marne, France

C. Schwarze
Centre de Recherches Appliquées aux Techniques de l'Eau de l'Agence de l'Eau Seine Normandie, France

RESUME : Une bonne connaissance des origines des désordres responsables des dégradations occasionnées dans les réseaux d'assainissement non accessibles à l'homme est aujourd'hui nécessaire pour optimiser leur gestion technique

Le Département du Val de Marne et le Centre de Recherche et d'Essais Appliqués aux Techniques de l'Eau de l'Agence de l'Eau Seine Normandie ont miniaturisé, robotisé et testé un certain nombre de techniques, couramment utilisées dans les ouvrages accessibles pour les rendre opérationnels dans des collecteurs de petites dimensions.

Ces essais ont pu être réalisés dans une "canalisation expérimentale" dont l'encaissant comporte des anomalies connues, indispensables à la validation des outils mis en oeuvre.

Cet article présente d'une part le site expérimental, d'autre part le protocole utilisé, enfin les résultats obtenus ainsi que la méthodologie suivie pour ausculter les ouvrages d'assainissement non accessibles à l'homme.

INTRODUCTION

Le Département du Val de Marne et le Centre de Recherches Appliquées aux Techniques de l'Eau de l'Agence de l'Eau Seine Normandie ont uni leurs efforts pour développer les techniques d'auscultation des ouvrages d'assainissement non accessibles à l'homme.

En effet, la fonction première des gestionnaires de réseaux étant d'assurer le transport d'effluents et d'en mesurer la qualité et la quantité. Les collecteurs urbains nécessaires aux rôles précédemment cités doivent donc être opérationnels et la connaissance de leur état physique est indispensable pour assurer une gestion de qualité.

Aussi la démarche suivie pour obtenir une bonne connaissance des origines des désordres, responsables des dégradations constatées dans certains ouvrages, passe par une méthodologie originale fondée sur la détection, l'analyse et la compréhension de leurs problèmes grâce à l'observation, aux mesures, aux traitements et aux interprétations s'y rattachant. Cette méthodologie impose le développement d'outils et leur validation dans les collecteurs non accessibles. Ceci est rendu possible grâce à la construction d'une canalisation expérimentale située à Champigny sur Marne dans le département du Val de Marne.

Le texte ci-dessous présente brièvement :

- le protocole expérimental,
- les phases d'expérimentation,
- les résultats.

I PROTOCOLE EXPERIMENTAL

1.1 *DESCRIPTION DU SITE EXPERIMENTAL*

La canalisation expérimentale se compose de dix éléments (ϕ 400 mm) en amiante ciment. Elle mesure 25 mètres de long et pour des raisons techniques, n'a pas pu être enterrée. Elle se trouve donc dans un support en forme de U composé de 13 éléments préfabriqués. Cet ensemble est clos aux deux extrémités par des plaques métalliques.

Les joints internes de cette structure en "U", recouverts de résine et de plaques en aluminium, assurent son étanchéité. Un drain installé au fond évacue l'eau emmagasinée dans "l'encaissant" lors des essais.

La canalisation repose sur un lit de sablon avec une pente de 5/1000.

Cette "tranchée" se subdivise en quatre compartiments séparés par des cloisons de bois :

- le premier, témoin et dépourvu d'anomalie, renferme des matériaux non compactés,
- le second, témoin et dépourvu d'anomalie, renferme des matériaux compactés,
- le troisième, pourvu d'anomalies, renferme des matériaux compactés,
- le quatrième, pourvu d'anomalies identiques aux précédentes, renferme des matériaux non compactés.

Les anomalies se composent : de blocs de béton, de polystyrène, de béton cellulaire , mais aussi de caisses

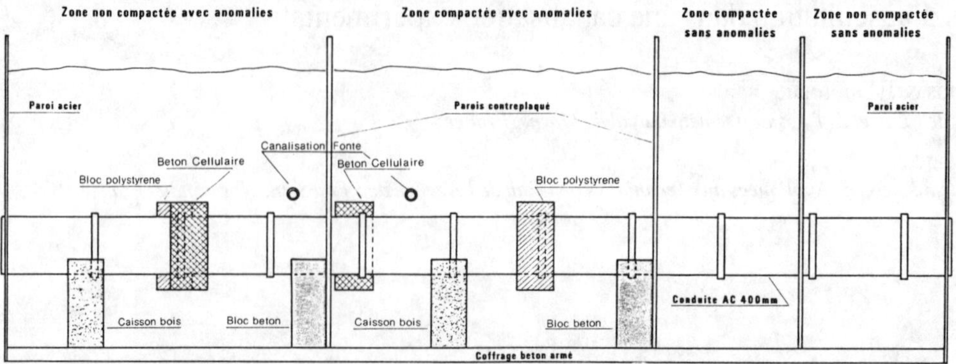

Fig.1 **CANALISATION EXPERIMENTALE de CHAMPIGNY sur MARNE -Vue de Profil**

simulant des vides, de morceaux de canalisation et de compactages différents (fig 1).

1.2 <u>*PHASES EXPERIMENTALES*</u>

L'expérimentation effectuée dans cette canalisation nécessite de faire varier l'état de l'environnement et les techniques d'auscultation en les décomposant en plusieurs phases.

1.2.1 *Première phase*

"L'encaissant" de la canalisation, composé de sablon, compacté ou non et muni de ses anomalies, est sec.

Les auscultations réalisées sont : la visualisation vidéo, le radar, les diagraphies nucléaires composées de sondes à neutrons et gamma/gamma.

1.2.2 *Deuxième phase*

"L'encaissant" reste sec mais la canalisation est mise en eau, sur une hauteur de 10 cm. La conductivité de cet "effluent" est celle de l'eau de consommation. Les auscultations sont ensuite effectuées selon un protocole établi.

1.2.3 *Troisième phase*

Les mêmes opérations que dans les phases un et deux sont réalisées mais une nappe est simulée dans l'encaissant.

Parallèlement, des mesures d'humidité du sol sont effectuées régulièrement afin de suivre le front d'humidité.

Ainsi dans chacune de ces phases, nous avons fait varier les conditions diélectriques des milieux utilisés.

1.3 <u>***LES ESSAIS***</u>

Pour connaître l'état du collecteur et de son environnement plusieurs techniques ont été adaptées et testées à partir de l'intérieur de la canalisation, mais également depuis la surface pour certaines d'entre elles.

En effet, si de nombreux outils sont couramment utilisés dans les réseaux accessibles à l'homme, les contraintes de mises en oeuvre ne permettent pas leur utilisation dans des ouvrages de petite dimension. Il a donc fallu adapter quelques uns d'entre eux pour les exploiter dans des conditions optimales.

Les outils utilisés dans cette expérimentation ont permis de :

- caractériser l'origine de certains désordres par une connaissance des terrains adjacents à l'ouvrage.
- mieux maîtriser certaines techniques d'auscultation et les méthodes de traitement qui les accompagnent.

1.3.1 *Mise en oeuvre*

Pour la plupart des techniques utilisées, la mise en oeuvre est sensiblement la même. Lorsque les inspections sont longitudinales ou ponctuelles, nous considérons la section de la canalisation comme une horloge de façon à pouvoir positionner nos mesures à des heures préalablement définies et effectuer la manipulation désirée. Les heures retenues sont :
0 h, 2 h, 4 h, 6 h, 8.h, 10 h.

Lorsque les investigations doivent se faire transversalement, leur position est définie sur la longueur de la canalisation.

Quelque soit la technique utilisée, le premier passage dans la canalisation se fait en aveugle.
Cela signifie qu'aucune information n'est communiquée aux opérateurs de l'outil aussi bien lors de la prise des mesures que pour le traitement. Un premier dépouillement est effectué et suivi d'une discussion des résultats. Lors de cette première manipulation, des mesures longitudinales sont mises en oeuvre toutes les deux heures mais également, des mesures transversales, au droit des joints et de tout autre position choisie par l'opérateur, si celui-ci le juge nécessaire après ses observations.

1.3.2 *Auscultation radar*

1.3.1.1 *Principe*

Cette auscultation, couramment utilisée dans les réseaux visitables permet d'une part d'approfondir les connaissances de la structure et de l'environnement d'un ouvrage d'autre part de détecter, et dans des conditions favorables de dimensionner et d'identifier, des anomalies.

Le principe consiste à envoyer une impulsion de très courte durée (quelques nanosecondes), émise à une cadence très élevée (50KHz) par une antenne (100 MHz à 1 GHz). Lorsque le signal rencontre une discontinuité, il est réfléchi et capté par un récepteur qui l'envoie vers un enregistreur. Les échos, correspondant aux différentes discontinuités sont observées successivement lors du déplacement régulier (1 à 3 km/h) des antennes le long de profils à ausculter dont le nombre dépend de la précision demandée. Une "coupe temps" est ainsi constituée, transformable en "coupe profondeur". Les profondeurs d'investigation s'échelonnent de 0,1 à 20 mètres pour des matériaux de construction. Ces profondeurs dépendent :

- de la capacité de l'appareil,
- de la fréquence des ondes émises,
- des propriétés diélectriques des matériaux.

L'interprétation peut être immédiate mais nécessite souvent un traitement adapté à l'objectif.

1.3.1.2 *Déroulement des essais.*

Un mobile miniaturisé (porte antenne), tracté ou poussé par une caméra, elle même autotractée, permet d'inspecter la canalisation depuis l'intérieur.

Les inspections sont réalisées dans un diamètre de 400 mm mais le matériel a été étudié pour accéder à des ouvrages de 200 mm de diamètre.

Photo 1 "Chariot-caméra" en place dans la canalisation avec l'antenne radar en place.

La prise des mesures est possible en tout point de la canalisation , c'est-à-dire sur des tracés longitudinaux ou transversaux.

En raison de la position de l'antenne radar de 1GHz, située devant l'objectif de la caméra, une visualisation permanente est effectuée à partir d'un écran de contrôle, embarqué dans un "camion caméra". Ceci offre la possibilité d'effectuer une correction permanente de la position de l'antenne surveillée au moyen d'un cadran indiquant les variations angulaires (photo 1). La mesure est donc faite avec une couche d'air, entre la paroi et l'antenne, la plus faible possible.

1.3.3 <u>*Auscultations par diagraphies nucléaires gamma-gamma et neutron-neutron.*</u>

Des essais ont été réalisés dans la canalisation expérimentale de Champigny sur Marne avec des sondes de diagraphies gamma-gamma et neutron-neutron du LCPC (Laboratoire Central des Ponts et Chaussées) et du BRGM (Bureau de Recherche Géologique et Minière).

L'objectif de ces essais était d'évaluer l'intérêt de ces outils, normalement employés en forage, pour l'auscultation non-destructive et continue de l'encaissant d'une conduite d'assainissement de faible diamètre sur une profondeur de 0 à 20 cm afin d'y détecter et d'y localiser des anomalies telles que cavités, points durs, zones décompactées.

1.3.3.1. *Principes et équipements*

* La diagraphie gamma-gamma mesure, par rétro-diffusion et à une distance donnée d'une source de ^{137}Cs, l'intensité des photons gamma diffusés par effet COMPTON. Nous montrons que l'intensité des photons diminue quand la densité électrique de la matière, elle-même fonction du poids volumique du terrain, augmente. La diagraphie gamma-gamma mesure donc la densité humide d'un matériau.

La sonde gamma-gamma du BRGM (Fabrication Mount-Sopris) contient une source de ^{137}Cs de 100

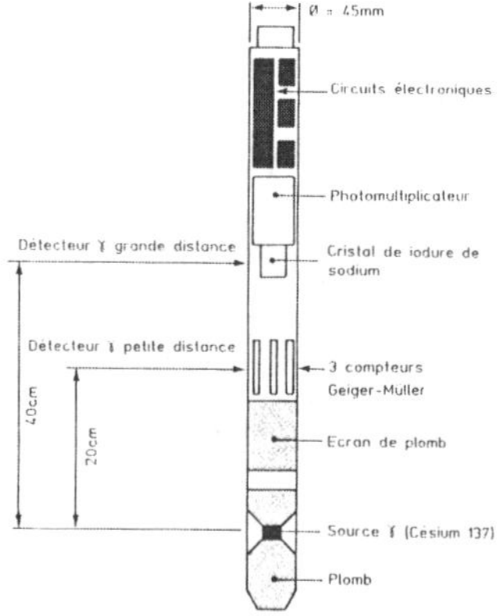

Fig. 2 schéma de principe de la sonde gamma-gamma développée par le LPC

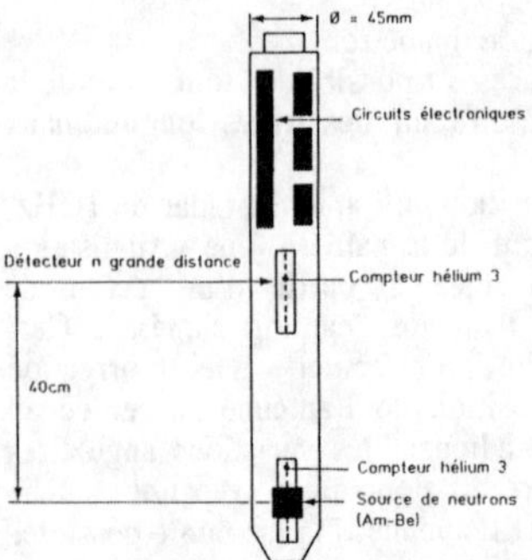

Fig.3 Schéma de principe de la sonde de diagraphie neutron-neutron

mCi d'intensité et possède 2 détecteurs, l'un situé à 10 cm (détecteur), l'autre à 35 cm (détecteur GD) de la source. Source et détecteurs sont collimatés suivant une même génératrice. En forage, un "bras diamètreur" permet de plaquer la sonde contre la paroi du trou. La longueur de l'outil est de 2,20 m pour un diamètre de 57 mm et un poids de 17 kg.

La sonde gamma-gamma du LCPC (fabrication LCPC, France) possède une source radioactive ^{137}Cs d'une intensité de 60 mCi. L'outil comporte 2 détecteurs, l'un à 20 cm (détecteur Petite Distance PD), l'autre à 40 cm (détecteur Grande Distance GD) de la source. la longueur de la sonde est de 1,50 m pour un diamètre de 46 mm et un poids de 8 kg (fig 2).

* La diagraphie nucléaire neutron-neutron enregistre, à une distance donnée d'une source de ^{241}AmBe, le flux rétro-diffusé de neutrons thermiques, c'est-à-dire de neutrons rapides ayant perdu une partie de leur énergie initiale par collision successives avec la matière, essentiellement avec les atomes d'hydrogène présents dans le milieu. Dans la pratique ceci revient à mesurer la teneur en eau volumique du terrain ausculté.

La sonde neutron-neutron du LCPC (fabrication LCPC) comporte un ensemble de 3 sources de neutrons rapides AmBe d'une intensité de 3x200 mCi. Le détecteur PD est centré autour de la source, tandis que le détecteur GD se situe à 40 cm de celle-ci (fig. 3). La longueur de la sonde est de 1,76 cm pour un diamètre de 46 mm et un poids de 5 kg.

Le volume restant dans les sondes est occupé par des circuits électroniques qui transforment les comptages en courant électrique, l'amplifient et l'envoient vers l'unité de traitement en surface via le câble de traction.

1.3.3.2 *Déroulement des essais*

Les différentes sondes ont été tractées dans la canalisation expérimentale le long de génératrices espacées toutes les 2 heures : à des vitesses allant de 1,5 à 20 m/min suivant les passages et la sonde utilisée.

Afin de pouvoir ausculter les génératrices autres que celles situées à 6 h (traction de l'outil en radier), les sondes ont été montées sur un chariot porte-sonde, de conception artisanale, ce qui a permis de les maintenir dans la position voulue à moins de 2 cm de la paroi de la conduite durant le passage.

D'autre passages ont été réalisés en radier (génératrice 6 h) dans la canalisation mise en eau, avec des hauteurs d'eau variant entre 7 cm au point haut et 30 cm au point bas.

II - LES RESULTATS

Ils concernent les auscultations radar et par diagraphies nucléaires.

2.1 *AUSCULTATION RADAR*

L'acquisition des données, effectuée par deux entreprises (Géoscan et la Compagnie Générale de Géophysique), est réalisée de façon continue pour chacun des profils longitudinaux ou transversaux. L'antenne utilisée, d'une fréquence de 1 GHz, a été choisie en raison de ses dimensions réduites. Les enregistrements ont été effectués simultanément suivant deux durées d'auscultation 15 et 30 nanosecondes sur chaque profil, ces temps correspondent à des profondeurs d'investigations moyennes de 75 et 150 cm derrière la paroi de l'ouvrage.

L'auscultation radar, après traitement du signal, montre de nombreuses discontinuités présentes dans l'environnement de l'ouvrage.

La figure 4 correspond à la scanérisation d'un profil longitudinal réalisé en voûte c'est-à-dire à 12 h depuis le joint n°5 au joint n°7, soit environ 5 m de longueur. Du haut vers le bas, nous reconnaissons :

- une interface continue correspondant à la surface de la paroi,
- des anomalies de faible extension situées tous les 2,5m correspondant aux joints entre chaque élément de la canalisation.
- une discontinuité constituée par une série de réflecteurs de forte amplitude encadrant le joint n°5 et dont l'extension horizontale est d'environ un mètre. Elle correspond à la présence d'une cavité artificielle de 0,6x1 m situé au dessus de la canalisation.
- entre les joints 6 et 7, l'ensemble des réflexions laisse apparaître une forme hyperbolique synonyme de la présence d'un tuyau. Ceci correspond effectivement à un tuyau de fonte placé à 12 h, à proximité et perpendiculairement à la canalisation expérimentale.
- à la base de la coupe, un ensemble de réflexions montre une forme de ligne correspondant à la surface extérieure de l'encaissant.

La technique d'auscultation radar a donc permis de localiser et d'identifier certaines anomalies (bloc de béton, vides, canalisation, etc...).

2.2 *DIAGRAPHIES NUCLEAIRES*

Les résultats sont présentés sous la forme de profils avec en ordonnées les comptages (en coups/seconde) et en abscisses la distance (en m), parcouru par la sonde dans la conduite.

Conformément à des expérimentations ultérieures, seuls ont été pris en compte les comptages :

- du détecteur GD de la sonde GG du LCPC,

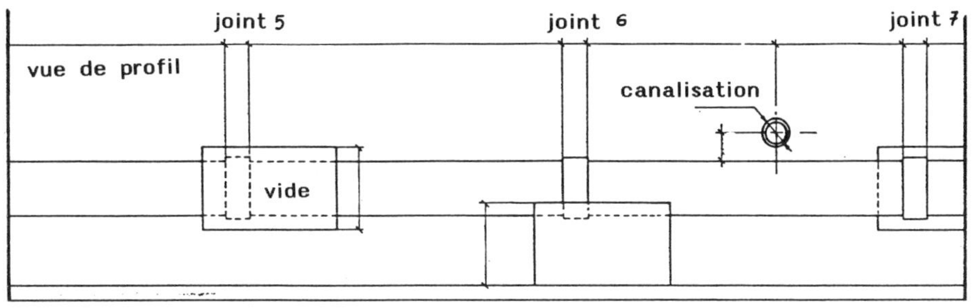

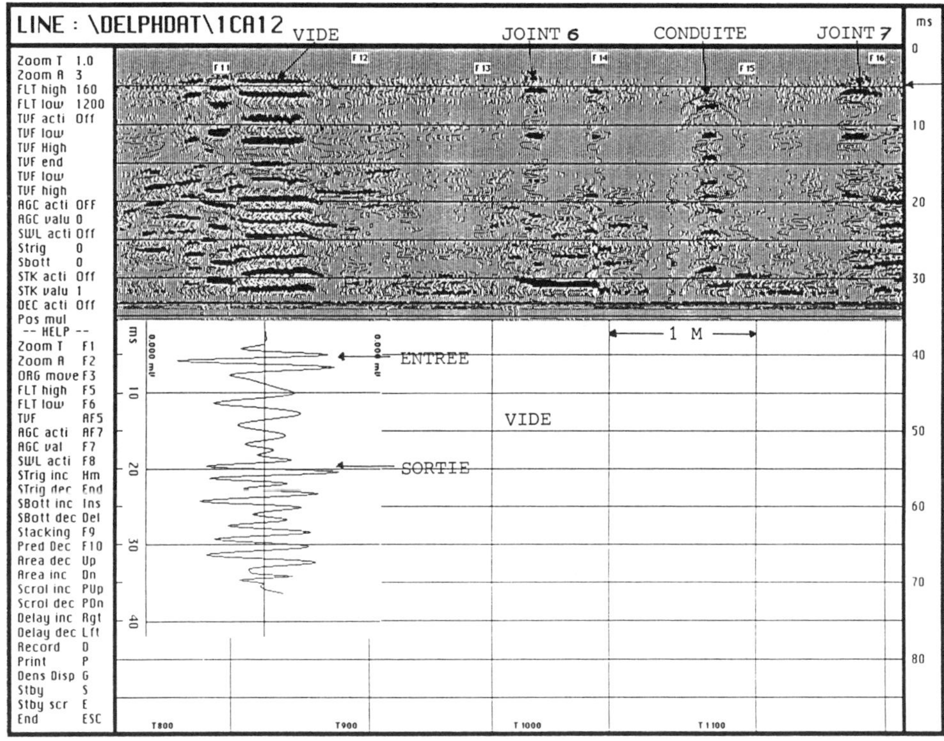

DETECTION D'EVENEMENTS DANS LE TERRAIN ENCAISSANT LA CANALISATION

PROFIL LONGITUDINAL A 12 h

Fig.4 CANALISATION EXPERIMENTALE A CHAMPIGNY SUR MARNE AUSCULTATION RADAR

- du détecteur PD de la sonde GG du BRGM,
- du détecteur PD de la sonde NN du LCPC.

La figure 5 présente les profils des comptages gamma de la sonde du BRGM, obtenus suivant les différentes génératrices auscultées. Une augmentation significative des comptages indique une diminution du poids volumique du milieu, tandis que leur diminution signale la présence de points de plus forte densité. Nous y observons :

* une augmentation significative des comptages gamma, signalant la présence probable d'un vide ou point de moindre densité à environ :

- 3,5 m sur les profils de 8 et 6 h, indiquant le vide (caisson de bois) situé à cet endroit sous la génératrice inférieure de la canalisation.
- 7,5 m sur les profils de 10 et 8 h, signalant la présence du bloc de béton cellulaire situé à 9 h.
- 6,5 m sur les profils de 4 et 2 h indiquant la présence du bloc de polystyrène enfoui à 3 h.
- 11,5 m sur les profils de 4 et 2 h (et

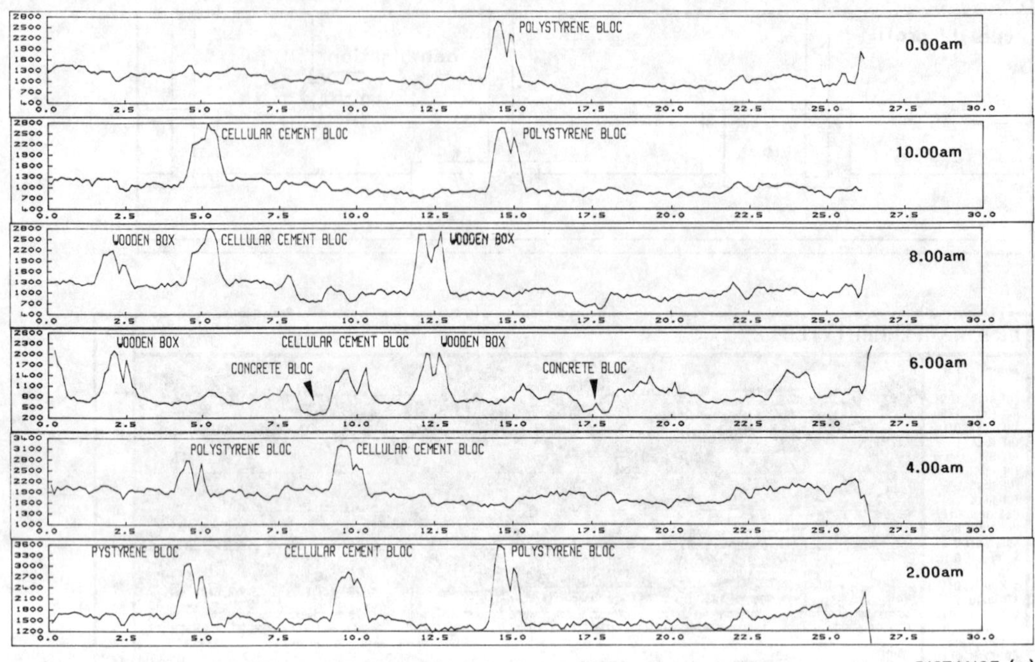

Fig 5 Profils des comptages gamma du BRGM

Tableau 1 Méthodologie d'auscultation mise en oeuvre dans le Val de Marne.

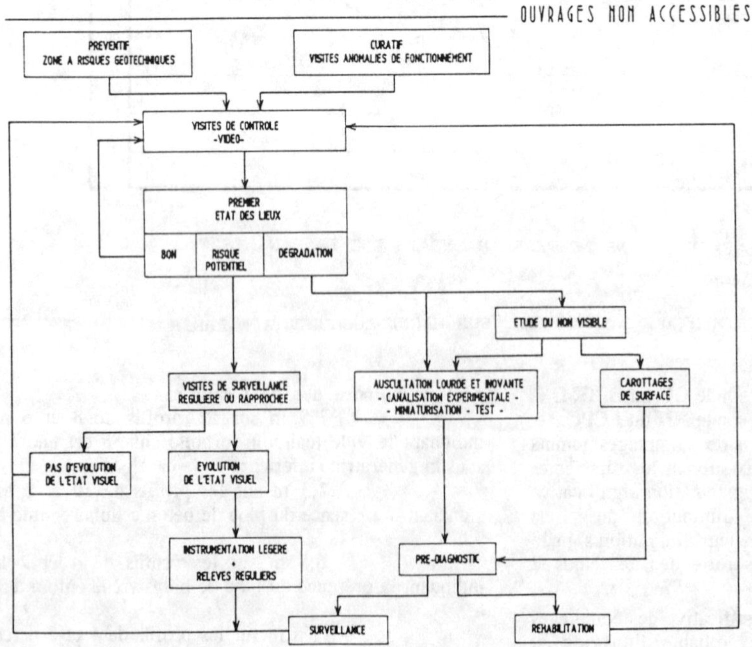

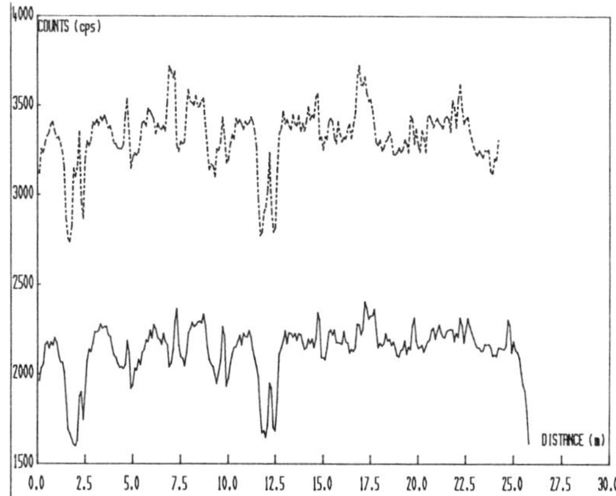

Fig 6 profils de comptages neutron de la sonde LCPC.

atténué sur celui de 6 h), décelant le bloc de béton cellulaire dissimulé à 3 h.

- 16 h sur les profils de 12, 10 et 2 h, ce qui correspond bien à la présence du bloc de polystyrène situé sur la génératrice supérieure de la canalisation.

* une diminution significative des comptages, indiquant la présence probable d'un point de plus forte densité (point dur) à 10,5 et 19,5 m sur les profils de 6 et 8 h, ce qui correspond aux position des 2 blocs de béton situés sous la génératrice inférieure de la canalisation.

Les zones compactées de la tranchée expérimentale, situées entre 9 et 22 m, se traduisent sur les profils de 12, 10, a et 2 h par une baisse du niveau moyen des comptages, quand on passe des zones non compactées aux zones compactées (flèches).

La figure 6 montre 2 profils de comptages neutron de la sonde LCPC réalisés sur la génératrice 6 h dans la canalisation à sec et en eau. on observe que la présence des vides à 1,7 m et 11, 7 m se traduit par une baisse significative des comptages neutron, tandis que les points plus denses (8 et 17 m) apparaissent sous la forme d'une diminution des comptages neutron.

CONCLUSION

Les critères de sélection des techniques adaptées et testées pour l'auscultation des canalisations non accessibles ont été leurs performances et leurs complémentarités.

Ces tests réalisés dans une canalisation expérimentale équipée de zones témoins et de zones présentant des anomalies connues ont permis de mieux maîtriser les outils utilisés mais aussi leurs traitements.

Les premiers résultats obtenus ont montré la possibilité d'adapter certains outils et d'envisager leur utilisation, à cours terme et après quelques modifications, dans des réseaux en service.

L'ensemble de ces résultats associés aux inspections visuelles vont permettre de mettre en application la méthodologie mise en oeuvre dans le Val de Marne résumé dans le tableau 1.

3 Strategic and legal aspects

No Trenches in Town, Henry & Mermet (eds) © 1992 Balkema, Rotterdam. ISBN 90 5410 085 0

Microtunnelling advances in the UK

Haydn White
Corporate Business, Yorkshire Water plc, Leeds, UK

Colin Tregoing
WRc, UK

ABSTRACT: Trenches in towns will be discouraged by recent UK legislation, but decisions to use trenchless technology will probably continue to be based on estimated direct engineering costs. The old legislation and notoriously variable ground conditions slowed the pace at which microtunnelling was expected to be adopted for routine sewer construction in the UK but, now that the legal, technical and economic obstacles have largely been overcome, there remains the last challenge of converting UK designers to regard the technique as routine. This challenge is being met and significant advances have been made by various UK utilities, manufacturers and contractors, who have also prompted a number of research and development initiatives. The scene is therefore set for the UK microtunnelling market to increase, both in terms of machine sales and installed sewer lengths.

1 INTRODUCTION

Microtunnelling has been defined (Yorkshire Water plc 1991) as "the use of a steerable remote controlled tunnel boring machine to allow installation of pipelines up to DN900 by pipe jacking". The technique as such is only just seven years old in the UK, but in that time much effort has gone into its development and marketing. Now that most of the legal, technical and economic difficulties found in the UK have been overcome, the time is ripe to assess the advances made so far and the prospects for future growth.

2 HISTORY

The first microtunnel in the UK was completed in July 1986, when a DN500 sewer was successfully installed at Gateshead using a Japanese Iseki Unclemole slurry machine to jack vitrified clay pipes into place through backfill material and stiff clay (ISTT 1986). The Unclemole was then moved on to tackle several other drives in the north-east of England over the next few months, with varying degrees of success.

The main problem faced by the early UK microtunnelling enthusiasts was that the utilities, and in particular the sewerage authorities, did not (and still do not) take social costs into account when deciding on methods of installation. Thus a contractor wishing to offer microtunnelling as an alternative to installing a sewer by trench excavation had to do so on the basis of no greater overall direct cost. This was difficult enough in those early days, but to base the offer on Japanese, or by then German, technology not developed for use in the notoriously variable UK ground was too much of a risk.

Yorkshire Water Authority noticed this dilemma and realised that it was in a position to help promote the use of microtunnelling in the UK. On 16 April 1987 it signed a Joint Venture Agreement with Decon Engineering and ARC Pipes to develop a complete microtunnelling system, including machines and unreinforced concrete pipes, over a two year period on jobs provided by Yorkshire Water in widely varying ground conditions. Success was to be judged on the basis of a detailed "user specification" (including soil and rock parameters), but the prime objective was to develop a system that could install pipelines at no more overall direct cost than doing so by conventional trench excavation.

During the discussions that led to the Joint Venture, those involved came to the conclusion that, on balance, the auger system was physically and economically better able to cope with wide variations in ground than was the slurry system when microtunnelling in the UK (White 1989). In the light of this and the "user specification" it was found that the RVS system developed by Dr Ing Soltau GmbH in Germany came closest to meeting the requirements of the prospective Joint Venture partners. Accordingly, Decon began negotiations with Soltau and eventually signed a Licence Agreement on 5 April 1987. This cleared the way for the Joint Venture and its subsequent success has been described elsewhere (White, Moss and Rowlands 1988).

The concept of client involvement in the development of microtunnelling in the UK was further taken up by North West Water Authority and in August 1989 it too launched a microtunnelling initiative. The idea was to specify various microtunnelling systems currently available in the UK, for use on several jobs throughout North West Water's area, and to assess their performance. The systems chosen were the Decon RVS (auger), the Okumura-Markham (slurry) and the Herrenknecht (slurry) and the results were subsequently reported (Glynn and Smith 1990).

Microtunnelling in the UK received another boost in 1989 for, towards the end of that year, a DN675 sewer was installed at Caerphilly in Wales using an Iseki TCC 675 Unclemole machine to jack reinforced concrete pipes into place. The Caerphilly site was particularly difficult, not only due to the bad ground, but also to the close proximity of the 13th century Caerphilly Castle, its foundations and water-filled moats. By avoiding driving a tunnel in compressed air the client was not only able to save over £0.5 million, but also reduced environmental disruption to a minimum (Russell 1991).

Over the past few years many more micro-tunnelling contracts have been carried out in the UK, though not as many as some had expected, and several required the technology to be advanced still further. For example, the last job for the Yorkshire Water/Decon/ARC Joint Venture in 1989 was chosen to try out the Decon RVS auger system in solid bunter sandstone. Although successful, no-one in the UK attempted a full-scale job in rock until the following year, when a Herrenknecht AVN700 slurry system was used to install DN675 reinforced concrete pipes for Northumbrian Water Ltd at Gateshead (Hayes 1992).

A common response by UK engineers when challenged to use microtunnelling was that guidance on how to go about designing and documenting such contracts was not available. This prompted Yorkshire Water to produce its Code of Practice for Microtunnelling (Yorkshire Water plc 1990) and this has been described elsewhere (White and Dew 1991).

Another reason why microtunnelling did not take off in the UK as quickly as some had hoped was the fact that the utilities had complete freedom to choose their methods of construction, even to the extent of being able to arrange for roads to be closed so as to allow the digging of trenches. In 1992 this freedom will be limited by a new Act of Parliament and the ways in which this will create a climate more conclusive to trenchless technology have been previously described (White and Dew 1991, Tregoing 1991).

3 PIPELINE MATERIALS AND STANDARDS

As indicated earlier, the first UK microtunnel was driven in July 1986, when Hepworth DN500 vitrified clay pipes were used. The following month another DN500 pipeline began to be installed, this time using unreinforced concrete pipes with a GRP outer skin applied, due to the particularly aggressive nature of the ground on that contract. These were manufactured by Spun Concrete Ltd.

The first UK reinforced concrete microtunnel-ling pipes were also DN500, supplied in January 1987, and again made by Spun Concrete Ltd. The first unprotected unreinforced concrete pipes began to be installed with the commencement of the

Yorkshire Water/Decon/ARC Joint Venture in April that year; ARC's DN600 pipes being used on each of the ten associated schemes.

Over the past few years other UK manufacturers have begun to offer microtunnelling pipes, but the materials have remained basically unreinforced concrete and reinforced concrete. Contrary to the situation in other parts of Europe, vitrified clay, fibre cement, ductile iron and GRP pipes have not figured largely in the UK microtunnelling scene.

There are no specific British Standards for microtunnelling pipes, but many of the requirements of BS5911: Part 120 "Specification for reinforced jacking pipes with flexible joints" can be applied to precast concrete pipes. In fact, that Standard covers only reinforced concrete pipes of DN900 and larger and an Amendment drafted to accommodate unreinforced concrete pipes and smaller diameters in both materials cannot now be published because of the Standstill procedure imposed by CEN, the European standards-making body. Similarly, vitrified clay microtunnelling pipes can be manufactured to BS EN295 (in UK terms, formerly BS65), though attributes peculiar to microtunnelling will not be covered.

A harmonised European Standard covering reinforced and unreinforced concrete microtunnelling pipes is currently being drafted and is due for publication by CEN at the end of 1993. Work is now also under way to widen the scope of EN295 to cover vitrified clay microtunnelling pipes, a new Standard being expected at about the same time as the one for concrete pipes.

4 THE CURRENT UK MICROTUNNELLING MARKET

4.1 Microtunnelling Manufacturers and Operators

At present there are eight manufacturers of microtunnelling systems who have supplied equipment for use in the UK (Table 1). Most of these companies have the resources and necessary licences to manufacture systems in the UK to original Japanese or German designs. However, one or two companies have designed and developed microtunnelling systems from first principles in this country. Complete microtunnelling systems are offered for sale to prospective purchasers and in addition four companies can provide machines and operators on a sub-contract or hire basis (Dosco, Euro-Iseki, Howden, DJB). The Westfalia machine is leased at the moment, with a purchase option.

Details of the 11 contractors who have bought and now operate microtunnelling systems are shown in Table 2. The four manufacturing companies who also act as operators of systems have been included to give a total of 15 microtunnelling operators in the UK.

4.2 Microtunnelling installation

A detailed analysis of UK microtunnelling activity was presented at the Pipe Jacking and Microtunnelling 91 Conference, held in London in October 1991 (Nicholas 1991). Information was given on total drive lengths, pipe diameters, machine numbers/types and annual productions. These details covered activities up to June 1991 and the intention in this present paper is to extend and broaden the analysis using new data. All the UK operators of systems were contacted and all readily cooperated by responding to a survey which covered the time period July 1991 to June 1992 (12 months). Details were collected for machines available and pipelines constructed using microtunnelling during this period.

The information contained in this paper relates to pipelines not greater than DN900. It is worth noting that at present there is not an accepted international definition for microtunnelling diameter limits. However, a European Standard covering safe man-entry requirements for tunnels is currently being prepared. Some operators also own machines in the 900 - 1200mm internal diameter range and tend to look upon these as microtunnelling systems, although such sizes have traditionally been regarded in the UK as pipe jacking. Data on these pipeline sizes has not been included although there is certainly a degree of activity in this area in the UK.

In the 12 months under review, an overall length of 14,540m of microtunnelled pipeline was installed, using 20 machines (Table 3). The pipe nominal sizes ranged from DN450 to DN875. Some short lengths of pipeline of below DN450 were installed but this is believed to be less than 300m in total. The data in Table 3 is as reported by the various operators but it should be noted that machine production figures relate to work secured during the year under review. This is not necessarily a reflection of the capabilities of individual machines.

Thus an average figure for annual usage per machine is 727m but this is somewhat distorted since a small number of machines have been under-utilised during the year. Slurry and auger machines are compared in Table 4 which shows that slurry machines are more numerous and have a higher average installed length.

Of the 14,540m total length installed, about 93% was for sewerage schemes and 95% of pipelines used concrete pipes. A small amount of glass reinforced plastic (GRP) pipe was used and it is believed that some vitrified clay pipes were used for a limited amount of work of less than DN450. The 7% of microtunnelled pipelines which were not sewerage were mainly ducts for telecommunications, gas mains and water or sewer pressure mains.

The most popular size for microtunnelled pipelines was DN600, with 6150m installed at this size (Figure 1).

5 UK RESEARCH AND DEVELOPMENT IN MICROTUNNELLING

A number of organisations continue to carry out research and development work in the UK which is directly related to trenchless technology and microtunnelling. These include national research bodies serving water utilities or highway authorities and University-based Science and Engineering Research Council projects.

WRc have for some time disseminated technical information on microtunnelling and other trenchless systems to UK and overseas water utility clients. In April 1992 WRc began work on a major three-year research project into trenchless technology systems, including microtunnelling. Part of this work will involve the preparation of a manual on trenchless technology equipment together with practical guidance on the selection of appropriate techniques for pipeline installation.

The UK's Construction Industry Research and Information Association (CIRIA) have commenced a one-year research project on the planning and investigation for the use of trenchless and minimum excavation technology. This multi-client study will focus on the benefits of good planning at an early stage in scheme preparation in order to maximise the potential offered by trenchless methods.

The University of Manchester Institute of Science and Technology is working on the social and environmental costs caused by utility works in highways (Read and Vickridge 1991, Ling, Read and Vickridge 1991, Vickridge, Ling and Read 1992). However, such costs are not currently taken into account when assessing schemes for the UK and a requirement to do so will only be a reserve power under new legislation.

Nevertheless, the Department of Transport is concerned that the fullest use is made of the existing roads infrastructure and that the closures or partial closures caused by utilities are minimised. To this end, the Transport Research Laboratory has commissioned Jason Consultants in a study into social costs and the potential savings to the road user from a wider use of trenchless methods. A further research project is assessing the cost of damage to the structural performance and durability of the highway caused by openings such as trenches.

Other relevant research in the UK includes work at the University of Oxford into various aspects of pipe jacking technology (Milligan and Ripley 1989, Milligan and Norris 1991, Milligan and Norris 1992). Sponsors for this work have included the Pipe Jacking Association, Concrete Pipe Association, Science and Engineering Research Council and several water service companies.

6 THE FUTURE

Microtunnelling activity in the UK has been steadily increasing each year since 1986 (Figure 2). This trend is continuing and the total annual microtunnelled length of pipeline and number of machines available are both at their highest ever levels.

However, the rate of increase of microtunnelling activity has not been as great as was predicted in the early years. There are several possible reasons why the pace of advance has slowed, including a continuing reluctance by designers and planners to specify installation techniques other than long established traditional trenching methods.

In the near future, certain factors are likely to make microtunnelling an even more attractive proposition, for routine lengths of urban sewers and ducts as well as the more obvious crossing situations.

The first of these is the effect of the New Roads and Street Works Act 1991, which has a number of provisions likely to make disruptive excavation potentially more difficult and uneconomic; also the capital expenditure programmes of the UK water service companies are shifting in emphasis and priority away from water and sewage treatment schemes towards pipeline infrastructure renewal. Attention has already been drawn to the significance of these two factors (White and Dew 1991). Over the next few years there is therefore likely to be far more new sewerage installation work than previously and this will have to be constructed in accordance with the new legislative framework.

ACKNOWLEDGEMENTS

The authors would like to thank their respective employers and those water undertakings sponsoring the current WRc research study into trenchless technology systems for allowing this paper to be written. Thanks are also due to the manufacturers and contractors who kindly supplied commercial information on UK microtunnelling achievements.

REFERENCES

Glynn M G and Smith S G. 1990. Microtunnelling in North West Water Ltd. Proc. ISTT NO-DIG 90. Rotterdam.

Hayes N F. 1992. Microtunnelling the Western Interceptor Sewer - Phase 5, Newcastle-Upon-Tyne, England. Proc. 2nd International Microtunnelling Symposium. Bauma '92, Munich.

ISTT. September 1986. *Underground Magazine*, p28.

Ling D J, Read G F and Vickridge I G. 1991. Road space rental - A structural incentive for the adoption of no-dig technologies. Proc. ISTT NO-DIG 91. Hamburg.

Milligan G W E and Norris P. 1991. Concrete jacking pipes, the Oxford research project. Proc. 1st Pipe Jacking and Microtunnelling Conference 91. London.

Milligan G W E and Norris P. 1992. Pipe end load transfer mechanisms during pipe jacking. Proc. ISTT NO-DIG 92. Washington.

Milligan G W E and Ripley K J. 1989. Packing material in jacked pipe joints. Proc. ISTT NO-DIG 89. London.

Nicholas P. 1991. Microtunnelling - Markets and Techniques. Proc. 1st Pipe Jacking and Microtunnelling Conference 91. London.

Read G F and Vickridge I G. 1991. The social and environmental benefits of microtunnelling and pipe jacking. Proc. 1st Pipe Jacking and Microtunnelling Conference 91. London.

Russell A T. 1991. An assessment of the performance of Iseki TCC 675 Unclemole microtunnelling equipment. WRc Report No. UM1173.

Tregoing C E. 1991. Trenchless technology and the new legislation. Proc. HAUC 91 Conference. Birmingham.

Vickridge I G, Ling D J and Read G F. 1992. Evaluating the social costs and setting the charges for road space occupation. Proc. ISTT N0-DIG 92. Washington.

White H. 1989. Range of applications and limits of auger systems dependent on soil conditions. Proc. 1st International Microtunnelling Symposium. Bauma '89, Munich.

White H and Dew R J. 1991. UK micro scope. Proc. ISTT NO-DIG 91. Hamburg.

White H, Moss A F and Rowlands E. 1988. Making micro work. Proc. ISTT NO-DIG 88. Washington.

Yorkshire Water plc. 1990. Code of Practice for Microtunnelling (First Edition).

Yorkshire Water plc, 1991. Code of Practice for Microtunnelling (Second Edition), Preface.

Table 1. Manufacturers of microtunnelling machines used in the UK

Company	Machine Name
Decon Engineering Company, Bridgwater, Somerset	Decon
DJB Trenchless Techniques Limited, Glasgow	Witte
Dosco Overseas Engineering Limited, Newark, Nottinghamshire	Dosco
Euro-Iseki Limited, Stratford-Upon-Avon, Warwickshire	Iseki
Herrenknecht International Limited, Seaham, County Durham	Herrenknecht
James Howden and Company Limited, Glasgow	Howden Sanwa
Markham and Company Limited, Chesterfield, Derbyshire	Okumura Super-Mini
Westfalia Becorit, Lunen, Germany	Westfalia

Table 2. Microtunnelling Operators in the UK

Company/Contact details	Machine/Model	Machine Type and Internal Diameter Range (mm)	Machine Ownership
M J Clancy and Sons Ltd Steve Woodger Tel: (0895) 823711 Fax: (0895) 825263	Decon RVS 250A Herrenknecht AVN600	Auger (450-720) Slurry (600)	Owned Owned
DCT Civil Engineering Ltd John Nuttall Tel: (0706) 842929 Fax: (0706) 882158	Herrenknecht AVN600	Slurry (600)	Owned
Delta Civil Engineering Co Ltd Albert Edwards Tel: (0223) 843912 Fax: (0223) 845335	Decon RVS 250A	Auger (450-600)	Owned
DJB Trenchless Techniques Ltd Peter Taylor Tel: (041) 7631234 Fax: (041) 7630333	Witte	Auger (450-675)	Hired to contractors
J F Donelon and Co Ltd Alan Boden Tel: (0204) 699222 Fax: (0204) 699333	Decon RVS 250A Herrenknecht AVN800	Auger (450-600) Slurry (900)	Owned Owned
Dosco Overseas Engineering Ltd David Peat Tel: (0777) 870621 Fax: (0777) 871580	Dosco DMB760 HS	Slurry (600)	Hired to contractors

...../continued

Table 2 (continued)

Company/Contact details	Machine/Model	Machine Type and Internal Diameter Range (mm)	Machine Ownership
James Howden and Co Ltd Bob Bonnar Tel: (041) 8866711 Fax: (041) 8863979	Howden-Sanwa 716	Auger (250-700)	Hired to contractors
Euro Iseki Ltd Andrew French Tel: (0789) 292227 Fax: (0789) 268350	Iseki TCC Unclemole (various models)	Slurry (500-675)	Hired to contractors
Kennedy Construction Ltd David Forrest Tel: (061) 8776300 Fax: (061) 8776301	Herrenknecht AVN700 Herrenknecht AVN800	Slurry (675) Slurry (900)	Owned Owned
Laserbore Ltd David Mansell Tel: (0246) 452000 Fax: (0246) 453860	Decon RVS 250A Herrenknecht AVN800	Auger (450-600) Slurry (900)	Owned Owned
McNicholas Construction Co Ltd Steven McNicholas Tel: (081) 9534144 Fax: (081) 9531860	Decon RVS 250H	Auger (450-600)	Owned
Microline Tunnelling Ltd Neil McLcod Tel: (0922) 24148 Fax: (0922) 38812	Iseki TCC780	Slurry (600)	Owned
Miller-Markham Ltd John McInally Tel: (061) 9414023 Fax: (061) 9269628	Super-Mini 500 (with reaming system)	Slurry (500-700)	Owned
Queghan Civil Engineering James Quinn Tel: (061) 6202115 Fax: (061) 6270181	Westfalia Becorit WBM-LF6	Slurry (600)	Leased (with purchase option)
Sillars Ltd Andrew Scott Tel: (0429) 268125 Fax: (0429) 268038	Decon RVS 250A	Auger (450-600)	Owned

Table 3. Details of microtunnelling activity in the UK (July 1991 - June 1992)

Operator	Machine	Type	Lengths installed (m) DN 450 Conc.	DN 500 Conc.	DN 600 Conc.	DN 675 Conc.	DN 720 GRP	DN 875 Conc.	Function (m) Sewer	Other	Totals (m) Machine	Overall
Clancy	Decon 250	Auger	170		525		150		655	190	845	1855
	Herren 600	Slurry			1010				720	290	1010	
DCT	Herren 600	Slurry			1000				1000		1000	1000
Delta	Decon 250	Auger			705				705		705	705
DJB	Witte	Auger	20			60			80		80	80
Donelon	Decon 250	Auger	45		1100				900	245	1145	2270
	Herren 800	Slurry	200					925	1125		1125	
Dosco	Dosco 760	Slurry			180				180		180	180
Howden	Sanwa 716	Auger				60			60		60	60
Iseki	Iseki TCC	Slurry		750					750		750	1450
	Iseki TCC	Slurry			110	590			700		700	
Kennedy	Herren 700	Slurry				1160			1160		1160	1660
	Herren 800	Slurry						500	500		500	
Laserbore	Decon 250	Auger	1065		520				1585		1585	1715
	Herren 800	Slurry						130	130		130	
McNicholas	Decon 250	Auger			270				140	130	270	270
Microline	Iseki 780	Slurry		210	500				710		710	710
Miller-Markham	Super-Mini 500	Slurry		1050		235			1285		1285	1285
Queghan	Westfalia	Slurry		600					600		600	600
Sillars	Decon 250	Auger	470		230				465	235	700	700
TOTALS			1970	2610	6150	2105	150	1555	13450	1090	14540	14540

Note: The information in this table is as reported by the various operators but it should be noted that machine production figures relate to work secured during the year under review. This is not necessarily a reflection of the capabilities of individual machines.

Table 4. Lengths installed by type of machine

	Number of machines	Total lengths (m)	Annual length per machine (m)
Slurry	12	9150	763
Auger	8	5390	674
TOTAL	20	14540	727

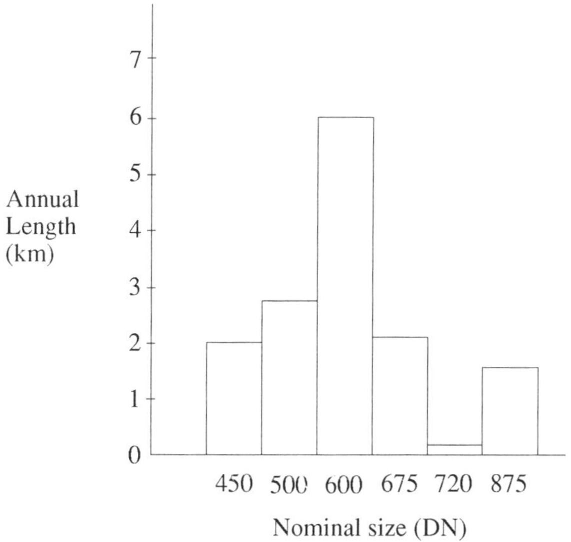

Figure 1. Annual installed length by pipe size

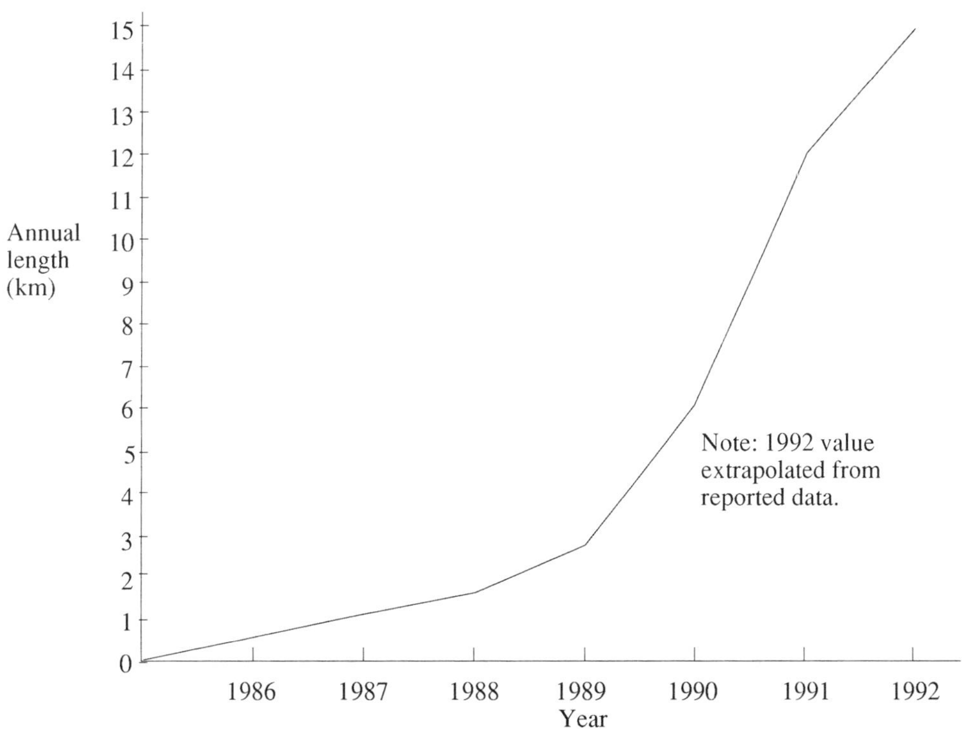

Figure 2. Annual microtunnelling installed length

No Trenches in Town, Henry & Mermet (eds) © 1992 Balkema, Rotterdam. ISBN 90 5410 085 0

Applying enlightened contracting practices to trenchless technology projects

Robert J. Smith
WG International, Washington, D.C. & Wickwire Gavin, P.C., Madison, Wisc., USA

ABSTRACT: Owners and their designers and constructors are increasingly recognizing and acknowledging the benefits of applying rational risk allocation to contracting practices utilized for underground work. This paper describes what owners can do to better allocate risk and reduce their total project costs and how those practices can be beneficially and effectively utilized for urban "no-dig" projects. These practices include better funded acquisition of geotechnical information, full disclosure of all known subsurface information, full compliance with legal requirements of many jurisdictions for preconstruction location of utilities by designers or utility owners or both, early identification of permitting requirements, and predefined equations and procedures for adjusting contract price when the unexpected is encountered. Thus, many of the legal/contractual problems endemic to trenchless construction in urban areas can be prevented or mitigated.

1 INTRODUCTION

Underground conduit construction in general is laden with risks to all parties to the process. Utilization of trenchless technology presents new opportunities for risk avoidance, but new challenges as well, particularly in urban areas. The purpose of this presentation is to discuss contracting practices and procedures that have been increasingly utilized with substantial success on pipeline and tunnel projects in the United States.

1.1 *Contracting philosophies for dealing with risk*

Current philosophies for major construction projects in the United States cover a very broad spectrum. On the one extreme is the risk allocation philosophy of the conservative owner who steadfastly believes that contractor acceptance of risk is inherent in the business of contracting and that the profit potentials apparently justify a contractor taking considerable risks. Some term this the "legalized gambling" approach to contracting. Numerous state departments of transportation and hundreds of municipalities both large and small adhere to this philosophy.

Towards the other end of the philosophical spectrum are major public owners with expansive (and expensive) construction programs who have concluded that the proper sharing and allocation of risk is cost effective.

1.2 *Generally accepted principles of risk allocation applicable to contracting practices*

Numerous papers and discussions at the 1979 Conference on Construction Risks and Liability Sharing, sponsored by the American Society of Civil Engineers, collectively concluded:

- Risks belong with those parties who are best able to evaluate, control, bear the cost, and benefit from the assumption of risks.

- Many risks and liabilities are best shared
- Every risk has an associated and unavoidable cost which must be assumed somewhere in the process. (ASCE, 1980).

The Construction Industry Institute has stated in a well-documented report:

> The ideal contract -- the one that will be most cost effective -- is one that assigns each risk to the party that is best equipped to manage and minimize that risk, recognizing the unique circumstances of the project. (Construction Industry Institute, 1986).

This paper supports the emerging view that better risk allocation is beneficial to all parties to the contracting process. It is through better contracting practices that better risk allocation is achieved.

2 BENEFITS OF ENLIGHTENED CONTRACTING PRACTICES

All parties to the design and construction process have benefitted where modern and innovative contracting practices have been used. The result is truly a "win-win" situation. Just as the project and the public benefit from modern trenchless technology, the project and the public can benefit from modern risk allocation.

2.1 Benefits to the owner

Owners will have a project that is much more likely to be completed on time, at a fair price without acrimony, and meeting their needs and expectations. Though this is what every owner strives and hopes for when planning a project, achievement of these goals is often prevented when misallocated risks result in costly and disruptive delays and disputes. Respondents to a Construction Industry Institute survey collectively concluded that better contracting practices could save an average of 5% of project costs with a 10:1 benefit to cost ratio. (Construction Industry Institute, 1990). One public owner reported hundreds of claims and dozens of lawsuits avoided due to its practice and philosophy of risk allocation.

2.2 Benefits to the contractor

Underground construction contractors want to complete a project on time and have a satisfied owner when they turn over the project. Certainly they want some return for their efforts, namely profit. However, when already thin profit margins are eroded by the occurrence of uncontrollable events and unrealistic or unfair contract practices, conflicts can occur. Projects employing modern contracting practices have a very low rate of such conflicts.

The fewer the uncertainties, and the more specific the allocation of risk, the easier it is for competing bidders to "sharpen their pencils" and provide more competitive bids without contingencies and without the need to use litigation to try to stay solvent.

2.3 Reduced cost of delays and disputes

The costs of improper risk allocation can be enormous. A contractor may lose thousands of dollars each day it is shut down. If either the contractor or owner formally asserts a claim, that usually triggers involvement of outside professionals like consultants and lawyers, again at costs often measured in thousands of dollars per day. Both parties are then spending huge sums of money for something other than the project, and it isn't being constructed. This has led some owners to conclude "there must be a better way." Indeed there is and they have found it: risk allocation.

2.4 Reducing language of questionable enforceability

Some owners use one sided contract language, such as "no damage for delay" clauses and disclaimers on soil borings. In some cases the tendency of the design professional is to include carefully worded contract clauses to protect the design professionals' interests. All of this can be costly to the owner. Even though such clauses may be clearly included in a contract, many courts will not enforce them. (Kovars and Sandler, 1987). Obviously, making decisions based on language that may ultimately not be enforced is not a sound practice.

3 ENLIGHTENED AND INNOVATIVE PRACTICES USED BY SOME U.S. OWNERS TO BETTER ALLOCATE RISKS AND REDUCE TOTAL PROJECT COSTS (ACEC/AGC, 1991)

Contracting procedures and contract documents are the keys to implementing improved risk allocation. Thus, to get in a position to reap the benefits of risk allocation, owners are performing systematic reviews of their existing procedures and documents and comparing them with modern practices. Existing procedures and documents are usually the result of incremental additions over the years, with few subtractions. Many times both procedures and documents are "handed down" from one person to another, the "we've always done it this way" approach.

In other words, owners are taking the time to sit down and ask themselves how they are allocating risk, why they are doing it that way, and if there is a better way.

3.1 Reviewing and revising "front end documents"

A principal tool for risk allocation and assignment is the part of each contract called the general conditions. Sometimes these are the front part of a "standard specifications" book, other times they are preprinted "boilerplate." Thorough reviews of these documents, testing them against checklists of construction risks, often indicate the opportunity for beneficial, cost-saving changes.

Among other things, one needs to ask if the general conditions are compatible with no-dig methods. Often they are not. Careful scrutiny may reveal that they were drafted for more traditional methods of construction in an earlier era.

3.2 Obtaining and providing geotechnical information

It should go without saying that the more information a trenchless contractor has about anticipated subsurface conditions, the more informed the bid. There is a decided trend towards (1) increasing geotechnical exploration and analysis budgets and (2) making all of the geotechnical information available to bidders. Some owners and engineers argue that this will lead to claims if the information is wrong. This misses the basic point that if the bid was based on the provided information, then that defined the bargain struck between the owner and the contractor. Accordingly, if the actual underground conditions are worse than the contractor expected, it is only fair that the owner pay, for if the contractor had known of the more severe conditions, it certainly would have bid higher. Economic benefits of trenchless methods will be enhanced when the designer and the contractor have as much information as reasonably possible.

The epitome of better risk allocation through fuller geotechnical disclosure is the Geotechnical Design Summary Report (GDSR). (ASCE, 1991). The GDSR sets forth the designer's anticipated subsurface conditions and their impact on design and construction. Thus the engineer and owner establish the geotechnical baseline for all anticipated conditions. Use of this single contractually established interpretation of conditions will generally result in more uniform bid prices, and less exposure to claims involving interpretation of subsurface data. It also fosters a cooperative climate since the emphasis is on openness and candor.

If conditions are materially different from the baseline, and the contractor can demonstrate a financial impact, he is entitled to additional compensation. Thus, the owner accepts the risk for conditions more difficult than the baseline.

Contractors can submit competitive bids based on efficient management, innovative construction methods, and more optimistic interpretation of conditions (with additional risk acceptance), knowing that the owner's geotechnical baseline will provide a clear basis for identifying differing conditions and limiting contractor exposure.

The following entities have used various types of geotechnical design summary reports for underground projects:

Alaska Power Authority
City of Colorado Springs, Colorado
City and County of Honolulu, Hawaii
City of Los Angeles, California
Colorado Department of Highways
Hawaii Department of Transportation
Massachusetts Water Resources Authority
Milwaukee Metropolitan Sewerage District, Wisconsin
Municipality of Anchorage, Alaska
Municipality of Metropolitan Seattle, Washington
Pennsylvania Turnpike Commission
Pierce County (Washington) Public Works Department
Rail Construction Corporation, Los Angeles, California
U.S. Bureau of Reclamation
U.S. Department of Energy
Washington State Department of Transportation
Washington (D.C.) Suburban Sanitary Commission
Washington (D.C.) Metropolitan Area Transit Authority

3.3 *Escrow bid documents* (ASCE, 1991)

Escrow bid documents (EBD) preserve the contractor's calculations and the information used in preparing his bid. These documents may be consulted whenever they might be helpful in avoiding or resolving disputes. The usual practice is to use them during negotiations for equitable adjustments to the contract price.

Escrow bid documents include all of the quantity take-offs, calculations, quotes, consultant's reports, notes, and other information that a bidder used to arrive at the bid price. No standard format is required, although the escrow documents should be in the language of the contact documents to avoid ambiguities in translation.

The typical procedure is for the three lowest bidders to submit EBD, sealed in a container, along with a certificate of authenticity, normally within one to three working days after the bid opening, depending on the size of the project and the complexity of the proposal. EBD of the unsuccessful bidders are returned after award of the contract.

The EBD are jointly examined by representatives of the contractor and owner prior to award of the contract. This examination is only to verify authenticity, completeness, and legibility. It is not for reviewing or evaluating the contractor's qualifications, estimating technique, or construction methods.

The EBD are consulted whenever either party believes this would facilitate resolution of an issue. They are examined only in the presence of representatives of both parties. If approved by both parties, they may be examined by the Disputes Review Board.

When not in use, the EBD are stored in the care of a mutually-agreed third party, at the owner's expense. The escrow documents are returned to the contractor once the contract is completed and closed out.

Specifications should contain language confirming that the documents remain the property of the contractor. In addition, the U.S. Freedom of Information Act contains several exceptions for business confidential information. The specification should include language intended to give EBD the benefits of these exemptions. Similarly, many states' Freedom of Information Acts exempt trade secrets from disclosure, and the provisions of the specification have been drafted to bring EBD under a trade secret exemption. The law of the particular jurisdiction should be consulted, however, to confirm the effectiveness of the language presented in the specifications.

To date there is no record of any complaints involving disclosure of confidential information found in the EBD.

3.4 *Using differing site conditions clauses*

Over the past 15-20 years there has been a

significant increase in the use of differing site conditions clauses in contracts in the United States. Such clauses assign the risk of unexpected and unknown subsurface conditions to the owner. In the absence of such a clause, contractors are forced to include a contingency amount in their bids. If there are unexpected conditions, they have to hope the contingency is sufficient. If it isn't, they are usually forced to sue anyway. And, when the contingency is not needed to deal with bad conditions, then the contractor gets a windfall. Differing site conditions clauses result in lower bids and result in payments to contractors only when the owner's site is actually more difficult to deal with than anticipated. This is consistent with the kind of risk that an owner ought to be taking. The administration and application of such clauses for small diameter trenchless projects can sometimes be challenging becuase of limitations on the ability to determine actually encountered conditions.

3.5 *Performing constructibility reviews*

Contractors sometimes complain that the designs they are required to follow are not practical. If this is the case, there may be delays and additional costs incurred to develop an alternative. Even if the design is buildable, the owner may pay more money to get the same results. By having the plans and specifications reviewed for "constructibility" before contractors bid on them, owners have been able to utilize suggestions for modifying the designs and making the job of construction easier. If planned for in advance, a constructibility review will not add to the time required to do the design, and the cost is relatively minimal. Constructibility reviews are particularly important for "no dig" projects where the designer and specifier have little or no experience with trenchless technology.

3.6 *Utilizing "real time" dispute resolution*

The risk of delayed resolution of disputes is sometimes overlooked, but it is very real and expensive, particularly on underground projects with expensive equipment which can't produce progress or dollars when idle. To address this risk, more and more owners and contractors are agreeing to use a panel of third-party experts to decide disputes as they occur, rather than via arbitration or litigation after construction is complete.

Sometimes called "project arbitrators" and other times called "disputes review boards," these dispute resolvers are sometimes available from the first day of the contract. If a dispute occurs, it is immediately referred for a recommendation or decision which comes out shortly thereafter. Local, state and U.S. government agencies are increasingly using these methods which have been used successfully on about $6 billion worth of public works construction completed or under way, much of it underground. ("Successful" means that not a single dispute has had to be arbitrated or litigated). The record on private sector work is equally impressive.

The recent American Society of Civil Engineers publication Avoiding and Resolving Disputes During Construction (ASCE, 1991) provides a more detailed description of the Disputes Review Board process. It also includes:

- A list of all known projects where the method has been utilized.
- Suggested specifications for implementation.

3.7 *Establishing realistic contract performance times*

In those situations where there are no firm legal or business requirements for completion of construction by a specific date, owners are learning what contractors have known all along. It is to the contractor's advantage to complete the job within a reasonable time. Extended time on a job costs money. Yet, if contract performance time is insufficient, then it will cost more to do the work or else the contractor will finish late. Either scenario is a disadvantage to the owner. Owners are avoiding these problems by getting contractor advice and input on what is a realistic time to allow for construction of a given project. Obviously an experienced trenchless technology contractor's input should be sought.

3.8 Recognizing the need for a budget contingency

Virtually every construction project will have some changes as it is being constructed. The owner's needs or desires may change as the work is being performed, or there may be omissions in the plans and specifications. (A mere omission doesn't mean anyone is at fault.) It is easier -- and faster and certainly less painful -- to implement changes if the necessary funding is set aside in advance. (If the owner needs to go through a formal procedure to get more funding every time a change is needed, this can only delay the process and also lead to "finger pointing" and "second guessing" as to the cause or need for the change.) Acknowledging that changes can and will occur -- and are a fact of life -- and setting aside a sufficient contingency helps the owner get what it truly needs with a minimum of delay and controversy. This is particularly important given the uncertainties of underground construction.

3.9 Planning for communications

To avoid the "paper wars" of letters making increasingly strong accusations back and forth between project participants many owners require frequent and regularly scheduled face-to-face meetings of project participants with decision-making authority. Such meetings are inexpensive ways of identifying and solving problems while they are still small, and anticipating and avoiding other problems. On trenchless projects it might be appropriate to have such meetings on a weekly basis.

3.10 Pre-planning for permits/utilities/zoning

There are typically dozens of various regulatory requirements that have to be complied with in the course of designing and constructing an underground project. It is equally obvious that if these requirements are not known and considered in advance, delays may result. These will be costly to both the owner and the contractor. To help avoid this, many owners and their engineers are putting forth an extra effort to specifically identify requirements for all of the various required permits, and the environmental constraints that might affect a contractor's performance. For example, on one project, it was learned during a precontract review that city noise regulations would not permit 24-hour trucking of tunnel spoil from a project as had been assumed. The contract schedule was changed to reflect this prior to bidding. As a result, the owner probably avoided a major contractor delay or acceleration claim, or a large bid contingency.

Similarly, today's urban environments often have a complex underground web of utilities. Absent careful pre-planning and construction phase coordination, costly re-routing and delays can occur. However, the cost of appropriate advance planning and coordination is but a fraction of what the cost of delays in re-routing would be. No-dig methodologies mandate a particularly thorough investigation of existing utilities. In addition, nearly all of the 50 United States have state-specific laws (1) establishing procedures and responsibilities for preconstruction location of existing underground utilities, and (2) assigning specific construction phase responsibilities.

3.11 Recognizing that design is a very small -- and often underfunded -- component of cost

When one adds the total cost of engineering, construction, financing, and operation and maintenance of a project together, it turns out that engineering is a small percentage of the total, typically 1-2%. Owners are discovering that a modest increase in engineering effort can be a highly leveraged investment. For example, alternative designs, materials, and construction methods can be more carefully considered. Life cycle costs of operating and maintaining the project can be reduced through more intensive engineering.

3.12 Delegating decisionmaking authority to owner's site representative

Untimely decisionmaking, often the result of either layers of bureaucracy or lack of authority has repeatedly been documented as a major problem in the construction process. To solve this problem more and more owners have delegated authority to appropriate levels so decisions on changes and claims can be made on the spot. Prompt decisions permit the work to go on without a loss of momentum. The risk of untimely decisionmaking is readily avoidable. A corollary of this practice is assigning a resident representative with trenchless experience.

3.13 Including predefined adjustment equations and procedures

In order to eliminate many sources of disagreement from the contract administration process, some owners specify clear and accurate provisions that establish formulae or methods to predetermine computation of sums for disputable items such as profit on changes, overhead, equipment rates, and force account labor. Home office overhead rates (G&A), although subject to wide variation within the industry, are also preset in a range acceptable to the owner and contractor. A contract provision is also included to establish a generally accepted manual for determining the equipment rates to be used in pricing any change orders.

It is equally important for the contract to contain very clear provisions with respect to how change orders will be processed and what information should be included in change order requests. The same is true for force account provisions, which would enable the contractor to be paid on a timely basis for disputed work, pending negotiation of a change order or modification.

Some consideration should be given to include, as a unit price, a per diem value for extended project time. In the event of an owner-caused delay, this amount would be added to any change order carrying with it entitlement to an extension of time.

4 CONCLUSIONS

An often used cliche in the United States is "necessity is the mother of invention." In the case of contracting practices for underground works, the necessity was the avoidance of the costly and acrimonious delays, disruptions, claims, and lawsuits. Various owners and engineers from throughout the United States over time fashioned innovations in contracting practices. While not a total panacea either individually or collectively, overall they have proven to be beneficial to the construction process. Because of some of the unique aspects of trenchless construction, they are particularly applicable to projects utilizing no-dig technology. While quantification of results of the use of such practices is imperfect and difficult at best, overall it has been shown that disputes have been prevented and more readily resolved. This in itself is quite an accomplishment for a litigious industry (construction) in a litigious country (the United States).

REFERENCES

American Consulting Engineers Council and Associated General Contractors of America (ACEC/AGC), 1991, *Owner's Guide to Saving Money by Risk Allocation.*

American Society of Civil Engineers, 1980, *Proc. Construction Risks and Liability Sharing, Volume II*, New York: ASCE.

American Society of Civil Engineers, 1991, *Avoiding & Resolving Disputes During Construction*, New York: ASCE.

Construction Industry Institute, 1990, *Assessment of Construction Industry Project Management Practices and Performance*, University of Texas: Austin.

Construction Industry Institute, 1986, *Im-*

pact of Various Construction Contract Types and Clauses on Project Performance, University of Texas: Austin.
Kovars and Sandler, 1987, "No Damage for Delay Clauses," 87-11 *Construction Briefings*, Washington, D.C.: Federal Publications, Inc.

No Trenches in Town, Henry & Mermet (eds) © 1992 Balkema, Rotterdam. ISBN 90 5410 085 0

L'Application de stratégies novatrices aux marchés de travaux sans tranchée

Robert J. Smith
Wickwire Gavin International, Washington, D.C., Etats-Unis

RESUME: Les propriétaires, leurs dessinateurs et leurs constructeurs reconnaissent de plus en plus les avantages de l'application d'un programme raisonnable d'allocation de risque aux stratégies employées pour les travaux sous terre. Cet article décrit d'une part ce que les propriétaires peuvent faire afin de mieux distribuer le risque et de réduire les coûts totaux du projet, et d'autre part, comment ces stratégies peuvent servir de manière avantageuse et efficace aux travaux urbains sans fossé. Ces stratégies comprennent l'acquisition mieux finincée de l'information géotechnique, de la révélation complète de toute information connue sur la sous-surface, de l'adhérance complète aux conditions légales de plusieurs juridictions pour la localisation avant la construction des services publics par les dessinateurs ou les propriétaires de ces services, ou les deux, l'identification à l'avance des conditions des permis de travail, et les équations et les procédures prédéterminées pour l'ajustement du prix du contrat quand l'imprévu se présente. Ainsi, plusieurs de ces problèmes légaux et/ou contractuels associés à la construction sans fossé aux zones urbaines peuvent être évités ou mitigés.

1 INTRODUCTION

La construction sous terre des conduits en général est chargée de risques à tous les participants au processus. L'utilisation d'une technologie de construction sans fossé présente non seulement de nouvelles occasions pour la dérobade de risque, mais aussi de nouvelles difficultés, surtout dans les zones urbaines. Le dessein de cette présentation est de discuter des stratégies et des procédures contractuelles utilisés de plus en plus avec un succès considérable aux Etats Unis dans la construction des pipe-lines et des tunnels.

1.1 Les philosophies du marché pour le traitement du risque

Les philosophies actuelles concernant les grands projets de construction aux Etats Unis comprennent toute une gamme d'opinions. D'un côté, il y a la philosophie du propriètaire conservateur au sujet de l'allocation de risque; celui-ci croit fermement que l'acceptation du risque de la part de l'entrepreneur de construction fait partie du marché et que les bénéfices éventuels justifient que l'entrepreneur prend des risques considérables. On peut regarder cette approche aux contrats comme un jeu de hasard. De nombreux départements de transportation et des centaines de municipalités, les grandes aussi bien que les petites, adhèrent à cette philosophie.

A l'autre côté, des grands propriétaires publics avec des programmes de construction expansifs (et coûteux) ont conclu que le partage et l'allocation convenable du risque est économique.

1.2 *Les principes généralement acceptés de l'allocation de risque concernant les stratégies du marché*

De nombreux articles et discussions à la "1979 Conference on Construction Risks and Liability Sharing" (la Conférence sur les Risques de Construction et le Partage des Responsabilités), organisée par l'American Society of Civil Engineers, ont conclu collectivement que:

- les risques appartiennent aux parties qui sont les mieux capables d'évaluer, de contrôler, de prendre les frais en charge, et de profiter de la prise des risques;
- il vaut mieux partager plusieurs risques et responsabilités;
- chaque risque comprend un coût associé et inévitable qui doit être assumé

quelque part dans le processus. (ASCE, 1980).

La "Construction Industry Institute" (L'Institut de l'Industrie de Construction) a constaté dans un rapport bien documenté que:

> le contrat idéal -- celui qui sera le plus éconmique -- est celui qui attribue chaque risque à la partie qui est la mieux préparée d'entreprendre et de minimiser ce risque, étant données les circonstances particulières du projet.

(Construction Industry Institute, 1986)

Cet article soutient l'opinion émergeant qu'un meilleur programme d'allocation de risque est avantageux à toutes les parties dans le processus du marché. Par suite de meilleures stratégies du marché, un meilleur programme d'allocation de risque est atteint.

2 LES AVANTAGES DES NOUVELLES STRATEGIES DU MARCHE

Toutes les parties qui s'occupent du processus de dessin et de construction ont profité lorsqu'on s'était servi de nouvelles stratégies du marché. Il en resulte une situation véritablement réussie. Le projet et le public profitent de cette technologie moderne de construction sans fossé; ils peuvent aussi profiter de ce programme moderne d'allocation de risque.

2.1 *Les avantages au propriétaire*

Les propriétaires auront un projet qui est beaucoup plus disposé á être terminé à l'heure, à un prix juste et sans acrimonie, et qui honore leurs besoins et leurs espérances. Bien que ce soit ce que tout propriétaire veut atteindre lorsqu'il entreprend un projet, la réalisation de ces buts est souvent empêchée par des conflits et des délais coûteux et paralysants. Les participants d'un sondage du "Construction Industry Institute" ont conclu collectivement que les meilleures stratégies de construction pourraient économiser en moyenne 5% le coût du projet avec un rapport de 10 pour 1 en ce qui concerne le profit vis-à-vis du coût. (Construction Industry Institute, 1990). Par exemple, il y a un propriétaire public qui croit avoir évité des centaines de revendications et des douzaines de procés judiciares grace à son pratique et à sa philosophie de l'allocation de risque.

2.2 *Les avantages à l'entrepreneur de construction*

Les entrepreneurs de construction sous terre veulent compléter un projet à l'heure, et il visent à satisfaire le propriétaire quand ils terminent la construction. Ils veulent, bien sûr, du récompense pour ce qu'ils ont fait, c'est-à-dire du profit. Cependant, quand les marges bénéficiaires déjà maigres sont même plus diminuées par l'existence des événements imprévus et quand les stratégies contractuelles irréalistes et injustes peuvent se présenter, il peut y avoir des conflits. De tels conflits soulevés au cours des projets qui se servent des nouvelles stratégies du marché ne sont pas très communs.

Moins les incertitudes, et plus l'allocation spécifique de risque, plus il est facile pour les enchérisseurs en concurrence de présenter des enchères plus compétitives sans éventualités et sans besoin de se porter en justice.

2.3 *Les coûts réduits des délais et des conflits*

Les coûts de l'allocation incorrecte peuvent être énormes. Un entrepreneur de construction peut perdre des milliers de dollars chaque jour que le travail est arrêté. Si l'entrepreneur ou le propriétaire demande une indemnité, ceci déclenche normalement la participation des prefessionnels de l'extérieur tels que des ingénieurs conseils ou des avocats, encore à des prix souvent mésurés dans les milliers de dollars par jour. Les deux parties dépensent alors des énormes sommes d'argent pour quelque chose autre que le projet, qui d'ailleurs reste inachevé. Cela a mené quelques propriétaires à conclure qu'il devrait y avoir un meilleur moyen de parvenir. En effet il y en a un et ils l'ont trouvé: l'allocation du risque.

2.4 *La Réduction du langage d'une nature contestable*

Quelques propriétaires se servent d'un langage inégal dans leurs contrats, par exemple les clauses comme "pas d'indemnité par suite d'un retard" et les dénégations de responsabilité pour les forages de terre. Dans certains cas, la tendance d'un professionel de dessin est d'inclure des clauses soigneusement écrites afin de protéger ses intérêts. Tout cela peut être couteux au propriétaire. Même si de telles clauses peuvent s'insérer d'une manière claire dans un contrat, beaucoup de tribunaux ne les appliqueront pas (Kovars and Sandler, 1987). Bien sûr, il n'est pas recommandé de prendre des décisions basées sur un langage contestable.

3 DES STRATEGIES INNOVATRICES EMPLOYEES PAR QUELQUES PROPRIETAIRES AMERICAINS AFIN DE MIEUX DISTRIBUER LES RISQUES ET DE REDUIRE LES COUTS TOTAUX DU PROJET (ACEC/AGC, 1991)

Les procédures du marché et les documents du contrat sont les clefs à l'exécution d'un programme amélioré d'allocation de risque. Ainsi, afin de se mettre en position pour retirer les avantages de l'allocation de risque, les propriétaires complètent des révisions systématiques de leurs procédures et de leurs documents actuels et ils les comparent aux stratégies modernes. Les procédures et les documents actuels sont d'habitude le résultat d'une augmentation de marches au cours des années, avec peu de marches supprimées. Souvent les procédures et les documents sont transmis d'une personne à une autre; c'est l'approche de "on l'avait toujours fait comme ça."

En d'autres termes, les propriétaires devraient prendre le temps de réfléchir et de se demander comment ils distribuent le risque, pourquoi ils le font de cette façon, et s'il existe un moyen plus efficace ou plus économique.

3.1 *La Révision des premiers documents*

Un instrument principal pour l'allocation du risque est la partie de chaque contrat appelé les conditions générales. Elles sont parfois la première partie d'un livre de "spécifications courantes," ou bien elles composent un "boilerplate" imprimé à l'avance. Des révisions en profondeur de ces documents, en les comparant avec des listes de contrôle des risques de construction, indiquent souvent une occasion pour des changements avantageux.

Parmi d'autres choses, on a besoin de se demander si les conditions générales sont compatibles avec les méthodes de construction sans fossé. Souvent elles ne le sont pas. L'examen minitieux peut révéler qu'elles ont été créées auparavant pour des méthodes traditionnelles de construction.

3.2 *L'Acquisition et la fourniture d'information géotechnique*

Il ne faut pas dire qu'une enchère bien renseignée dépend de la qualité et de la profondeur de l'information que possède un entrepreneur de construction sans fossé en ce qui concerne les conditions anticipées sur la sous-surface. Il existe une tendance marquée vers (1) une augmentation des travaux de recherche géotechniques et des analyses budgétaires et (2) la distributation de l'information géotechnique aux enchérisseurs. Quelques propriétaires et ingénieurs constatent que cela mènera à des réclamations d'indemnité si l'information n'est pas correcte. Mais ils ne reconnaissent pas l'idée de base: si l'enchére a été basée sur l'information fournie, ceci définit l'accord entre le propriétaire et l'entrepreneur de construction.

Donc, si les conditions actuelles sous terre sont pires que ce à quoi l'entrepreneur s'attendait, ce n'est que juste que le propriétaire paie puisque si l'entrepreneur avait été au courant des conditions plus sèvéres, il aurait enchéri une offre plus élévée. Les avantages économiques des méthodes de construction sans fossé s'amélioreront quand le dessinateur <u>et</u> l'entrepreneur possèderont autant d'information que possible.

Les meilleures méthodes d'allocation de risque à travers une plus grande révélation géotechnique sont incarnées dans le "Geotechnical Design Summary Report" (GDSR) (ASCE, 1991). Le GDSR présente les conditions anticipées sur la sous-surface et leur impact sur les études de l'avant-projet et sur la construction. Ainsi l'ingénieur et le propriétaire établissent la ligne de base géotechnique pour toutes conditions anticipées. L'emploi de cette seule interprétation des conditions établie dans le contrat ménera à des prix plus uniformes des enchères et à moins d'exposition aux demandes d'indemné en ce qui concerne l'interprétation des données sur la sous-surface. Cela encourage aussi un climat de coopération puisque l'accent est mis sur la franchise et la candeur.

Si les conditions diffèrent en matière de la ligne de base, et l'entrepreneur de construction démontre un impact économique, il a droit à une indemnité supplémentaire. Donc, le propriétaire accepte les risques pour les conditions plus difficiles que celles de la ligne de base.

Les entrepreneurs de construction peuvent soumettre des enchères compétitives basées sur une organisation efficace, des méthodes de construction innovatrices, et une interprétation plus optimiste des conditions (avec l'acceptation de risque supplémentaire), avec l'assurance que la ligne de base géotechnique du propriétaire fournira une base claire pour l'identification des diverses conditions et pour l'établissement des limites d'exposition de l'entrepreneur.

Les entités suivantes ont employés plusieurs types de rapports sommaires du dessin g

Alaska Power Authority
City of Colorado Springs, Colorado
City and County of Honolulu, Hawaii
City of Los Angeles, California
Colorado Department of Highways
Hawaii Department of Transportation
Massachusetts Water Resourses Authority
Milwaukee Metropolitan Sewerage District, Wisconsin
Municipality of Anchorage, Alaska
Municipality of Metropolitan Seattle, Washington
Pennsylvania Turnpike Commission
Pierce County (Washington) Public Works Department
Rail Construction Corporation, Los Angeles, California
U.S. Bureau of Reclamation
U.S. Department of Energy
Washington State Department of Transportation
Washington (D.C.) Suburban Sanitary Commission
Washington (D.C.) Metropolitan Area Transit Authority

3.3 Les documents des enchères en dépôt légal (ASCE, 1991)

Les documents des enchères en dépôt légal ("Escrow Bid Documents") préservent les calculations de l'entrepreneur et l'information employée dans la préparation de son enchère. On peut consulter ces documents quand ils peuvent aider à éviter ou à résoudre des conflits. Normalement on les consulte pendant les négociations afin d'arriver à des ajustements équitables du prix du contrat.

Ces documents comprennent tous les enlèvements de quantité, les calculations, les devis, les rapports de l'ingénieur conseil, les notes, et d'autre information employée par un enchérisseur pour arriver au prix de l'enchère. Il n'y a pas de format standardisé, mais les documents devraient être dans le langage des documents du contrat afin d'éviter des traductions ambigues.

Selon la procédure typique, les trois enchérisseurs les plus bas soumettent leurs documents des enchères en dépôt légal, cachetés dans un récipient, avec un certificat d'authenticité, normalement dans les premiers trois jours ouvrables après l'ouverture des enchères. Tout ceci dépend de l'importance du projet et de la complexité de la proposition. Les ducuments en dépôt légal des enchérisseurs sans succès sont remis après l'attribution du contrat.

Ces documents sont examinés par les représentants de l'entrepreneur de construction et par le propriétaire avant l'attribution du contrat. Cette examination sert seulement à vérifier l'authenticité, l'état complet, et la lisibilité. Elle ne sert ni à revoir ou à évaluer les qualifications de l'entrepreneur, ni à rendre jugement sur la technique ou sur les méthodes de construction.

On consulte ces documents quand une partie croit que ceci rendrait plus facile la résolution d'un conflit. Ils ne sont examinés qu'en présence des représentants des deux parties. Si les deux parties sont d'accord, ils peuvent être examinés par le Comité de Révision des Conflits ("Disputes Review Board").

Quand on n'a pas besoin de ces documents, ils sont confiés aux soins d'une tierce personne au coût du propriétaire. Les documents en dépôt légal sont remis à l'entrepreneur une fois que le contrat est complété et terminé.

Les spécifications devraient contenir un langage qui confirme que les documents restent la propriété de l'entrepreneur. En outre, l'Acte de Libre Accès à l'Information des Etats Unis ("U.S. Freedom of Information Act") contient plusieurs exceptions quant à l'information confidentielle des affaires. La spécification devrait inclure un langage qui a l'intention de donner à ces documents les avantages de ces excemptions. De plus, l'Acte de Libre Accès à l'Information de plusieurs états exclut les secrets de commerce de la révélation, et les provisions de la spécification ont été rédigées afin d'inclure les documents des enchères en dépôt légal dans ces excemptions. Cependant, on devrait consulter la loi des juridictions particulières afin de conformer à l'efficacité du langage présenté dans ces spécifications.

Jusqu'à nos jours, il n'y a pas d'indication d'aucune plainte en ce qui concerne le révélation de l'information confidentielle trouvée dans un de ces documents.

3.4 L'Emploi des clauses au sujet des conditions différentes du site

Au cours des dernières 15-20 années, il y avait une augmentation considérable de l'emploi des clauses au sujet des conditions différentes du site dans les contrats aux Etats Unis. De telles clauses attribuent au propriétaire le risque d'une condition imprévue et inconnue de la sous-surface. Dans l'absence d'une telle clause, les entrepreneurs de construction sont forcés d'inclure des frais divers dans leur enchère afin de parer à l'imprévu. Si des conditions inattendues se présentent, ils ne peuvent qu'espèrer que la somme ajoute est suffisante. Si ce n'est pas le cas, ils sont obligés de toute façon de poursuivre le propriétaire en justice. Et, quand ces frais divers ne sont pas nécessaires, l'entrepreneur reçoit des bénéfices inattendus. Les clauses au sujet des conditions différentes du site entraînent des enchères plus bas et des paiements à l'entrepreneur seulement quand il est plus difficile que ce qu'on avait anticipé de traiter du site du propriétaire. Cela est un exemple de l'espèce de risque que le propriétaire devrait prendre. Quelquefois l'administration et l'application de telles clauses pour des projets sans fossé d'un petit diamètre peuvent être difficiles à cause des limites sur la capacité de déterminer les véritables conditions rencontrées.

3.5 Les études sur la "constructabilité"

Les entrepreneurs de construction se plaignent parfois du fait que les dessins qu'ils doivent suivre ne sont pas pratiques. Si c'est le cas, il peut y avoir des retards et aussi des coûts supplémentaires encourus afin de développer une alternative. Même si le dessin est faisable, le propriétaire peut payer plus d'argent pour arriver aux mêmes résultats. Les propriétaires, en étudiant la "constructabilité" avant l'ouverture des enchères, sont plus prêts à accueillir des suggéstions pour apporter des modifications aux dessins et pour rendre le travail de construction plus facile. Si c'est envisagée en avance, une étude sur la constructabilité n'ajoutera pas de temps nécessaire pour compléter le dessin, et les frais ne sont pas relativement chers. Les études sur la consrtuctabilité sont importantes surtout pour les projets de construction sans fossé où le dessinateur et celui qui est responsable des spécifications n'ont pas beaucoup d'expérience dans le domaine de la technologie de construction sans fossé.

3.6 La Résolution des conflits en termes de "temps réels"

On oublie parfois le risque d'une résolution retardée des conflits, mais cela arrive est c'est cher, surtout pour les travaux sous terre avec une installation considérable qui n'effectue aucun progrès pendant une période d'inactivité. Au lieu de soumettre un conflit à l'arbitrage ou de le porter en justice après l'achèvement de la construction afin d'adresser ce risque, un nombre croissant de propriétaires et d'entrepreneurs se mettent d'accord sur l'emploi d'un comité indépendant d'experts qui résout les conflits quand ils se présentent.

Ces comités, appelés parfois "médiateurs du projet" ou bien "les comités des révisions des conflits," sont souvent disponibles dès le premier jour du contrat. Si un conflit se soulève, il est soumis immédiatement pour une décision à laquelle le comité arrive en peu de temps. Des agences régionales, de l'état, et des Etats Unis utilisent de plus en plus ces méthodes avec succès pour $6 billion de construction des services publics, terminée ou en cours, dont beaucoup est sous terre. ("Avec succès" veut dire que pas un seul conflit n'a été soumis à l'arbitrage ou porté en justice.) Le bilan du secteur privé est aussi impressionnant.

La publication récente de l'American Society of Civil Engineers, Avoiding and Resolving Disputes During Construction (Comment éviter et résoudre des conflits au cours de la construction) (ASCE, 1991) offre une description plus détaillée du processus de ces comités de révision des conflits. Elle fournit aussi:

- une liste de tous les projets connus qui emploient cette méthode;
- quelques stipulations suggérées pour exécution.

3.7 L'établissement d'un délai pratique de bonne exécution des contrats

Dans les situations où il n'existe pas d'exigences fermes, soit juridiques soit commerciales, pour achever la construction à une date prédéterminée, les propriétaires apprennent ce que les entrepreneurs de construction savaient déjà. C'est à l'avantage de l'entrepreneur de compléter son projet bien à l'heure. Les projets qui avancent lentement sont coûteux. Cependent, si le progrès d'une performance contractuelle est insuffisant, le travail va coûter plus cher ou bien l'entrepreneur va le compléter en retard. Tous les deux scénarios sont désavantageux au propriétaire. Les propriétaires évitent ces problèmes en demandant conseil et d'autres informations en retour aux entrepreneurs en ce qui concerne ce que c'est qu'un délai pratique qui permet la construction d'un projet qeulconque. Il est évident qu'il faut chercher le conseil d'un entrepreneur de technologie de construction sans fossé.

3.8 En reconnaissance du besoin d'un budget de prévoyance

Presque tout projet de construction aura quelques changements au cours de son exécution. Les besoins ou les désirs du propriétaire mènent toujours à des changements pendant que la construction se déroule, ou bien il se peut qu'il y ait certaines omissions dans les plans et les spécifications. (Une simple omission ne veut pas dire que quelqu'un est fautif.) Il est plus facile, plus rapide, et moins pénible bien sûr, d'exécuter les changements si les fonds requis sont mis de côté à l'avance. (Si le propriétaire a besoin de se soumettre à une procédure formelle pour obtenir plus de fonds chaque fois qu'il faut changer quelque chose, cela ne peut qu'empêcher la procédure aussi bien que mener aux accusations et aux anticipations de ce qu'on va dire quant à la cause ou à la raison du changement.) Il est de fait que les changements sont des réalités de la vie qui se présentent d'un moment à autre. Un budjet de prévoyance réduit au minimum les délais et les controverses, et il aide aussi le propriétaire à obtenir tout ce dont il a vraiment besoin. Ceci est particulièrement important étant donné l'existence des incertitudes dans la construction sous terre.

3.9 Les préparations pour les correspondances

Afin d'éviter les "guerres de papier" des lettres qui produisent des accusations de plus en plus fortes entre les participants du projet, un grand nombre de propriétaires organise régulièrement des réunions pour les participants qui ont l'autorité de prendre des décisions. De telles réunions offrent un moyen économique pour identifier et résoudre des problèmes tout en restant une affaire modérée. Elles peuvent offrent aussi un moyen par lequel on peut prévoir et circomvenir d'autres problèmes. Aux projets sans fossé il pourrait être approprié d'organiser de telles réunions une fois par semaine.

3.10 Les préparations à l'avance pour les permits, les services publics, et la réparation de zones

Il y a en général des douzaines de diverses exigences régulatrices qu'il faut observer au cours de la préparation et la construction d'un projet sous terre. Il est également evident que certains délais peuvent se présenter si toutes les exigences ne sont pas considérées à l'avance. Ces retards seront coûteux au propriétaire aussi bien qu'à l'entrepreneur. Afin d'éviter ce problème, plusieurs propriétaires et leurs ingénieurs font un effort pour identifier de maniére spécifique les exigences pour tous les divers permis nécessaires, et de comprendre toutes les contraintes qui ont rapport à l'environnement et qui peuvent empêcher le progrès de l'entrepreneur. Par exemple, au cours d'un projet on a appris pendant une enquête préliminaire que, selon le règlement municipal concernant le bruit, il n'était pas permis de comionner 24h/24 les déblais du tunnel d'un projet comme on avait espéré. Alors le propriétaire a probablement évité un grand délai de la part de l'entrepreneur, une demande d'accélération, ou des frais supplémentaires dans les enchères.

De plus, les environnements urbains de nos jours ont souvent un réseau complexe de services publics. L'absence des préparations à l'avance et de la coordination des marches de la construction, des changements couteux et des retards peuvent se présenter. Cependant, le coût des préparations à l'avance et la coordination n'est qu'une petite portion de ce que le coût serait avec ces délais et ces changements. Les méthodes de construction sans fossé demandent une investigation en profondeur des services publics existants. De plus, presque tous les 50 Etats Unis ont des lois particulières à l'état (1) qui établissent les procédures et les responsabilités pour trouver, avant la construction, les services publics sous terre existants, et (2) l'attribution des responsabilités spécifiques des marches de la construction.

3.11 La reconnaissance que le dessin est une petite partie -- souvent sans fonds suffisants -- des frais de construction

Quand on ajoute ensemble le coût total de l'engineering, de la construction, de l'opération et de l'entretien d'un projet, il arrive que l'engineering représente un pourcentage peu important, normalement 1-2%. Les propriétaires découvrent qu'une augmentation modeste de l'effort de l'engineering peut être un investissement à grand crédit. Par exemple, des dessins, des matériaux, et des méthodes de construction alternatifs peuvent être considérés de façon plus facile. Les frais du cycle de vie de l'opération et d'entretien du projet peuvent se réduire par suite d'un programme d'engineering plus intensif.

3.12 L'attribution des pouvoirs de la prise des décisions aux représentants du site du propriétaire

La prise inopportune de décisions, souvent le résultat des délais de bureaucratie ou la manque d'autorité, a été documentée à maintes reprises comme un problème majeur au cours de la construction. Pour résoudre ce problème, de plus en plus de propriétaires ont délégué l'autorité aux niveaux appropriés pour que les décisions au sujet des changements et des demandes puissent être prises sur place. Les décisions immédiates permettent que le travail continue sans arrêt. On peut éviter facilement le risque d'une prise inopportune de décisions.

3.13 L'inclusion des équations d'ajustement et des procédures prédéterminées

Afin d'éviter plusieurs sources de désaccord pendant le processus de l'administration contractuelle, quelques propriétaires signalent des provisions claires et exactes qui établissent des formules ou des méthodes pour prédéterminer le calcul des sommes pour les éléments disputés tel que le profit réalisé sur les changements, les frais généraux, le prix des matériaux, et le prix de la main-d'oeuvre. Les frais généraux de l'administration centrale (G&A), bien qu'ils puissent varier de manière considérable dans l'industrie en général, sont prédéterminés sur une échelle acceptée par le propriétaire et l'entrepreneur. On comprend aussi une provision du contrat pour établir un manuel généralement accepté pour déterminer le prix des matériaux si des changements de commande sont nécessaires.

C'est également important que le contrat contienne des provisions très claires en ce qui concerne comment les changements de commande seront effectués et quelle information devrait être comprise dans la demande des changements de commande. C'est aussi le cas pour les provisions des comptes de force, ce qui permettra à l'entrepreneur d'être payé à temps pour un travail contesté, en attendant la négotiation d'un changement de commande ou d'une modification.

On devrait considérer d'inclusion, comme un prix unitaire, d'une valeur de "per diem" pour une durée prolongée du projet. Au cas où le délai est la faute du propriétaire, cette somme serait ajoutée au changement de commande en comprennant l'allocation d'un prolongement de temps.

4 CONCLUSIONS

Une expression courante aux Etats Unis nous dit que "necessity is the mother of invention" (la nécessité est la mère de l'invention). En ce qui concerne les stratégies contractuelles des travaux sous terre, la nécessité est la raison pour laquelle on s'est mis à éviter les délais couteûx et acrimonieux, les interruptions, les demandes d'indemnité et les procès juridiques. Plusieurs propriétaires et ingénieurs de tous les Etats Unis ont développé au cours des années certaines innovations pour les stratégies contractuelles. Bien que cela ne signifie pas une espèce de panacée individuelle ou collective, en général ces innovations se sont démontrés favorables au processus de la construction. On peut les appliquer surtout aux projets qui se servent des méthodes technologiques de construction sans fossé parce qu'elles sont uniques. Bien que l'évaluation quantative des résultats de telles stratégies soit imparfaite sinon difficile, il est vrai que les conflits ont été empêchés et résolus de façon plus efficace. C'est un véritable accomplissement en soi, surtout dans une industrie litigieuse (la construction) et dans un pays litigieux (les Etats Unis).

REFERENCES

American Consulting Engineers Council and Associated General Contractors of America (ACEC/AGC), 1991, *Owner's Guide to Saving Money by Risk Allocation.*

American Society of Civil Engineers, 1980, *Proc. Construction Risks and Liability Sharing, Volume II*, New York: ASCE.

American Society of Civil Engineers, 1991, *Avoiding & Resolving Disputes During Construction*, New York: ASCE.

Construction Industry Institute, 1990, *Assessment of Construction Industry Project Management Practices and Performance*, University of Texas: Austin.

Construction Industry Institute, 1986, *Impact of Various Construction Contract Types and Clauses on Project Performance*, University of Texas: Austin.

Kovars and Sandler, 1987, "No Damage for Delay Clauses," 87-11 *Construction Briefings*, Washington, D.C.: Federal Publications, Inc.

No Trenches in Town, Henry & Mermet (eds) © 1992 Balkema, Rotterdam. ISBN 90 5410 085 0

Coordination of utility networks as a first step towards underground town planning

Pierre Duffaut & Monique Labbé
France

ABSTRACT : The necessary coordination of wires and pipes networks should be the first step towards a new brand of town-planning, fully including the underground (in french, Urbanisme souterrain). Transportation networks are the most concerned but also facilities for any kind of social activity and for any extension of sites for storage, and production. The paper calls for a prospective thought on subsurface use, taking account very early of the space needed, the construction methods and the operation of vital networks.

1 INTRODUCTION

The extraordinary interlacing of pipes and wires networks embedded within the subsurface of the streets is for long felt as the result of a total lack of foresight on behalf of city authorities. It may well have come also from a lack of savoir-vivre *(good manners)* on behalf of utilities: the first one has taken the best place and has spread its facilities at one's will, without any concern for those likely to follow, and so on to the present! Should such attitudes be understandable at the time of the first step, when no one did think that others could soon follow, they are no longer so, since many decades, as so many new networks have developed since and will develop ahead.

The same occurs on the surface: the first road in a valley takes the best place, without consideration for railways and motorways to come in the future; even the late duplication of a railway track shows that the first built bridges and tunnels are at the best place and the second best are sometimes on one side, sometimes on the other side; the same occurs for the penstocks of hydroelectric plants. It is more or less easy to displace some of those surface facilities to obtain the optimum coordination of competitors, even if their owner is the same. It is the more difficult when the networks are embedded in the soil. True underground tunnels and caverns are quite impossible to displace and even very difficult to modify.

Of course the layout of any network is not fixed for ever; each day someone needs to be strengthened or extended, not to speak of maintenance and repairs; new customers have to be connected, new networks are added to the former ones without any rethinking about clarification of the tangle. On the contrary the confusion does not cease to aggravate.

Projects of deeper works (sewers, metros, car parks, motorway tunnels, etc.) provide as many opportunities for ascertaining the state of the shallow subsurface, but unfortunately not to set things to right (unless sometimes very locally).

Should the use of underground space be more general, it will be mandatory to clear the table of all these anarchic networks (fig. 1) and to redesign the whole of them on a new basis (see below the Boston central Artery case history).

An interesting example has been given when Nice city, about fifteen years ago, restricted the narrow streets in its center to pedestrian use: all the shallow utilities have been reorganized along a simple layout, with better "branching" of buildings on both sides.

2 CASE HISTORIES

The Paris metro provides a significant example of a network developed over a long period, as the 13 lines have been built from 1898 and are yet extending in the suburbs: many stations provide commuting facilities between 2 to 4 lines, but the lines have not been planned at the same time and the pedestrians endure difficult corridors with too many stairs up and down. Conversely the crossing of RER (express regional metro) lines A and B at Les Halles provides change on the same platform for the most of commuters, as they have been designed together.

The Central Artery Project, now in construction in Boston, Massachussets, will transfer a major motorway from viaduct to tunnel inside the true center of the city; more than 5000 wires and ducts were to be crossed or displaced : the first work has been to reorganize those networks in new corridors, as well along the project as across it. So, thanks to the project, the whole set of utilities in the center has been both updated and rationalized.

Early in 1992 ten kilometers of streets and most of bordering buildings in Guadalajara (Mexique) have been destroyed by a series of blasts inside a trunk sewer. Upstream along the sewer an oil pipeline was crossed by a water main, in close contact (the water pipe was second to the oil one); the cathodic protection of the oil pipe being impaired, oil was seeping since years and part of it was recovered in the sewer. Halfway a factory for alimentation products would have released hexane gas in the same sewer. Last, downstream, the sewer has been temporarily occluded before being displaced at crossing a future light metro line. In spite of many complaints from inhabitants alarmed by gas odours, nothing has been done up to the sequence of blasts (about 200 casualties, 2000 wounded, 10000 unlodged).

The Tolbiac quarter of Paris along the Seine left bank, upstream of Austerlitz railway Station, is to be soon fully redeveloped from industrial and railway estates, around the so-called Grande Bibliothèque (a new French national library to be built there). Together with a new metro line (an automated one named METEOR), the whole sewerage network will be redesigned. For the first time in Paris a binary system will be used, separating rain and waste water.

Electricité de France has recently decided to bury all the MT lines of its network (medium tension, up to 60 kV), beginning with any new lines, then from 1994 on, the lines operated today. Of course this policy applies to lines in the country; inside cities most of the lines are yet buried, contrary to what appears in America and some other parts of the world. It could be a good opportunity for thinking to the location of wires to be buried: they will introduce new "cuts" inside the shallow subsurface, they will have to cross other branches of utilities, they will impede future underground works of any kind. They have to follow the examples of new road- or railways which include provisions for crossings of other networks not yet established.

3 UTILIDORS

Without sharing of a few common galleries (galeries techniques in french, utilidors in North America), the space occupied by utilities is far greater than necessary, due to the distance reserved between successive works of different concessionaries. The same is true for transit tunnels: inside Paris, from Gare de Lyon to les Halles four tunnels of conventional and RER metros run rather parallel and three more will be soon added.The space they use is many times the space of a common box likely to host all the tracks, as those designed for immersed tunnels.

The sewers of Paris, built with a big size to cope with storm water and to allow for cleaning by workers, have long been used to host other networks such as water, electricity and compressed air (today water ducts are restricted to non drinkable water, electricity wires to public lighting and traffic lights control).

A century ago, Chicago, Illinois, had an underground network of tunnels for freight trains, abandoned before 1900, and reused later by a set of utilities. Only one city in the world has a large network of technical galleries, Fairbanks, Alaska. Other examples in the USA are given by some university campuses and leisure parks (Disneyworld, Florida).

A few quarters in a few towns have been built with a false ground : streets and utilities run below the slab according to the Charte d'Athènes (as in Disneyworld) : Front de Seine and la Défense in Paris, Lyon Part-Dieu, Bordeaux-Meriadek, Bobigny, but this model has not fully succeeded up to now. Earlier, architect Perret had proposed to rebuild the whole Le Havre city, destroyed by bombing during WW2, at a + 3m level in order to benefit a general service storey ; his project has not been considered.

Among the many benefits of man size technical galleries are easy interventions for maintenance, repairs and extension, without opening the streets and pavements. They could also provide a protected path to shelters in case of chemical or nuclear fall out. But the well-known drawbacks often balance the benefits : much money has to be invested early, and many solutions become quickly obsolete. Some needs disappear, as the coal delivery for domestic heating and cooking, some new technologies are no longer compatible with the space allowed.

4 COVERING THE CUTS AND ASSOCIATING DIFFERENT WORKS, for better use of underground space and of the surface.

On the surface, railways and motorways cut the territory into two parts (Duffaut and Labbé, 1991), just as do rivers and canals, needing special crossings for any other ways and utilities. Bridges have long been the conventional answer, but they are no longer perceived as an enrichment to any landscape ; an increasing number of bridges aggravates the obstruction of a site, the more why they differ in style for being built at different times, for different functions and with different techniques (think of the crossings over East River inside New York, over la Marne about Nogent).

In order to limit the drawbacks of too many cuts, long sections of such ways should be covered, as has happened for many small rivers inside cities (la Bièvre in Paris, etc ...) Examples of short sections are widespread along urban rail -and motorways but their length is growing fastly and will grow ahead.

The number of cuts may be limited by setting two or more side by side : so, many pipelines or transmission lines follow the same routes, so the TGV North rail track has just been built along motorway A1, etc ... But a better coupling could be obtained by superposing two or more "cuts", as is often made for rail and road on bridges. As elevated viaducts are no longer accepted even out of the city streets, be they narrow or wide, one of the ways at less should be entrenched. Should a motorway cover a railtrack, the latter will be protected against any fall-out, either from a hillside or from the sky. So it has been proposed to locate a motorway below the Seine bed all along its course through Paris (along with plenty of space for car parks and other functions).

Associating different works and functions is a general way of sharing costs of underground works, and the more to maximise the benefits of underground space. Many examples are given by metro works worldwide. Two may be quoted from Paris RER : the car park along Boulevard Haussmann, on six storeys from the train tunnel to the pavement ; and the les Halles station.

Conversely la Défense development did not succeed to find other useful uses of its subsurface than underground transport and car parks. It could have been an excellent opportunity to coordinate utilities and to locate more services.

5 WHAT SHOULD BE DONE.

5.1 What should be avoided :

From examples above it is clear that many problems should be better addressed and many faults should be avoided :

i/ Separate tunnels take a lot of space, separate ducts and wires waste underground space, the more when they are designed separately ; only coordination can save both space and future.

ii/ Crossing points of different utilities often are dangerous points (Guadalajara is an extreme case, many occur with less damage and casualties).

iii/ Ground being opaque, no embedded things are visible, their defects do not appear. Secret wires may be embedded (a through telephone wire between public buildings happened to be cut by roadworks on the Concorde place, Paris).

iiii/ The more : trenching across road pavements and footpaths as has been done for most of sewers, early metro lines, car park Avenue George V in Paris, and is yet being done for most of repairs and extensions of utilities, will no longer be tolerated. Trenchless methods of working have to be used.

5.2 The site concept (Duffaut, 1977):

Nature provides more or less favourable sites

for harbours and for dams, to only quote two examples, thanks to Earth surface morphology. Taking account of 3-D structures inside the ground, Nature also provides sites for underground space uses, for instance natural caves for domestic storage and sometimes housing. Horizontal strata of rock suitable for building stone (or even for concrete aggregate) may provide precious underground space as in Kansas City (Illinois and Kansas); in Paris such quarries are too irregular in size and stability to be useful, except some gypsum quarries, one of them harbouring a major defence center.

The rock hills inside any city provide big volumes of underground space, escaping the needs for power to bring to surface water, goods and people. Many examples of such use is given by Salzburg, Austria (without evidence that any general plan has been proposed to maximise the benefits of underground facilities for the community).

Such underground sites are a major part of the natural supply of underground space; they deserve to be clearly recognized, and often preserved for the most demanding uses:

5.3 Rehabilitation opportunities :

After a period of random development, many features need to be erased including the hazardous nearness of housing to industrial and potentionally dangerous sites. Up to now many areas have been redevelopped inside Paris and in the suburbs, without any more consideration for sudsurface use than a few car parks (except in les Halles case).

5.4 Liaison of infrastructures with structures above the ground, as well at the city level as at the street or building level, in order to fully benefit of the nodes of networks :

Only some metro stations (and pedestrian subways in many eastern cities) benefit from association with commercial areas ; this model should be improved and followed by other networks ; nodes of networks often include special equipment, switch and transformer rooms, valves and control rooms, etc ... and could be associated with volumes devoted to other activities.

In the same way as most of ducts and wires are associated to the streets, the nodes should be designed as points of development of underground space, the more for rail stations and car parks which favour access to commercial and cultural activities.

More generally, the subsurface storey (below ground surface), should not be thought only as a sanitary void space as under any small house ; neither it should be only a passage for ducts and wires as the space over a false ceiling. Instead it should provide various volumes for various uses, in order to meet more needs of inhabitants, bypassers and users of transportation systems.

5.5 Need for accurate recording :

As geological and geotechnical data from any excavation or borehole has to be made available to public, data on utilities have to be kept by the local authority, including records on the way works have been performed and all the unexpected discoveries.

5.6 Need for planning :

Planning underground is far more than only day-by-day coordinating the networks of infrastructures and utilities ; it is necessary to foresee the long term needs of the city and to take advantage of any work, public or private, to improve the overall layout instead of adding to the confusion.

5.7 Three examples of underground projects are proposed :

East of Paris, the subsurface of Bois de Vincennes (Lépine, 1957), should be used for storage development (cars, wines and fuels), river quay, thanks to a favourable geological strata (calcaire grossier, extensively quarried under Paris and suburbs). Similar to the use of Kansas City quarries this project dates back to the period of early developments there.

Projects for underground motorways (Doublet and Duffaut, 1992) below Paris and close western suburbs, LASER and MUSE, have been published during last years (and are closely considered by authorities in charge) ; unfortunately they are restricted to car traffic (or mass transit like the EOLE RER liaison) and they do not imply any underground space use for other purposes.

The Rungis national market, 10 km south of Paris, dates back to the demolition of les Halles pavillions in the center of the city. No extension is possible now as the whole surface has been used ; only an underground extension is likely. Thanks to the plateau site, it could easily be served by level underground roads and tracks from the Seine valley floor, and together benefit unexpensive cold storage rooms.

6 CONCLUSIONS

Three sets of conclusions are proposed at the end of this review :

i/ Plenty of subsurface space is available under cities, provided some care is taken when beginning to use it (not to prevent access to a deeper space ...)

ii/ Trenches and shafts inside a city are felt like traumatizing surgery and no longer accepted by inhabitants ; in the hospital more and more surgeons use new techniques, thanks.

Trenchless technologies may be compared to microsurgery and cutless surgery, for the benefit of inhabitants and patients.

iii/ Coordinating utilities is not only a necessity for today, it will provide a model and a base for a full underground town planning for tomorrow.

Underground more than aboveground, a prospective thought of surface and volume use is necessary, taking account very early of space needed, of construction methods, and of the networks vital for the life of the city.

REFERENCES

Doublet M. and Duffaut P., 1992, "State of art of underground space use in France", Proc. 5th ICUSESS, DELFT (in press).

Duffaut P., 1977, "Site reservation policies for large underground openings", Rockstore Symposium, Bergman M. ed. Stockholm; reprinted *Underground Space*, 3 (4): 187-194.

Duffaut P.and Labbé M. 1991, Le sous-sol au service du Val-de-Marne - Report to Conseil Général(D.S.E.A. 94).

Lépine R., 1957, Aménagement souterrain du plateau de Gravelle, Le Monde Souterrain n° 103, Juin PP 502-4.

No Trenches in Town, Henry & Mermet (eds) © *1992 Balkema, Rotterdam. ISBN 90 5410 085 0*

Trenchless excavation methods for the installation of electrical distribution cables – A strategic overview

Alwyn Morgan
Midlands Electricity plc, UK

Abstract: The paper describes the strategy developed for the introduction of trenchless excavation methods. In the last two years the company has steadily built up its range of moleing machines, presently some 85 moles are regularly being used for cable installation work. These range from the Grundomat 45mm machine to the Ditch Witch Jet Track.

Well over half of all road crossings as well as a significant proportion of service replacements are being carried out using trenchless excavation techniques. The recent arrival of the steerable mole will allow longer lengths of low voltage mains cable to be installed.

Future developments are discussed including the companies response to the New Roads and Street Works Act.

1. BACKGROUND

Midlands Electricity plc (MEB) is one of the United Kingdoms recently privatised Regional Electricity Companies. Situated in the heart of England the Company distributes electricity to more than 2.1 million customers.

MEB operates extensive underground cable networks (35,000km) and overhead power lines (27,000km). The Company installs approximately 1900km of underground cables per year, the split by voltage level and typical sizes of cables is shown in Table 1 below:-

TABLE 1

VOLTAGE	LENGTH	OUTSIDE DIAMETER
Services	(1250km)	19mm - 35mm
LV	(350km)	35mm - 60mm
11kV	(300km)	60mm - 85mm

Approximately 60% of the cables are installed in unmade ground (pasture, new housing developments etc) the remaining 40% being made up ground with the majority of cables being installed in footpaths and the remainder being associated with road crossings.

Traditionally the majority of work has been carried out by hand excavation, with only a small proportion being actioned using mechanical excavators.

MEB employ either direct labour or contractors for this activity.

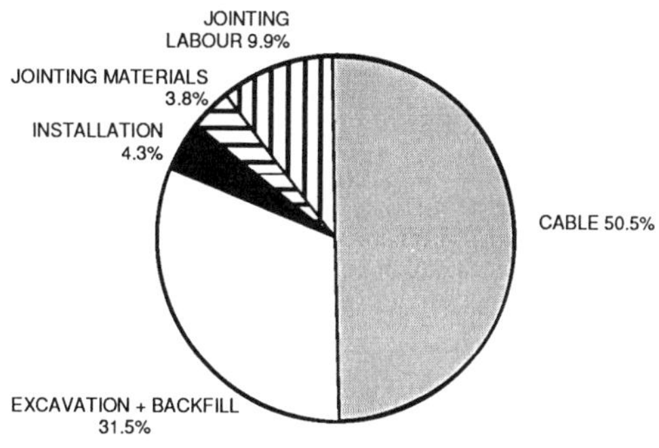

Fig 1

Contractors being traditionally paid by cost per metric volume of trench.

The pie chart (Fig 1) provides an indication of the relative costs associated with the installation of a unit one metre length of power cable in unmade ground. Installation, excavation and back-filling include all costs, ie labour, material etc.

The pie chart below (Fig 2) again shows the installation of a unit length of cable. This time it is being installed in made up ground. In this case the costs are much higher, as the costs of temporary and permanent reinstatement of the footpath/road surface have been included.

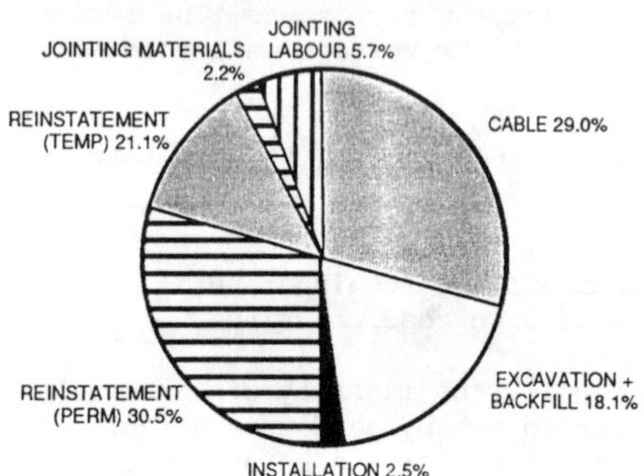

Fig 2

Traditionally reinstatement work is divided into two distinct parts. After the cable is installed MEB is responsible for the temporary reinstatement of the trench, once settlement occurs, the local authority carries out the permanent reinstatement.

In December of this year the 'New Roads and Street Works Act' will come into force, the principal effects of the Act are:-

- Utilities to have full responsibility for reinstating the street following excavation and completing the reinstatement as soon as reasonably practical (ideally total one-shot reinstatement).

- The change in the reinstatement specification will require all persons with responsibility for work in the highway to be assessed as capable of carrying out this work in a satisfactory manner.

- The Highway Authority will be enpowered to inspect and be entitled to payment for inspection of 6% of the utilities work. Inspections which fail to meet the prescribed standards will be subject to three additional inspections at the utilities expense. Should the utility fail to pass the final inspection then they become responsible for the re-excavation and reinstatement of the road or footpath repair.

2. STRATEGIC APPRAISAL

Early in 1990 the Company set up a small team chaired by the author to review:-

- The opportunities/threats imposed by the impending 'New Roads and Street Works Act'.

- Opportunities associated with the mechanisation of the installation of cables.

The results of the review are summarised below:-

SWOT Analysis

Strengths of Traditional Method
• Uses well established methods. • Relatively low skill level.
Weakness of Traditional Method
• Uses most expensive excavation method. • High level of disruption to road and footpath users. • Quality of reinstatements
Opportunities
• Trenchless excavation technology. • Narrow Trenching. • Increased use of traditional Mechanical excavation. • Reduced 'traffic disruption. • Move to single-shot reinstatement (possible 20% saving)
Threats
• 'Cost of failure' associated with the 'New Roads and Street Work Act'. • Additional training required to comply with new 'Act'. • Traditional 'poor image' of trenchless technology.

Our analysis revealed that, in made up ground, the ideal solution was to minimise the negative aspects associated with the 'Act' by reducing the amount of reinstatement work required. In priority, and potential lowest cost order, the options available were:-

- Trenchless Technology

  Reinstatement limited to the launch and receive pits and associated 'check' holes.

- Narrow Trenching

  Reduces reinstatement by limiting the surface area disturbed.

- Combination of Machine and Hand Dig

  Whilst the reinstatement costs are not minimised using this method, savings in overall installation costs are made via the increased use of mechanization.

  Additional research and extensive trials showed that trenchless technology offered the optimum solution for the installation of cable across roadways. This had the dual benefits of saving substantial amounts of money whilst at the same time drastically reducing the disruption caused to traffic flow and the general public.

  Unfortunately the picture was not so clear in the case of footpaths, the limiting factors here being:-

(i) The level of congestion caused by the amount of utilities equipment, see Fig 3

(ii) The fact that existing electrical cables do not always lie in a dead straight line.

(iii) The degree of accuracy of cable location equipment.

(iv) The level of accuracy of utility records.

(v) The accuracy achieved by trenchless machines.

As regards the use of moleing equipment in other activities several opportunities were identified:-

(i) Service Replacement work traditionally involved the opening of a trench across the gardens of customers properties, this led to costly reinstatement work such as relaying lawns, provision of replacement plants and the resurfacing of driveways. Trenchless technology offered the perfect solution to this problem.

(ii) In unmade ground opportunities also existed in areas such as: river and canal crossings and areas of outstanding natural beauty.

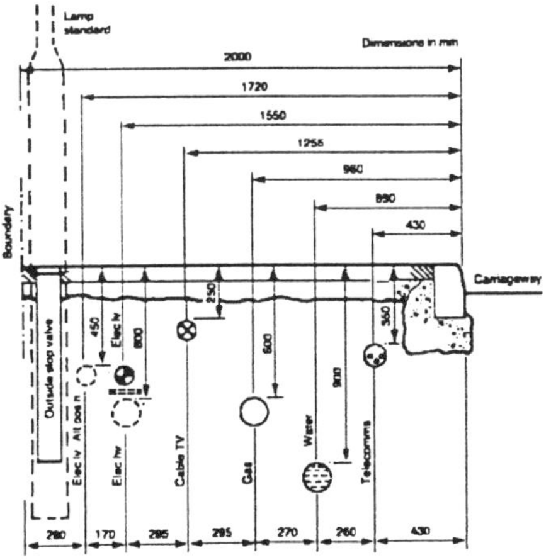

Fig 3

3. STRATEGY JANUARY 1990 ONWARDS

Our first concern was to identify a key individual in each of the companies seven Divisions who would have the role of championing the use of trenchless technology.

Next we equipped a team in each location with 45mm and 65mm impact moles their initial area of operation being the installation of service replacement cables (see Fig 4) and the

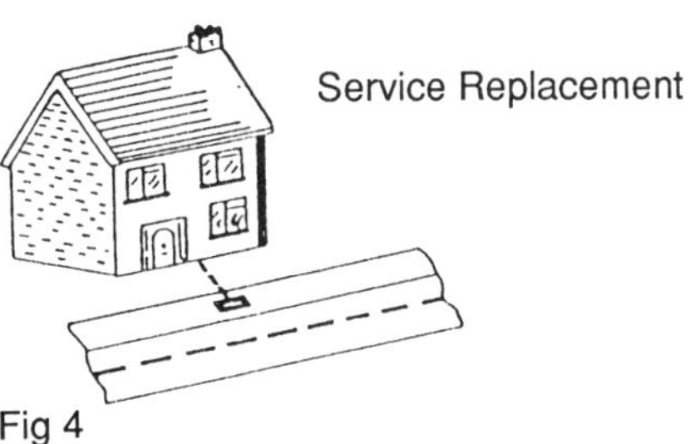

Fig 4

crossing of small roadways with LV cables.

By July of the first year we were operating the following:-

17 45mm soil displacement hammers
8 65mm soil displacement hammers
1 95mm soil displacement hammers

As the success rate grew so did the enthusiasm of the companies manual staff for trenchless technology.

At the start of the second year we began to experiment with the use of tracking equipment. Our aim being to extend the range of operation to cater for the more complex road crossings and to experiment with laying cables directly under footpaths.

During this period we purchased:-

5 79mm pneumatic machines
1 95mm pneumatic machines
1 102mm pneumatic machines
2 140mm pneumatic machines

'Stitching' The use of the electronic transmitters in the head of the larger impact moles such as the Pierce Arrow 140 has allowed LV cables to be installed directly in the footpath. For this type of operation two options are available.

(i) The mole is fired from the launch pit and is tracked along a pre-determined straight line, once the mole is seen to be deviating (via the use of the location equipment), it is stopped and a new launch pit is excavated around the buried mole. The technique is repeated until the mole reaches the final receive pit (see Fig 5).

Fig 5

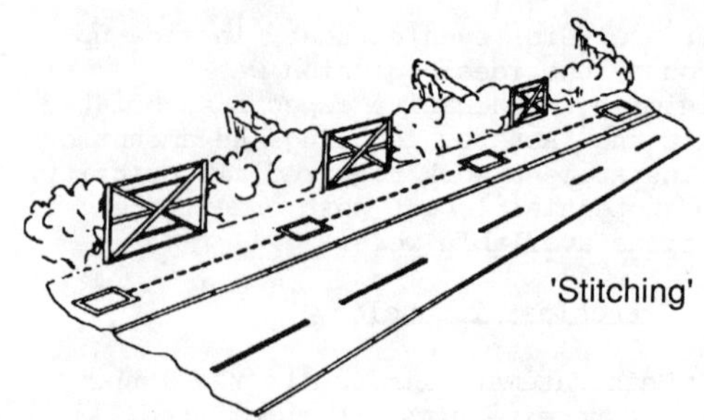

Fig 6

(ii) As for (i) above but in this case the mole is fired between a set of previously excavated pilot holes (used to identify the location of utilities equipment) (see Fig 6).

Of the order of 14km of cable has been installed using this method.

In the later part of 1991 our attention became focused on two distinct areas: The provision of small bore moles for the companies cable jointing teams, and the acquisition of a guided boring system.

By this phase of the operation trenchless technology was becoming a 'way of life' within the Company, with well over half of all road crossings being successfully carried out using this technique.

In 1992 we purchased our first Guided Boring System and we are pleased to report that the break-even point was reached in only three months. Initially the Jet Track was used on the more complex jobs i.e. railway, river and motorway crossings. At the time of writing we are now using the system to install LV cables in the footpath.

As regards the cable jointers as of this year all of the companies new jointers vans will come fitted with an underfloor compressor unit and associated moles (45mm and 65mm).

4. Results To-date

By the end of 1992 it is estimated that the total savings associated with the companies move to trenchless technology will be of the order of £800,000. Added to this must be the considerable reduction to the disruption to traffic and the general public that has been achieved by the use of this technique.

We now regularly employ 85 moles and it is our intention to significantly increase this number in the near future.

To-date we have carried out the following operations using trenchless technology:-

3000 Service Replacements
1200 Road Crossings

and in addition we have laid some 14km of LV cables.

5. The Future

It is intended to further extend the companies 'fleet' of moles during the next six month period. This will centre on the provision of up to two additional guided boring systems along with the purchase of a range of non-steerable machines.

The Company intends to comply with the 'New Roads and Street Works Act' three months ahead of schedule (October 1992).

The revised 'Cable Installation Contract' has been programmed to commence as of October 1992. Under the terms of this new contract a significant proportion of the work is planned to be carried out using trenchless techniques i.e. use of moles at road crossings is expected to reach 95%.

It is our planned intention to raise the overall profile of trenchless technology in the area of electrical distribution.

To this end we held in July the first annual trenchless technology workshop for Electrical Distribution Companies. Our aim is to provide a common forum for the sharing of experience in this field and to raise the awareness of manufacturers and suppliers of our particular needs. As part of this event we actively encouraged participants to join ISTT.

Finally what of MEB's next strategic phase. This will focus on two key areas:-

(i) The location of moleing equipment:-
   - At the high tech end this means having the capability of remotely detecting the exact location of a mole as it crosses a roadway, that is without the need for an operator to walk into the road
   - At the other end of the market we require a robust and accurate electronic 'sond' which can operate from the head of a small mole (45mm or 65mm).

(ii) Ground Radar, required to allow us to operate efficiently and effectively in the footpath.

No Trenches in Town, Henry & Mermet (eds) © 1992 Balkema, Rotterdam. ISBN 90 5410 085 0

Méthodes d'excavation sans tranchée pour l'installation de câbles de distribution électriques – Plan stratégique

Alwyn Morgan
Midlands Electricity plc, Royaume-Uni

Abstrait: Le document décrit la stratégie développée pour l'introduction de méthodes d'excavation sans tranchées. Au cours des deux dernieres années, l'entreprise a régulierement augmenté sa panoplie de machines a creuser des galeries. 85 "taupes" sont actuellement utilisées pour des travaux d'installation de câbles, allant de la Grundomat 45mm a la Ditch Witch Jet Track.

Plus de la moitié des passages routiers ainsi qu'une proportion importante des opérations de remplacements de raccords utilisent des techniques d'excavation sans tranchées. La venue récente de la taupe dirigeable permettra la mise en plâce de cables basse tension plus longs.

Les développements futurs sont discutés ainsi que la réponse de l'entreprise envers la nouvelle législation routiere.

1. HISTORIQUE

La Midlands Electricity plc (MEB) est une des entreprises régionales d'électricité britannique récemment privatisées. Située au coeur de l'Angleterre, la société alimente plus de 2,1 millions de clients en électricité.

MEB exploite d'importants réseaux souterrains de câbles (35000km) et de lignes aériennes (27000km). Chaque année, elle installe environ 1900km de câbles souterrains, dont les différentes tensions et dimensions de câbles sont indiquées dans le tableau ci-dessous:

TENSION	LONGUEUR	DIAMETRE HORS TOUT
Raccords	(1250km)	19mm - 35mm
Basse tension	(350km)	35mm - 60mm
11kV	(300km)	60mm - 85mm

Environ 60% des câbles sont installés en terrain naturel (prés, nouveaux lotissements etc), et les 40% restants en terrain artificiel avec la majorité installée sous des trottoirs et le reste associé aux passages routiers.

Traditionnellement, la plus grande partie des travaux a été effectuée en creusant a la main, avec seulement une petite proportion effectuée a l'aide d'engins mécaniques.

Pour ces travaux, MEB emploie soit une main-d'oeuvre directe soit des sous-traitants, ces derniers étant généralement payés selon le volume métrique de la tranchée.

COUT DE CABLE BASSE TENSION WAVECON
Excavation à la main-terrain naturel
Coûts de raccords exclus

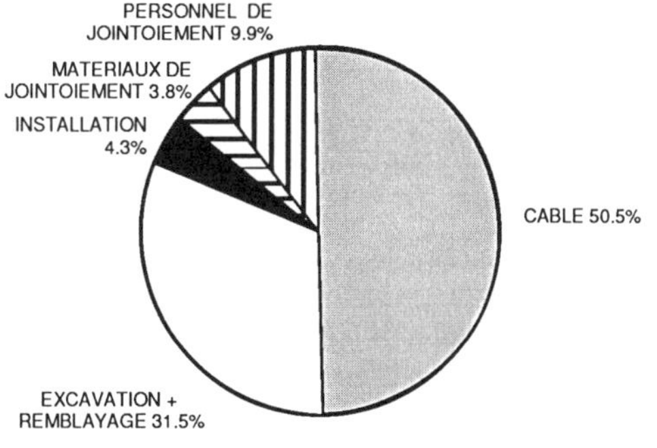

Fig 1

Le schéma (Fig 1) indique les coûts relatifs associés a l'installation de 1 metre de câble électrique en terrain naturel. L'installation, l'excavation et le remblayage comprennent la totalité des frais, c'est-a-dire la main-d'oeuvre, les matériaux etc.

Le schéma ci-dessous (Fig 2) illustre également l'installation de 1 metre de câble, cette fois en terrain artificiel. Les coûts sont ici bien plus élevés car sont inclus les frais de remise en place temporaire et permanente de la couverture du trottoir ou de la route.

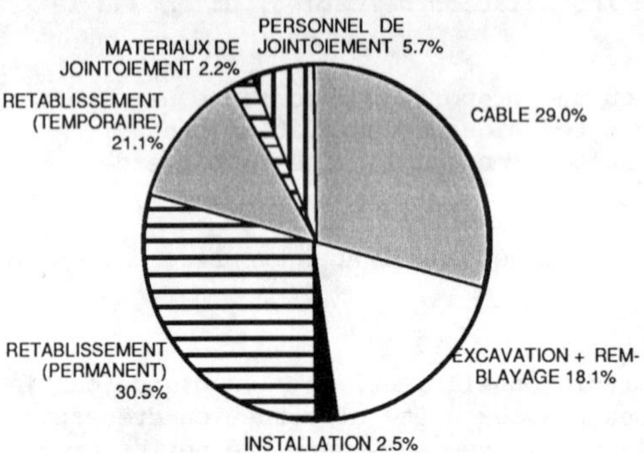

Fig 2

Traditionnellement, les travaux de rétablissement sont divisés en deux parties bien distinctes. Une fois le câble mis en place, MEB est responsable de la reconstruction temporaire de la tranchée, puis des que le tassement a eu lieu, les autorités locales entreprennent la reconstruction permanente.

La nouvelle législation routiere sera appliquée cette année en décembre, dont les effets principaux seront les suivants:

- Suite a une excavation, les entreprises assurant les services publics seront entierement responsables de la reconstruction de la rue et devront achever les travaux de reconstruction dans la mesure du possible dans les temps les plus brefs (au mieux en une fois).
- Les changements apportés au cahier des charges concernant la reconstruction exigent que toutes personnes responsables des travaux routiers devront certifier d'une maniere satisfaisante leur compétence dans l'entreprise de ces travaux.
- L'Office routier aura le pouvoir d'inspecter et recevra en échange de ce contrôle 6% des travaux entrepris. La non-conformité des travaux aux normes prescrites entraînera trois contrôles supplémentaires aux frais de l'entreprise de travaux publics. La non-conformité au contrôle final entraînera la responsabilité de l'entreprise et la ré-excavation et la reconstruction de la route ou du trottoir.

2. Estimation stratégique

Début 1990, la société a mis en place une petite équipe présidée par l'auteur du présent exposé, pour la revue:

- Des opportunités/des dangers présentés par la législation routiere imminente.
- Des opportunités associées a la mécanisation des travaux d'installation de câbles.

Les résultats de cette remise a jour sont résumés ci-dessous:

Avantages des méthodes traditionnelles
• Utilisent des méthodes bien établies. • Requierent un niveau de compétence relativement bas.
Opportunités
• Technologie d'excavation sans tranchées. • Tranchées étroites. • Davantage d'excavations classiques par engins mécaniques. • Moins de dérangement aupres de la circulation. • Reconstruction éventuelle en une seule opération (20% d'économies possibles).
Inconvénients des méthodes traditionnelles
• Utilisent les méthodes d'excavation les plus coûteuses. • Bouleversement important aupres des utilisateurs routiers et des piétons. • Qualité de la reconstruction.
Dangers
• Frais de non-conformité associés a la nouvelle législation routiere. • Formation supplémentaire exigée par la nouvelle législation. • Mauvaise image traditionnellement présentée par la technologie sans tranchées.

Notre analyse a révélé qu'en terrain artificiel la meilleure solution était de minimiser les aspects négatifs associés a la législation en réduisant le montant des travaux de reconstruction requis. En priorité et dans l'ordre des coûts potentiels les moins élevés, les options disponibles étaient les suivantes:

- Technologie sans tranchée

  Reconstruction limitée aux puits de lancement et de réception et aux regards d'inspection connexes.

- Tranchées étroites

  Réduisent les travaux de reconstruction en limitant la surface de la zone dérangée.

- Combinaison d'excavation machine et main

  Bien que cette méthode n'entraîne pas la réduction des coûts de reconstruction, on peut réaliser des économies dans les coûts globals d'installation en utilisant davantage les engins mécaniques.

  Les études supplémentaires et les nombreux essais effectués ont montré que la technologie sans tranchée offrait la meilleure solution pour l'installation de câbles en travers des routes. Elle présentait le double avantage de faire de grandes économies tout en réduisant considérablement les bouleversements occasionnés aupres de la circulation et du public.

  Malheureusement, la situation n'était pas si claire concernant les trottoirs, dû principalement aux facteurs suivants:

  (i) Le niveau de congestion causée par la quantité du matériel, voir Fig 3

  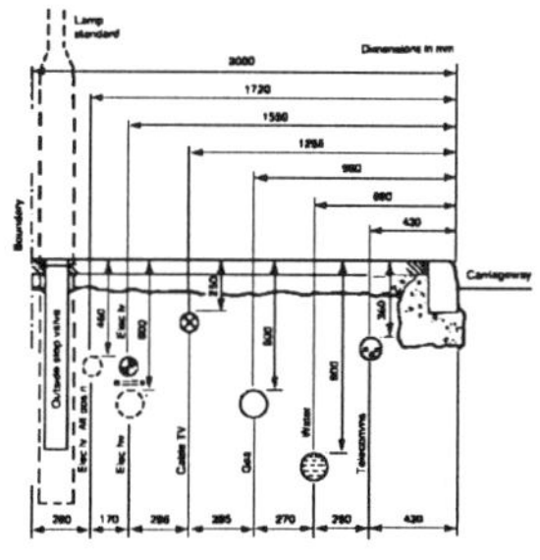

  Fig 3

  (ii) Le fait que les câbles électriques existants ne sont pas toujours placés en ligne parfaitement droite.

  (iii) Le degré d'exactitude offert par le matériel de repérage des câbles.

  (iv) Le degré d'exactitude des plans publics.

  (v) L'exactitude offerte par les machines a creuser les galeries.

En ce qui concerne l'utilisation du matériel taupe dans différentes activités, plusieurs opportunités ont été identifiées, a savoir:

(i) Les travaux de remplacement des raccords signifiaient généralement l'ouverture d'une tranchée dans les jardins privés, ce qui signifiait des travaux de reconstruction coûteux tels que le rétablissement des pelouses, le remplacement des plantes et la reconstruction des allées. La technologie sans tranchée offrait la parfaite solution a ce probleme.

(ii) En terrain naturel, des possibilités existaient également, a savoir traversées de rivieres et de canaux et parcs naturels.

3. STRATEGIE A PARTIR DE JANVIER 1990

Notre premiere préoccupation a été d'identifier dans chacune des sept divisions de la société la personne qui prendrait fait et cause de l'adoption de la technologie sans tranchée.

Pour chaque zone, nous avons ensuite équipé une équipe de taupes a percussion responsables initialement de l'installation de raccords de remplacement (voir Fig 4) et de la traversée de petites routes avec des câbles de basse tension.

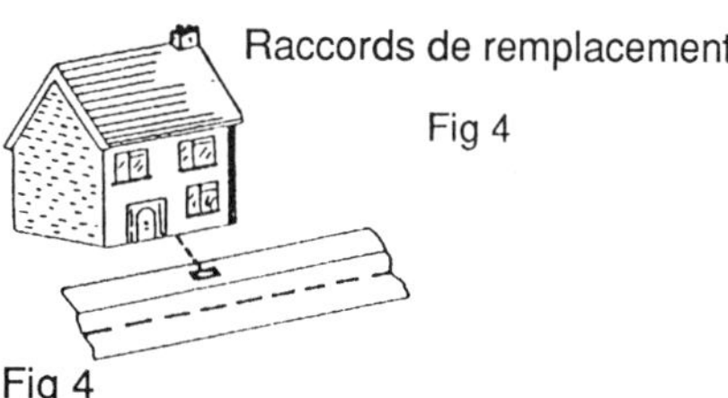

Fig 4

En juillet la premiere année, nous utilisions le matériel suivant:

17 marteaux de terrassement de 45mm
8 marteaux de terrassement de 65mm
1 marteau de terrassement de 95mm

A mesure que le taux de réussite s'améliorait, la main-d'oeuvre des entreprises montrait de plus en plus d'enthousiasme pour la technologie sans tranchée.

Au début de la deuxieme année, nous avons démarré les essais avec les engins extracteurs. Notre objectif était d'étendre les possibilités d'exploitation pour répondre aux besoins des traversées de routes plus complexes et pour la pose de câbles sous les trottoirs.

Au cours de cette période, nous avons acheté:

5 engins pneumatiques de 79mm
1 engin pneumatique de 95mm
1 engin pneumatique de 102mm
2 engins pneumatiques de 140mm

"Pigûre" - L'utilisation de transmetteurs électroniques dans la tête des taupes les plus grosses comme le modele Pierce Arrow 140, a permis la pose de câbles de basse tension directement dans le trottoir. Pour ce genre d'opération, deux options sont disponibles.

(i) La taupe est lancée a partir de la fosse de lancement et est dirigée le long d'une ligne droite pré-déterminée. Dès que la taupe dévie (cộntrole effectué par le matériel de repérage), elle est arretée et on creuse une nouvelle fosse de lancement autour de la taupe enterrée. La technique est renouvelée jusqu'a ce que la taupe atteigne la fosse finale de réception (voir Fig 5).

Fig 5

(ii) Comme pour (i) précédemment mais cette fois la taupe est lancée entre des trous pilotes préal ablement creusés (utilisés pour identifier l'emplacement du matériel) (voir Fig 6).

Fig 6

Environ 14 km de câbles ont été posés avec cette méthode.

Vers la fin de l'année 1991, notre attention s'est tournée vers deux domaines distincts: La fourniture de deux taupes de petit calibre pour les équipes de jointoiement, ainsi que l'acquisition d'un matériel d'alésage guidé,

Dorénavant, la technologie sans tranchée faisait parie du quotidien au sein de notre entreprise, avec plus de la moitié des traversées de routes effectuées avec succés avec cette technique.

En 1992, nous avons acquis notre premier matériel d'alésage guidé et nous sommes heureux d'annoncer qu'il n'a fallu que trois mois pour atteindre le seuil de rentabilité. Initialement, le Jet Track fut utilisé pour les travaux les plus complexes tels que les traversées de voies ferrées, de riviéres et d'autoroutes. A l'heure de la rédaction, nous utilisons actuellement ce matériel pour la pose de câbles de basse tension dans les trottoirs.

Quant aux équipes de jointoiement, toutes les nouvelles camionnettes seront équipées d'un compresser sous plancher et de taupes connexes (45 mm et 65 mm).

4. Résultats actuels

En fin 1992, il est estimé que le total des économies liées a la décision de l'entreprise d'adopter la technologie sans tranchée sera de l'ordre de 800,000

Livres Sterling. On peut ajouter a cela la réduction considérable des dérangements auprés de la circulation et du public grâce a cette technique.

Aujourd'hui, nous utilisons régulierement 85 taupes et notre intention est d'augmenter considérablement bientôt ce chiffre.

En utilisant la technologie sans tranchée, nous avons effectué jusqu'a présent les travaux suivants:-

3000 remplacements de raccords
1200 traversées de routes

et nous avons en sus posé environ 14 km de câbles de basse tension.

5. <u>L'avenir</u>

Nous avons l'intention d'étendre notre série de taupes au cours des six prochains mois. Notre préoccupation principale sera la provision de deux engins d'alésage guidé ainsi que l'achat d'une série d'engins non-dirigeables.

Nous avons l'intention de conformer a la nouvelle législation routiére 3 mois plus tot que prévu (Octobre 1992).

Un contrat révisé concernant la pose de câbles est prévu pour Octobre 1992. Selon les clauses de ce nouveau contrat, une proportion significant des travaux utiliserait les techniques sans tranchées (l'utilisation de taupes pour les traversées de routes devrait atteindre 95%).

Notre objectif est d'améliorer le profil de la technologie sans tranchée dans le domaine de la distribution en électricité.

A cette fin a eu lieu en juillet la premiere réunion annuelle de technologie sans tranchée regroupant les compagnies d'électricité. Notre but est d'offrir un forum commun des expériences vécues dans ce domaine et de présenter davantage nos besoins particuliers aupres des fabricants et des fournisseurs. Au cours de cette réunion, nous avons encouragé les participants a devenir membre de la ISTT.

Finalement, qu'en est-il de notre prochaine phase stratégique. Nous nous concentrerons sur deux domaines clé, a savoir:-

(i) Le repérage du matériel de taupe:-

- En ce qui concerne la haute technologie, cela signifie pou voir détecter a distance l'em placement exact d'une taupe lors de la traversée d'une route sans que l'opérateur ait besoin de marcher sur la route.
- A l'autre extrémité du marché, il nous faut une sonde robuste et précise qui puisse être mon tée sur la tête d;une petite taupe (45 mm ou 65 mm).

(ii) Un radar au sol nous permettant de travailler avec efficacité dans les trottoirs.

No Trenches in Town, Henry & Mermet (eds) © 1992 Balkema, Rotterdam. ISBN 90 5410 085 0

The pipe renovation market: Inventory of regulations

Bergue
Service Technique de l'Urbanisme, France

Joussin
Société Eternit, France

Lagubeau
Société Denys Telerep, France

Ferrier
Société des Tuyaux Bonna, France

ABSTRACT : This paper :
• first, makes an inventory of the regulations related to potable water, gas and industrial fields, based on the results of an international survey,
• second, lists and comments the certification or approval procedures covering processes and contractors,
• lastly, proposes the ground work for a draft French charter.

We should like to focus on the provisions made in the various countries most active in renovating their existing pipe systems using trenchless processes or techniques, to bring order to this market. It would seem difficult, to say the least, to organize such an effervescent field, with a great many ideas coming and going all the time. But we all feel that principals interested in finding out about these techniques, attractive for the environment, would undoubtedly shape their desires into orders more often if, one way or another, recognized specialists sorted them out and exercised some supervision over them. Since our country cannot claim to be a forerunner in becoming aware of the ecological aspects of pipe systems management, it is important to take a look at what our neighbours are doing, to see if this need has been taken into account and whether there are ways of meeting it. A quick survey was made of the partners affiliated with our ISTT organization, and likewise of organizations known to have undertaken such action.

List of organizations and associations surveyed : ISTT, GSTT, NSTT, NASTT, WRc in England, RAL in Germany and for France : CSTB, CEBTP, AGHTM, GDF, etc.

Our text is based on the results of this survey and also on documents and knowledge obtained by the authors. During the oral presentation, you may note certain differences, resulting from further knowledge obtained between the end of June and the date of the No Dig session in Paris.

The following table (table 1) sums up the knowledge we possess.

Rules are of widely different natures. They are of course laid down in accordance with standards where standards exist, and are applicable to renovation in each country.

In Great Britain, the WRc is the technical reference for the assessment of renovating processes for water and sewage disposal networks. It has also developed specifications for materials and for their implementation procedures. However, only Certifying Agencies, such as BSI, WICS, etc are entitled to approve the techniques. The WRc also furnishes a list of specialized contractors, but does not give the contractor a seal of approval.

In Germany, standards are proposed for each family of renovation process by the ATV. Furthermore, specialized contractors have been certified by the " Gueteschutz Kanalbau" association since 1990. Some contractors have this seal of approval and a great many files are being examined.

For water and gas, the operators have got together to organize the DVGW. This

Table 1. Certifying agency

Country	Field	Concerning processes	Concerning contractors
Date	June 26, 1992		
United Kingdom	Drinking water	WICS C/O WRc	WRc
	Sewage disposal	WICS C/O WRc	WICS C/O WRc
	Gas		
	Industry		
Germany	Drinking water		
	Sewage disposal		Guetschutz Kanalbau
	Gas		
	Industry		
France	Drinking water		
	Sewage disposal	AT of the CSTB	
	Gas	AGHTM	
	Industry		

organization publishes valid recommendations on renovation matters, and gives official approval for processes.

In other countries, regulation has not gone as far as in the countries mentioned.

In France, several organizations or associations have contributed to laying the groundwork for regulations.

These are :

- AGHTM, with a collection of technical recommendations applicable to the work of renovating sewers,
- CSTB, with the Technical Agreements that are designed to approve the capability of the process to be used, when this process or its use is new and has not been standardized yet,

or operators, such as :

- GDF, with procedures for authorizing use of gas conduits.

It must be specified that in France, the law compels the contractor to be responsible for the operation of the work laid down for 10 years. Since this provision applies to renovation work, it may easily be understood why we are interested in completing, or even institutionalizing this regulatory groundwork.

The overall situation of the French renovation market being what it is, it would be helpful to think about setting up an ambitious quality policy in this field, to convince the public that these young solutions provide a mature alternative to out-and-out replacement of pipe systems. It is first in the well understood interest of the renovators, second of the prime contractors, to answer the legitimate concerns and doubts of the owner of the works. The action may take the form of setting up a seal of approval given to the renovator-process tandem.

The seal of approval would be based on the voluntary adhesion of contractors to a " Pipe renovator charter" the main ideas of which would be the following, illustrated by extracts from the project being set up by the A 9 SG 2 shop of the FSTT :

A - Contractors specializing in pipe renovation in France agree to respect the following quality rules, in all fields and for the processes and techniques that they use :

"Criterion 3 - Appendix 1 : Shape of the file
As an indication, the application file for the seal of approval must specify in particular :

• for the contractor's firm and the industrial plant

1 - The contractor's quality organization plan

2 - The quality assurance plan of the plant operating in the pipe renovation field.
3 - The list of names of the specialized operators.

- for techniques and processes

1 - List of the renovation processes and techniques used and also a descriptive file for each of them, of operating procedures.
2 - The test procedures given to laboratories or test centres concerning : the materials used and the work accomplished.
3 - The quality assurance plan appertaining to each process or technique, noting, for example, the critical phases for quality and the measures taken to avoid these risks.
4 - The references acquired in France by the contractor for each family of process :

- (TC) short tube tubing,
- (TL) long tube tubing,
- (TE) tubing after burster,
- (MT) "mange tube" (tube eater),
- (TS) spiral strip tubing,
- (CH) sheathing,
- (MI) inside sleeve,
- (PMe) machine spraying of mortar or resin,
- (PMa) hand spraying of concete, mortar or resin,
- (IJ) point by point injection of joints or cracks,
- (EP) prefabricated unit,
- (RP) point by point repairs,
- (AU) other."

B - The certifying agency is of mixed composition with several disciplines represented, a group of individuals concerned with renovation.

"Article 3 : Seal of approval
Any contractor or plant meeting the criteria spelled out in Article 1 may request the France pipe renovator seal of approval from the certifying agency. This seal of approval will be issued by the certifying agency on the basis of the criteria spelled out in Appendix 1 and on the technical criteria designed to check the appropriateness for use of each process or technique subject to grant of the seal of approval. The commission constituting the certifying agency must include at least 10 members chosen among the public or semi-public owners, private and public prime contractors, industrial suppliers of pipes and of the chemistry of composite materials, and pipe renovators."

C - This seal of approval is subject to inspection and may be withdrawn.

"Criterion 1 - Appendix 4 : Competency
The certifying agency alone is competent to examine withdrawals or suspensions of the seal of approval. The Agency may obtain confirmation from a commission of surveyors, designate an auditor, etc., for the purpose of informing its discussions. Opinions expressed after hearing the contractor concerned may not be appealed. The certifying agency only acts for the purpose of providing users of renovated pipes with a high level of reliability and quality. Its role is therefore mainly pedagogical where sanctioned contractors are concerned, so as to promote an even level of quality and competence within the profession."

D - Examining and arbitrating problems met with. :

"Criterion 2 - Appendix : Inquiry application
Any owner or prime contractor may apply to the certifying agency for an inquiry to be made into any technical problem or dispute between himself and a contractor with the seal of approval .If necessary, the certifying agency names one of its members or designates one or more surveyors to track down the causes and propose an arbitrated solution for the purpose of settling the dispute amicably. This procedure avoids isolating the principal faced with a difficulty and allows pertinent information drawn from an incident to be taken into account in the procedures worked out by the contractor."

E - Problems yet to be solved

The unanswered questions that we submit to our national audience are :
- Who will be the certifying agency ?
- What means might be available when it is a question of having an audit made to check a contractor's statement, or to settle the expenses tied to arbitration ?
- Which authority can take the liberty of

promoting or sanctioning a commercial enterprise in a sector enjoying free competition ?
- How are manufacturing secrets and procedures to be protected among contractors belonging to the certifiying agency ?
- How are the actions of certifying agencies to be coordinated in Europe ?

To conclude this talk with a list of problems yet to be solved may sound pessimistic. We shall therefore emphasize the fact that work is under way, and that contractors have understood that it will be useful to convince principals that pipe renovation techniques, used by competent renovators, are today part of the efficient and ecologically valid solutions to the problem of maintenance work on the vast pipe systems inherited from over past generations.

Appendix : Addresses of the organizations mentioned

RAL - Gueteschutz Kanalbau, W-Gartenstr. 9, 5340 Bad Honnef (Germany)
DVGW (Deutscher Verein des Gas und Wasserfaches), Mergenthalerallee 27-29, W-6236 Eschborn 1 (Germany)
ATV (Abwasser Technische Vereinigung), Markt 71, Postfach 11-60, W-5205 St Augustin 1 (Germany)
AGHTM (Association Générale des Hygiénistes et Techniciens Municipaux), 9, rue de Phalsbourg, 75017 Paris (France)
CSTB (Centre Scientifique et Technique du Bâtiment), 84, avenue Jean-Jaurès, BP 2, Champs-sur-Marne, 77421 Marne la Vallée Cedex 2 (France)
GDF/DETN (Gaz de France - Direction des Etudes et Techniques Nouvelles), 361, avenue du Président Wilson, BP 33, 93211 La Plaine Saint Denis Cedex (France)
WRc Water Research Centre, Swindon,P.O. Box 85, Frankland Road, Blagrove, Swindon, Wiltshire SN 5 8YR (Great Britain)

No Trenches in Town, Henry & Mermet (eds) © 1992 Balkema, Rotterdam. ISBN 90 5410 085 0

Le marché de la réhabilitation: État de la réglementation

Bergue
Service Technique de l'Urbanisme, France

Joussin
Société Eternit, France

Lagubeau
Société Denys Telerep, France

Ferrier
Société des Tuyaux Bonna, France

ABSTRACT : La communication :

- dans un premier temps, fait le constat de l'état de la réglementation concernant le domaine de la réhabilitation des conduites assainissement, eau potable, gaz et industrie en s'appuyant sur les résultats d'une enquête internationale,
- dans un deuxième temps, détaille et commente les procédures de certification ou d'agrément couvrant les procédures et les entreprises,
- enfin, sont proposées les bases pour un projet de charte française.

La communication se propose de faire le point, dans les différents pays les plus actifs dans le domaine de la réhabilitation de leur patrimoine de réseaux par des procédés ou techniques sans tranchées, des dispositions prises pour organiser ce marché. L'organisation, d'un domaine aussi bouillonnant, avec l'apparition et la disparition de nombreuses idées, semble une gageure. Mais, nous ressentons tous que les décideurs, intéressés et curieux de ces techniques séduisantes pour l'environnement, concrétiseraient sans doute plus souvent leurs désirs en ordres si, d'une manière ou d'une autre, un tri et un contrôle étaient faits par des spécialistes reconnus. Notre pays n'étant pas précurseur dans cette prise de conscience écologique de la gestion des réseaux, il est important de jeter un oeil sur nos voisins, afin de vérifier si ce besoin a été pris en compte et s'il existe des réponses. Une enquête rapide a été réalisée auprès des partenaires affiliés à notre organisation ISTT, et, parallèlement, auprès d'organismes notoirement connus pour avoir entrepris une telle démarche.

Liste des organismes et associations interrogés : ISTT, GSTT, NSTT, NASTT, WRc en Angleterre, RAL en Allemagne et, pour la France, CSTB, CEBTP, AGHTM, GDF, etc.

Le texte que nous vous communiquons est issu des résultats de cette enquête et aussi des documents et connaissances obtenus par les auteurs. Il sera peut être constaté, lors de l'exposé oral, certaines différences qui seront le fruit d'un enrichissement de ces connaissances intervenu entre fin juin et la date de la cession No Dig à Paris.

Nous résumons dans le tableau 1 qui suit les informations que nous possédons.

De nature très différentes, les règles sont édictées, bien sûr, en accord avec les normes, lorsque celles-ci existent, et sont applicables à la réhabilitation dans chaque pays.

En Grande-Bretagne, le WRc est la référence technique pour l'évaluation des techniques et procédés de réhabilitation des réseaux dans le domaine de l'eau et de l'assainissement. Il a développé de nombreuses spécifications pour les matériaux et leurs procédures de mise en oeuvre. Ce sont cependant des agences de certification telles que BSI, WICS,... qui approuvent les techniques. Le WRc fournit aussi une liste d'entreprises spécialisées, mais sans fournir de label à l'entreprise.

En Allemagne, les normes sont proposées pour chaque famille de procédés de réhabilitation par l'ATV. Par ailleurs, la certification des entreprises spécialisées est effectuée par l'association "Gueteschutz

Tableau 1. Organismes "certificateurs"

Pays	Domaines	Concernant procédés	Concernant entreprises
Date	26/06/92		
Royaume Uni	Eau potable	WICS C/o WRc	WRc
	Assainissement	WICS C/o WRc	WICS C/o WRc
	Gaz		
	Industrie		
Allemagne	Eau potable		
	Assainissement		Gueteschutz Kanalbau
	Gaz		
	Industrie		
France	Eau potable		
	Assainissement	AT du CSTB	
	Gaz	AGHTM	
	Industrie		

Kanalbau" depuis 1990. Quelques entreprises ont ce label et beaucoup de dossiers sont en cours d'instruction.

Pour l'eau et le gaz, les exploitants se sont réunis dans un organisme DVGW. Cet organisme édite des recommandations valables en matière de réhabilitation et donne des agréments pour les procédés.

Dans les autres pays, la réglementation n'est pas aussi avancée que pour les pays cités.

En France, plusieurs organismes ou associations ont contribué à poser les bases d'une réglementation.

Il s'agit de :

- AGHTM avec un recueil de recommandations techniques applicables aux travaux de réhabilitation des réseaux d'assainissement,
- CSTB avec la procédure des avis techniques qui concerne l'aptitude à l'emploi de procédés lorsque leur nouveauté ou celle de l'emploi qui en est fait n'en permet pas encore la normalisation,

ou les exploitants comme :

- GDF avec ses procédures d'autorisation d'emploi pour les conduites de gaz.

Il faut préciser qu'en France, la loi oblige l'entreprise à être responsable de la fonctionnalité de l'ouvrage construit pendant 10 ans. Cette disposition s'appliquant aux travaux de réhabilitation, on comprend l'intérêt de compléter, voire d'institutionnaliser, ces bases réglementaires.

Il s'avère qu'au vue de la situation globale du marché français de la réhabilitation, il est utile de réfléchir à la mise en place d'une ambitieuse politique de qualité dans ce domaine afin de convaincre de la maturité de ces jeunes solutions alternatives au remplacement pur et simple des conduites. C'est l'intérêt bien compris d'abord des réhabiliteurs, ensuite des maîtres d'oeuvre, de répondre aux interrogations et aux doutes légitimes des maîtres d'ouvrage. La démarche peut se concrétiser par la création d'un label qui serait décerné pour le couple : réhabiliteur-procédé.

Le label s'appuierait sur l'adhésion volontaire des entreprises à une "Charte du réhabiliteur de canalisations" dont les procédés seraient les suivants, illustrés par des extraits du projet en cours de réalisation au sein de l'atelier A9 SG2 de la FSTT.

A - Les entrepreneurs spécialisés dans la réhabilitation des conduites en France, dans tous les domaines pour les procédés et techniques qu'ils mettent en oeuvre, s'engagent à respecter les règles de qualité suivantes :

"Critère 3 - Annexe 1 : Forme du dossier.
Le dossier de demande d'obtention du label, à tire indicatif, devra préciser notamment :
• pour l'entreprise et l'établissement :
1. Le plan d'organisation de la qualité dans l'entreprise.
2. Le plan d'assurance qualité de l'établissement exerçant dans le domaine de la réhabilitation des conduites.
3. La liste nominative des opérateurs spécialisés.
•pour les techniques ou procédés :
1. La liste des procédés et techniques de réhabilitation mis en oeuvre ainsi qu'un dossier descriptif pour chacun d'eux des procédures de mise en oeuvre.
2. Les protocoles d'essais confiés à des laboratoires ou centres d'essais concernant les matériaux employés et l'ouvrage réalisé.
3. Le plan d'assurance qualité afférent à chaque procédé ou technique en notant par exemple les phases critiques pour la qualité et les dispositions prises pour éviter les risques.
4. Les références acquises en France par l'entreprise pour chaque famille de procédés :

- (TC) tubage par tube court,
- (TL) tubage par tube long,
- (TE) tubage après éclateur,
- (MT) mange tube
- (TS) tubage par bande spirale,
- (CH) chemisage,
- (MI) manchette intérieure,
- (PMe) projection mécanique de mortier ou de résine,
- (PMa) projection manuelle de béton, mortier ou résine,
- (IJ) injection ponctuelle d'étanchement de joint ou fissure,
- (EP) élément préfabriqué,
- (RP) réparation ponctuelle,
- (AU) autre."

B - Le certificateur est mixte et pluridisciplinaire regroupant des acteurs concernés par la réhabilitation.

"Article 3 : Label
Toute entreprise ou établissement répondant aux critères exposés à l'article 1 peut demander au certificateur l'obtention du label : Réhabiliteur de canalisations de France.
Ce label sera délivré par le certificateur sur la base des critères détaillés dans l'annexe 1 et sur des critères techniques destinés à vérifier l'aptitude à l'emploi de chaque procédé ou technique soumis à labellisation. La commission constituant le certificateur devra compter un minimum de 10 membres choisis parmi des maîtres d'ouvrage publics et semi-publics, des maîtres d'oeuvre privés et publics, des fournisseurs industriels de canalisations et de la chimie des matériaux composites et de réhabiliteurs."

C - Le label est contrôlé et peut être retiré

"Critère 1 - Annexe 4 : Compétence
Le certificateur est seul compétent pour l'examen des retraits ou des suspensions du label. Il peut s'appuyer sur une commission d'experts, désigner un cabinet d'audit, etc, en vue d'éclairer ses délibérations. Les avis signifiés après l'écoute de l'entreprise concernée ne sont pas susceptibles d'appel. Le certificateur n'agit que dans le but d'assurer aux utilisateurs des conduites réhabilitées un niveau de fiabilité et de qualité élevé. Son rôle est donc essentiellement pédagogique envers les entreprises sanctionnées afin de promouvoir, au sein de la profession, un même niveau de qualité et de compétence."

D - Examen et arbitrage des problèmes rencontrés

"Critère 2 - Annexe 4 : Saisie
Tout maître d'ouvrage ou maître d'oeuvre peut saisir le certificateur de tout problème technique ou litige l'opposant à une entreprise labellisée. Si nécessaire, le certificateur nomme parmi ses membres ou désigne un ou des experts chargés de rechercher les causes et proposer une solution d'arbitrage en vue d'aboutir à un règlement amiable du conflit. Cette procédure permet de ne pas isoler le donneur d'ordres face à une difficulté et de permettre de tenir compte des enseignements utiles à tirer d'un incident dans les procédures mise au point par l'entreprise."

E - Problèmes à résoudre

Les questions non résolues que nous soumettons à notre auditoire national sont :

- Qui sera le certificateur ?
- Quels pourront être ses moyens quand il s'agira de charger un cabinet d'audit de vérifier les déclarations d'une entreprise, ou de régler les frais liés à l'arbitrage ?
- Quelle autorité peut se permettre de promouvoir ou de sanctionner une entreprise commerciale dans un secteur de libre concurrence ?
- Comment protéger les secrets de fabrication et les procédures entre les entrepreneurs membres de l'organisme certificateur ?
- Comment coordonner les actions des certificateurs en Europe ?

Conclure cette communication sur l'énoncé des problèmes à résoudre peut paraître pessimiste, nous mettrons donc l'accent sur le fait que la démarche est en route et que les entreprises en ont compris tout l'intérêt pour convaincre les donneurs d'ordres que les techniques de réhabilitation des conduites, mises en oeuvre par des réhabiliteurs compétents, font aujourd'hui partie des solutions performantes et écologiques pour l'entretien du vaste patrimoine de réseaux constitué au cours des générations précédentes.

Annexe : Adresses des organismes mentionnés

RAL - Gueteschutz Kanalbau, W-Gartenstr. 9, 5340 Bad Honnef (Allemagne)

DVGW (Deutscher Verein des Gas und Wasserfaches), Mergenthalerallee 27-29, W-6236 Eschborn 1 (Allemagne)

ATV (Abwasser Technische Vereinigung), Markt 71, Postfach 11-60, W-5205 St Augustin 1 (Allemagne)

AGHTM (Association Générale des Hygiénistes et Techniciens Municipaux), 9, rue de Phalsbourg, 75017 Paris (France)

CSTB (Centre Scientifique et Technique du Bâtiment), 84, avenue Jean-Jaurès, BP 2, Champs-sur-Marne, 77421 Marne la Vallée Cedex 2 (France)

GDF/DETN (Gaz de France - Direction des Etudes et Techniques Nouvelles), 361, avenue du Président Wilson, BP 33, 93211 La Plaine Saint Denis Cedex (France)

WRc Water Research Centre, Swindon,P.O. Box 85, Frankland Road, Blagrove, Swindon, Wiltshire SN 5 8YR (Grande Bretagne)

No Trenches in Town, Henry & Mermet (eds) © 1992 Balkema, Rotterdam. ISBN 90 5410 085 0

Rehabilitation of sewerage systems in Scandinavia – New objectives and new technology

Erling Holm
I. Krüger AS, Copenhagen, Denmark

ABSTRACT: In the following, we will be reviewing the rehabilitation and sewerage systems in Scandinavia in the nineteen eighties and looking at our plans for the nineties.

Today, we are more prepared to make an effort to increase the number of rehabilitations in all fields. The extent of rehabilitation work will undoubtedly increase even more in the nineties than it did in the eighties. This applies to examination methods, model computations and implementation.

Additionally, a recent Danish survey indicates that sewerage condition is not as serious as previously thought: thus, there is time to set up specific objectives for rehabilitation efforts. These objectives should be adjusted to the resources, to the wishes of the population, and to political issues.

A NEW PERIOD

Many sewerage systems in Denmark, Norway and Sweden have been extended with interceptors; and during the last ten years, parts of the sewerage systems have been rehabilitated because their performance was no longer satisfactory. However, these rehabilitations have not been nearly as extensive as predicted at the beginning of the eighties.

There are many reasons for this. In the early eighties, very little was known about the condition of the pipes. CCTV (closed circuit TV-inspection) in old sewerage systems had just begun. - The possibilities for a total evaluation of the sewerage system's capacity and of pollution loads were very moderate - The use of computers had only just made its entry - Renovation techniques for rehabilitation without inconvenient excavation had just been introduced on the market.

The surveys which led to the prophecies that, "we are facing a mountain of rehabilitation work", "threatened by breakdown of the infra-structure" and "enormous investments" were not good or comprehensive enough.

Today, the situation is different. Many techniques have overcome "beginner difficulties" and the use of computer-based data bases and models has become widespread. Moreover, it is more and more common for both public and private companies to work towards fixed maintenance and economic objectives.

NEW OBJECTIVES

The first years of rehabilitation were dominated by measures to rectify acute problems such as pipe collapse and chronic basement flooding.

In several municipalities and especially in the larger ones, investigation and planning of more comprehensive rehabilitation began early. This work has been most revealing as to what should - and should not - be done. Thus, this pioneer work has also been very profitable - in regards to new work methods as well as computational developments (computer-based).

A very important issue has been the need to assess priorities and whether they should relate to the limiting of financial or personnel ressources.

In many places, assessing priorities was very difficult until objectives were defined. General as well as detailed objectives were defined and used to determine the order in which to solve the sewerage system problems. For municipalities, the advantages of this method are that overall evaluation is a natural consequence and that sewerage system's priority can be ranked relative to the municipality's other responsibilities. This also applies to their financial situation and to their choice of solutions. Moreover, it is then natural to integrate the needs and wishes of the inhabitants, and to use the experience gathered from implementation in the planning.

Thus, the planning becomes a tool - not an objective in itself.

The objectives are listed below in the most general terms:

* Clean recipients

Obtain a clean water environment in streams, lakes and coastal areas, and, for bathing, at the sea shore. Specific objectives for the recipients are decisive factors in the designing of basins and overflows.

* Safe drainage of wastewater and storm water

Minimize the chance of operation failure in the sewerage system during rain or thawing periods and the consequences thereof such as basement and terrain flooding. Objectives, which are ranked according to the desired degree of security determine, among other things, when to rehabilitate a sewer pipe.

* Safe work environment

Solutions which provide a satisfactory and safe work environment - and in which operation is best and easiest.

* Good overall economy

Achieve optimum solutions with minimal system and maintenance costs and simultaneously obtain a satisfactory environment, choice of technical solutions, choice of service level to the inhabitants and a sound economic situation.

Within each of these objectives (and other more general objectives), it is possible to set up detailed, measurable objectives. For example: flooding / damming (of 0.2 meters under basement level/under manhole cover) can be accepted at most once every five years. Such an objective can be planned for by means of, for example, the computerized Mouse pipe models.

DANISH SURVEY OF NEED FOR REHABILITATION

In Denmark, a comprehensive survey to quantify the need for rehabilitation of sewerage systems has recently been

Figure 1: Setting objectives is an important process.

completed. The survey also presents the economic aspects of improving today's sewerage system to achieve a satisfactory level based on the current knowledge of the condition of pipes and structures.

The survey is based on information as to the extent and condition of the sewerage system, and the realized and expected rehabilitation investments from about half of Denmark's municipalities.

Moreover, the calculations, which are performed with successive calculation models, are based on the fact that the survey has shown that many municipalities have only recently begun using their objectives and demands in order to obtain the desired standard.

The level of demand entirely determines the economic situation, both at the individual municipality and at country level. For example, consider the following objectives for choosing pipe tightness:

1) All sewage pipes must be tight for waterpressure

2) Pipes must be watertight where leaking could cause crucial loading on the wastewater treatment plant/recipient, or where leaking could cause pollution of groundwater used for drinking.

The general objective is about 75 per cent more expensive than the objective with limitations.

The survey, which includes a summary of the number of pipes, overflow structures, pumping stations, basins, separators etc. proves that financial key numbers and key numbers for the sewer condition and quantities are of decisive importance for sound and effective rehabilitation. This applies to each individual municipality as well as to the country as a whole.

Regarding the condition of sewerage systems, it is essential to note that, as yet only a very small portion of local sewerage systems have been investigated. About 15 percent of the pipes have been examined with CCTV and there are very few computation models to determine the possibilities for the abatement of flooding and pollution.

Here in Denmark, there are very few problems serious enough to require immediate attention. Thus, field investigations and rehabilitation planning take high priority for the next few years; the basis of decision will thereby be improved both for the individual municipalities and for the country as a whole.

Today, several Danish municipalities face problems with pollution, overloading of recipients, maintenance and excessive expenses for overflow structures, basins, pumping stations and separators.

* Costs

Danish municipalities have used an average 55 DKK per inhabitant per year during the last five years for rehabilitation of sewerage systems (this amount does not include expenses for new constructions and extensions of wastewater treatment plants). Currently the municipalities have budgeted increasing this amount to the range of 130-155 DKK per inhabitant per year in the coming ten years.

The survey however, indicates that current needs require investments of 200-400 DKK per inhabitant per year for the next 20 years. This estimate is based on using the first five years primarily for planning and, thereafter, in the following 15 years, carrying out extensive rehabilitation.

Figure 2 shows the disposition of investments on

* pipes and manholes
* pumping stations
* overflow structures
* basins
* separators
* SCADA (Supervisory Control And Data Acquisition)
* service pipes (municipal part)
* planning and investigations
* design and inspection

The investments in basins are necessary to fulfil demands on the quality of the recipient. The investments in service pipes are necessary as more and more Danish municipalities are responsible for the service pipes lying in public roads and pavements.

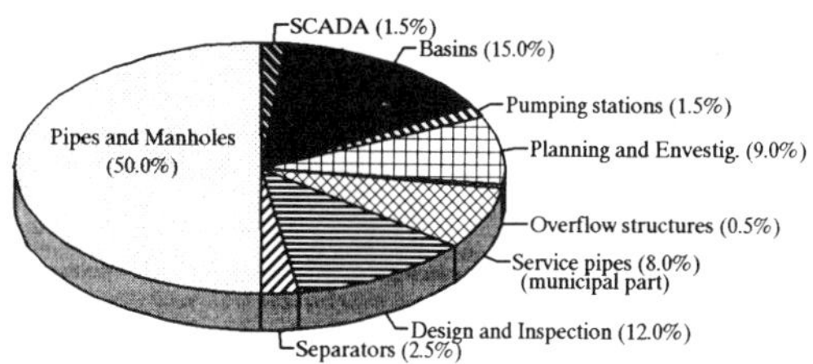

Figure 2: Disposition of expenses for rehabilitation in Denmark.

NEW TECHNOLOGY

* Computer models

Computer models have significantly improved the possibilities for carrying out a plan based on demands for improved protection of recipients, and for a better, more economical maintenance of the sewerage systems.

The "Mouse" computer model makes it possible to evaluate the extent of flooding /damming and the extent of different kinds of pollution in extreme rain situations as well as on a yearly basis. There are also computer models that simulate operations, calculate loadings of wastewater treatment plants during rain and then calculate oxygen depletion in streams.

Thus, it is possible to estimate the entire sewerage systems' operation, especially during rain and thawing periods.

* Surveys

Closed circuit TV inspection (CCTV) has developed considerably during recent years with, among other things, improved cameras, colour filming, video techniques, and better reports which are well-suited to direct application and storage in data bases.

The possibilities for data recording and co-ordination of large amounts of data have made it possible to employ a variety of data for planning, without loosing sight of the general picture. Risk analyses can also be conducted to compare the many survey results, model calculations and experience, and thereby rank plans according to probabilities and consequences of damages. (Not loosing sight of the required objectives and demands.)

* Pipe Renovation

The market for renovation has changed drastically during the last few years. In the Scandinavian countries, the number of companies specializing in these methods has

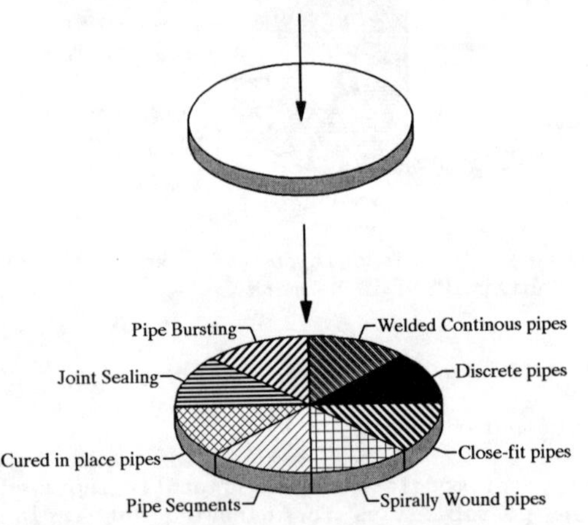

Figure 3: Today, there are many different renovation methods.

more than tripled - but the demand has not noticeably grown.

In Norway and especially in Sweden, lining with discrete pipes and welded continuous pipes of polythylen dominates the market. This is because a large part of the renovations are carried out by the municipalities themselves.

In Denmark and Norway, cured in place pipes (flexible linings) and repairs of pipes are performed to a great extent via joint sealing. In recent years Denmark has made many renovations by pipe bursting.

The Danish municipalities with the most extensive rehabilitation needs are also the most willing to use the renovation methods.

It is interesting to note that most of the renovated pipes were constructed during the period 1930 to 1975. Thus, it is not only older pipes but, to a great extent, also newer pipes which require rehabilitation.

There are many reasons for using renovation methods. The most frequent reason is the avoidance of digging and thereby minimizing inconvenience.

People who use the renovation methods are very uncertain of the quality. So even if costs is a very decisive issue when choosing between renovation and replacement, more objective knowledge regarding the methods' durability, functionality, and material parameters is necessary.

THE FUTURE

Today's Scandinavian market has a great variety of renovation methods. In years to come, quality assurance demands will be tightened and there will not be a market for many methods. The demand for improved methods to renovate service pipes and manholes will increase, in order to carry out renovation as a whole.

Today, experience with cured in place pipes (flexible linings), pipe bursting, lining with discrete pipes and welded continuous pipes is extensive in Scandinavia: The Scandinavian guidelines for choosing and designing renovation methods, written in 1989, can be used advantageously in other countries.

Likewise, many of the Scandinavian computer models for the calculation of hydraulic capacity of pipe systems and of pollution to recipients through overflow structures and wastewater treatment plants, can be advantageously applied elsewhere.

The possibilities for evaluating the need for rehabilitations throughout an entire sewerage system and for evaluating the consequences of different solutions are much greater today than just ten years ago.

No Trenches in Town, Henry & Mermet (eds) © 1992 Balkema, Rotterdam. ISBN 90 5410 085 0

Rénovation des systèmes d'assainissement en Scandinavie – Nouvelle époque, nouveaux objectifs, nouvelle technique

Erling Holm
I. Krüger AS, Copenhagen, Danemark

RÉSUMÉ: Revue des travaux de rénovation des systèmes d'assainissement pendant les années quatre-vingts et vue sur ceux prévus pendant les années quatre-vingt-dix.

Nous sommes aujourd'hui, sur toutes les lignes, mieux préparés pour l'enjeu - les innovations dans le domaine d'études, de simulation/modélisation et d'exécution de rénovations seront sans doute plus nombreuses pendant les années quatre-vingt-dix que pendant les années quatre- vingt.

Il ressort par ailleurs d'une récente étude décrivant l'état actuel des systèmes d'assainissement - qu'il nous reste encore du temps pour fixer des objectifs concrets nécessaires quant aux travaux de rénovation. Ces objectifs doivent pourtant être adaptés aux ressources disponibles et aux désirs des habitants et des hommes politiques.

NOUVELLE EPOQUE

Pendant ces derniers 10 ans de nombreux systèmes d'assainissement au Danemark, en Norvège et en Suède ont été complétés par des canalisations coupantes et certaines parties des systèmes ont été modernisées vu que le fonctionnement n'a pas été satisfaisant.

Les rénovations réalisées ces derniers 10 ans n'ont cependant pas pris les dimensions prévues au début des années quatre-vingt.

Il y a plusieurs raisons - au début des années quatre-vingt l'état de ces canalisations ne recevra que peu d'attention. La surveillance télévisée des anciens systèmes d'assainissement n'avait que vu le jour et les possibilités d'obtenir une évaluation d'ensemble de la capacité des systèmes et des décharges étaient limitées. En effet, l'application des moyens informatiques de même que des techniques permettant une rénovation des systèmes d'assainissement sans de grandes travaux de déterrement avait justement commencé.

Les prédictions déclarant que nous nous trouvions face à une "montagne de rénovations" menaçant de détruire l'infrastructure existante et face aux énormes investissements n'étaient donc pas bien fondées.

La situation d'aujourd'hui est bien différente. La plupart des diverses techniques introduites ont prouvé leur viabilité et l'application de moyens informatiques sous forme de bases de données et de simulation/-modélisation est devenue une partie intégrée et naturelle des projets. Il sera d'ailleurs de plus en plus courant que les institutions publiques et les entreprises privées suivent des objectifs fixes quant aux services offerts et aux divers aspects financiers.

NOUVEAUX OBJECTIFS

Les premières rénovations étaient bien naturellement caractérisées par des actions d'urgence rendues nécessaires suite à des ruptures de canalisations et des inondations répétées de caves.

Plusieurs municipalités (surtout les grandes) ont tôt constaté le besoin d'études et de planification au cas de travaux de renovation considérables. Par l'intermédiaire des travaux de rénovation on a d'ailleurs pu acquérir des expériences utiles concernant ce qu'on doit faire et ce qu'on ne doit pas faire. Ces travaux pionnier ont donc contribué au développement de nouvelles méthodes pour l'exécution des rénovations et à l'application des techniques informatiques.

La hiérarchisation des priorités est un aspect essentiel dans cet ordre d'idées soit à cause des fonds limités disponibles soit à cause des ressources de personnel limitées.

Dans de nombreuses municipalités cependant la hiérarchisation a cependant été très compliquée jusqu'à ce qu'on ait commencé de fixer des objectifs concrets - càd fixer les objectifs généraux ainsi que les objectifs plus détaillés indiquant l'ordre selon lequel une solution devra être apportée aux divers problèmes du système d'assainissement. Pour les municipalités ce procédé est avantageux parce qu'il permet une évaluation d'ensemble et une hiérarchisation des priorités quant au système d'assainissement faisant partie intégrante de tous les domaines d'activités de la municipalité. Cela en ce qui concerne les aspects économiques et le choix de solutions. Il deviendra par ailleurs plus naturel de considérer les besoins et les désirs des habitants et d'inclure à la planification les expériences d'exploitation acquises.

La planification sera ainsi plutôt l'outil et non pas l'objectif même.

De manière général les objectifs pourront être formulés comme:

* Milieu récepteur pur

Càd des eaux pures (cours d'eau, lacs et zones littorales) y compris l'eau des plages. Il est essentiel de fixer des objectifs concrets quant au milieu récepteur pour permettre le dimensionnement de bassins et de trop pleins.

* Système de décharge non-défaillant des eaux pluviales

Càd minimiser les risques de défaillance du système d'assainissement lors de pluies, de fontes de neige évitant ainsi des inondations inacceptables de caves et de terrains. Les objectifs fixant le degré de la non-défaillance doivent entre autres indiquer quand il est temps de remplacer/rénover une canalisation.

* Sécurité de Travail

Càd appliquer des solutions accordant une haute priorité à la sécurité de travail de même qu'à une exploitation efficace et facile.

* Bonne Economie

Càd voir dans un contexte commun la protection de l'environnement, le choix de solutions techniques, l'étendue des services offerts aux habitants et l'économie afin d'identifier des solutions optimales de même que réduire autant que possible les frais liés à la construction de stations de traitement et à l'exploitation de celles-ci.

Pour chacun de ces objectifs (et les autres objectifs généraux) il est possible de fixer des objectifs plus détaillés et plus concrets. On pourra p.ex. dire que des inondations/accumulations d'eau ne sont acceptées qu'une fois chaque cinquième année et seulement jusqu'à un niveau de 0,2 m dessous le niveau de sous-sol. Un tel objectif pourra être incorporé dans la planification par la simulation MOUSE sur ordinateur.

ETUDE DANOISE DES BESOINS DE RENOVATION

Une étude vient d'être achevée au Danemark quant aux besoins de rénovation des systèmes d'assainissement. Cette étude prévoit les frais nécessaires pour mettre les systèmes d'assainissement existants au niveau des connaissances actuelles dans le domaine.

L'étude a d'ailleurs été faite sur la base de données venant de 50% des municipalités danoises quant à l'étendue et l'état du système d'assainissement ainsi que les investissements déjà encourus et ceux envisagés.

Le calcul des frais, fait par méthode

Figure 1. C'est essentiel de fixer des objectifs

successive, se base d'ailleurs sur les résultats de l'étude - le plus grand nombre des municipalités n'ont que commencé de fixer des objectifs et des exigences quant au standard souhaité.

Tant au niveau municipal qu'au niveau national les exigences choisies sont décisives. On peut par exemple choisir deux objectifs différents quant à l'étanchéité des canalisations:

1) toutes les canalisations doivent être étanches

2) Les canalisations doivent être étanches là où l'infiltration pourra avoir un effet nuisible important à la station de traitement/au milieu récepteur et là où le suintement donne lieu à une pollution des eaux souterraines utilisées pour de l'eau potable. L'objectif général sera 75% plus cher que l'objectif comprenant des limitations.

A part de nous donner pour la première fois une vue d'ensemble du nombre de canalisations, d'installations de trop-plein, de stations de pompage, de bassins, de séparateurs etc. cette étude montre que les chiffres clés en ce qui concerne l'économie, l'état des canalisations existantes et l'envergure des travaux à exécuter sont essentiels afin d'assurer une rénovation utile et efficace. Cela s'applique tant au niveau des municipalités qu'au niveau national.

En ce qui concerne l'état des systèmes d'assainissement il s'est avéré que même au début des années quatre-vingt-dix seulement une partie très limitée des systèmes d'assainissement municipaux a fait l'objet d'études.

Seulement 15% environ des canalisations sont pourvus de surveillance télévisée et la simulation informatique pouvant prévenir les risques d'inondations et de décharges polluantes éventuelles est seulement utilisée dans une mesure limitée.

Heureusement ce n'est que dans assez peu de cas que la gravité de la situation est telle qu'une action urgente est nécessitée.

Il est donc bien fondé de donner pendant les prochaines années une haute priorité aux études sur site et à la planification. Cela contribuera aussi à rendre plus facile la prise de décision dans les municipalités mais aussi au niveau national.

Un grand nombre des municipalités danoises se trouve aujourd'hui face aux difficultés liées à des décharges polluantes énormes au milieu récepteur, et il y a plusieurs municipalités qui se trouvent à la face des irrégularités d'exploitation et des frais d'exploitation importants des installations de trop plein, des bassins, des stations de pompage et des séparateurs.

* Economie

Ces derniers 5 ans les municipalités danoises ont en général affecté 55 CD/habitant/an à la rénovation des systèmes d'assainissement (nouvelles installations, extensions de stations existantes non compris). Pendant les prochains 10 ans les municipalités prévoient au budget des augmentations de cette somme - s'élevant ainsi à 130 à 150 CD/habitant/an.

Cependant l'étude déjà mentionnée montre aussi que pour couvrir les besoins réels d'aujourd'hui il faudra plutôt affecter 200 CD ou 400 CD/habitant/an au cours des 20 prochaines années. Ces chiffres assureraient la planification pendant les premiers 5 ans ainsi que les vastes travaux de rénovation pendant les 15 ans suivents.

La répartition des investissement quant aux:

* canalisations et puits
* stations de pompage
* installations de trop plein
* bassins
* séparateurs
* moyens d'instrumentation, de commande et de mesure
* branchements municipaux
* planification et études
* conception et supervision

ressort de la figure montrée ci-dessous

Les investissements liés aux bassins sont rendus nécessaires à cause d'exigences quant à la qualité de l'eau du milieu récepteur tandis que les investissements liés aux branchements sont envisagés parce qu'une partie de plus en plus grande des municipalités danoises se chargent des branchements situés en voie publique.

NOUVELLE TECHNIQUE

* Simulation/modélisation sur ordinateur

L'apparition de la simulation/modélisation sur ordinateur a fortement favorisé la planification quant aux exigences visant une meilleure protection du milieu récepteur et aux exigences d'une exploitation plus économique des systèmes d'assainissement.

Le paquet de simulation MOUSE permettra d'évaluer les inondations/accumulations d'eau ainsi que les décharges polluantes au cas de pluies extrêmement fortes et sur base

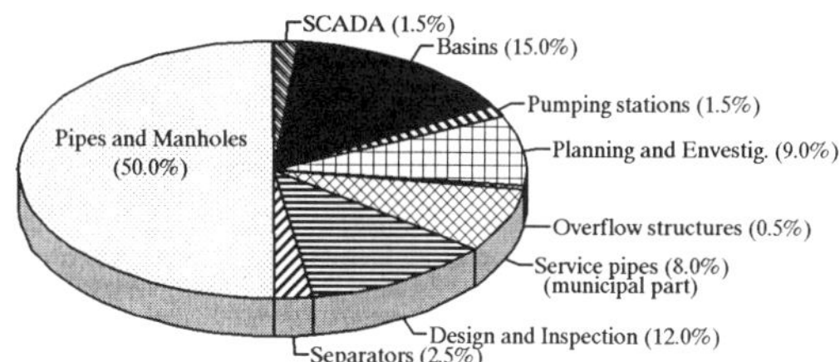

Figure no. 2: Répartition des frais de rénovation au Danemark.

annuelle. Il sera aussi possible de faire une simulation de l'exploitation et des décharges à partir de la station de traitement lors de pluies - la raréfaction d'oxygène dans les cours d'eau peuvent d'ailleurs aussi être calculée.

Cela nous permettra donc d'obtenir une évaluation d'ensemble du fonctionnement du système d'assainissement surtout lors de pluies et de fontes de neige.

* Études

En ce qui concerne les études des systèmes d'assainissement la surveillance télévisée a été considérablement perfectionnée pendant ces derniers ans, entre autres à cause de l'utilisation de nouveaux caméras, tournages en couleurs, nouvelles techniques vidéo et à cause de l'amélioration portée au collecte de données appropriées au traitement et au stockage dans des bases de données.
Les diverses possibilités d'enregistrement informatique et de coordination de grandes quantités de données ont rendu possible l'utilisation d'une diversité de données dans la planification sans ainsi perdre la vue d'ensemble. On peut p.ex. effectuer des analyses de risques comparant les nombreuses données des études, des simulations et des expériences. Cela permettra d'établir les priorités en considérant les dégâts les plus probables et les conséquences les plus néfastes y relatives tout en les comparant aux objectifs et aux exigences fixés.

* Rénovation des canalisations

Le "marché de rénovation" a subi des changements importants ces dernières années - en Scandinavie les maisons spécialisées dans le domaine de rénovation ont ainsi triplé ou même quadruplé. Et cela en dépit du fait qu'on n'a pas pu constater en même temps une augmentation considérable de la demande.

* Méthodes de rénovation

En Norvège et surtout en Suède la rénovation préférée est: le revêtement par segments de tuyaux et le revêtement par des tuyaux en polyéthylène assemblés par soudage. La raison en est que dans ces pays une très grande partie des rénovations sont réalisées par les municipalités elles-mêmes.

Au Danemark et en Norvège les rénovations sont dans une large mesure faites par revêtement des tuyaux ainsi que par des réparations faites par l'injection aux assemblages de matière de revêtement. Récemment il a en outre été exécuté de nombreuses rénovations par craquement de tuyaux au Danemark.

Il s'est avéré qu'au Danemark ce sont les municipalités les plus innovatrices qui sont aussi prêtes à utiliser les méthodes de rénovation.

Il est intéressant de constater que la plus grande partie des canalisations rénovées sont construites dans la période 1930-1975. Ce ne donc pas les canalisations les plus anciennes mais dans une large mesure les canalisations construites plus récemment.

Il y a plusieurs raisons différentes pour choisir la rénovation des canalisations - parmi celles-ci les raisons les plus communes sont:

* les travaux de déterrement sont évités
* les gênes relatifs à la rénovation sont minimisés autant que possible.

Parmi les usagers des méthodes de rénovation il y a toujours une certaine incertitude quant à la qualité de ces méthodes - l'économie restant l'aspect décisif pour choisir la rénovation, ils expriment un profond souhait pour pouvoir obtenir des informations neutres quant à la durée des méthodes, leur fonctionnement et les paramètres quant à la matière utilisée.

LE FUTUR

Dans les pays scandinaves les méthodes de rénovation offrent aujourd'hui une grande diversité. Lors des prochaines années les exigences quant au contrôle de qualité de celles-ci seront plus strictes et plusieurs méthodes seront donc sans doute abandonnées. Il est prévu que les exigences quant aux méthodes de rénovation des branchements et des puits seront plus rigoureuses aussi - de cette manière les rénovations pourront être réalisées comme un ensemble.

En Scandinavie nous possédons aujourd'hui de vastes expériences quant aux méthodes de rénovation par revêtement des tuyaux, craquement des tuyaux, revêtement par segments de tuyaux ainsi que revêtement par tuyaux assemblés par soudage. Les lignes directives quant au choix de méthodes et au dimensionnement, élaborées en 1989, peuvent donc, avec avantage, être appliquées dans d'autres pays.

C'est d'ailleurs aussi le cas pour la simulation développée en Scandinavie pour le calcul des capacités des canalisations hydrauliques ainsi que des décharges polluantes au milieu récepteur par l'intermédiaire des installations de trop-plein et des stations de traitement.

Nous pouvons donc en conclure que nous sommes mieux préparés aujourd'hui qu'il y a 10 ans pour évaluer les besoins de rénovation dans tout le système d'assainissement et pour évaluer les effets des diverses solutions.

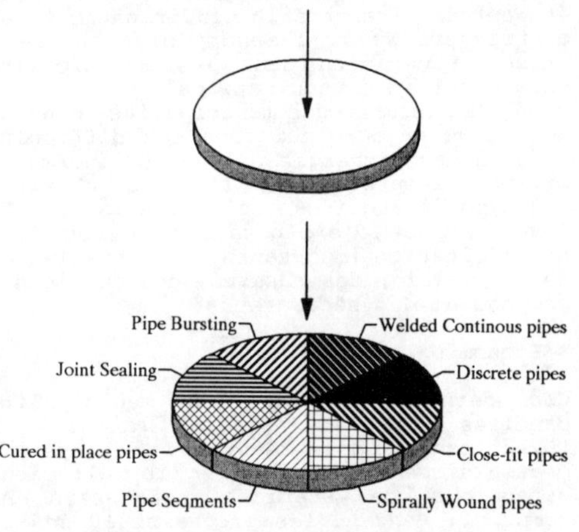

Figure no. 3: Aujourd'hui il y a plusieurs méthodes divers de renovation.

Late papers

No Trenches in Town, Henry & Mermet (eds) © 1992 Balkema, Rotterdam. ISBN 90 5410 085 0

New developments in microtunnelling techniques

Steve Orchard
Euro Iseki Limited, Stratford-upon-Avon, UK

Summary of Current Status of Microtunnelling Techniques

The microtunnelling market is only around 20 years old. Much of its history has been devoted to identifying the benefits to potential users and building a track record so as to provide confidence for the market to proceed with employing this new technology.

There is now a strong track record for the technology in Japan and in most Western countries. Most countries are enthusiastically taking to the technology albeit with differing degrees of enthusiasm.

Clients are now interested in expanding microtunnelling's use in more difficult ground conditions and in new industries.

Current microtunnelling equipment available

Currently in the microtunnelling market, the range of microtunnelling systems is based upon the use of auger and slurry shields. The basic concepts which can be employed for microtunnelling are limited but new ideas are appearing.

Both the conventional slurry and auger systems employ a tunnelling shield to excavate ground in front of the shield. The alternative to this has been to ram or force a head through the ground. This principle is not always well received. It also generally suffers from the inability to steer accurately.

The most recent genuinely new tunnelling shield principle is the Perimole shield. This shield employs the same principle of eccentric movement as the well known Unclemole shield. However, it does it in a new and novel way. Instead of employing the eccentric movement as a crusher within a cone, the eccentric movement is employed to compress the ground away from the passage of the shield. It is done with the use of two split cones which push the ground in opposite directions to ensure the shield does not vibrate or oscillate.

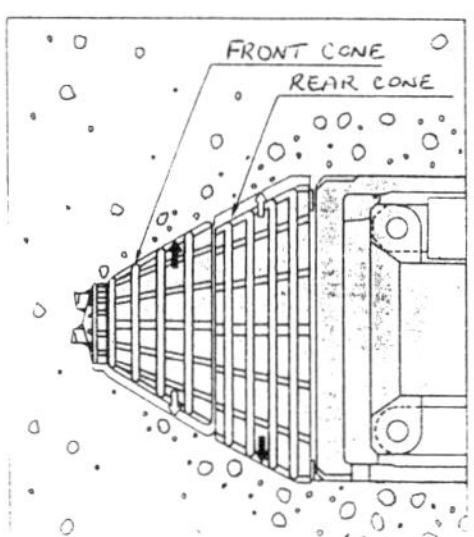

The shield has been tested in a range of grounds and has been found to be effective in ground with an N value of up to 30. It has been proven to create no heave or vibration when operating up to 2 pipe diameters from a neighbouring pipe or duct.

This shield offers the potential of installing pipes from 200mm id upwards and todate has been built in sizes up to 350mm id. Iseki believes that the principle has an application in larger sized shields and will continue expanding the range.

Limitations of ground Conditions

End users require to be able to cope with both a wide range of ground conditions as well as to tunnel through rock and hard ground. The convinced users of microtunnelling are no longer deliberating over the straight forward use of the technique but demanding that system capabilities are enhanced in order to tunnel through common rock strengths perceived to be up to 200 Mpa.

Currently most equipment is limited in its ability to handle a wide range of ground conditions with one shield or type of head. In the future if large volumes of work become available, it is possible the specialist shield will become more acceptable. However, at the present time, the need to effectively amortise an investment as soon as possible means that investors in this technology want to be able to use their equipment across a wide range of ground conditions. A shield capable of working over a wide range of ground conditions also provides security when highly variable ground is being tunnelled.

In order to overcome hard or rock conditions, it is generally perceived that disc cutters, roller bits or some combination of both or required. However, there are disadvantages to the use of these systems not the least being that there is a need to have either a specialist shield or that it is necessary to continually change the head of the shield.

There are a number of shields in the market with disc cutters including the Rockmole manufactured by Iseki which can tunnel through up to 200 Mpa rock. However, the ideal shield is still perceived as the shield which can proceed through a wide range of ground conditions including the difficult combination of clay and rock. In order to do this, it is necessary to incorporate a conventional system, such as the Unclemole slurry system, with the ability to tackle rock. Such a shield is being investigated by Iseki with both the use of a new design of cutter bit and a mixture of roller bits operating on a conventional shield.

Targets for Technical Development

The requirement to provide better instrumentation on equipment is another end user requirement. This arises from the need to have accurate records for the work for both project evaluation and later record filing. This increase in the instrumentation of equipment ties in with the user's need to automate equipment. With more modern systems this automation offers both more accurate driving and the ability to minimise costly training and operation.

The new automatic guidance systems employ the sophistication of sophisticated computer software to provide not just "reactive" but "proactive" guidance instructions. The way this has been managed by the AS system is by programming a computer with a history of actual guidance records. These are based upon the experience of fully qualified operators over a wide range of projects. The data used is sorted by mean deviation assessment to include only the most successful and best supported guidance responses. The computer is then equipped with a memory like a human from which it can make a range of decisions best suited to control the shield.

Field tests of this system show that the shield actually responds faster and better than under human guidance. This means that drive results are more accurate. To date, the ground variables are so great that it is impossible to have a guidance system that can analyse what to do if the shield hits a totally unforeseen obstruction like a steel pipe or a concrete block. The human operator is therefore not yet an extinct species!

Alternative Microtunnelling Applications

If we step aside from considering only the "dirty water" industry and consider the full potential range of microtunnelling users, the range would be considerable increased if the choice of suitable jacking pipes is expanded. The most recent development is the pressure jacking pipe. Various types exist already. However, a recent breakthrough is the installation of polyethylene pipe by microtunnelling.

The LLP system developed by Keido construction in Japan employs a central pushing shaft. This is capable of accepting the full jacking load and transfers the load to the shield at the head of the pipe line.

Along the shaft are specially designed inflatable carrier bags which grip the inside of the polyethylene pipe. In this manner, the polyethylene pipe is pulled into the tunnel internally. Tests carried out to date show this method can be successfully employed over a wide range of ground conditions.

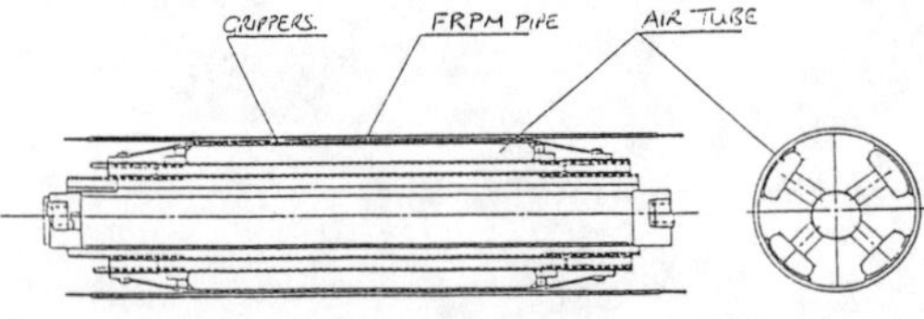

The LL B method opens up the opportunity of installing polyethylene pipe in conditions which are not suitable for conventional jacking pipe due to effluent quality or aggressive ground condition. It also means that lower pressure gas lines or other lines which must be leak proof can be installed by a non disruptive method.

Reducing The Size of Access Shafts

Another frequent requirement of the end user in the European market is the need to reduce the size of shafts. Currently shafts are already so small that a compromise between pipe length and shaft size has to be reached. When shafts are very small, they enforce the need to employ small pipe lengths with the consequential loss of speed of installation. However there are occasions when there is no choice but to look for a small shaft.

The first answer to this is the development of small jacking systems such as the T bar jacking system. However, Iseki has now also introduced the possibility of splitting its shields shaft so that the exit shafts can be existing manholes as small as 1,00m diameter. This has a particularly important implication with regard to pipe replacement as the client is often strongly motivated to utilise an existing manhole to avoid cost and disruption.

Conclusion

All the machine manufacturers are looking to meet market needs as they are identified. However, there is a growing awareness of the need for the manufacturers, the contractors and the end users to combine forces to explore the realistic targets for the microtunnelling market. It is an exciting and challenging area of development. The goal is the ability to carry out a wider range of work in a manner which the general public is going to find attractive.

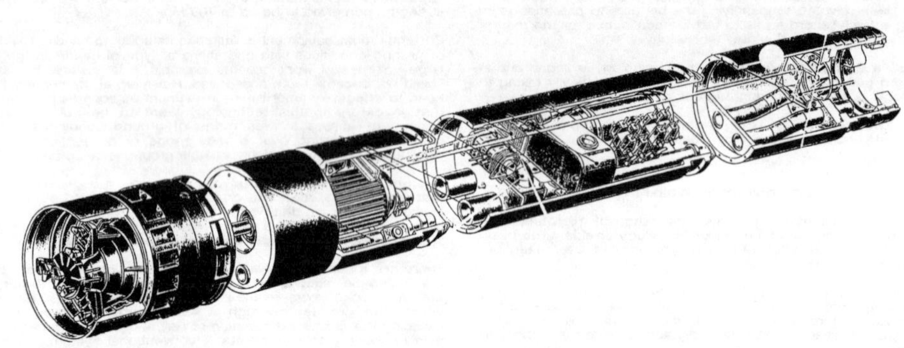

No Trenches in Town, Henry & Mermet (eds) © 1992 Balkema, Rotterdam. ISBN 90 5410 085 0

Practical survey of experiments in no-dig techniques conducted on the SEDIF area

Michel Mercier
Ile de France Water Board (SEDIF), France

ABSTRACT: The Ile de France Water Board (SEDIF) is a public institution in charge of supplying drinking water to 144 municipalities in the Parisian region, which involves catering for a population of about 4 million. Deeply aware of the importance of the job assigned to it by the municipal authorities, SEDIF has always made a point of creating a minimum of inconvenience for road users and its own subscribers when effecting works on pipe lines. Agreements on quality have been signed with contractors working regularly for the Water Board to ensure that the aspects of environmental conservation are taken into account and, in addition, SEDIF decided to pioneer the development of trenchless techniques.

1 INTRODUCTION

The Ile de France Water Board is a public institution whose duty it is to supply 144 Communes in the Parisian region housing a total population of 4 million. This job can only be done by production units proportionate to the commitment (1 million m^3/day) and by a very large pipe network (more than 8,500 linear metres) ranging from the feeder with a diameter of 2m to the local branch line, 100 mm in diameter.
It is obvious that, in these circumstances, the number of sites where pipes are being laid or are undergoing various maintenance operations at the same time, is also impressive.

As a result, the SEDIF, deeply aware of its responsibility towards the municipal authorities, has always made a point of creating the least inconvenience possible for road users and subscribers to the water utility when working on pipe lines, however indispensable and cumbersome these jobs may be. In this perspective, quality agreements were signed with the contractors working regularly for the water Board so that environmental conservation aspects would be taken into account. Particular care is therefore taken in beaconing the work sites, installing protective barriers, storage areas, materials, passages for pedestrians and in using low

noise level road-working machinery, etc.
It naturally follows that the Ile de France Water Board felt in duty bound to promote no-dig techniques.
Studies and trials undertaken in cooperation with the operator of the network, Compagnie Générale des Eaux and the contractors concerned namely: in situ replacement of cast-iron pipes by pipe-extraction techniques, sliplining of existing pipes, guided drilling, pneumatic driving of side connection pipes, soft insertion lining methods, microtunnelers.
Let us notice that most of the jobs are realized on a network under operation, what influences strongly the organization of the works (setting up of temporary mains, rapidity of intervention, etc.).

1.1 PIPE CRACKING BY EXTRACTION OF GREY CAST-IRON PIPE SECTIONS

1 Principle:
The grey cast-iron pipe is extracted in section of 20 to 30 meters long by exerting a strength of 30 to 60 tons transmitted by cable. As the pipe is extracted, it is broken up and the new PEHD pipe is pulled into position.
2 Field of application:
Used for replacing worn out pipes of a nominal diameter (ND) 60 to 100mm, by a PEHD main, dia 125 mm. As this is an in situ replacement the operation usually requires laying a temporary main.
This is a process that enables the diameter of the existing main to be enlarged; it uses the well mastered electrically weldable PEHD technology: seals, side connections, repairs.
3 Improvements being studied:
This concerns mainly the machinery for cracking the cast-iron pipe section, especially ND 60mm, the machine for cutting the steel or ductile cast iron replacement sections, grouting the annular space if the fitment diameter of the grey cast-iron section is above 125 mm, and the possibility of extracting dia 150 mm cast iron in order to replace it with dia. 180 mm and 250 mm PEHD in coils.
4 Economic aspect:
The five jobs carried out until now show that the cost of the pipe, including the temporary mains, is identical to, or slightly more , than laying the new sections in an excavated trench. However, the transfer of side connections should result in a slight saving and the process should reach better productivity once it has been broken in.

1.2 SLIP LINING WITHOUT ANNULAR SPACE - SWAGELINING

1 Principle:
The diameter of the PEHD pipe is first compressed by 7 to 15% of its original diameter, by being forced through a profile. The pipe is then inserted in the section being renovated. When traction is released, the pipe tends to snap back to its original diameter and expands until it is composite with the wall in the old main.
2 Field of application:
For slip lining obsolete pipes of nominal diameters between 150 and 400 mm with PEHD elements, but welded into sections of 200 linear metres. As this is a lining process a temporary main may be required. The process uses

PEHD 63 a well mastered material, both as regards its mechanical characteristics and assemblies: seal joints, side connections.
Measurements taken on the sites processed show that, despite the reduced cross-section, the excellent hydraulic coefficient of PEHD enables the flow rate to be increased.
3 Development:
This process is under controlled development (in France). 1000 metres have been laid up till now (cast-iron ND 200mm lined with PEHD ND 200mm, NP 10bar. The objective for 1992 is to install 2000 metres in sizes ND 150 to 400 mm as well as a site using nominal diameters other than 200 mm.
4 Improvements being studied:
The idea is to use firstly PEHD pipe made of resin PE 100 in order to increase the inside cross-section and, secondly, dia 180 and 225 mm PEHD pipe on reels to reduce the constraints on the laying operation caused by the use of cut pipe lengths.
5 Economic aspects:
The three sites completed today in dia 200 mm pipe show that we can count on a saving of 5 to 15% compared with the conventional replacement method, the exact amount depending on the quality of the soil restoration work which is no longer necessary.

1.3 GUIDED DRILLING BY HYDROJET
1 Principle:
The basis is the drilling of a pilot hole between two excavations, 100 m to 200 m apart, using a specific tool which pushes a string of rods equipped with a high pressure bentonite injection head. On the return run, the hole is bored while the high density polyethylene lining is being simultaneously pulled through.
2 Fields of application:
We lay mainly: polyethylene pipes dia 50 to 225 mm, between 1 and 3 metres deep; PEHD pipes dia 125 mm NP 16bar in coils; 800 mm of PEHD dia 180 mm NP 16 in mirror welded assemblies, in suitable ground: clay, silt, sand.
3 Development:
The scheduled operations for 1991 were for the laying of 5300 Lmt. The objective has been reached. The aim for 1992 is 9000 Lmt.
4 Improvements under study:
These aim especially at using PE resin pipes in diameters of 100, 180 and 225 mm, delivered in coils.
5 Economic aspect:
For sites where the comparison between laying with or without excavation is possible, there is a saving of 5 to 15%, depending on the length of the project and the nature of the soil covering.
In addition, the process allows singular points to be executed in places where it is impossible to open the trench: rivers, motorway slip roads, railway lines etc.

1.4 LAYING MAINS AND SIDE CONNECTIONS BY PNEUMATIC SHAFT BORING
1 Principle:
The idea is to create a hole by compacting the soil using a pneumatically propelled horizontal bore-head (rocket). The PEHD pipe (dia 25 to 125 mm) is pulled into position by the power line feeding the bore-head.
2 Field of application:
For laying PEHD side

connections, dia 25 to 125 mm, and for PEHD mains, dia 63 to 125 mm. The material rarely allows more than 20 Lmt to be laid per blast.

3 Development:

This process has been operative for ten years. 20 to 25 km of pipe laying is done by this method each year.

4 Improvements under study:

Over the last ten years the reliability of machinery has increased. Reversed drive systems are frequent. Some equipment can be fitted with a transmitting probe allowing to monitor the path.

5 Economic aspect:

This process is mainly used to carry out PE side connections, diameter 25mm, 32mm and 50 mm. The cost is identical to that of laying in open trench conditions for a normal type of road. It is a process that has largely contributed towards reducing inconvenience to traffic by laying side connections pipes and modernizing networks.

1.5 LINING WITH FLEXIBLE INNER MEMBRANE

1 Principle:

This consists in positioning a lining, inserted by the pulling method, which is then blown up and heated. After being hardened by polymerization, it forms a new pipe leaving no annular space.

The lining is made up of four components:

. an inner polyethylene membrane for water tightness;

. a central glass fibre reinforcement containing heating elements.

. external PVC sheathing to protect the foregoing components during installation.

. impregnation with epoxy resin of the central core and the inner felt material composite with the polyethylene membrane.

2 Field of application:

For lining old cast-iron pipes with diameters of 300 mm minimum. It is important to note that pipe bends and tee connections must be processed in the traditional manner, using components of ductile cast iron, steel or manufactured by BONNA.

A project was studied by the SEDIF including structural lining to absorb the entire pressure stress. If necessary, the COPEFLEX lining can be designed to absorb external loads.

3 Development:

These are additional certification tests on the SEDIF drinking water network, i.e.

. negative pressure tests

. burst due to impact tests

. ageing tests.

The means of assembly at pipe ends will be tested simultaneously.

4 Improvements:

This lining is primarily designed for the rehabilitation of worn out cast-iron feeders minimum diameter 300 mm, and with side connections up to 40 mm in diameter only. These connections can however be transferred during the lining operation. A device to allow the realisation of under pressure connection of 40 mm side connections is at the present studied.

5 Economic aspect:

The economic comparisons made at the moment based on diameters of 300 to 500 mm authorize us to say that about 10% saving is possible compared with the

conventional system of replacement. Saving should be even greater for larger diameters.
6 Experiments - References in France:
35,000 Lmt laid at the end of 1991 representing an area of 55,000 m^2, mainly in sewerage.

1.6 INVERSION LINING (Paltem or Phoenix systems)
1 Principle:
A woven textile lining, coated on the outside with resin, is positioned by inversion in the existing pipe. The lining, turned inside out, is glued to the internal pipe wall by the resin placed inside the lining before its insertion.
2 Fields of application:
For lining worn cast-iron pipes with diameters of 200 to 500 mm maximum. The pipe bends and tee connections must be processed in the traditional manner using components of cast iron, steel or manufactured by BONNA.
A version recently tested on SEDIF's network, the Kevlar reinforced Paltem absorbs all pressure stresses.
3 Development:
Manufacturing the lining in Europe instead of in Japan should enable costs to be reduced.
The ferrules presented by a supplier need to be tested in order to ascertain the reliability of the realisation under pressure of side connections.
4 Improvements:
Technologies are now under study, to improve the treatment of branch connections, pipe bends and to bear heavy weight of the ground.
5 Economic aspect:
The lack of visibility on the market does not enable us to forecast definitely the economic position.
6 Experiments - References:
200 Lmt dia 300mm processed until now on behalf of SEDIF.

1.7 MICRO- AND MINI-TUNNELERS
1 Principle:
The basis is the positioning of prefabricated pipes by a system of horizontal shaft boring, using a machine which is remote controlled from outside. The string of pipes is jacked into position by cylinders in the working shaft. The tunneller is guided by an interactive laser beam system, and the cuttings are removed mechanically or by hydraulic means.
2 Field of application:
This is a system for laying sewage or drinking water networks in pipe lengths of 150 m maximum, in homogeneous soil without rocks.
The microtunneller is suitable for diameters of 400 to 800 mm.
The minitunneler is used for dia 900 à 1200 mm pipes.
For drinking water, we use BONNA steel or concrete lined pipes.
3 Development:
Its use is mainly for doing away with open trenches in sensitive urban surroundings.
4 Improvements under study:
They concern the testing of concrete steel-core pressure pipes with flexible joints.
5 Economic aspect:
The economic interest can only find justification in view of local conditions: preliminary removal of networks prior to laying in the open air, laying at great depths.

6 Experiments - Reference:
Except for one site, the most frequent work done on drinking water systems consists in lining operations.
In sewage, the references are more numerous: fifty sites in France over the last three years, executed by more or less ten contractors, half of whom own their own equipment.

No Trenches in Town, Henry & Mermet (eds) © 1992 Balkema, Rotterdam. ISBN 90 5410 085 0

Point concret des expériences menées sur l'ensemble du territoire du SEDIF en matière de techniques sans tranchée

Michel Mercier
Syndicat des Eaux d'Ile de France (SEDIF), France

RESUME : Le Syndicat des Eaux d'Ile de France (SEDIF) est un établissement public chargé de l'alimentation en eau potable de 144 communes de la région parisienne soit 4 millions d'habitants à desservir. Conscient du mandat que lui ont conféré les communes, le SEDIF s'est toujours attaché à ce que ses travaux d'installation de canalisations présentent un minimum de gêne pour les usagers de la voie publique et pour ses abonnés. Des protocoles de qualité ont été signés avec les entreprises qui travaillent régulièrement pour le compte du Syndicat afin que les notions d'environnement soient prises en compte et préservées et le SEDIF a décidé d'être pionnier pour le lancement et la promotion des techniques sans tranchée.

1 INTRODUCTION

Le Syndicat des Eaux d'Ile de France est un établissement public chargé de l'alimentation en eau potable de 144 communes de la région parisienne soit 4 millions d'habitants à desservir.
Cette mission ne peut se faire sans des unités de production à la taille de l'enjeu (1 million de m3/jour) et sans un réseau très important de canalisations (plus de 8 500 kms) allant du feeder de 2 mètres de diamètre à la conduite de desserte locale de 100 mm de diamètre.
Il est clair dans ce contexte que le nombre de chantiers de pose de canalisations ou d'interventions diverses sur le réseau menés en concomitance sur son territoire est également impressionnant.
Dès lors, conscient du mandat que lui ont conféré les communes, le SEDIF s'est toujours attaché à ce que ses travaux d'installation de canalisation pour indispensables et encombrants qu'ils soient présentent un minimum de gêne pour les usagers de la voie publique et pour ses abonnés.
Dans cet esprit, des protocoles de qualité ont été signés avec les entreprises qui travaillent régulièrement pour le compte du Syndicat afin que les notions d'environnement soient prises en compte et préservées. Un soin tout particulier est donc apporté aux balisages des chantiers, aux barrières de protection, aux lieux de stockage des matériaux, aux passages réservés aux piétons, au caractère silencieux des matériels de voirie, etc...

Il était donc naturel que le Syndicat des Eaux d'Ile de France se donne comme devoir d'être pionnier pour le lancement et la promotion des techniques sans tranchée.
Les études et expérimentations qu'il a menées de concert avec l'exploitant du réseau, la Compagnie Générale des Eaux, et les entreprises portent notamment sur : la pose place sur place par extraction de conduites en fonte grise, le tubage de canalisations existantes, les forages dirigés, le fonçage pneumatique des branchements, le chemisage par membrane intérieure souple et par reversion, les micro-tunneliers.
il est important de noter que la plupart de ces travaux sont conduits sur un réseau en exploitation et que cela conditionne très fortement l'organisation des chantiers.

1.1 LA POSE PLACE SUR PLACE PAR EXTRACTION DE CONDUITES EN FONTE GRISE

1 Principe :
La conduite en fonte grise est extraite par tronçons de 20 à 30 mètres, par application d'une force de 30 à 60 tonnes transmise par des câbles, au fur et à mesure de l'extraction, le tuyau en fonte grise est brisé et la nouvelle conduite en tubes PEhd est tractée.
2 Domaine d'emploi :
Il réside dans le remplacement de conduites vétustes DN 60 à 100 mm par une canalisation

en PEhd Ø 125. S'agissant d'un renouvellement "place pour place" l'opération nécessite le plus souvent la pose d'un communicateur.
Le procédé permet d'augmenter le diamètre nominal de la conduite existante ; il utilise par ailleurs la technologie maîtrisée du PEhd électrosoudable :raccordements, branchements, réparations.
3 Améliorations à l'étude :
Il s'agit essentiellement de l'appareillage d'éclatement de la fonte grise surtout en DN 60, l'appareillage du découpage des pièces de réparation en acier ou en fonte ductile, du remplissage éventuel du vide annulaire si le diamètre de l'emboîtement de la fonte grise est supérieur à 125 mm et de la possibilité d'extraire de la fonte grise de DN 150 pour la remplacer par du PEhd en couronne de Ø 180 et 225 mm.
4 Aspect économique :
Les cinq chantiers réalisés à ce jour font ressortir des coûts pour la canalisation (y compris communicateur) identiques ou légèrement supérieurs à ceux d'une pose avec ouverture de tranchée. Cependant, le poste reports de branchements devrait présenter une légère économie et le rodage du procédé devrait induire une meilleure productivité.

1.2 TUBAGE SANS VIDE ANNULAIRE - SWAGELINING
1 Principe :
Le diamètre du tuyau en PEhd est réduit de 7 à 15% en le forçant par traction à travers une filière ; il est ainsi inséré dans la canalisation à rénover. Lorsque l'effort de traction est relâché, le tuyau tend à reprendre son diamètre initial et vient alors se plaquer sur la paroi dans l'ancienne canalisation.
2 Domaine d'emploi :
Il procède au tubage de conduites vétustes de DN 150 à 400 mm par des éléments PEhd assemblés par soudure au miroir par tronçons de 200 ml. S'agissant d'un tubage, l'opération peut nécessiter la pose d'un communicateur.
Le procédé utilise un matériau bien maîtrisé, le PEhd 6,3, aussi bien pour ses caractéristiques mécaniques que pour ses assemblages : raccordements, branchements.
Les mesures faites sur les chantiers réalisés démontrent que malgré la réduction de section, l'excellent coefficient hydraulique du PEhd permet un accroissement de débit.
3 Développement :
Ce procédé est en développement contrôlé (en France). 1000 ml ont été réalisés à ce jour (fonte DN 200 tubée par du PEhd 200 PN 10. L'objectif 92 est de 3000 m DN 150 à 400 ainsi que la réalisation de chantier en DN autres que 200 mm.
4 Améliorations à l'étude :
Il s'agit de l'utilisation d'une part de tube PEhd en résine PE 100 afin d'augmenter la section interne et d'autre part de tube PEhd Ø 180 et 225 sur tourets pour diminuer les contraintes de mise en oeuvre induites par l'utilisation de tubes en barre.
5 Aspect économique :
Les trois chantiers réalisés à ce jour en Ø 200 mm permettent d'escompter une économie de 5 à 15% par rapport au remplacement traditionnel le gain étant fonction de la qualité de la réfection de sols économisée.

1.3 FORAGES DIRIGES PAR HYDROJETS
1 Principe :
Il réside dans le forage d'un trou pilote entre deux fouilles distantes de 100 à 120 m à l'aide d'un outil spécifique poussant un train de tiges équipé d'une tête d'injection à haute pression de solution de bentonite. Au retour, il est procédé à l'alésage du trou avec traction simultanée du tube de polyéthylène haute densité.
2 Domaine d'emploi :
Nous posons essentiellement : des tubes polyéthylène Ø 50 à 225 mm entre 1 et 3 m de profondeur, du PEhd Ø 125 mm PN 16 en couronne, 800 mm de PEhd Ø 180 mm PN 16 en tubes assemblés au miroir, terrains appropriés. Sols homogènes : argile, limon, sable.
3 Développement :
Le programme opérationnel était de 5300 ml exécutés en 1991. L'objectif 1992 est de 9000 ml.
4 Améliorations à l'étude :
Elles visent essentiellement, à utiliser des tubes en résine PE 100 Ø 180 et 225 mm en couronne.
5 Aspect économique :
Pour les chantiers où la comparaison entre la pose avec ou sans tranchée est possible, l'économie de 5 à 15% dépend de la longueur du projet et de la nature du revêtement de sol. Le procédé permet par ailleurs l'exécution de points singuliers pour lesquels une ouverture de tranchée est impossible : rivière, bretelles d'autoroute, voies SNCF, etc...

1.4 POSE DE CANALISATIONS ET BRANCHEMENTS PAR FONCAGE PNEUMATIQUE
1 Principe :
Il consiste à créer un trou par compactage du sol à l'aide d'un fonceur pneumatique (fusée). Celui-ci est aiguillé à l'aide du flexible d'alimentation du fonceur afin de permettre le tirage du tube en PEhd Ø 25 à 125 mm.
2 Domaine d'emploi :
La réalisation de branchements en PEhd de Ø

25 à 125 mm. et de canalisations en PEhd de Ø 63 et 125 mm.Le matériel permet rarement de dépasser 20 ml par tir.
3 Développement :
Le procédé est opérationnel depuis dix années, 20 à 25 kms sont réalisés chaque année.
4 Améliorations à l'étude :
Depuis dix ans, la fiabilité des matériels s'est accrue. Les dispositifs d'inversion de marche sont courants. Certains matériels peuvent être équipés de sonde émettrice permettant de suivre la trajectoire.
5 Aspect économique :
Ce procédé est essentiellement utilisé pour l'exécution des branchements en PE de 25,32 et 50 mm. Le coût est identique à la pose avec ouverture de tranchée, pour une chaussée courante ; procédé ayant largement contribué à la diminution de la gêne causée par l'exécution des branchements et des modernisations.

1.5 CHEMISAGE PAR MEMBRANE INTERIEURE SOUPLE
1 Principe :
Mise en place d'une chemise introduite par traction, puis gonflée et chauffée ; après durcissement par polymérisation, elle devient une nouvelle conduite, sans aucun vide annulaire.
La chemise est constituée de quatre éléments :
. une membrane intérieure en polyéthylène assurant l'étanchéïté,
. une armature centrale en tissus de fibres de verre comportant des résistances chauffantes,
. une enveloppe extérieure en PVC assurant la protection des éléments précédents lors de la mise en oeuvre,
. une imprégnation de résine époxy de l'armature centrale et du feutre interne solidaire de la membrane en polyéthylène.
2 Domaine d'emploi :
Le chemisage de conduite fonte vétuste de Ø ≥ 300 mm. A noter que les condes et les tés doivent être traités de façon classique par pièce fonte acier ou BONNA.
Un projet a été étudié pour le SEDIF qui comprenait un chemisage structurant reprenant seul les contraintes de pression. Si cela s'avère nécessaire, le chemisage COPEFLEX peut être calculé pour reprendre les charges extérieures.
3 Développement :
Il s'agit de compléments d'essais pour l'agrément sur le réseau d'eau potable du SEDIF, à savoir :
. essais de dépression
. essais d'éclatement sous choc mécanique
. essais de vieillissement.
Ce mode de raccordement aux extrémités sera testé simultanément.
4 Améliorations à l'étude :
Le chemisage est prioritairement destiné à la réhabilitation des feeders fonte vétustes de Ø ≥ 300 mm qui ne comportent pas de prise de branchement de Ø ≤ 40 mm ; le report de ces branchements est cependant possible lors de l'exécution du chemisage. Un dispositif de prise en charge pour le Ø 40 mm est actuellement à l'étude.
5 Aspect économique :
Les comparaisons économiques faites à ce jour des diamètres de 300 et 500 mm permettent d'escompter une économie d'environ 10 % dans ces diamètres par rapport à un remplacement traditionnel. Dans les diamètres supérieurs l'économie devrait être accrue.
6 Expériences - Références en France :
35 000 ml posés à fin 91 représentent une surface de 55 000 m2 essentiellement en assainissement.

1.6 CHEMISAGE PAR REVERSION
1 Principe :
Une gaine tissée, enduite de résine à l'extérieur, est mise en oeuvre par réversion à l'intérieur de la conduite existante. Le collage de la gaine sur la paroi interne de la conduite est donc assuré par la résine mise en place à l'intérieur de la gaine avant introduction.
2 Domaine d'emploi :
Le chemisage de conduites fonte vétustes de Ø ≤ 200 ≤ 500 mm. Les coudes et les tés doivent être traités de façon classique par pièces fonte, acier ou Bonna.
Une version récemment testée sur le territoire du SEDIF, avec une gaine KEVLAR reprend seul les contraintes de pression, mais ne reprendrait pas systémtiquement le poids des terres.
3 Amélioration à l'étude :
Des perfectionnements sont en cours pour améliorer le traitement des branchements, des coudes, et des fortes surcharges extérieures.
La fabrication de la gaine en Europe au lieu de l'importer du Japon devrait permettre de diminuer les coûts. Il conviendra également de tester les viroles présentées par un fabricant pour fiabiliser les prises en charge.
5 Aspect économique :
Le manque de visibilité du marché ne permet pas encore de cerner définitivement le positionnement économique.
6 Expériences - Références :
200 ml Ø 300 mm ont été réalisés à ce jour pour le compte du SEDIF.

1.7 MICRO ET MINI-TUNNELIERS.
1 Principe :
Il réside dans la mise en place par fonçage horizontal des tuyaux préfabriqués en utilisant

une machine de forage télécommandée de l'extérieur. Le train de tubes est poussé par des vérins installés dans un puits de travail. Le guidage du tunnelier est assuré par un système laser interactif et les déblais sont évacués mécaniquement ou par voie hydraulique.

2 Domaine d'emploi :

Il s'agit de la réalisation de réseaux d'assainissement ou d'eau potable par tronçons unitaires de longueur maximum 150 m, dans des terrains homogènes non rocheux.

Le micro-tunnelier s'emploie pour des Ø 400 à 800 mm.

Le mini-tunnelier s'emploie pour des Ø 900 à 1200 mm.

En eau potable, les tuyaux sont du type BONNA ou en acier ou en gaine béton.

3 Développement :

Son utilisation vise essentiellement à supprimer les tranchées en zone urbanisée sensible.

4 Améliorations à l'étude :

Elles consistent dans le domaine de l'eau potable, à tester des tuyaux pression âme-tôle à joints souples.

5 Aspect économique :

L'intérêt économique ne peut se justifier qu'en fonction des conditions locales : déplacement préliminaire de réseaux avant pose à ciel ouvert, grande de profondeur de pose.

6 Expériences - Références :

En eau potable, à l'exception d'un chantier, il s'agit le plus souvent de la mise en oeuvre de gaines.

En assainissement, les références sont plus nombreuses : cinquante chantiers en France depuis trois ans par environ dix entreprises différentes dont la moitié possèdent un matériel.

No Trenches in Town, Henry & Mermet (eds) © 1992 Balkema, Rotterdam. ISBN 90 5410 085 0

Author index